Student's Solutions Manual to Accompany

Student's Solutions Manual to Accompany

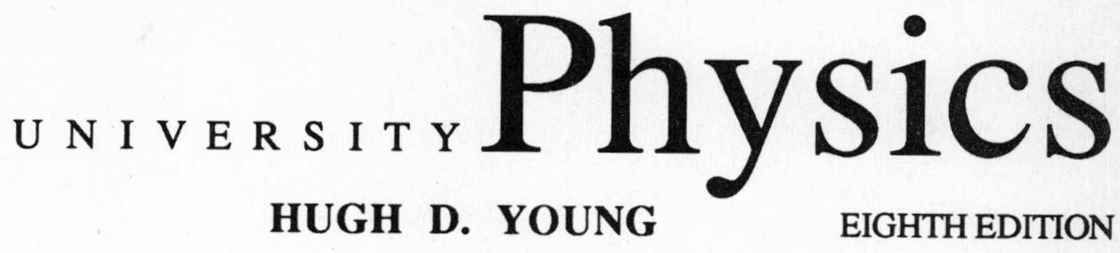

UNIVERSITY Physics

HUGH D. YOUNG EIGHTH EDITION

A. Lewis Ford
Texas A&M University

Addison-Wesley Publishing Company
Reading, Massachusetts • Menlo Park, California • New York
Don Mills, Ontario • Wokingham, England • Amsterdam • Bonn
Sydney • Singapore • Tokyo • Madrid • San Juan • Milan • Paris

Reproduced by Addison-Wesley from camera-ready copy supplied by the author.

ISBN 0-201-19653-0

1 2 3 4 5 6 7 8 9 10-BA-95949392

PREFACE

This student's solution manual contains detailed solutions for approximately one-third of the Exercises and Problems in the 8th edition of *University Physics*. The Exercises and Problems included here were selected solely from the odd-numbered Exercises and Problems in the text, for which the answers are tabulated in the back of the textbook. The Exercises and Problems to be included were not selected at random, but rather have been carefully selected so as to include at least one representative example of each problem type. This solution manual greatly expands the set of worked-out examples that go along with the presentation of the physics laws and concepts in the text. The remaining Exercises and Problems, for which solutions are not given, constitute an ample set of problems for the students to tackle on their own. In addition, there are Challenge Problems for which no solutions are given here.

This solution manual is written for student use. A primary function of the manual is to provide the student with models to follow in working physics problems. The problems are worked out in the manual in the manner and style in which the students should carry out their own problem solutions.

The author will gratefully receive comments as to style, points of physics, errors, or anything else relating to the manual.

A. L. F.

Physics Department
Texas A and M University
College Station, TX 77843
June, 1992

CONTENTS

Chapter			Page
15	Exercises	3, 11, 13, 17, 19, 25, 27, 29, 31, 33, 35, 39, 45, 49, 51, 53, 57	158
	Problems	59, 61, 63, 67, 69, 71, 75, 77, 81, 83, 85, 87	
16	Exercises	3, 7, 11, 17, 19, 23, 25, 27, 29	168
	Problems	35, 37, 39, 41, 43, 47, 49, 51	
17	Exercises	3, 9, 13, 15, 19, 21	174
	Problems	23, 25, 29, 31, 33	
18	Exercises	1, 7, 9, 13, 15, 19, 21, 25, 27	179
	Problems	29, 31, 33, 35	
19	Exercises	3, 5, 7, 11, 13, 15, 21, 23, 25	187
	Problems	27, 31, 33	
20	Exercises	3, 5, 7, 11, 13, 17	191
	Problems	19, 21, 23, 25	
21	Exercises	1, 7, 9, 11, 13, 15, 17	196
	Problems	19, 23, 27	
22	Exercises	1, 5, 9, 11, 17, 19, 21, 25, 29	201
	Problems	35, 37, 39, 43, 45, 47, 49	
23	Exercises	1, 5, 7, 9	211
	Problems	13, 17, 21, 23, 25	
24	Exercises	1, 3, 5, 9, 11, 13, 15, 19, 21, 27, 31	218
	Problems	37, 39, 41, 45, 47, 49, 53, 55	
25	Exercises	1, 5, 7, 13, 15, 17, 19, 21	229
	Problems	25, 27, 29, 31, 35	
26	Exercises	3, 7, 9, 13, 15, 17, 21, 25, 27, 29, 31	237
	Problems	33, 35, 37, 39, 43	
27	Exercises	3, 5, 7, 9, 11, 13, 15, 21, 25, 27	242
	Problems	29, 33, 35, 37, 39, 41, 43, 49, 51	
28	Exercises	1, 3, 7, 9, 11, 15, 19, 23, 25, 27, 29	254
	Problems	31, 33, 39, 41, 43, 47, 49	

Chapter			Page
29	Exercises	1, 3, 5, 11, 15, 17, 21, 23, 27, 29, 31	263
	Problems	33, 37, 39, 43, 45, 47, 49, 53	
30	Exercises	1, 3, 7, 9, 13, 15, 17, 21, 25	273
	Problems	27, 29, 35, 37, 39	
31	Exercises	3, 5, 9, 11, 13, 19, 21, 25, 27	281
	Problems	29, 35, 37, 39, 41, 43, 45	
32	Exercises	3, 7, 9, 11, 13, 17, 21, 23, 25	289
	Problems	29, 31, 33, 35, 39, 41, 45	
33	Exercises	3, 5, 9, 13, 15, 17	297
	Problems	21, 25, 27, 29, 31, 33, 35	
34	Exercises	3, 7, 9, 11, 17, 19	302
	Problems	27, 29, 31, 33, 35, 37, 39	
35	Exercises	1, 5, 9, 13, 15, 17, 21, 23, 25, 29	307
	Problems	33, 35, 37, 39, 41, 43, 47, 49, 51, 53, 57, 61, 63, 65	
36	Exercises	3, 5, 7, 9, 15, 17, 19, 23	319
	Problems	25, 29, 31	
37	Exercises	1, 3, 5, 9, 13, 17, 19, 25	323
	Problems	27, 29, 33, 35, 37	
38	Exercises	1, 3, 5, 7, 9, 11, 15, 17, 21	328
	Problems	29, 33, 37, 39	
39	Exercises	1, 3, 7, 9, 11, 15, 17, 21, 25, 31, 33	333
	Problems	37, 39, 43, 45, 47, 49	
40	Exercises	1, 5, 7, 9, 11, 13, 15, 19, 21, 23, 25, 29, 35	339
	Problems	39, 41, 45, 47, 51, 53, 55, 59	
41	Exercises	3, 5, 7, 11, 13, 15	346
	Problems	21, 23, 25, 29, 31, 33	
42	Exercises	3, 5, 9, 13, 15, 17, 19	350
	Problems	21, 23, 27, 31, 33	

Chapter			Page
43	Exercises	3, 5, 11, 13, 15, 19, 21, 25	355
	Problems	27, 29, 31, 35, 39	
44	Exercises	7, 9, 11, 13, 15, 17, 19, 23, 25, 27	360
	Problems	29, 33, 35, 37, 41	
45	Exercises	3, 5, 9, 11, 13, 15, 19, 21, 25, 27	368
	Problems	31, 33, 35, 37, 41, 43	
46	Exercises	3, 5, 7, 9, 13, 15, 19, 23, 25	374
	Problems	27, 29, 35, 37	

Student's Solutions Manual to Accompany

UNIVERSITY Physics

HUGH D. YOUNG

EIGHTH EDITION

CHAPTER 1

Exercises 1, 3, 5, 11, 15, 17, 27, 29, 31, 33, 37, 41

Problems 43, 45, 47, 49, 51, 53

Exercises

1-1

$1.00 \text{ in.} = 2.54 \text{ cm}$

$$1 \text{ km} = 1 \cancel{\text{km}}\left(\frac{1000 \cancel{\text{m}}}{1 \cancel{\text{km}}}\right)\left(\frac{100 \cancel{\text{cm}}}{1 \cancel{\text{m}}}\right)\left(\frac{1.00 \cancel{\text{in.}}}{2.54 \cancel{\text{cm}}}\right)\left(\frac{1 \cancel{\text{ft}}}{12 \cancel{\text{in.}}}\right)\left(\frac{1 \text{ mi}}{5280 \cancel{\text{ft}}}\right) = 0.621 \text{ mi}$$

$1 \text{ km} = \underline{0.621 \text{ mi}}$

1-3

a) $1450 \text{ mi/h} = 1450 \text{ mi}/\cancel{\text{h}}\left(\frac{1 \cancel{\text{h}}}{60 \cancel{\text{min}}}\right)\left(\frac{1 \cancel{\text{min}}}{60 \text{ s}}\right) = \underline{0.403 \text{ mi/s}}$

b) $0.403 \text{ mi/s} = 0.403 \cancel{\text{mi}}/\text{s}\left(\frac{1.609 \cancel{\text{km}}}{1 \cancel{\text{mi}}}\right)\left(\frac{1000 \text{ m}}{1 \cancel{\text{km}}}\right) = \underline{648 \text{ m/s}}$

1-5

$$19.0 \text{ km/L} = (19.0 \cancel{\text{km}}/\cancel{\text{L}})\left(\frac{0.6214 \text{ mi}}{1 \cancel{\text{km}}}\right)\left(\frac{3.788 \cancel{\text{L}}}{1 \text{ gal}}\right) = \underline{44.7 \text{ mi/gal}}$$

1-11

$$\text{density} = \frac{m}{V}$$

For a sphere $V = \frac{4}{3}\pi r^3$, so

$$\text{density} = \frac{m}{\frac{4}{3}\pi r^3} = \frac{3m}{4\pi r^3} = \frac{3(5.98 \times 10^{24} \text{ kg})}{4\pi(6.38 \times 10^6 \text{ m})^3} = \underline{5.50 \times 10^3 \text{ kg/m}^3}$$

1-15

I <u>estimate</u> that my scalp's area is about that of a 10 in. diameter circle:

$$d = 10 \cancel{\text{in.}}\left(\frac{2.54 \cancel{\text{cm}}}{1 \cancel{\text{in.}}}\right)\left(\frac{10 \text{ mm}}{1 \cancel{\text{cm}}}\right) = 250 \text{ mm}$$

The estimated area is thus $A = \pi r^2$ with $r = \frac{d}{2} = 125 \text{ mm}$:

$$A = \pi(125 \text{ mm})^2 = 5 \times 10^4 \text{ mm}^2$$

I further <u>estimate</u> that on my scalp there are 5 hairs per mm^2.

The number of hairs on my head is thus estimated to be about $(5 \text{ hairs/mm}^2)(5 \times 10^4 \text{ mm}^2) = \underline{2 \times 10^5 \text{ hairs.}}$

1-17

Estimate that the pile is 18 in. X 18 in. X 5 ft. 8 in., so the volume of gold in the pile is $V = 18\text{ in.} \times 18\text{ in.} \times 68\text{ in.} = 22,000\text{ in.}^3$

Convert to cm^3: $V = 22,000\ \text{in.}^3 \left(\frac{1000\ cm^3}{61.02\ \text{in.}^3}\right) = 3.6 \times 10^5\ cm^3$

From Example 1-4 the density of gold is $19.3\ g/cm^3$, so the mass of this volume of gold is $m = (19.3\ g/cm^3)(3.6 \times 10^5\ cm^3) = 7 \times 10^6\ g$.

Convert to ounces: $m = 7 \times 10^6\ g \left(\frac{1\ \text{ounce}}{30\ g}\right) = 2 \times 10^5$ ounces

Also from Example 1-4, the monetary value of one ounce is \$400, so the gold has a value of $(\$400/\text{ounce})(2 \times 10^5\ \text{ounces}) = \$8 \times 10^7$, or about $\$100 \times 10^6$ (one hundred million dollars).

1-27

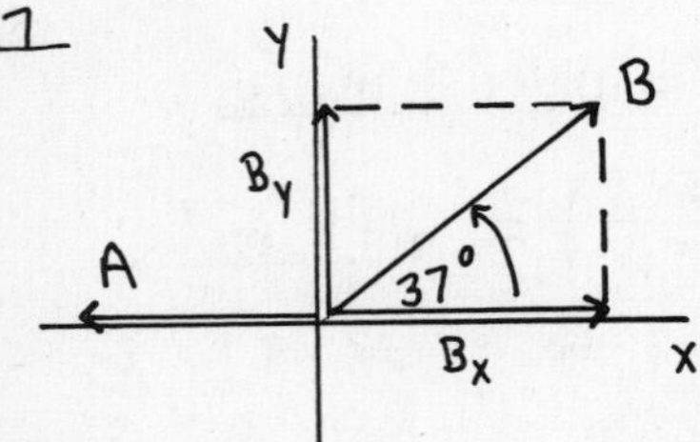

$A_x = -12.0\ m,\ A_y = 0$

$B_x = B\cos 37° = (18.0\ m)\cos 37° = 14.38\ m$

$B_y = B\sin 37° = (18.0\ m)\sin 37° = 10.83\ m$

Note that both B_x and B_y are positive. A_x is negative because $\vec{A}$ is in the negative x-direction.

a) Let $\vec{R} = \vec{A} + \vec{B}$

$R_x = A_x + B_x = -12.0\ m + 14.38\ m = +2.38\ m$

$R_y = A_y + B_y = 0 + 10.83\ m = +10.83\ m$

$R = \sqrt{R_x^2 + R_y^2} = \sqrt{(2.38\ m)^2 + (10.83\ m)^2} = \underline{11.1\ m}$

$\tan\theta = \frac{R_y}{R_x} = \frac{10.8\ m}{2.38\ m} = 4.538 \Rightarrow \theta = \underline{77.6°}$,

measured counterclockwise from the +x-axis.

b) Now let $\vec{R} = \vec{A} - \vec{B}$

$R_x = A_x - B_x = -12.0\ m - 14.38\ m = -26.38\ m$

$R_y = A_y - B_y = 0 - 10.83\ m = -10.83\ m$

Now both R_x and R_y are negative.

1-27 (cont)

$$R = \sqrt{R_x^2 + R_y^2} = \sqrt{(-26.38\,m)^2 + (-10.83\,m)^2} = \underline{28.5\,m}$$

$$\tan\theta = \frac{R_y}{R_x} = \frac{-10.83\,m}{-26.38\,m} = +0.4105 \Rightarrow \underline{\theta = 202^\circ},$$

measured counterclockwise from the +x-axis

Note: My calculator gives arctan(+0.4105) = 22.3°. But the sketch shows that $\vec{R}$ is in the third quadrant. The angle 22.3° + 180° = 202° also has a tangent of +0.4105 and the sketch shows that this is the correct answer.

1-29

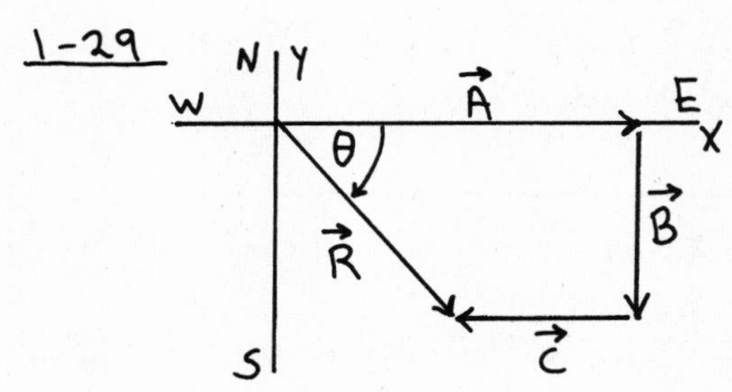

$A = 4.92$ km
$B = 3.95$ km
$C = 1.80$ km

Select a coordinate system where +x is east and +y is north. Let $\vec{A}$, $\vec{B}$, and $\vec{C}$ be the three displacements of the professor. Then the resultant displacement $\vec{R}$ is given by $\vec{R} = \vec{A} + \vec{B} + \vec{C}$. By the method of components, $R_x = A_x + B_x + C_x$ and $R_y = A_y + B_y + C_y$. Find the x and y components of each vector; add them to find the components of the resultant. Then the magnitude and direction of the resultant can be found from its x and y components that we have calculated. As always, it is essential to draw a sketch.

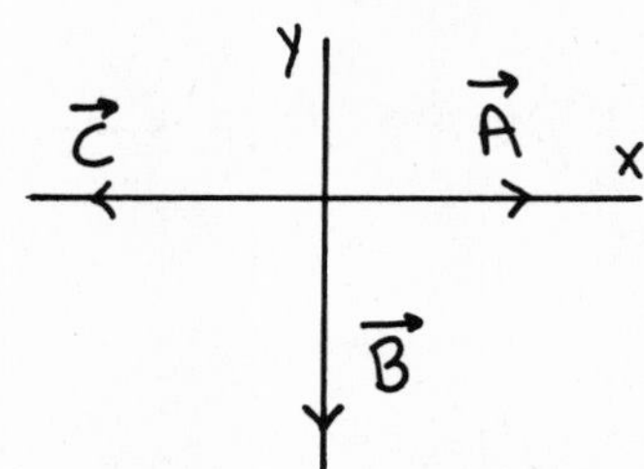

$A_x = +4.92$ km, $A_y = 0$

$B_x = 0$, $B_y = -3.95$ km

$C_x = -1.80$ km, $C_y = 0$

$$R_x = A_x + B_x + C_x = 4.92\,km + 0 - 1.80\,km = +3.12\,km$$

$$R_y = A_y + B_y + C_y = 0 - 3.95\,km + 0 = -3.95\,km$$

$$R = \sqrt{R_x^2 + R_y^2} = \sqrt{(3.12\,km)^2 + (-3.95\,km)^2} = \underline{5.03\,km}$$

$$\tan\theta = \frac{R_y}{R_x} = \frac{-3.95\,km}{3.12\,km} = -1.266 \Rightarrow \theta = -51.7^\circ$$

The angle θ is clockwise from the +x-axis since it is negative. In terms of compass directions, the resultant displacement is in the direction 51.7° S of E, or <u>38.3° E of S.</u>

1-31

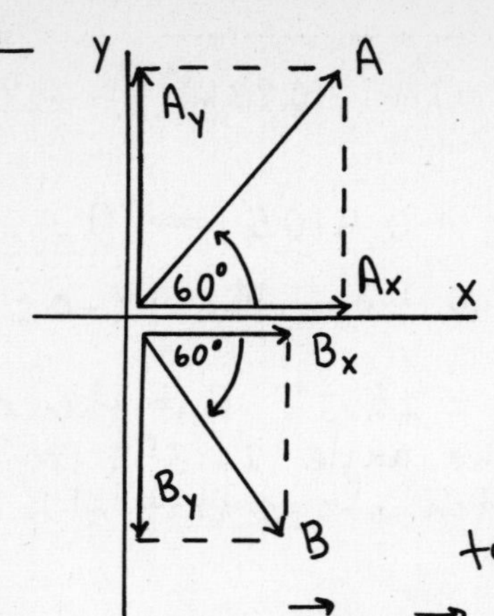

$$A_x = A\cos 60° = (2.80\text{ cm})\cos 60° = +1.40\text{ cm}$$

$$A_y = A\sin 60° = (2.80\text{ cm})\sin 60° = +2.425\text{ cm}$$

$$B_x = B\cos(-60°) = (1.90\text{ cm})\cos(-60°) = +0.95\text{ cm}$$

$$B_y = B\sin(-60°) = (1.90\text{ cm})\sin(-60°) = -1.645\text{ cm}$$

Note that the signs of the components correspond to the directions of the component vectors.

a) Let $\vec{R} = \vec{A} + \vec{B}$.

Then $R_x = A_x + B_x = +1.40\text{ cm} + 0.95\text{ cm} = +2.35\text{ cm}$

$R_y = A_y + B_y = +2.425\text{ cm} - 1.645\text{ cm} = +0.779\text{ cm}$

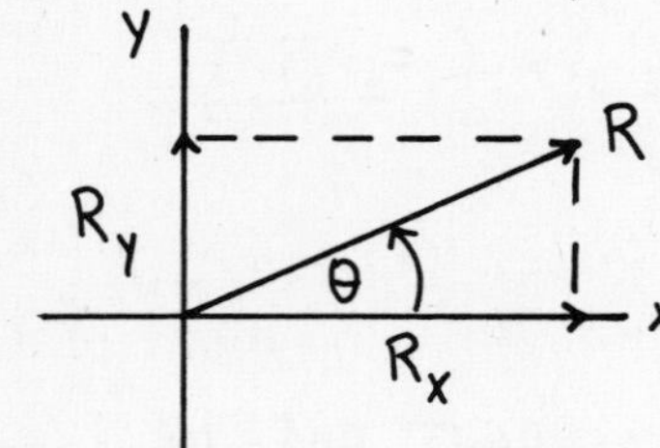

$$R = \sqrt{R_x^2 + R_y^2} = \sqrt{(2.35\text{ cm})^2 + (0.779\text{ cm})^2} = \underline{2.48\text{ cm}}$$

$$\tan\theta = \frac{R_y}{R_x} = \frac{+0.779\text{ cm}}{+2.35\text{ cm}} = +0.3315 \Rightarrow \theta = \underline{18.3°}$$

b) Let $\vec{R} = \vec{A} - \vec{B}$.

Then $R_x = A_x - B_x = 1.40\text{ cm} - 0.95\text{ cm} = 0.45\text{ cm}.$

$R_y = A_y - B_y = 2.425\text{ cm} + 1.645\text{ cm} = 4.070\text{ cm}.$

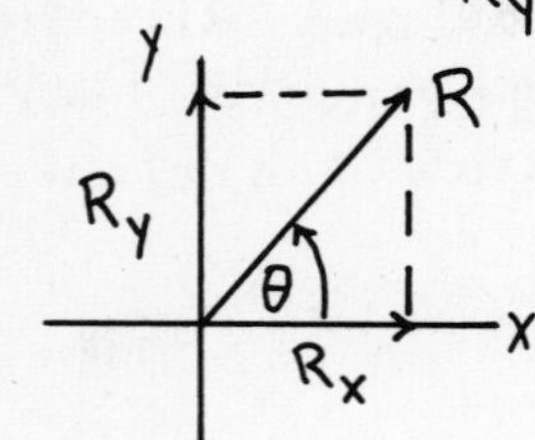

$$R = \sqrt{R_x^2 + R_y^2} = \sqrt{(0.45\text{ cm})^2 + (4.070\text{ cm})^2} = \underline{4.09\text{ cm}}$$

$$\tan\theta = \frac{R_y}{R_x} = \frac{4.070\text{ cm}}{0.45\text{ cm}} = 9.044 \Rightarrow \theta = \underline{83.7°}$$

c) Let $\vec{R} = \vec{B} - \vec{A}$.

Then $R_x = B_x - A_x = +0.95\text{ cm} - 1.40\text{ cm} = -0.45\text{ cm}$

$R_y = B_y - A_y = -1.645\text{ cm} - 2.425\text{ cm} = -4.070\text{ cm}$

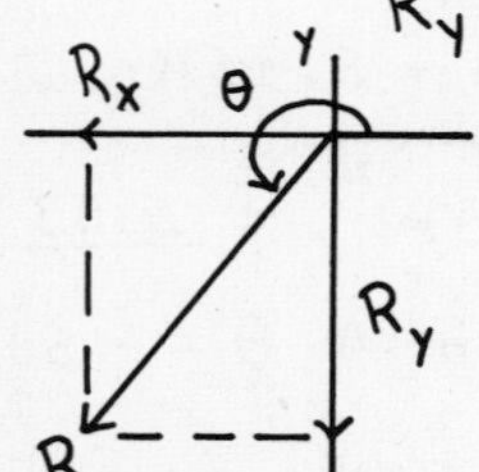

$$R = \sqrt{R_x^2 + R_y^2} = \sqrt{(-0.45\text{ cm})^2 + (-4.070\text{ cm})^2} = \underline{4.09\text{ cm}}$$

$$\tan\theta = \frac{R_y}{R_x} = \frac{-4.070\text{ cm}}{-0.45\text{ cm}} = 9.044 \Rightarrow \theta = 83.7° + 180° = \underline{263.7°}$$

Note: $\vec{B} - \vec{A} = -(\vec{A} - \vec{B})$; $\vec{B} - \vec{A}$ and $\vec{A} - \vec{B}$ are equal in magnitude and opposite in direction.

1-33

a) $\vec{A} = 4\vec{i} + 3\vec{j} \Rightarrow A_x = +4, \quad A_y = +3$

$$A = \sqrt{A_x^2 + A_y^2} = \sqrt{(4)^2 + (3)^2} = \underline{5}$$

$\vec{B} = \vec{i} - 5\vec{j} \Rightarrow B_x = 1, \quad B_y = -5$

$$B = \sqrt{B_x^2 + B_y^2} = \sqrt{(1)^2 + (-5)^2} = \underline{5.10}$$

b) $\vec{A} - \vec{B} = 4\vec{i} + 3\vec{j} - (\vec{i} - 5\vec{j}) = (4-1)\vec{i} + (3+5)\vec{j} = \underline{3\vec{i} + 8\vec{j}}$

c) Let $\vec{R} = \vec{A} - \vec{B} = 3\vec{i} + 8\vec{j}$. Then $R_x = +3$, $R_y = +8$.

$$R = \sqrt{R_x^2 + R_y^2} = \sqrt{(3)^2 + (8)^2} = \underline{8.54}$$

$$\tan\theta = \frac{R_y}{R_x} = \frac{+8}{+3} = +2.667 \Rightarrow \theta = \underline{69.4^\circ}$$

1-37

$\vec{A} = 4\vec{i} + 3\vec{j}, \quad \vec{B} = \vec{i} - 5\vec{j}$

$$\vec{A}\cdot\vec{B} = (4\vec{i} + 3\vec{j})\cdot(\vec{i} - 5\vec{j}) = (4)(1) + (3)(-5) = 4 - 15 = \underline{-11}$$

1-41

$\vec{A} = 4\vec{i} + 3\vec{j}, \quad \vec{B} = \vec{i} - 5\vec{j}$

$$\vec{A}\times\vec{B} = (4\vec{i} + 3\vec{j})\times(\vec{i} - 5\vec{j}) = 4\vec{i}\times\vec{i} - 20\vec{i}\times\vec{j} + 3\vec{j}\times\vec{i} - 15\vec{j}\times\vec{j}$$

But $\vec{i}\times\vec{i} = \vec{j}\times\vec{j} = 0$, and $\vec{i}\times\vec{j} = \vec{k}$, $\vec{j}\times\vec{i} = -\vec{k}$ so

$$\vec{A}\times\vec{B} = -20\vec{k} + 3(-\vec{k}) = -23\vec{k}$$

The magnitude of $\vec{A}\times\vec{B}$ is $|\vec{A}\times\vec{B}| = 23$.

Note: Sketch the vectors $\vec{A}$ and $\vec{B}$ in a coordinate system where the xy-plane is the plane of the paper and the z-axis is directed out toward you.

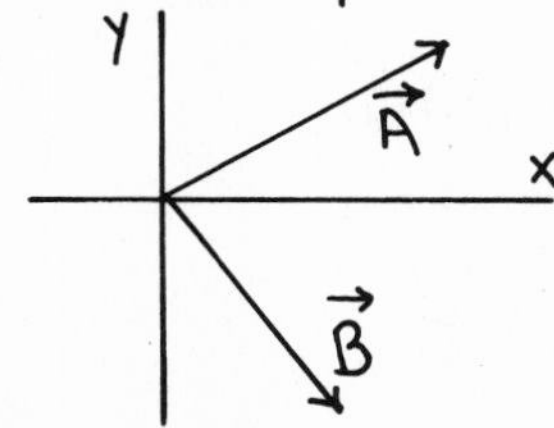

By the right-hand rule $\vec{A}\times\vec{B}$ is directed into the plane of the paper, in the $-z$-direction. This agrees with the above calculation that used unit vectors.

Problems

1-43

a) $f = 1.420 \times 10^9$ cycles/s $\Rightarrow \frac{1}{1.420 \times 10^9}$ s $= \underline{7.04 \times 10^{-10}\text{ s}}$ for one cycle

b) $\frac{3600\text{ s/h}}{7.04 \times 10^{-10}\text{ s/cycle}} = \underline{5.11 \times 10^{12}\text{ cycles/h}}$

c) Calculate the number of seconds in 4600 million years $= 4.60 \times 10^9$ y and divide by the time for 1 cycle:

$$\frac{(4.60 \times 10^9 \cancel{y})(3.156 \times 10^7 \cancel{s}/\cancel{y})}{7.04 \times 10^{-10}\ \cancel{s}/\text{cycle}} = \underline{2.06 \times 10^{26}\text{ cycles}}$$

d) The clock is off by 1 s in 100,000 y $= 1 \times 10^5$ y, so in 4.60×10^9 y it is off by $1\text{ s}\left(\frac{4.60 \times 10^9}{1 \times 10^5}\right) = \underline{4.6 \times 10^4\text{ s}}$ (about 13 h)

1-45

Let $\vec{R}$ be the resultant; $\vec{R} = \vec{M} + \vec{N}$.

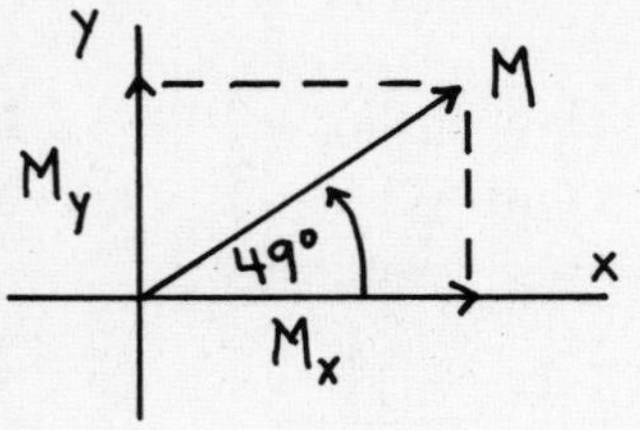

$M_x = M \cos 49° = (5.20\text{ cm}) \cos 49° = 3.412\text{ cm}$

$M_y = M \sin 49° = (5.20\text{ cm}) \sin 49° = 3.924\text{ cm}$

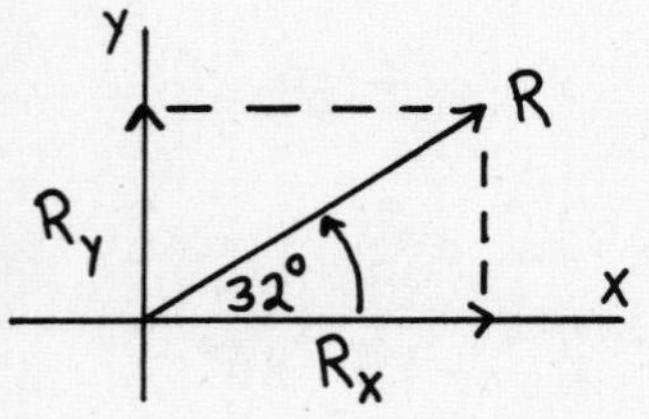

$R_x = R \cos 32° = (5.20\text{ cm}) \cos 32° = 4.410\text{ cm}$

$R_y = R \sin 32° = (5.20\text{ cm}) \sin 32° = 2.756\text{ cm}$

a) $\vec{R} = \vec{M} + \vec{N} \Rightarrow \vec{N} = \vec{R} - \vec{M}$

$N_x = R_x - M_x = 4.410\text{ cm} - 3.412\text{ cm} = \underline{1.00\text{ cm}}$

$N_y = R_y - M_y = 2.756\text{ cm} - 3.924\text{ cm} = \underline{-1.17\text{ cm}}$

b) $N = \sqrt{N_x^2 + N_y^2} = \sqrt{(1.00\text{ cm})^2 + (-1.17\text{ cm})^2} = \underline{1.54\text{ cm}}$

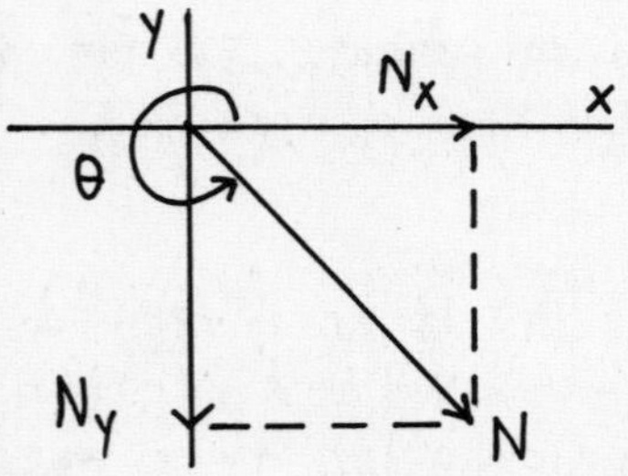

$\tan\theta = \frac{N_y}{N_x} = \frac{-1.17\text{ cm}}{1.00\text{ cm}} = -1.17$

$\Rightarrow \theta = 49.5° + 180° = \underline{310.5°}$

1-47

Use a coordinate system where east is in the +x-direction and north is in the +y-direction.

Let $\vec{A}$, $\vec{B}$, and $\vec{C}$ be the three displacements that are given and let $\vec{D}$ be the fourth unmeasured displacement. Then the resultant displacement is $\vec{R} = \vec{A} + \vec{B} + \vec{C} + \vec{D}$. And since he ends up back where he started, $\vec{R} = 0$.

$$0 = \vec{A} + \vec{B} + \vec{C} + \vec{D} \Rightarrow \vec{D} = -(\vec{A} + \vec{B} + \vec{C}); \quad D_x = -(A_x + B_x + C_x)$$
$$D_y = -(A_y + B_y + C_y)$$

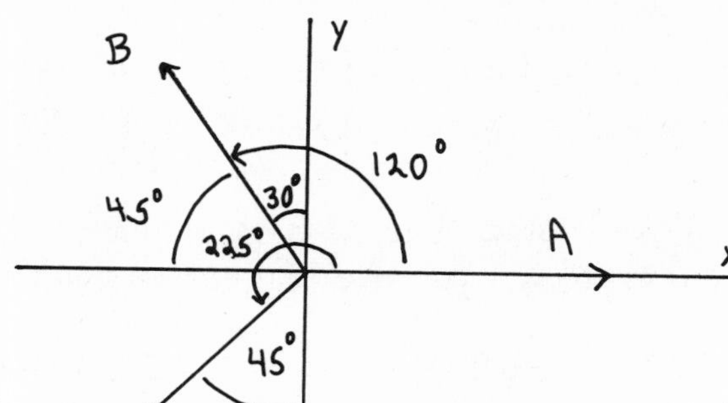

$A_x = +180\text{ m}, \quad A_y = 0$

$B_x = B\cos 120° = (80\text{ m})\cos 120° = -40.0\text{ m}$
$B_y = B\sin 120° = (80\text{ m})\sin 120° = +69.28\text{ m}$

$C_x = C\cos 225° = (150\text{ m})\cos 225° = -106.1\text{ m}$
$C_y = C\sin 225° = (150\text{ m})\sin 225° = -106.1\text{ m}$

$D_x = -(A_x + B_x + C_x) = -(180\text{ m} - 40\text{ m} - 106.1\text{ m}) = -33.9\text{ m}$
$D_y = -(A_y + B_y + C_y) = -(0 + 69.3\text{ m} - 106.1\text{ m}) = -(-36.8\text{ m}) = +36.8\text{ m}$

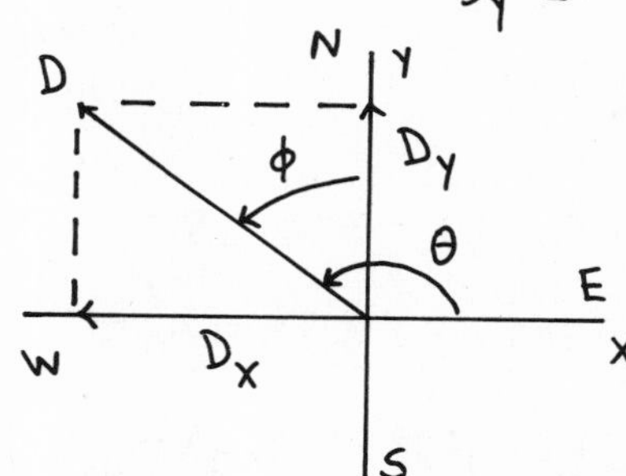

$D = \sqrt{D_x^2 + D_y^2} = \sqrt{(-33.9\text{ m})^2 + (36.8\text{ m})^2} = \underline{50.0\text{ m}}$

$\tan\theta = \dfrac{D_y}{D_x} = \dfrac{+36.8\text{ m}}{-33.9\text{ m}} = -1.086$

$\Rightarrow \theta = 180° - 47.3° = 132.7°$

($\vec{D}$ is in the second quadrant since D_x is negative and D_y is positive.)

The direction of $\vec{D}$ can also be specified in terms of $\phi = \theta - 90° = 42.7°$; $\vec{D}$ is <u>42.7° W of N.</u>

1-49

a)

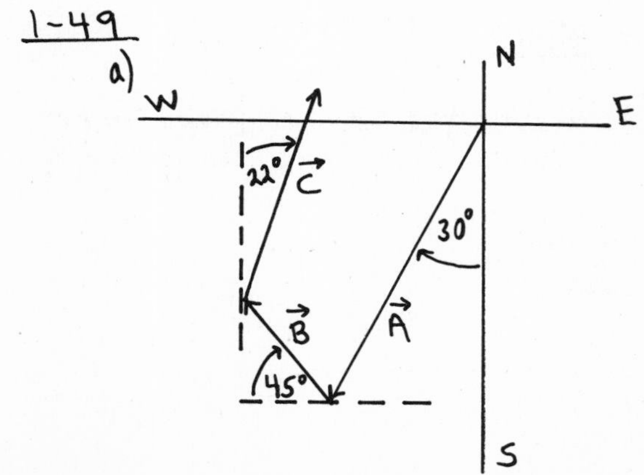

Let the three displacements that are given in the problem be called $\vec{A}$, $\vec{B}$, and $\vec{C}$.

Let the resultant displacement be $\vec{R}$; $\vec{R} = \vec{A} + \vec{B} + \vec{C}$.

Part (b) of the problem is asking for the magnitude of $\vec{R}$.

1-49 (cont)

b) Use a coordinate system where the +x-direction is east and the +y-direction is north.

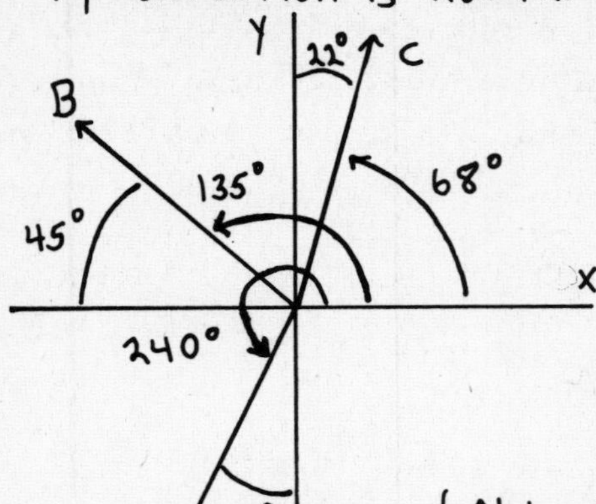

$A_x = A\cos 240° = (7.40\text{ km})\cos 240° = -3.70\text{ km}$
$A_y = A\sin 240° = (7.40\text{ km})\sin 240° = -6.409\text{ km}$

$B_x = B\cos 135° = (2.80\text{ km})\cos 135° = -1.980\text{ km}$
$B_y = B\sin 135° = (2.80\text{ km})\sin 135° = +1.980\text{ km}$

$C_x = C\cos 68° = (5.20\text{ km})\cos 68° = +1.948\text{ km}$
$C_y = C\sin 68° = (5.20\text{ km})\sin 68° = +4.821\text{ km}$

(Note that in each case the signs of the components correspond to the directions of the vectors.)

$R_x = A_x + B_x + C_x = -3.70\text{ km} - 1.980\text{ km} + 1.948\text{ km} = -3.732\text{ km}$
$R_y = A_y + B_y + C_y = -6.409\text{ km} + 1.980\text{ km} + 4.821\text{ km} = +0.392\text{ km}$
$R = \sqrt{R_x^2 + R_y^2} = \sqrt{(-3.732\text{ km})^2 + (0.392\text{ km})^2} = \underline{3.75\text{ km}}$

1-51

$\vec{A} = -\hat{i} + 2\hat{j} - 5\hat{k}\ ,\quad \vec{B} = 3\hat{i} + 2\hat{j} - 2\hat{k}$

a) $A = \sqrt{A_x^2 + A_y^2 + A_z^2} = \sqrt{(-1)^2 + (2)^2 + (-5)^2} = \sqrt{30} = \underline{5.48}$

$B = \sqrt{B_x^2 + B_y^2 + B_z^2} = \sqrt{(3)^2 + (2)^2 + (-2)^2} = \sqrt{17} = \underline{4.12}$

b) $\vec{A} - \vec{B} = (-\hat{i} + 2\hat{j} - 5\hat{k}) - (3\hat{i} + 2\hat{j} - 2\hat{k}) = (-1-3)\hat{i} + (2-2)\hat{j} + (-5+2)\hat{k}$
$\vec{A} - \vec{B} = -4\hat{i} - 3\hat{k}$

c) Let $\vec{C} = \vec{A} - \vec{B}$, so $C_x = -4$, $C_y = 0$, $C_z = -3$

$C = \sqrt{C_x^2 + C_y^2 + C_z^2} = \sqrt{(-4)^2 + (-3)^2} = \underline{5.00}$

$\vec{B} - \vec{A} = -(\vec{A} - \vec{B})$, so $\vec{B} - \vec{A}$ and $\vec{A} - \vec{B}$ have the same magnitude but opposite directions

1-53

Use $\cos\phi = \dfrac{\vec{A}\cdot\vec{B}}{AB}$

a) $\vec{A} = \vec{k}$ (along line ab)

$\vec{B} = \vec{i} + \vec{j} + \vec{k}$ (along line ad)

$A = 1, \quad B = \sqrt{1^2+1^2+1^2} = \sqrt{3}$

$\vec{A}\cdot\vec{B} = \vec{k}\cdot(\vec{i}+\vec{j}+\vec{k}) = 1$

So $\cos\phi = \dfrac{\vec{A}\cdot\vec{B}}{AB} = \dfrac{1}{\sqrt{3}} \Rightarrow \phi = \underline{54.7^\circ}$

b) $\vec{A} = \vec{i} + \vec{j} + \vec{k}$ (along line ad)

$\vec{B} = \vec{j} + \vec{k}$ (along line ac)

$A = \sqrt{1^2+1^2+1^2} = \sqrt{3} \quad ; \quad B = \sqrt{1^2+1^2} = \sqrt{2}$

$\vec{A}\cdot\vec{B} = (\vec{i}+\vec{j}+\vec{k})\cdot(\vec{j}+\vec{k}) = 1+1 = 2$

So $\cos\phi = \dfrac{\vec{A}\cdot\vec{B}}{AB} = \dfrac{2}{\sqrt{3}\,\sqrt{2}} = \dfrac{2}{\sqrt{6}} \Rightarrow \phi = \underline{35.3^\circ}$

CHAPTER 2

Exercises 3, 5, 7, 9, 15, 17, 21, 23, 27, 29, 31, 33, 37

Problems 41, 43, 47, 49, 51, 53, 57, 59

Exercises

2-3

$$v_{av} = \frac{\Delta x}{\Delta t} \Rightarrow \Delta x = v_{av}\,\Delta t \text{ and } \Delta t = \frac{\Delta x}{v_{av}}$$

Use the information given for normal driving conditions to calculate the distance between the two cities:

$$\Delta x = v_{av}\,\Delta t = (96\ \text{km/h})\left(\frac{1\ \text{h}}{60\ \text{min}}\right)(130\ \text{min}) = 208\ \text{km}$$

Now use v_{av} for a rainy day to calculate Δt; Δx is the same as before.

$$\Delta t = \frac{\Delta x}{v_{av}} = \frac{208\ \text{km}}{72\ \text{km/h}} = 2.89\ \text{h} = 2\ \text{h and } 53\ \text{min}$$

The trip takes an additional 43 min.

2-5

$$x = \alpha t^2 + \beta t^3, \text{ with } \alpha = 1.80\ \text{m/s}^2,\ \beta = 0.250\ \text{m/s}^3$$

$$v_{av} = \frac{\Delta x}{\Delta t} = \frac{x_2 - x_1}{t_2 - t_1}$$

We can use the equation for x as a function of t to calculate Δx.

a) $t_1 = 0 \Rightarrow x_1 = 0$

$t_2 = 2.00\ \text{s} \Rightarrow x_2 = (1.80\ \text{m/s}^2)(2.00\ \text{s})^2 + (0.250\ \text{m/s}^3)(2.00\ \text{s})^3 = 9.20\ \text{m}$

Then $v_{av} = \dfrac{x_2 - x_1}{t_2 - t_1} = \dfrac{9.20\ \text{m} - 0}{2.00\ \text{s} - 0} = 4.60\ \text{m/s}$.

b) $t_1 = 0 \Rightarrow x_1 = 0$

$t_2 = 4.00\ \text{s} \Rightarrow x_2 = (1.80\ \text{m/s}^2)(4.00\ \text{s})^2 + (0.250\ \text{m/s}^3)(4.00\ \text{s})^3 = 44.8\ \text{m}$

$v_{av} = \dfrac{x_2 - x_1}{t_2 - t_1} = \dfrac{44.8\ \text{m} - 0}{4.00\ \text{s} - 0} = 11.2\ \text{m/s}$

c) $t_1 = 2.00\ \text{s} \Rightarrow x_1 = 9.20\ \text{m}$

$t_2 = 4.00\ \text{s} \Rightarrow x_2 = 44.8\ \text{m}$

2-5 (cont)

$$v_{av} = \frac{x_2 - x_1}{t_2 - t_1} = \frac{44.8\text{ m} - 9.20\text{ m}}{4.00\text{ s} - 2.00\text{ s}} = \frac{35.6\text{ m}}{2.00\text{ s}} = \underline{17.8\text{ m/s}}$$

Note that v_{av} depends on the Δt interval.

2-7

$$x = bt^2, \text{ where } b = 12.0\text{ cm/s}^2$$

$$v = \frac{dx}{dt} = 2bt$$

$$t = 3.00\text{ s} \Rightarrow v = 2(12.0\text{ cm/s}^2)(3.00\text{ s}) = \underline{72.0\text{ cm/s}}$$

2-9

$$a_{av} = \frac{\Delta v}{\Delta t} \Rightarrow \Delta t = \frac{\Delta v}{a_{av}}$$

$$\Delta v = v_2 - v_1 = 96\text{ km/h} - 120\text{ km/h} = -24\text{ km/h}$$

$$\Delta v = -24\text{ km/h}\left(\frac{10^3\text{ m}}{1\text{ km}}\right)\left(\frac{1\text{ h}}{3600\text{ s}}\right) = -6.67\text{ m/s}$$

$$\Delta t = \frac{\Delta v}{a_{av}} = \frac{-6.67\text{ m/s}}{-4.0\text{ m/s}^2} = \underline{1.7\text{ s}}$$

2-15

a) v_0 ——— 80 m ———→ $\vec{v} = 15.0$ m/s, along x

$x_0 = 0$, $t = 0$ $\qquad$ $x = 80$ m, $t = 6.00$ s

$x - x_0 = 80$ m
$t = 6.00$ s
$v = 15.0$ m/s
$v_0 = ?$

Use $x - x_0 = \left(\frac{v_0 + v}{2}\right)t \Rightarrow v_0 = \frac{2(x - x_0)}{t} - v$

$$v_0 = \frac{2(80\text{ m})}{6.00\text{ s}} - 15.0\text{ m/s} = \underline{11.7\text{ m/s}}$$

b) Use $v = v_0 + at \Rightarrow a = \frac{v - v_0}{t} = \frac{15.0\text{ m/s} - 11.7\text{ m/s}}{6.00\text{ s}} = \underline{0.55\text{ m/s}^2}$

2-17

a) The acceleration a at time t is the slope of the tangent to the v versus t curve at time t.

At $t = 3$ s, the v versus t curve is a horizontal straight line, with zero slope. Thus $\underline{a = 0}$.

2-17 (cont)

At $t = 7s$, the v versus t curve is a straight-line segment with slope $\frac{45\ m/s - 20\ m/s}{9s - 5s} = 6.25\ m/s^2$.

Thus $a = \underline{6.25\ m/s^2}$.

At $t = 11s$ the curve is again a straight-line segment, now with slope $\frac{0 - 45\ m/s}{13s - 9s} = -11.2\ m/s^2$. Thus $a = \underline{-11.2\ m/s^2}$.

b) For the time interval $t = 0$ to $t = 5s$ the acceleration is constant and equal to zero. For the time interval $t = 5s$ to $t = 9s$ the acceleration is constant and equal to $6.25\ m/s^2$. For the interval $t = 9s$ to $t = 13s$ the acceleration is constant and equal to $-11.2\ m/s^2$.

During the first 5 seconds the acceleration is constant, so the constant acceleration kinematic formulas can be used.

$v_0 = 20\ m/s$

$a = 0$

$t = 5s$

$x - x_0 = ?$

$x - x_0 = v_0 t + \frac{1}{2} a t^2$ (the $\frac{1}{2}at^2$ term is crossed out)

$x - x_0 = (20\ m/s)(5\ s) = \underline{100\ m}$; this is the distance the officer travels in the first 5 seconds

During the interval $t = 5s$ to $9s$ the acceleration is again constant. The constant acceleration formulas can be applied to this 4 second interval. It is convenient to restart our clock so the interval starts at time 0 and ends at time 4s. (Note that the acceleration is <u>not</u> constant over the whole $t = 0$ to $t = 9s$ interval.)

$v_0 = 20\ m/s$

$a = 6.25\ m/s^2$

$t = 4s$

$x_0 = 100\ m$

$x - x_0 = ?$

$x - x_0 = v_0 t + \frac{1}{2} a t^2$

$x - x_0 = (20\ m/s)(4s) + \frac{1}{2}(6.25\ m/s^2)(4s)^2$

$x - x_0 = 80\ m + 50\ m = 130\ m$

Thus $x = x_0 + 130\ m = 100\ m + 130\ m = 230\ m$.

At $t = 9s$ the officer is at $x = 230\ m$, so she has traveled $\underline{230\ m}$ in the first 9 seconds.

During the interval $t = 9s$ to $t = 13s$ the acceleration is again constant. The constant acceleration formulas can be applied for this 4 second interval but <u>not</u> for the whole $t = 0$ to $t = 13s$ interval. To use the equations restart our clock so this interval begins at time 0 and ends at time 4s.

$v_0 = 45\ m/s$ (at the start of the time interval)

$a = -11.2\ m/s^2$

$t = 4s$

$x_0 = 230\ m$

$x - x_0 = ?$

$x - x_0 = v_0 t + \frac{1}{2} a t^2$

$x - x_0 = (45\ m/s)(4s) + \frac{1}{2}(-11.2\ m/s^2)(4s)^2$

$x - x_0 = 180\ m - 89.6\ m = 90.4\ m$

2-17 (cont)

Thus $x = x_0 + 90.4\,m = 230\,m + 90.4\,m = 320\,m$. At $t = 13\,s$ the officer is at $x = 320\,m$, so she traveled <u>320m</u> in the first 13 seconds.

<u>2-21</u>

a) The maximum speed occurs at the end of the initial acceleration period.

$a = 15.0\ m/s^2$
$t = 10\ min = 600\,s$
$v_0 = 0$
$v = ?$

$v = v_0 + at$
$v = 0 + (15.0\,m/s^2)(600\,s) = 9000\ m/s$

b) The motion consists of three constant acceleration intervals. In the middle segment of the trip $a = 0$ and $v = 9000\ m/s$, but we can't directly find the distance traveled during this part of the trip because we don't know the time. Instead, find the distance traveled in the first ($a = +15.0\ m/s^2$) part of the trip and in the last ($a = -15.0\ m/s^2$) part of the trip. Subtract these two distances from the total distance of 4.0×10^5 km to find the distance traveled in the middle ($a=0$) part of the trip.

<u>first segment</u>

$x - x_0 = ?$
$t = 10\ min = 600\,s$
$a = +15.0\ m/s^2$
$v_0 = 0$

$x - x_0 = \cancel{v_0 t}^{\,0} + \frac{1}{2}at^2$
$x - x_0 = \frac{1}{2}(15.0\ m/s^2)(600\,s)^2$
$x - x_0 = 2.70 \times 10^6\ m = 2.70 \times 10^3\ km$

<u>third segment</u>

$x - x_0 = ?$
$t = 10\ min = 600\,s$
$a = -15.0\ m/s^2$
$v_0 = 9000\ m/s$

$x - x_0 = v_0 t + \frac{1}{2}at^2$
$x - x_0 = (9000\ m/s)(600\,s) + \frac{1}{2}(-15.0\ m/s^2)(600\,s)^2$
$x - x_0 = 5.40 \times 10^6\ m - 2.70 \times 10^6\ m = 2.70 \times 10^6\ m = 2.70 \times 10^3\ km$
(The same distance as in the first segment.)

Therefore the distance traveled at constant speed is
$4.00 \times 10^5\ km - 2.70 \times 10^3\ km - 2.70 \times 10^3\ km = 3.946 \times 10^5\ km = 3.946 \times 10^8\ m$.

The fraction this is of the total distance is $\dfrac{3.946 \times 10^5\ km}{4.00 \times 10^5\ km} = \underline{0.986}$

c) Find the time for the constant speed segment:

$x - x_0 = 3.946 \times 10^8\ m$
$v = 9000\ m/s$
$a = 0$
$t = ?$

$x - x_0 = v_0 t + \cancel{\frac{1}{2}at^2}^{\,0}$

$t = \dfrac{x - x_0}{v_0} = \dfrac{3.946 \times 10^8\ m}{9000\ m/s} = 4.384 \times 10^4\ s$

2-21 (cont)

The total time for the whole trip is thus

$$600\,s + 4.384\times10^4\,s + 600\,s = 4.504\times10^4\,\cancel{s}\left(\frac{1\,h}{3600\,\cancel{s}}\right) = \underline{12.5\,h}$$

2-23

Take the origin at the ground and the positive direction to be upward.

a) At the maximum height $v = 0$. Thus

$v = 0$
$y - y_0 = 0.640\,m$
$a = -9.80\,m/s^2$
$v_0 = ?$

$$\cancel{v^2}^{\,0} = v_0^2 + 2a(y-y_0)$$
$$v_0 = \sqrt{-2a(y-y_0)}$$
$$v_0 = \sqrt{-2(-9.80\,m/s^2)(0.640\,m)} = \underline{3.54\,m/s}$$

b) When the flea has returned to the ground $y - y_0 = 0$.

$y - y_0 = 0$
$v_0 = +3.54\,m/s$
$a = -9.80\,m/s^2$
$t = ?$

$$\cancel{y-y_0}^{\,0} = v_0 t + \tfrac{1}{2}at^2$$
$$0 = v_0 t + \tfrac{1}{2}at^2$$
$$t = -\frac{2v_0}{a} = -\frac{2(3.54\,m/s)}{-9.80\,m/s^2} = \underline{0.722\,s}$$

2-27

a) $v_{av} = \dfrac{\Delta y}{\Delta t} = \dfrac{y-y_0}{t}$

We need to use the constant acceleration formulas to find t. Take the origin of coordinates at the roof and take the positive y-direction to be upward.

$v_0 = +8.00\,m/s$
$y - y_0 = -12.0\,m$
$a = -9.80\,m/s^2$
$t = ?$

First find v:

$$v^2 = v_0^2 + 2a(y-y_0)$$
$$v = -\sqrt{v_0^2 + 2a(y-y_0)}$$

(We know that v is negative since the rock is traveling downward when it reaches the ground.)

$$v = -\sqrt{(8.00\,m/s)^2 + 2(-9.80\,m/s^2)(-12.0\,m)} = -17.30\,m/s$$

Now use $v = v_0 + at$ to solve for t:

$$t = \frac{v - v_0}{a} = \frac{-17.3\,m/s - 8.00\,m/s}{-9.80\,m/s^2} = 2.582\,s$$

2-27 (cont)

Finally, $v_{av} = \frac{y-y_0}{t} = \frac{-12.0\text{ m}}{2.582\text{ s}} = -4.65\text{ m/s}$.

The average velocity has magnitude 4.65 m/s and from the minus sign we see that it is directed downward.

b) $a_{av} = \frac{\Delta v}{\Delta t} = \frac{v - v_0}{t} = \frac{-17.30\text{ m/s} - 8.00\text{ m/s}}{2.582\text{ s}} = -9.80\text{ m/s}^2$

The average acceleration has magnitude 9.80 m/s² and is directed downward. (The acceleration is <u>constant</u> and equal to −9.80 m/s², so the average value must equal this constant value.)

<u>2-29</u>

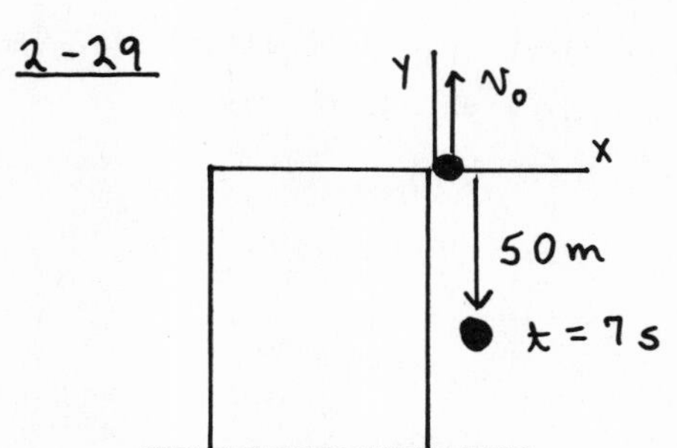

Take the positive y-direction to be upward, and the origin of coordinates at the cornice.

a) Consider the motion from the initial point to 7.0 s later:

$v_0 = ?$
$t = 7.00\text{ s}$
$y - y_0 = -50.0\text{ m}$
$a = -9.80\text{ m/s}^2$

$$y - y_0 = v_0 t + \tfrac{1}{2} a t^2$$

$$v_0 = \frac{y-y_0}{t} - \tfrac{1}{2} a t = \frac{-50.0\text{ m}}{7.00\text{ s}} - \tfrac{1}{2}(-9.80\text{ m/s}^2)(7.00\text{ s})$$

$$v_0 = -7.143\text{ m/s} + 34.30\text{ m/s} = \underline{27.2\text{ m/s}}$$

b) Consider the motion from the initial point to the highest point. At the maximum height $v = 0$.

$v = 0$
$a = -9.8\text{ m/s}^2$
$v_0 = +27.2\text{ m/s}$
$y - y_0 = ?$

$$v^2 = v_0^2 + 2a(y - y_0)$$

$$y - y_0 = \frac{v^2 - v_0^2}{2a} = \frac{0^2 - (27.2\text{ m/s})^2}{2(-9.80\text{ m/s}^2)} = \underline{+37.6\text{ m}}$$

c) $v = 0$ at the maximum height.

d) In free-fall $a = 9.80\text{ m/s}^2$, downward at <u>all</u> points in the motion.

<u>2-31</u>

a) $v = 447\text{ m/s}$
$v_0 = 0$
$t = 1.80\text{ s}$
$a = ?$

$$v = v_0 + at$$

$$a = \frac{v - v_0}{t} = \frac{447\text{ m/s} - 0}{1.80\text{ s}} = \underline{248\text{ m/s}^2}$$

2-31 (cont)

b) $\frac{a}{g} = \frac{248\text{ m/s}^2}{9.80\text{ m/s}^2} = \underline{25.3}$

c) $x - x_0 = v_0 t + \frac{1}{2}at^2 = 0 + \frac{1}{2}(248\text{ m/s}^2)(1.80\text{ s})^2 = \underline{402\text{ m}}$

d) Calculate the acceleration, assuming that it is constant:

$t = 1.40\text{ s}$

$v_0 = 283\text{ m/s}$

$v = 0$ (stops)

$a = ?$

$v = v_0 + at$

$a = \frac{v - v_0}{t} = \frac{0 - 283\text{ m/s}}{1.40\text{ s}} = -202\text{ m/s}^2$

$\frac{a}{g} = \frac{-202\text{ m/s}^2}{9.80\text{ m/s}^2} = -20.6 \Rightarrow a = -20.6g$

If the acceleration while the sled is stopping is constant then the magnitude of the acceleration is only 20.6g. But if the acceleration is not constant, it is certainly possible that at some point the instantaneous acceleration could be as large as 40g.

2-33

$a = At - Bt^2$ with $A = 1.90\text{ m/s}^3$ and $B = 0.120\text{ m/s}^4$

a) $v = v_0 + \int_0^t a\,dt = v_0 + \int_0^t (At - Bt^2)\,dt = v_0 + \frac{1}{2}At^2 - \frac{1}{3}Bt^3$

At rest at $t = 0 \Rightarrow v_0 = 0$, so

$v = \frac{1}{2}At^2 - \frac{1}{3}Bt^3 = \frac{1}{2}(1.90\text{ m/s}^3)t^2 - \frac{1}{3}(0.120\text{ m/s}^4)t^3$

$$\boxed{v = (0.95\text{ m/s}^3)t^2 - (0.040\text{ m/s}^4)t^3}$$

$x = x_0 + \int_0^t v\,dt$

Use $v = \frac{1}{2}At^2 - \frac{1}{3}Bt^3 \Rightarrow x = x_0 + \int_0^t \left[\frac{1}{2}At^2 - \frac{1}{3}Bt^3\right]dt$

$x = x_0 + \frac{1}{6}At^3 - \frac{1}{12}Bt^4$

At the origin at $t = 0 \Rightarrow x_0 = 0$.

$x = \frac{1}{6}At^3 - \frac{1}{12}Bt^4 = \frac{1}{6}(1.90\text{ m/s}^3)t^3 - \frac{1}{12}(0.120\text{ m/s}^4)t^4$

$$\boxed{x = (0.317\text{ m/s}^3)t^3 - (0.010\text{ m/s}^4)t^4}$$

b) At the time t when v is maximum, $\frac{dv}{dt} = 0$. (Since $a = \frac{dv}{dt}$, the maximum velocity is when $a = 0$. For earlier times a is positive so v is still increasing. For later times a is negative and v is decreasing.)

$a = \frac{dv}{dt} = 0 \Rightarrow At - Bt^2 = 0$

$t = \frac{A}{B} = \frac{1.90\text{ m/s}^3}{0.120\text{ m/s}^4} = 15.83\text{ s}$

Then $v = (0.95\text{ m/s}^3)t^2 - (0.040\text{ m/s}^4)t^3$ gives

$v = (0.95\text{ m/s}^3)(15.83\text{ s})^2 - (0.040\text{ m/s}^4)(15.83\text{ s})^3 = 238.06\text{ m/s} - 158.67\text{ m/s}$

$v = \underline{79.4\text{ m/s}}$

2-37

Let W stand for the woman, G for the ground, and S for the sidewalk. Take the positive direction to be the direction in which the sidewalk is moving.

The velocities are $v_{W/G}$ (woman relative to the ground), $v_{W/S}$ (woman relative to the sidewalk), and $v_{S/G}$ (sidewalk relative to the ground).

Eq. (2-24) becomes $v_{W/G} = v_{W/S} + v_{S/G}$

The time to reach the other end is given by $t = \dfrac{\text{distance traveled relative to ground}}{v_{W/G}}$.

a) $v_{S/G} = 1.0 \text{ m/s}$

$v_{W/S} = 2.0 \text{ m/s}$

$v_{W/G} = v_{W/S} + v_{S/G} = 2.0 \text{ m/s} + 1.0 \text{ m/s} = 3.0 \text{ m/s}$

$$t = \frac{80.0 \text{ m}}{v_{W/G}} = \frac{80.0 \text{ m}}{3.0 \text{ m/s}} = \underline{26.7 \text{ s}}$$

b) $v_{S/G} = 1.0 \text{ m/s}$

$v_{W/S} = -2.0 \text{ m/s}$

$v_{W/G} = v_{W/S} + v_{S/G} = -2.0 \text{ m/s} + 1.0 \text{ m/s} = -1.0 \text{ m/s}$

$$t = \frac{-80.0 \text{ m}}{v_{W/G}} = \frac{-80.0 \text{ m}}{-1.0 \text{ m/s}} = \underline{80.0 \text{ s}}$$

(Now the woman travels opposite to the direction the sidewalk is moving; she gets on at the opposite end than she did in part (a).)

Problems

2-41 $a_{av} = \dfrac{\Delta v}{\Delta t} = \dfrac{v - v_0}{t}$

$v_0 = 0$ since the runner starts from rest.

$t = 4.0$ s, but we need to calculate v, the speed of the runner at the end of the acceleration period.

For the last 9.1 s − 4.0 s = 5.1 s the acceleration is zero and the runner travels a distance of $d_1 = (5.1 \text{ s})\, v$ (from $x - x_0 = v_0 t + \frac{1}{2} a t^2$).

During the acceleration phase of 4.0 s, where the velocity goes from 0 to v, the runner travels a distance

$$d_2 = \left(\frac{v_0 + v}{2}\right) t = \frac{v}{2}(4.0 \text{ s}) = (2.0 \text{ s})\, v$$

The total distance traveled is 100 m, so $d_1 + d_2 = 100$ m. This gives $(5.1 \text{ s})\, v + (2.0 \text{ s})\, v = 100 \text{ m}$

2-41 (cont) $v = \frac{100\,m}{7.1\,s} = 14.08\ m/s$

Now we can calculate a_{av}:

$$a_{av} = \frac{v - v_0}{t} = \frac{14.08\ m/s - 0}{4.0\ s} = \underline{3.5\ m/s^2}$$

2-43

Take the origin to be at Seward and the positive direction to be west.

a) $\text{average speed} = \frac{\text{distance traveled}}{\text{time}}$

The distance traveled (different from the net displacement $x - x_0$) is $76\,km + 34\,km = 110\,km$.

Find the total elapsed time by using $v_{av} = \frac{\Delta x}{\Delta t} = \frac{x - x_0}{t}$ to find t for each leg of the journey.

Seward to Auora: $t = \frac{x - x_0}{v_{av}} = \frac{76\ km}{88\ km/h} = 0.8636\ h$

Auora to York: $t = \frac{x - x_0}{v_{av}} = \frac{-34\ km}{-79\ km/h} = 0.4304\ h$

Total $t = 0.8636\,h + 0.4304\,h = 1.294\,h$.

$$\text{average speed} = \frac{110\ km}{1.294\ h} = \underline{85\ km/h}$$

b) $v_{av} = \frac{\Delta x}{\Delta t}$, where Δx is the displacement, not the total distance traveled.

For the whole trip he ends up $76\,km - 34\,km = 42\,km$ west of his starting point.

Thus $v_{av} = \frac{42\,km}{1.294\ h} = \underline{32\ km/h}$.

2-47

a)

Y

truck: $a = 0$, $v_0 = 14.0\ m/s$

X

auto: $v_0 = 0$, $a = 2.00\ m/s^2$

At $t = 0$ the auto and truck are at the same position. "The auto overtakes the truck" means that after some time, call it T, the auto and truck will have undergone the same displacement d.

truck	auto
$a = 0$ (constant speed)	$a = 2.00\ m/s^2$
$v_0 = 14.0\ m/s$	$v_0 = 0$ (starts from rest)
$t = T$	$t = T$
$x - x_0 = d$	$x - x_0 = d$

2-47 (cont)

Apply the equation $x - x_0 = v_0 t + \frac{1}{2}at^2$ to the motion of each object.

truck: $d = (14.0\ \text{m/s})\, T$

auto: $d = \frac{1}{2}(2.00\ \text{m/s}^2)\, T^2$

Combine these equations $\Rightarrow$ $(14.0\text{m/s})\, T = \frac{1}{2}(2.00\text{m/s}^2)\, T^2$

$$T = \frac{2(14.0\text{m/s})}{2.00\ \text{m/s}^2} = 14.0\,\text{s}$$

Now that we have T we can use either equation to calculate d:

$$d = (14.0\text{m/s})\,T = (14.0\text{m/s})(14.0\,\text{s}) = \underline{196\ \text{m}}$$

$$\left(\text{or } d = \tfrac{1}{2}(2.00\text{m/s}^2)\,T^2 = \tfrac{1}{2}(2.00\ \text{m/s}^2)(14.0\text{s})^2 = 196\text{m}\right)$$

b) For the auto $v = v_0 + at = 0 + (2.00\text{m/s}^2)(14.0\text{s}) = \underline{28.0\ \text{m/s}}$

c)

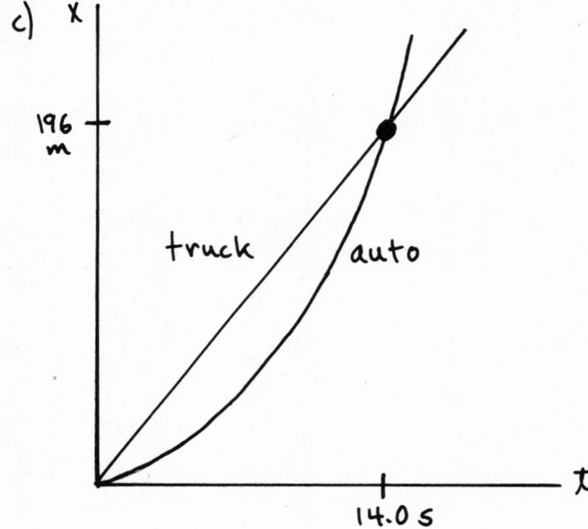

The graph shows that for $t < 14\,\text{s}$ the truck has traveled a greater distance than the auto, so is ahead.

At $t = 14\,\text{s}$ the displacements are the same, the two curves intersect, and the auto and truck are alongside each other. The graph also shows that at $t = 14\,\text{s}$ the speed of the auto (slope of the tangent to the curve) is greater than that of the truck, and that for $t > 14\,\text{s}$ the auto gets farther and farther ahead of the truck.

2-49

auto, d, truck, $a = 3.50\text{m/s}^2$, $v_0 = 0$; truck $a = 2.20\text{m/s}^2$, $v_0 = 0$; axes x, y

Take the origin of coordinates to be at the initial position of the truck. Let d be the distance that the auto initially is behind the truck, so $x_0(\text{auto}) = -d$ and $x_0(\text{truck}) = 0$. Let T be the time it takes the auto to catch the truck. Thus at time T the truck has undergone a displacement $x - x_0 = 75\ \text{m}$, so is at $x = x_0 + 75\ \text{m} = 75\text{m}$. The auto has caught the truck so at time T it is also at $x = 75\,\text{m}$.

2-49 (cont)

a) Use the motion of the truck to calculate T:

$x - x_0 = 75.0\ \text{m}$

$v_0 = 0$ (starts from rest)

$a = 2.20\ \text{m/s}^2$

$t = T$

$x - x_0 = \cancel{v_0 t}^{\,0} + \frac{1}{2}at^2$

$t = \sqrt{\frac{2(x-x_0)}{a}}$

$T = \sqrt{\frac{2(75.0\ \text{m})}{2.20\ \text{m/s}^2}} = \underline{8.26\ \text{s}}$

b) Use the motion of the auto to calculate d:

$x - x_0 = 75.0\ \text{m} + d$

$v_0 = 0$

$a = 3.50\ \text{m/s}^2$

$t = 8.26\ \text{s}$

$x - x_0 = \cancel{v_0 t}^{\,0} + \frac{1}{2}at^2$

$d + 75.0\ \text{m} = \frac{1}{2}(3.50\ \text{m/s}^2)(8.26\ \text{s})^2$

$d = 119.3\ \text{m} - 75\ \text{m} = \underline{44.3\ \text{m}}$

c) auto: $v = v_0 + at = 0 + (3.50\ \text{m/s}^2)(8.26\ \text{s}) = \underline{28.9\ \text{m/s}}$

truck: $v = v_0 + at = 0 + (2.20\ \text{m/s}^2)(8.26\ \text{s}) = \underline{18.2\ \text{m/s}}$

d)

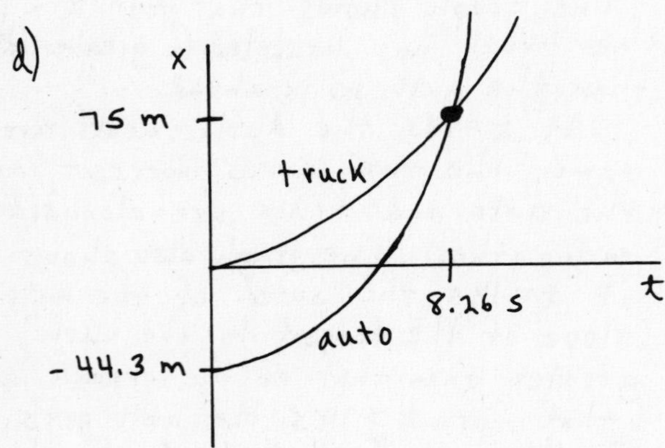

2-51

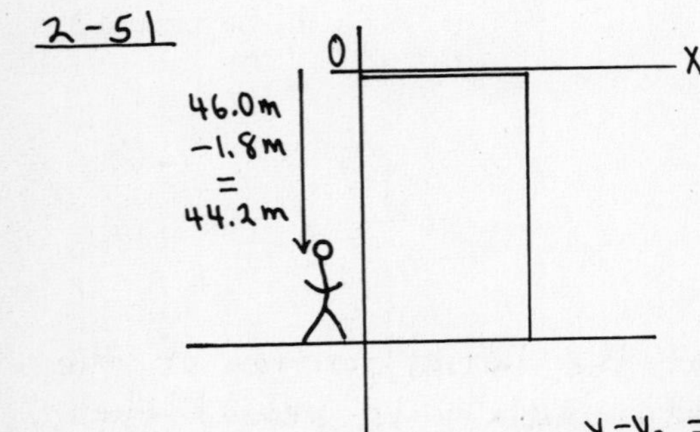

Take downward to be the +y-direction, and take the origin of coordinates to be at the roof.

First find the time for the egg to fall from the roof to the height of the top of the prof's head.

$y - y_0 = +(46.0\text{m} - 1.8\text{m}) = +44.2\ \text{m}$

$v_0 = 0$ (released from rest)

$a = +9.8\ \text{m/s}^2$

$t = ?$

$y - y_0 = \cancel{v_0 t}^{\,0} + \frac{1}{2}at^2$

$t = \sqrt{\frac{2(y-y_0)}{a}} = \sqrt{\frac{2(44.2\text{m})}{9.8\text{m/s}^2}}$

$t = 3.00\ \text{s}$

The distance the prof walks during this time is

$x - x_0 = v_0 t + \cancel{\frac{1}{2}at^2} = (1.20\text{m/s})(3.00\text{s}) = 3.60\ \text{m}$

Therefore you should release the egg when the professor is 3.60m from the point directly below you.

2-53

a)

Calculate the speed of the diver when she reaches the water. Take the origin of coordinates to be at the platform, and take the +y-direction to be downward.

$y - y_0 = +24.4\text{ m}$

$a = +9.80\text{ m/s}^2$

$v_0 = 0$ (diver just steps off)

$v = ?$

$v^2 = \cancel{v_0^2} + 2a(y - y_0)$

$v = +\sqrt{2a(y-y_0)}$

$v = +\sqrt{2(9.80\text{ m/s}^2)(24.4\text{ m})}$

$v = +21.9\text{ m/s}$

We know that v is positive because the diver is traveling downward when she reaches the water.

Convert to km/h to compare to the announcer's claim:

$v = (21.9\text{ m/s})\left(\frac{1\text{ km}}{10^3\text{ m}}\right)\left(\frac{3600\text{ s}}{1\text{ h}}\right) = 78.7\text{ km/h}$; the announcer has exaggerated the final speed of the diver.

b) Use the same coordinates as in part (a). Calculate the initial upward velocity needed to give the diver a speed of 105 km/h when she reaches the water.

$v_0 = ?$

$v = +(105\text{ km/h})\left(\frac{10^3\text{ m}}{1\text{ km}}\right)\left(\frac{1\text{ h}}{3600\text{ s}}\right) = 29.17\text{ m/s}$

$a = +9.80\text{ m/s}^2$

$y - y_0 = +24.4\text{ m}$

$v^2 = v_0^2 + 2a(y - y_0)$

$v_0 = -\sqrt{v^2 - 2a(y-y_0)} = -\sqrt{(29.17\text{ m/s})^2 - 2(9.80\text{ m/s}^2)(24.4\text{ m})} = \underline{-19.3\text{ m/s}}$

(v_0 is negative since the direction of the initial velocity is upward.)

One way to decide if this speed is physically reasonable is to calculate the maximum height above the platform it would produce:

$v_0 = -19.3\text{ m/s}$

$v = 0$ (at maximum height)

$a = +9.80\text{ m/s}^2$

$y - y_0 = ?$

$v^2 = v_0^2 + 2a(y - y_0)$

$y - y_0 = \frac{v^2 - v_0^2}{2a} = \frac{0 - (-19.3\text{ m/s})^2}{2(+9.80\text{ m/s}^2)} = -19.0\text{ m}$; this is not physically attainable.

2-57

a) It is very convenient to work in coordinates attached to the truck. Note that these coordinates move at constant velocity relative to the earth. In these coordinates the truck is at rest, and the initial velocity of the car is $v_0 = 0$. Also, the car's acceleration in these coordinates is the same as in coordinates fixed to the earth.

First, let's calculate how far the car must travel relative to the truck:

2-57 (cont)

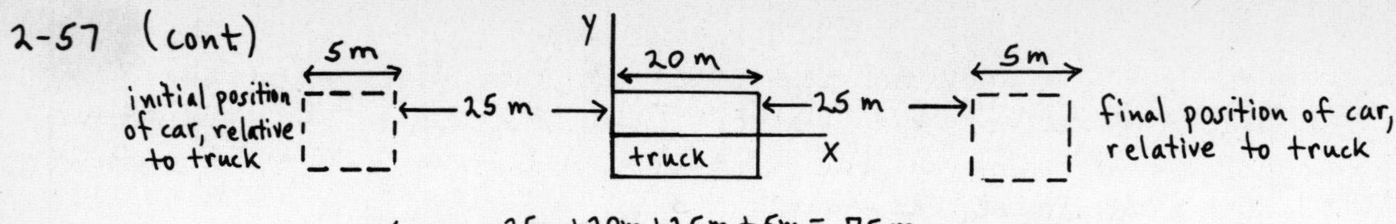

The car goes from $x_0 = -25.0\text{ m}$ to $x = 50.0\text{ m}$, so $x - x_0 = 75.0\text{ m}$.
Calculate the time that it takes the car to travel this distance:

$a = 0.800\text{ m/s}^2$
$v_0 = 0$
$x - x_0 = 75.0\text{ m}$
$t = ?$

$$x - x_0 = \cancel{v_0 t} + \tfrac{1}{2}at^2$$

$$t = \sqrt{\frac{2(x-x_0)}{a}} = \sqrt{\frac{2(75.0\text{ m})}{0.800\text{ m/s}^2}} = \underline{13.7\text{ s}}$$

b) Need how far the car travels relative to the earth, so go now to coordinates fixed in the earth. In these coordinates $v_0 = 20.0\text{ m/s}$ for the car. Take the origin to be at the initial position of the car.

$v_0 = 20.0\text{ m/s}$
$a = 0.800\text{ m/s}^2$
$t = 13.7\text{ s}$
$x - x_0 = ?$

$$x - x_0 = v_0 t + \tfrac{1}{2}at^2$$
$$x - x_0 = (20.0\text{ m/s})(13.7\text{ s}) + \tfrac{1}{2}(0.800\text{ m/s}^2)(13.7\text{ s})^2$$
$$x - x_0 = 274\text{ m} + 75\text{ m} = \underline{349\text{ m}}$$

c) In coordinates fixed to the earth:

$$v = v_0 + at$$
$$v = 20.0\text{ m/s} + (0.800\text{ m/s}^2)(13.7\text{ s}) = 31.0\text{ m/s}$$

2-59

The trick is to find the time it takes for the twins to meet. Since the pigeon flies at constant speed the total distance the pigeon flies is just this speed times the time it flies.

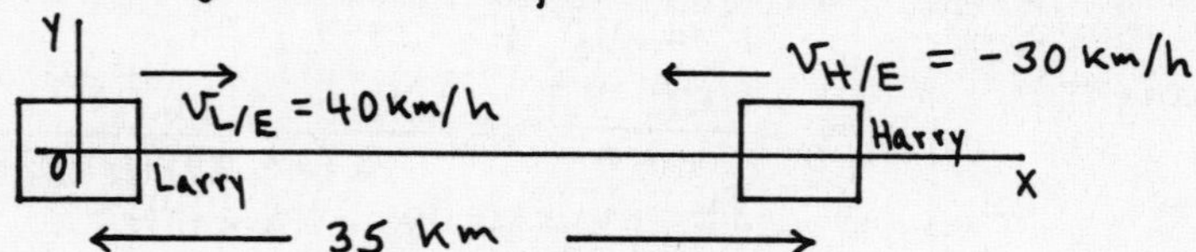

Work in a coordinate system attached to Larry's car. In these coordinates Harry undergoes a displacement of -35 km and the velocity of Harry is $v_{H/L}$, the velocity of Harry relative to Larry. Let $v_{L/E} = +40\text{ km/h}$ be the velocity of Larry relative to the earth and let $v_{H/E} = -30\text{ km/h}$ be the velocity of Harry relative to the earth (in the opposite direction). Then the relative velocity formula (Eq. 2-24) gives $v_{H/E} = v_{H/L} + v_{L/E}$, or
$v_{H/L} = v_{H/E} - v_{L/E} = -30\text{ km/h} - 40\text{ km/h} = -70\text{ km/h}$.

Then $x - x_0 = vt \Rightarrow t = \dfrac{x - x_0}{v} = \dfrac{-35\text{ km}}{-70\text{ km/h}} = +0.500\text{ h}$.

The distance the pigeon flies in this time is
distance = (speed)(time) = $(50\text{ km/h})(0.500\text{ h}) = \underline{25\text{ km}}$.

CHAPTER 3

Exercises 1, 3, 5, 7, 11, 15, 17, 21, 23, 25

Problems 33, 35, 37, 39, 43, 45, 47, 51

Exercises

3-1

a) $(v_x)_{av} = \dfrac{\Delta x}{\Delta t} = \dfrac{x_2 - x_1}{t_2 - t_1} = \dfrac{-4.5\text{ m} - 2.7\text{ m}}{4.0\text{ s} - 0} = \underline{-1.80\text{ m/s}}$

$(v_y)_{av} = \dfrac{\Delta y}{\Delta t} = \dfrac{y_2 - y_1}{t_2 - t_1} = \dfrac{8.1\text{ m} - 3.8\text{ m}}{4.0\text{ s} - 0} = \underline{+1.08\text{ m/s}}$

b) v_{av}, $(v_y)_{av}$, $(v_x)_{av}$, θ, x, y

$\tan\theta = \dfrac{(v_y)_{av}}{(v_x)_{av}} = \dfrac{1.08\text{ m/s}}{-1.80\text{ m/s}} = -0.600$

$\theta = \underline{149^\circ}$ (measured counterclockwise from +x-axis)

$v_{av} = \sqrt{(v_x)_{av}^2 + (v_y)_{av}^2} = \sqrt{(-1.80\text{m/s})^2 + (1.08\text{m/s})^2}$

$v_{av} = \underline{2.1\text{ m/s}}$

3-3

a) $(a_x)_{av} = \dfrac{\Delta v_x}{\Delta t} = \dfrac{v_{2x} - v_{1x}}{t_2 - t_1} = \dfrac{110\text{ m/s} - 190\text{ m/s}}{20.0\text{ s} - 0} = \underline{-4.00\text{ m/s}^2}$

$(a_y)_{av} = \dfrac{\Delta v_y}{\Delta t} = \dfrac{v_{2y} - v_{1y}}{t_2 - t_1} = \dfrac{60\text{ m/s} - (-120\text{ m/s})}{20.0\text{s} - 0} = \underline{+9.00\text{ m/s}^2}$

b)

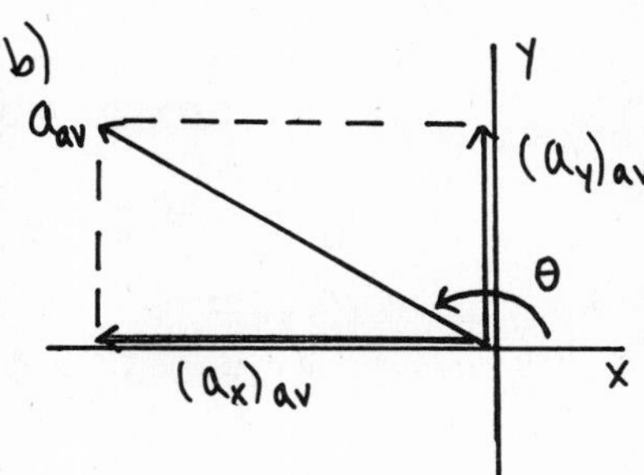

$\tan\theta = \dfrac{(a_y)_{av}}{(a_x)_{av}} = \dfrac{9.00\text{ m/s}^2}{-4.00\text{ m/s}^2} = -2.25$

$\theta = \underline{114^\circ}$ (measured counterclockwise from +x-axis)

$a_{av} = \sqrt{(a_x)_{av}^2 + (a_y)_{av}^2} = \sqrt{(-4.00\text{m/s}^2)^2 + (9.00\text{m/s}^2)^2}$

$a_{av} = \underline{9.85\text{ m/s}^2}$

3-5

a) $x = 2\text{m} - \alpha t, \quad \alpha = 3.6\text{ m/s}$

$y = \beta t^2, \quad \beta = 2.8\text{ m/s}^2$

$v_x = \dfrac{dx}{dt} = -\alpha = -3.6\text{ m/s}$

$v_y = \dfrac{dy}{dt} = 2\beta t = (5.6\text{m/s}^2)t$

$\Rightarrow \vec{v} = (-3.6\text{ m/s})\vec{i} + (5.6\text{ m/s}^2)t\vec{j}$

3-5 (cont)

$$a_x = \frac{dv_x}{dt} = 0$$

$$a_y = \frac{dv_y}{dt} = 2\beta = 5.6\,\text{m/s}^2 \qquad \Rightarrow \vec{a} = (5.6\,\text{m/s}^2)\,\vec{j}$$

b) velocity

$t = 3.0\,\text{s} \Rightarrow v_x = -3.6\,\text{m/s}\,,\ v_y = (5.6\,\text{m/s}^2)(3.0\,\text{s}) = 16.8\,\text{m/s}$

$$\tan\theta = \frac{v_y}{v_x} = \frac{16.8\,\text{m/s}}{-3.6\,\text{m/s}} = -4.67$$

$\theta = \underline{102^\circ}$ (measured counterclockwise from +x-axis)

$$v = \sqrt{v_x^2 + v_y^2} = \sqrt{(-3.6\,\text{m/s})^2 + (16.8\,\text{m/s})^2} = \underline{17.2\,\text{m/s}}$$

acceleration

$t = 3.0\,\text{s} \Rightarrow a_x = 0,\ a_y = 5.6\,\text{m/s}^2$ (both acceleration components are independent of time)

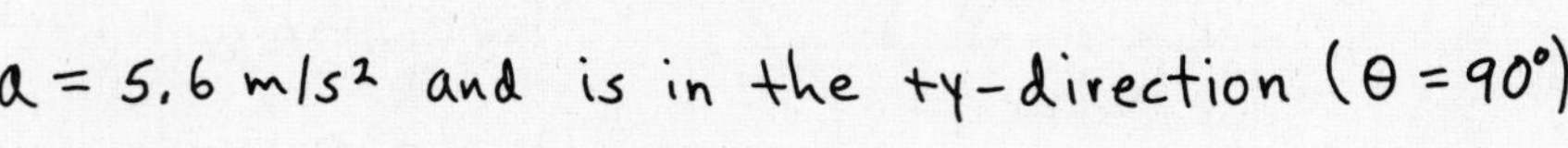

$a = 5.6\,\text{m/s}^2$ and is in the +y-direction ($\theta = 90^\circ$)

3-7

Take the positive y-direction to be upward. Take the origin of coordinates at the initial position of the book, at the point where it leaves the table top.

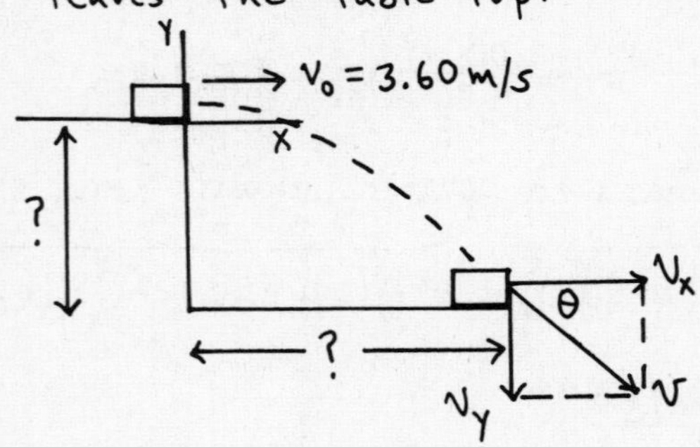

x-component	y-component
$a_x = 0$	$a_y = -9.80\,\text{m/s}^2$
$v_{0x} = 3.60\,\text{m/s}$	$v_{0y} = 0$
$t = 0.500\,\text{s}$	$t = 0.500\,\text{s}$

a) $y - y_0 = ?$

$$y - y_0 = v_{0y}t + \tfrac{1}{2}a_y t^2 = 0 + \tfrac{1}{2}(-9.80\,\text{m/s}^2)(0.500\,\text{s})^2 = -1.22\,\text{m}$$

The table top is $\underline{1.22\,\text{m}}$ above the floor.

b) $x - x_0 = ?$

$$x - x_0 = v_{0x}t + \tfrac{1}{2}a_x t^2 = (3.60\,\text{m/s})(0.500\,\text{s}) + 0 = \underline{1.80\,\text{m}}$$

c) $v_x = v_{0x} + \cancel{a_x t} = \underline{3.60\,\text{m/s}}$ (The x-component of the velocity is constant, since $a_x = 0$.)

$v_y = v_{0y} + a_y t = 0 + (-9.80\,\text{m/s}^2)(0.500\,\text{s}) = \underline{-4.90\,\text{m/s}}$

3-7 (cont)

$$\tan\theta = \frac{v_y}{v_x} = \frac{-4.90\ \text{m/s}}{3.60\ \text{m/s}} = -1.36 \Rightarrow \theta = -53.7^\circ$$

Direction of $\vec{v}$ is 53.7° below the horizontal.

$$v = \sqrt{v_x^2 + v_y^2} = \sqrt{(3.60\ \text{m/s})^2 + (-4.90\ \text{m/s})^2} = 6.08\ \text{m/s}$$

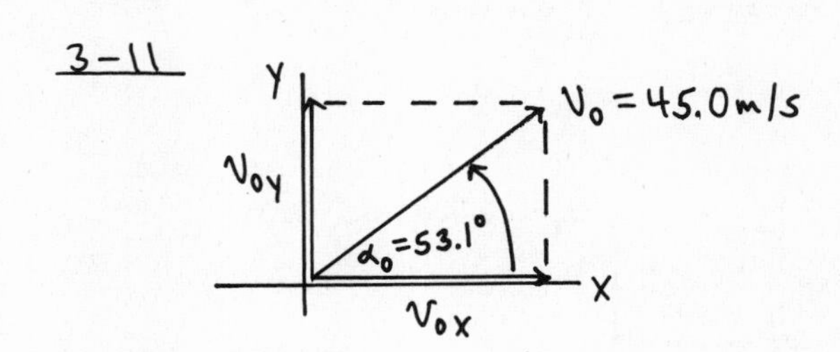

3-11

$v_0 = 45.0\ \text{m/s}$, $\alpha_0 = 53.1^\circ$

$$v_{0x} = v_0 \cos 53.1^\circ = +27.0\ \text{m/s}$$

$$v_{0y} = v_0 \sin 53.1^\circ = +36.0\ \text{m/s}$$

a) y-component (vertical motion)

$y - y_0 = +25.0\ \text{m}$
$v_{0y} = +36.0\ \text{m/s}$
$a_y = -9.80\ \text{m/s}^2$
$t = ?$

$$y - y_0 = v_{0y}t + \tfrac{1}{2}a_y t^2$$

$$25.0\ \text{m} = (36.0\ \text{m/s})t - (4.9\ \text{m/s}^2)t^2$$

$$(4.9\ \text{m/s}^2)t^2 - (36.0\ \text{m/s})t + 25.0\ \text{m} = 0$$

Apply the quadratic formula:

$$t = \frac{1}{9.8}\left[36.0 \pm \sqrt{(36.0)^2 - 4(4.9)(25)}\right]\ \text{s} = (3.672 \pm 2.896)\ \text{s}$$

The ball is at a height of 25.0 m 0.78 s and 6.57 s after being thrown.

b) $t = 0.78\ \text{s}$

$v_x = v_{0x} = +27.0\ \text{m/s}$ (same at any t, since $a_x = 0$)

$v_y = v_{0y} + a_y t = 36.0\ \text{m/s} + (-9.80\ \text{m/s}^2)(0.78\ \text{s}) = +28.4\ \text{m/s}$

($v_y > 0$ means the baseball is traveling upwards at this time)

$t = 6.57\ \text{s}$

$v_x = v_{0x} = +27.0\ \text{m/s}$

$v_y = v_{0y} + a_y t = +36.0\ \text{m/s} + (-9.8\ \text{m/s}^2)(6.57\ \text{s}) = -28.4\ \text{m/s}$

($v_y < 0$ means the baseball is traveling downward)

c) $v_x = v_{0x} = 27.0\ \text{m/s}$

solve for v_y:
$v_y = ?$
$y - y_0 = 0$ (when ball returns to level from which thrown)
$a_y = -9.80\ \text{m/s}^2$
$v_{0y} = +36.0\ \text{m/s}$

$$v_y^2 = v_{0y}^2 + 2a_y(y - y_0)^{\,0}$$

$v_y = -v_{0y} = -36.0\ \text{m/s}$
(minus since must be traveling downward at this point)

3-11 (cont)

Now that have the components can solve for the magnitude and direction of $\vec{v}$.

$$\tan\alpha = \frac{v_y}{v_x} = \frac{-36.0\ \text{m/s}}{+27.0\ \text{m/s}} = -1.33 \Rightarrow \alpha = -53.1^\circ$$

$$v = \sqrt{v_x^2 + v_y^2} = \sqrt{(27.0\ \text{m/s})^2 + (-36.0\ \text{m/s})^2} = 45.0\ \text{m/s}$$

The velocity of the ball when it returns to the level from which it was thrown has magnitude 45.0 m/s (the same as the initial speed) and is directed at an angle of 53.1° below the horizontal.

3-15

Take the origin of coordinates at the roof and let the +y-direction be upward.

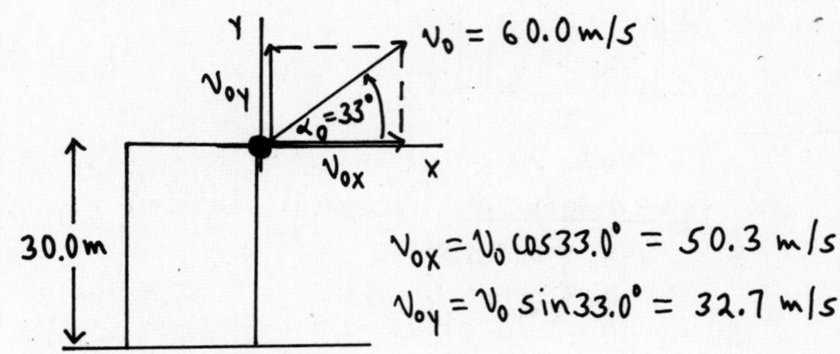

$v_{0x} = v_0 \cos 33.0° = 50.3$ m/s

$v_{0y} = v_0 \sin 33.0° = 32.7$ m/s

a) At the maximum height $v_y = 0$.

y-component

$a_y = -9.80\ \text{m/s}^2$

$v_y = 0$

$v_{0y} = +32.7$ m/s

$y - y_0 = ?$

$$v^2 = v_0^2 + 2a_y(y - y_0)$$

$$y - y_0 = \frac{v^2 - v_{0y}^2}{2a_y} = \frac{0^2 - (32.7\ \text{m/s})^2}{2(-9.80\ \text{m/s}^2)} = +54.6\ \text{m}$$

(measured from the roof)

b) $v_x = v_{0x} = 50.3$ m/s (since $a_x = 0$)

y-component

$v_y = ?$

$a_y = -9.80\ \text{m/s}^2$

$y - y_0 = -30.0$ m (minus because at ground rock is below its initial position)

$v_{0y} = +32.7$ m/s

$$v_y^2 = v_{0y}^2 + 2a_y(y - y_0)$$

$$v_y = -\sqrt{v_{0y}^2 + 2a_y(y - y_0)}$$

(v_y is negative because at the ground the rock is traveling downward)

$$v_y = -\sqrt{(32.7\ \text{m/s})^2 + 2(-9.80\ \text{m/s}^2)(-30.0\ \text{m})}$$

$$v_y = -40.7\ \text{m/s}$$

Then $v = \sqrt{v_x^2 + v_y^2} = \sqrt{(50.3\ \text{m/s})^2 + (-40.7\ \text{m/s})^2} = 64.7\ \text{m/s}$

3-15 (cont)

c) Use the vertical motion (y-component) to find the time the rock is in the air:

$t = ?$
$v_y = -40.7 \text{ m/s}$ (from part (b))
$a_y = -9.80 \text{ m/s}^2$
$v_{oy} = +32.7 \text{ m/s}$

$$v_y = v_{oy} + a_y t$$
$$t = \frac{v_y - v_{oy}}{a_y} = \frac{-40.7 \text{ m/s} - 32.7 \text{ m/s}}{-9.80 \text{ m/s}^2} = +7.49 \text{ s}$$

Can use this t to calculate the horizontal range:

$t = 7.49 \text{ s}$
$x - x_0 = ?$
$a_x = 0$
$v_{ox} = 50.3 \text{ m/s}$

$$x - x_0 = v_{ox} t + \cancel{\tfrac{1}{2} a_x t^2}$$
$$x - x_0 = (50.3 \text{ m/s})(7.49 \text{ s}) = \underline{377 \text{ m}}$$

3-17

Take the origin of coordinates at the point where the quarter leaves your hand, and take positive y to be upward.

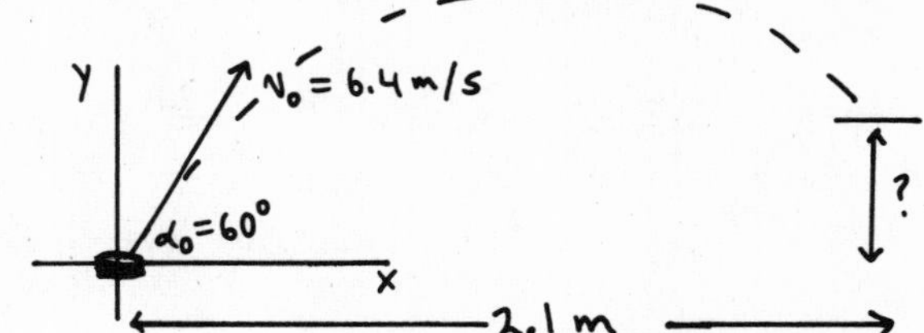

$v_{ox} = v_0 \cos\alpha_0 = (6.4 \text{ m/s}) \cos 60° = 3.2 \text{ m/s}$

$v_{oy} = v_0 \sin\alpha_0 = (6.4 \text{ m/s}) \sin 60° = 5.54 \text{ m/s}$

a) Use the horizontal (x-component) of motion to solve for t, the time the quarter travels through the air:

X-component

$t = ?$
$x - x_0 = 2.1 \text{ m}$
$v_{ox} = 3.2 \text{ m/s}$
$a_x = 0$

$$x - x_0 = v_{ox} t + \cancel{\tfrac{1}{2} a_x t^2}$$
$$t = \frac{x - x_0}{v_{ox}} = \frac{2.1 \text{ m}}{3.2 \text{ m/s}} = 0.656 \text{ s}$$

Now find the vertical displacement of the quarter after this time:

y-component

$y - y_0 = ?$
$a_y = -9.80 \text{ m/s}^2$
$v_{oy} = +5.54 \text{ m/s}$
$t = 0.656 \text{ s}$

$$y - y_0 = v_{oy} t + \tfrac{1}{2} a_y t^2$$
$$y - y_0 = (5.54 \text{ m/s})(0.656 \text{ s}) + \tfrac{1}{2}(-9.80 \text{ m/s}^2)(0.656 \text{ s})^2$$
$$y - y_0 = 3.64 \text{ m} - 2.11 \text{ m} = \underline{1.53 \text{ m}}$$

b) $v_y = ?$
$t = 0.656 \text{ s}$
$a_y = -9.80 \text{ m/s}^2$
$v_{oy} = +5.54 \text{ m/s}$

$$v_y = v_{oy} + a_y t$$
$$v_y = 5.54 \text{ m/s} + (-9.80 \text{ m/s}^2)(0.656 \text{ s})$$
$$v_y = -0.89 \text{ m/s}$$

(The minus sign indicates that the y-component of $\vec{v}$ is downward; at this point the quarter has passed through the highest point in its path and is on its way down.)

3-21

a) $|\vec{v}|$ constant $\Rightarrow a_{tan} = \frac{d|\vec{v}|}{dt} = 0$

$\vec{a}_{rad}$, $\vec{a}_{tan}$, $\vec{v}$

$$a_{rad} = \frac{v^2}{R} = \frac{(9.00\ m/s)^2}{14.0\ m} = 5.79\ m/s^2$$

The resultant acceleration is $\underline{5.79\ m/s^2}$, upward.

b) $v = \frac{2\pi R}{T}$ (Eq. 3-31) $\Rightarrow T = \frac{2\pi R}{v} = \frac{2\pi\,(14.0\,m)}{9.00\ m/s} = \underline{9.77\,s}$

3-23

a) N, E, W, S

The velocity vectors in the problem are:

$\vec{v}_{P/E}$, the velocity of the plane relative to the earth

$\vec{v}_{A/E}$, the velocity of the air relative to the earth (the wind velocity)

$\vec{v}_{P/A}$, the velocity of the plane relative to the air

The rule for adding these velocities is $\vec{v}_{P/E} = \vec{v}_{P/A} + \vec{v}_{A/E}$.

The problem tells us that $v_{P/A} = 290.0$ km/h. It tells us that $\vec{v}_{A/E}$ has magnitude 80.0 km/h and direction to the west. If the plane is to travel due north it must be that $\vec{v}_{P/E}$ is directed to the north.

The vector addition diagram is then

$v_{A/E} = 80.0$ km/h, $v_{P/E}$, $v_{P/A} = 290.0$ km/h, ϕ

(This diagram shows the vector addition $\vec{v}_{P/E} = \vec{v}_{P/A} + \vec{v}_{A/E}$ and also has $\vec{v}_{P/E}$ and $\vec{v}_{A/E}$ in their specified directions. Note that the vector addition diagram forms a right triangle.)

The direction the plane should head is the direction of $\vec{v}_{P/A}$.

From the diagram $\sin\phi = \frac{v_{A/E}}{v_{P/A}} = \frac{80.0\ km/h}{290.0\ km/h} = 0.2759 \Rightarrow \phi = 16.0°$

The pilot should head $\underline{16.0°\ \text{E of N}}$.

b) The Pythagorean theorem applied to the vector addition diagram gives $v^2_{P/A} = v^2_{A/E} + v^2_{P/E}$.

Thus $v_{P/E} = \sqrt{v^2_{P/A} - v^2_{A/E}} = \sqrt{(290.0\ km/h)^2 - (80.0\ km/h)^2} = \underline{279\ km/h}$

3-25

a) View the motion from above:

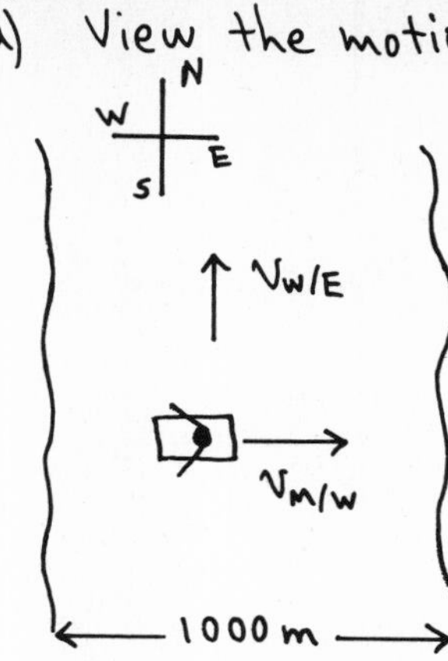

The velocity vectors in the problem are:
$\vec{v}_{M/E}$, the velocity of the man relative to the earth.
$\vec{v}_{W/E}$, the velocity of the water relative to the earth.
$\vec{v}_{M/W}$, the velocity of the man relative to the water.

The rule for adding these velocities is $\vec{v}_{M/E} = \vec{v}_{M/W} + \vec{v}_{W/E}$.
The problem tells us that $\vec{v}_{W/E}$ has magnitude 2.4 m/s and direction due north. It also tells us that $\vec{v}_{M/W}$ has magnitude 3.5 m/s and direction due east.

The vector addition diagram is then

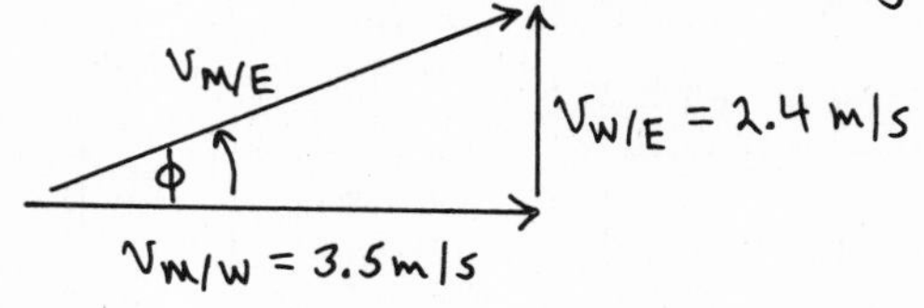

(This diagram shows the vector addition $\vec{v}_{M/E} = \vec{v}_{M/W} + \vec{v}_{W/E}$ and also has $\vec{v}_{M/W}$ and $\vec{v}_{W/E}$ in their specified directions. Note that the vector addition diagram forms a right triangle.)

The Pythagorean theorem applied to the vector addition diagram gives

$$v_{M/E}^2 = v_{M/W}^2 + v_{W/E}^2$$

$$v_{M/E} = \sqrt{v_{M/W}^2 + v_{W/E}^2} = \sqrt{(3.5\ m/s)^2 + (2.4\ m/s)^2} = 4.24\ m/s$$

$$\tan\phi = \frac{v_{W/E}}{v_{M/W}} = \frac{2.4\ m/s}{3.5\ m/s} = 0.6857 \Rightarrow \phi = 34.4^\circ$$

The velocity of the man relative to the earth has magnitude 4.24 m/s and direction 34.4° N of E.

b) This is a tricky question. To cross the river the man must travel 1000 m due east relative to the earth. The man's velocity relative to the earth is $\vec{v}_{M/E}$. But, from the vector addition diagram the eastward component of $\vec{v}_{M/E}$ equals $v_{M/W} = 3.5$ m/s.

Thus $t = \dfrac{x - x_0}{v_x} = \dfrac{1000\ m}{3.5\ m/s} = \underline{286\ s}$

c) The northward component of $\vec{v}_{M/E}$ equals $v_{W/E} = 2.4$ m/s. Therefore, in the 286 s it takes him to cross the river the distance north the man travels relative to the earth is

$$y - y_0 = v_y t = (2.4\ m/s)(286\ s) = \underline{686\ m}$$

Problems

3-33

$\vec{v} = (\alpha - \beta t^2)\vec{i} + \gamma t\vec{j}$; $v_x = \alpha - \beta t^2$, $v_y = \gamma t$

a) $a_x = \frac{dv_x}{dt} = \frac{d}{dt}(\alpha - \beta t^2) = -2\beta t$

$a_y = \frac{dv_y}{dt} = \frac{d}{dt}(\gamma t) = \gamma$

$$\boxed{\vec{a} = a_x\vec{i} + a_y\vec{j} = (-2\beta t)\vec{i} + \gamma\vec{j}}$$

From Eq. (2-21)

$x = x_0 + \int_0^t v_x\,dt = x_0 + \int_0^t (\alpha - \beta t^2)\,dt = x_0 + \alpha t - \frac{1}{3}\beta t^3$

$y = y_0 + \int_0^t v_y\,dt = y_0 + \int_0^t \gamma t\,dt = y_0 + \frac{1}{2}\gamma t^2$

Bird at the origin at $t = 0 \Rightarrow x_0 = y_0 = 0$.

Thus $\boxed{\vec{r} = x\vec{i} + y\vec{j} = (\alpha t - \frac{1}{3}\beta t^3)\vec{i} + (\frac{1}{2}\gamma t^2)\vec{j}}$

b) Find the value of t that gives $x = 0$:

$x = \alpha t - \frac{1}{3}\beta t^3 = 0 \Rightarrow t = 0$ or $\alpha - \frac{1}{3}\beta t^2 = 0$

$$t = \sqrt{\frac{3\alpha}{\beta}} = \sqrt{\frac{3(2.1\,m/s)}{3.6\,m/s^3}} = 1.32\,s$$

Now find y at this t:

$y = \frac{1}{2}\gamma t^2 = \frac{1}{2}(5.0\,m/s^2)(1.32\,s)^2 = \underline{4.4\,m}$

3-35

Take the origin of coordinates at the point where the cannister is released. Take +y to be upward. The initial velocity of the cannister is the velocity of the plane, 54.0 m/s horizontally.

Use the vertical motion to find the time of fall:

<u>y-component</u>

$t = ?$

$v_{0y} = 0$

$a_y = -9.80\,m/s^2$

$y - y_0 = -70.0\,m$

(when the cannister reaches the ground it is 70.0 m <u>below</u> the origin.)

$y - y_0 = \cancel{v_{0y}t} + \frac{1}{2}a_y t^2$

$$t = \sqrt{\frac{2(y-y_0)}{a_y}} = \sqrt{\frac{2(-70.0\,m)}{-9.80\,m/s^2}} = 3.78\,s$$

3-35 (cont)

Find out how far the cannister travels horizontally in this time:

x-component

$x - x_0 = ?$

$a_x = 0$

$v_{0x} = 54.0 \text{ m/s}$

$t = 3.78 \text{ s}$

$x - x_0 = v_{0x} t + \cancel{\tfrac{1}{2} a_x t^2}$

$x - x_0 = (54.0 \text{ m/s})(3.78 \text{ s}) = \underline{204 \text{ m}}$

3-37

a)

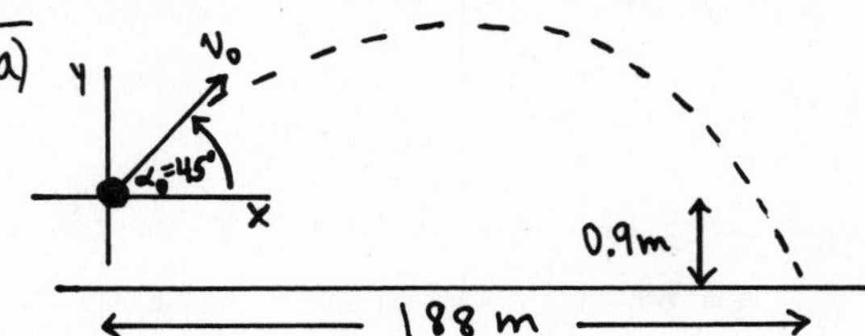

Take the origin of coordinates at the point where the ball leaves the bat, and take +y to be upward.

$v_{0x} = v_0 \cos\alpha_0$

$v_{0y} = v_0 \sin\alpha_0$, but we don't know v_0

Write down the equation for the horizontal displacement when the ball hits the ground and the corresponding equation for the vertical displacement. The time t is the same for both components, so this will give us two equations in two unknowns (v_0 and t).

y-component

$a_y = -9.80 \text{ m/s}^2$

$y - y_0 = -0.9 \text{ m}$

$v_{0y} = v_0 \sin 45°$

$y - y_0 = v_{0y} t + \tfrac{1}{2} a_y t^2$

$-0.9 \text{ m} = (v_0 \sin 45°) t + \tfrac{1}{2}(-9.80 \text{ m/s}^2) t^2$

x-component

$a_x = 0$

$x - x_0 = 188 \text{ m}$

$v_{0x} = v_0 \cos 45°$

$x - x_0 = v_{0x} t + \cancel{\tfrac{1}{2} a_x t^2}$

$t = \dfrac{x - x_0}{v_{0x}} = \dfrac{188 \text{ m}}{v_0 \cos 45°}$

Put this result into the y-component equation to eliminate t, and solve for v_0. (Note that $\sin 45° = \cos 45°$.)

$$-0.9 \text{ m} = (v_0 \sin 45°)\left(\frac{188 \text{ m}}{v_0 \cos 45°}\right) - (4.90 \text{ m/s}^2)\left(\frac{188 \text{ m}}{v_0 \cos 45°}\right)^2$$

$$4.90 \text{ m/s}^2 \left(\frac{188 \text{ m}}{v_0 \cos 45°}\right)^2 = 188 \text{ m} + 0.9 \text{ m} = 188.9 \text{ m}$$

$$\left(\frac{v_0 \cos 45°}{188 \text{ m}}\right)^2 = \frac{4.90 \text{ m/s}^2}{188.9 \text{ m}} \Rightarrow v_0 = \frac{188 \text{ m}}{\cos 45°}\sqrt{\frac{4.90 \text{ m/s}^2}{188.9 \text{ m}}} = \underline{42.8 \text{ m/s}}$$

b) Use the horizontal motion to find the time it takes the ball to reach the fence:

x-component

$x - x_0 = 116 \text{ m}$

$a_x = 0$

$v_{0x} = v_0 \cos 45° = (42.8 \text{ m/s}) \cos 45° = 30.3 \text{ m/s}$

$t = ?$

$x - x_0 = v_{0x} t + \cancel{\tfrac{1}{2} a_x t^2}$

$t = \dfrac{x - x_0}{v_{0x}} = \dfrac{116 \text{ m}}{30.3 \text{ m/s}} = 3.83 \text{ s}$

3-37 (cont)

Find the vertical displacement of the ball at this t:

y-component

$y - y_0 = ?$

$a_y = -9.80\ m/s^2$

$v_{oy} = v_0 \sin 45° = 30.3\ m/s$

$t = 3.83\ s$

$y - y_0 = v_{oy}t + \frac{1}{2}a_y t^2$

$y - y_0 = (30.3\ m/s)(3.83s) + \frac{1}{2}(-9.80\ m/s^2)(3.83\ s)^2$

$y - y_0 = 116.0\ m - 71.9\ m = +44.1\ m$, above point where ball was hit

The height of the ball above the ground is $44.1\,m + 0.9\,m = 45.0\ m$. It's height then above the top of the fence is $45.0\,m - 3.0\ m = \underline{42.0\ m}$.

3-39

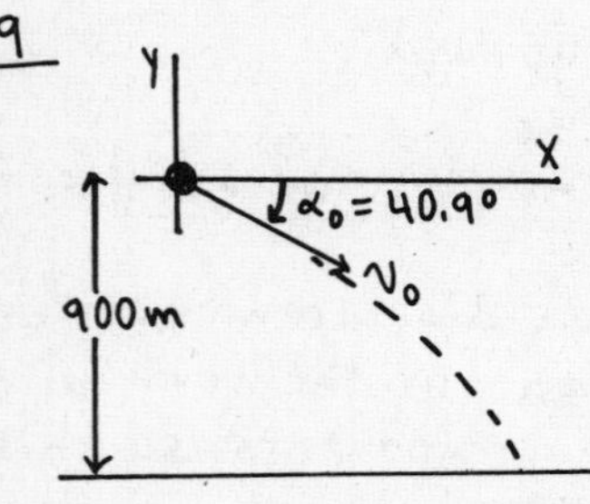

$v_{ox} = v_0 \cos \alpha_0$

$v_{oy} = -v_0 \sin \alpha_0$

Take the origin of coordinates to be at the point where the mailbag is released. Take +y to be upward. The initial velocity of the mailbag equals the velocity of the plane.

a) Use the y-component. Consider the motion from the origin to where the bag strikes the ground.

$a_y = -9.80\ m/s^2$

$y - y_0 = -900\ m$

$t = 6.00s$

$v_{oy} = -v_0 \sin 40.9°$

$y - y_0 = v_{oy}t + \frac{1}{2}a_y t^2$

$v_{oy} = \frac{y - y_0}{t} - \frac{1}{2}a_y t$

$v_{oy} = \frac{-900\ m}{6.00\ s} - \frac{1}{2}(-9.80\ m/s^2)(6.00s) = -150\ m/s + 29.4\ m/s$

$v_{oy} = -120.6\ m/s$

$v_0 = -\frac{v_{oy}}{\sin 40.9°} = -\frac{-120.6\ m/s}{\sin 40.9°} = 184\ m/s$

b) x-component

$a_x = 0$

$x - x_0 = ?$

$v_{ox} = v_0 \cos 40.9° = (184\ m/s) \cos 40.9°$

$v_{ox} = 139\ m/s$

$t = 6.00s$

$x - x_0 = v_{ox}t + \cancel{\frac{1}{2}a_x t^2}$

$x - x_0 = (139\ m/s)(6.00s) = \underline{834\ m}$

c) x-component

$a_x = 0 \Rightarrow v_x = v_{ox} = +\underline{139\ m/s}$

y-component

$v_y = ?$

$v_{oy} = -120.6\ m/s$

$a_y = -9.80\ m/s^2$

$t = 6.00\ s$

$v_y = v_{oy} + a_y t$

$v_y = -120.6\ m/s + (-9.80\ m/s^2)(6.00s)$

$v_y = \underline{-179\ m/s}$

3-43

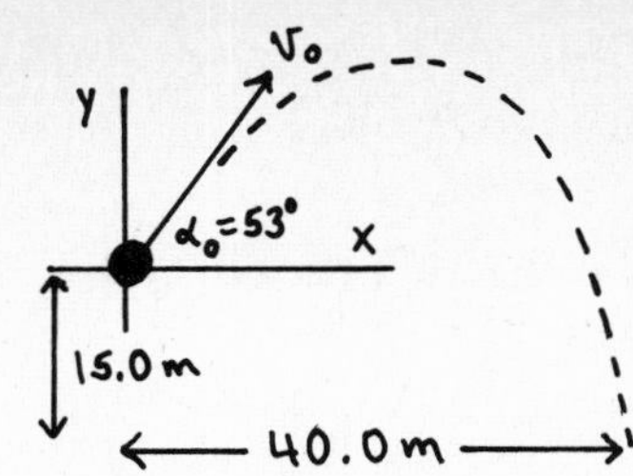

$v_{0x} = v_0 \cos\alpha_0$
$v_{0y} = v_0 \sin\alpha_0$

Take the origin of coordinates at the top of the ramp and take +y to be upward. The problem specifies that the object is displaced 40.0 m to the right when it is 15.0 m below the origin.

We don't know t, the time in the air, and we don't know v_0. Write down the equations for the horizontal and vertical displacements. Combine these two equations to eliminate one unknown.

y-component

$y - y_0 = -15.0$ m
$a_y = -9.80$ m/s²
$v_{0y} = v_0 \sin 53.0°$

$$y - y_0 = v_{0y}t + \tfrac{1}{2}a_y t^2$$
$$\boxed{-15.0\ \text{m} = (v_0 \sin 53°)t - (4.90\ \text{m/s}^2)t^2}$$

x-component

$x - x_0 = 40.0$ m
$a_x = 0$
$v_{0x} = v_0 \cos 53.0°$

$$x - x_0 = v_{0x}t + \cancel{\tfrac{1}{2}a_x t^2}$$
$$\boxed{40.0\ \text{m} = (v_0 t)\cos 53.0°}$$

The second eq. says $v_0 t = \dfrac{40.0\ \text{m}}{\cos 53.0°} = 66.47$ m.
Use this to replace $v_0 t$ in the first equation:

$$-15.0\ \text{m} = (66.47\ \text{m})\sin 53° - (4.90\ \text{m/s}^2)t^2$$

$$t = \sqrt{\frac{(66.47\ \text{m})\sin 53° + 15.0\ \text{m}}{4.90\ \text{m/s}^2}} = \sqrt{\frac{68.09\ \text{m}}{4.90\ \text{m/s}^2}} = 3.728\ \text{s}$$

Now that we have t we can use the x-component equation to solve for v_0:

$$v_0 = \frac{40.0\ \text{m}}{t \cos 53.0°} = \frac{40.0\ \text{m}}{(3.728\ \text{s})\cos 53.0°} = \underline{17.8\ \text{m/s}}$$

3-45

a)

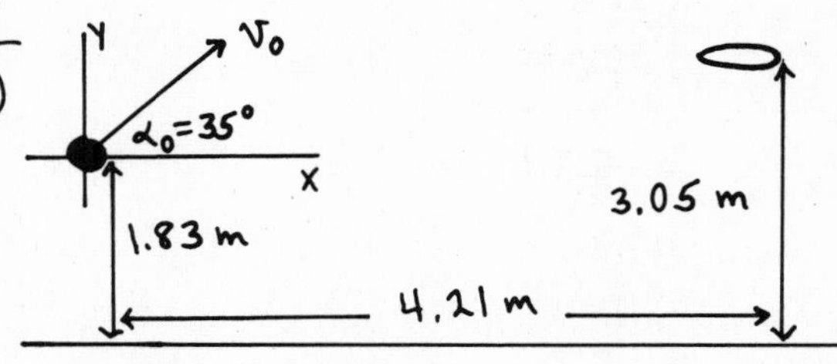

Take the origin of coordinates at the point where the player releases the ball.

$v_{0x} = v_0 \cos\alpha_0$
$v_{0y} = v_0 \sin\alpha_0$

y-component

$v_y = 0$ (at the maximum height)
$v_{0y} = v_0 \sin\alpha_0 = (4.88\ \text{m/s})\sin 35°$
$v_{0y} = 2.80$ m/s
$a_y = -9.80$ m/s²
$y - y_0 = ?$

$$v_y^2 = v_{0y}^2 + 2a_y(y - y_0)$$

$$y - y_0 = \frac{v_y^2 - v_{0y}^2}{2a_y} = \frac{0^2 - (2.80\ \text{m/s})^2}{2(-9.80\ \text{m/s}^2)} = +0.400\ \text{m}$$

The height above the floor then is
$0.400\ \text{m} + 1.83\ \text{m} = \underline{2.23\ \text{m}}$.

3-45 (cont)

b) Use the vertical motion to find the time the ball is in the air:

<u>y-component</u>

$y - y_0 = -1.83$ m (vertical displacement when the ball reaches the floor)

$v_{0y} = 2.80$ m/s

$a_y = -9.80$ m/s²

$t = ?$

$$y - y_0 = v_{0y} t + \tfrac{1}{2} a_y t^2$$

$$-1.83\ \text{m} = (2.80\ \text{m/s})t - (4.9\ \text{m/s}^2)t^2$$

$$4.9t^2 - 2.80t - 1.83 = 0$$

Quadratic formula gives

$$t = \frac{1}{9.8}\left(2.80 \pm \sqrt{(2.80)^2 + 4(4.9)(1.83)}\right)\text{s} = 0.286\ \text{s} \pm 0.675\ \text{s}$$

t must be positive, so $t = 0.286\ \text{s} + 0.675\ \text{s} = 0.961\ \text{s}$

Find the horizontal displacement for this t:

<u>x-component</u>

$x - x_0 = ?$

$v_{0x} = v_0 \cos\alpha_0 = (4.88\ \text{m/s})\cos 35°$

$v_{0x} = 4.00$ m/s

$a_x = 0$

$t = 0.961$ s

$$x - x_0 = v_{0x} t + \cancel{\tfrac{1}{2} a_x t^2}$$

$$x - x_0 = (4.00\ \text{m/s})(0.961\ \text{s}) = \underline{3.84\ \text{m}}$$

c) We don't know either v_0 or the time t for the ball to reach the basket. Write down the equations for the horizontal and vertical displacements to get two equations for these two unknowns.

<u>y-component</u>

$a_y = -9.80$ m/s²

$v_{0y} = v_0 \sin 35°$

$y - y_0 = 3.05\ \text{m} - 1.83\ \text{m} = 1.22\ \text{m}$

$$y - y_0 = v_{0y} t + \tfrac{1}{2} a_y t^2$$

$$\boxed{1.22\ \text{m} = (v_0 \sin 35°)t - (4.90\ \text{m/s}^2)t^2}$$

<u>x-component</u>

$a_x = 0$

$v_{0x} = v_0 \cos 35°$

$x - x_0 = 4.21$ m

$$x - x_0 = v_{0x} t + \cancel{\tfrac{1}{2} a_x t^2}$$

$$\boxed{4.21\ \text{m} = (v_0 \cos 35°)t}$$

$$v_0 t = \frac{4.21\ \text{m}}{\cos 35°} = 5.139\ \text{m}$$

Use this result in the y-component equation:

$$1.22\ \text{m} = (5.139\ \text{m})\sin 35° - (4.90\ \text{m/s}^2)t^2$$

$$(4.90\ \text{m/s}^2)t^2 = 1.728\ \text{m}$$

$$t = \sqrt{\frac{1.728\ \text{m}}{4.90\ \text{m/s}^2}} = 0.594\ \text{s}$$

Then $v_0 = \dfrac{5.139\ \text{m}}{t} = \dfrac{5.139\ \text{m}}{0.594\ \text{s}} = \underline{8.65\ \text{m/s}}$.

3-45 (cont)

d) We know that $v_0 = 8.65$ m/s from part (c)

<u>y-component</u>

$v_y = 0$ (at maximum height)

$a_y = -9.80 \text{ m/s}^2$

$v_{0y} = v_0 \sin 35° = (8.65 \text{ m/s}) \sin 35°$

$v_{0y} = 4.96$ m/s

$$v_y^2 = v_{0y}^2 + 2a_y(y - y_0)$$

$$y - y_0 = \frac{v_y^2 - v_{0y}^2}{2a_y} = \frac{0 - (4.96 \text{ m/s})^2}{2(-9.80 \text{ m/s}^2)} = +1.26 \text{ m}$$

The maximum height above the floor is given by $1.83 \text{ m} + 1.26 \text{ m} = \underline{3.09 \text{ m}}$.

Also use the y-component to find the time t to reach the maximum height; can then use this t to find the horizontal displacement at this point.

$$v_y = v_{0y} + a_y t$$

$$t = \frac{v_y - v_{0y}}{a_y} = \frac{0 - 4.96 \text{ m/s}}{-9.80 \text{ m/s}^2} = +0.506 \text{ s}$$

<u>x-component</u>

$x - x_0 = ?$

$a_x = 0$

$v_{0x} = v_0 \cos 35° = (8.65 \text{ m/s}) \cos 35°$

$v_{0x} = 7.09$ m/s

$t = 0.506$ s

$$x - x_0 = v_{0x} t + \cancel{\tfrac{1}{2} a_x t^2}$$

$$x - x_0 = (7.09 \text{ m/s})(0.506 \text{ s}) = 3.59 \text{ m}$$

This is the distance of the ball from the point where it was released. Its distance from the basket is then $4.21 \text{ m} - 3.59 \text{ m} = \underline{0.62 \text{ m}}$.

<u>3-47</u>

Use the equation that precedes Eq. (3-29):

$$R = (v_0 \cos \alpha_0) \frac{2 v_0 \sin \alpha_0}{g}$$

Call the range R_1 when the angle is α_0 and R_2 when the angle is $90° - \alpha$.

$$R_1 = (v_0 \cos \alpha_0) \frac{2 v_0 \sin \alpha_0}{g}$$

$$R_2 = (v_0 \cos(90° - \alpha_0)) \frac{2 v_0 \sin(90° - \alpha_0)}{g}$$

The problem asks us to show that $R_1 = R_2$.

We can use the trig identities in Appendix B to show:

$$\cos(90° - \alpha_0) = \cos(\alpha_0 - 90°) = \sin \alpha_0$$

$$\sin(90° - \alpha_0) = -\sin(\alpha_0 - 90°) = -(-\cos \alpha_0) = +\cos \alpha_0$$

Thus $R_2 = (v_0 \sin \alpha_0) \dfrac{2 v_0 \cos \alpha_0}{g} = (v_0 \cos \alpha_0) \dfrac{2 v_0 \sin \alpha_0}{g} = R_1$.

3-51

Select a coordinate system where +y is north and +x is east.

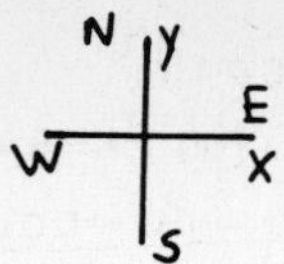

The velocity vectors in the problem are:

$\vec{v}_{P/E}$, the velocity of the plane relative to the earth

$\vec{v}_{P/A}$, the velocity of the plane relative to the air ($v_{P/A}$ is the airspeed of the plane and the direction of $\vec{v}_{P/A}$ is the compass course set by the pilot)

$\vec{v}_{A/E}$, the velocity of the air relative to the earth (the wind velocity)

The rule for combining relative velocities gives $\vec{v}_{P/E} = \vec{v}_{P/A} + \vec{v}_{A/E}$.

a) We are given the following information about the relative velocities:

$\vec{v}_{P/A}$ has magnitude 220 km/h and its direction is west. In our coordinates it has components $(v_{P/A})_x = -220$ km/h and $(v_{P/A})_y = 0$.

From the displacement of the plane relative to the earth after 0.500 h, we find that $\vec{v}_{P/E}$ has components in our coordinate system of

$$(v_{P/E})_x = -\frac{150\text{ km}}{0.500\text{ h}} = -300\text{ km/h (west)}$$

$$(v_{P/E})_y = -\frac{40\text{ km}}{0.500\text{ h}} = -80\text{ km/h (south)}$$

With this information the diagram corresponding to the velocity addition equation is

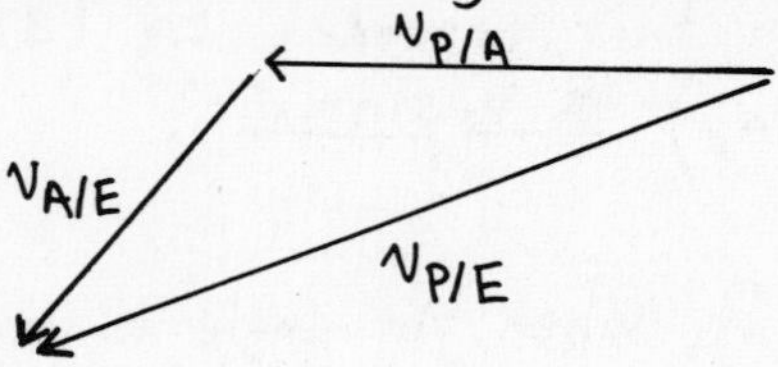

We are asked to find $\vec{v}_{A/E}$, so solve for this vector:

$$\vec{v}_{P/E} = \vec{v}_{P/A} + \vec{v}_{A/E} \Rightarrow \vec{v}_{A/E} = \vec{v}_{P/E} - \vec{v}_{P/A}$$

The x-component of this equation gives

$$(v_{A/E})_x = (v_{P/E})_x - (v_{P/A})_x = -300\text{ km/h} - (-220\text{ km/h}) = -80\text{ km/h}$$

The y-component of this equation gives

$$(v_{A/E})_y = (v_{P/E})_y - (v_{P/A})_y = -80\text{ km/h}$$

Now that we have the components of $\vec{v}_{A/E}$ we can find its magnitude and direction.

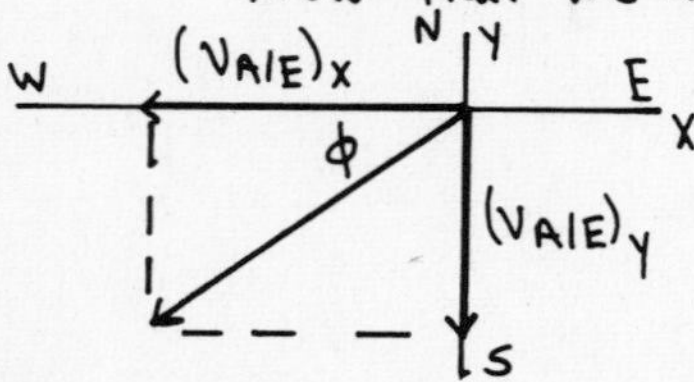

$$v_{A/E} = \sqrt{(v_{A/E})_x^2 + (v_{A/E})_y^2} = \sqrt{(-80\text{ km/h})^2 + (-80\text{ km/h})^2} = \underline{113\text{ km/h}}$$

$$\tan\phi = \frac{80\text{ km/h}}{80\text{ km/h}} = 1.0 \Rightarrow \phi = 45°$$

The direction of the wind velocity is 45° S of W.

3-51 (cont)

b) The rule for combing the relative velocities is still $\vec{v}_{P/E} = \vec{v}_{P/A} + \vec{v}_{A/E}$, but some of these velocities have different values than in part (a).

$\vec{v}_{P/A}$ has magnitude 220 km/h but its direction is to be found.

$\vec{v}_{A/E}$ has magnitude 120 km/h and its direction is due south.

The direction of $\vec{v}_{P/E}$ is west; its magnitude is not given.

The vector diagram for $\vec{v}_{P/E} = \vec{v}_{P/A} + \vec{v}_{A/E}$ and the specified directions for the vectors is

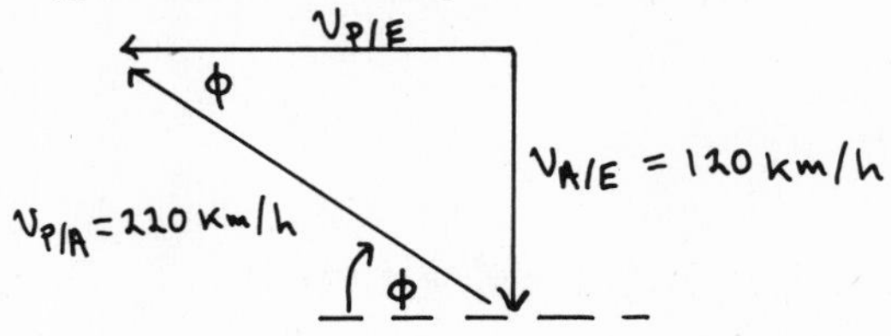

The vector addition diagram forms a right triangle.

$$\sin\phi = \frac{v_{A/E}}{v_{P/A}} = \frac{120\text{ km/h}}{220\text{ km/h}} = 0.545 \Rightarrow \phi = 33.1°$$

The pilot should set her course <u>33.1° N of W.</u>

CHAPTER 4

Exercises 3, 11, 13, 15, 17, 19, 21, 23

Problems 27, 29, 31, 35, 37, 39, 41

Exercises

4-3

Use a coordinate system where the +x-axis is in the direction of $\vec{F}_A$, the force applied by dog A.

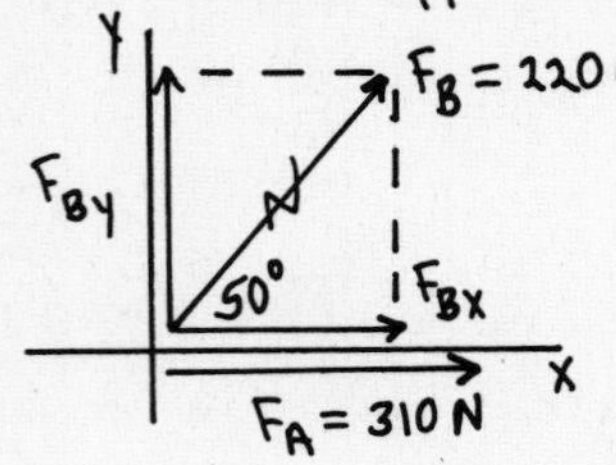

$F_{Ax} = +310\,N$

$F_{Ay} = 0$

$F_{Bx} = F_B \cos 50° = (220\,N)\cos 50° = +141\,N$

$F_{By} = F_B \sin 50° = (220\,N)\sin 50° = +169\,N$

$\vec{R} = \vec{F}_A + \vec{F}_B$

$R_x = F_{Ax} + F_{Bx} = +310\,N + 141\,N = +451\,N$

$R_y = F_{Ay} + F_{By} = 0 + 169\,N = +169\,N$

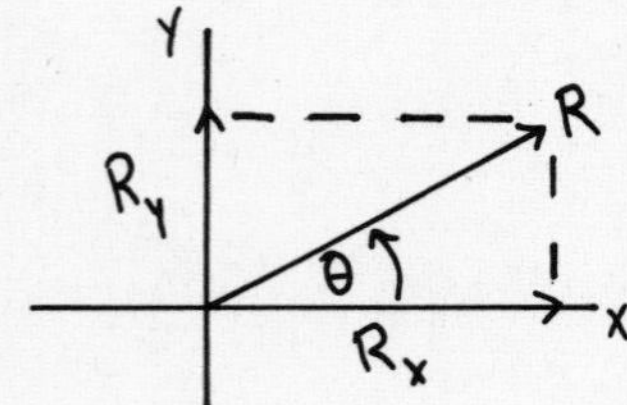

$R = \sqrt{R_x^2 + R_y^2} = \sqrt{(451N)^2 + (169N)^2} = \underline{482\,N}$

$\tan\theta = \frac{R_y}{R_x} = \frac{169N}{451N} = 0.375 \Rightarrow \theta = \underline{20.5°}$

4-11

a) During this time interval the acceleration is constant and equal to

$a_x = \frac{F_x}{m} = \frac{0.400\,N}{0.160\,kg} = 2.50\,m/s^2$

We can use the constant acceleration kinematic equations from Chapter 2.

$x - x_0 = v_{0x}t + \frac{1}{2}a_x t^2 = 0 + \frac{1}{2}(2.50\,m/s^2)(2.00s)^2 = 5.00\,m$, so the puck is at $x = \underline{5.00m}$

$v_x = v_{0x} + a_x t = 0 + (2.50\,m/s^2)(2.00s) = \underline{5.00\,m/s}$

b) In the time interval from $t = 2.00s$ to $5.00s$ the force has been removed so the acceleration is zero. The speed stays constant at $v_x = 5.00\,m/s$. The distance the puck travels is $x - x_0 = v_{0x}t = (5.00\,m/s)(5.00s - 2.00s) = 15.0\,m$. At the start of this interval the puck is at $x_0 = 5.00m$, so at the end of the interval it is at $x = x_0 + 15.0m = 20.0m$.

In the time interval from $t = 5.00s$ to $7.00s$ the acceleration is again $a_x = 2.50\,m/s^2$. At the start of this interval $v_{0x} = 5.00\,m/s$ and $x_0 = 20.0m$.

4-11 (cont)

$$x - x_0 = v_{0x}t + \tfrac{1}{2}a_x t^2 = (5.00\,m/s)(2.00\,s) + \tfrac{1}{2}(2.50\,m/s^2)(2.00\,s)^2 = 10.0\,m + 5.0\,m = 15.0\,m.$$

Therefore, at $t = 7.00\,s$ the puck is at $x = x_0 + 15.0\,m = 20.0\,m + 15.0\,m = \underline{35.0\,m}$.

$$v_x = v_{0x} + a_x t = 5.00\,m/s + (2.50\,m/s^2)(2.00\,s) = \underline{10.0\,m/s}$$

4-13

$w = mg$

The mass of the watermelon is constant, independent of its location. Its weight differs on earth and Mars. Use the information about the watermelon's weight on earth to calculate its mass:

$$w = mg \Rightarrow m = \frac{w}{g} = \frac{64.0\,N}{9.80\,m/s^2} = 6.53\,kg$$

On Mars, $\underline{m = 6.53\,kg}$, the same as on earth. Thus the weight on Mars is $w = mg = (6.53\,kg)(3.7\,m/s^2) = \underline{24.2\,N}$.

4-15

$F = ma$

We must use $w = mg$ to find the mass of the boulder:

$$m = \frac{w}{g} = \frac{2800\,N}{9.80\,m/s^2} = 285.7\,kg$$

Then $F = ma = (285.7\,kg)(24.0\,m/s^2) = \underline{6860\,N}$.

4-17

a). The free-body diagram for the bottle is ↓ $w = mg$

The only force on the bottle is gravity.

b) ↓w ↑w' earth

w is the force of gravity that the earth exerts on the bottle. The reaction to this force is w', the gravity force that the bottle exerts on the earth. Note that these two equal and opposite forces produce very different accelerations because the bottle and the earth have very different masses.

4-19

The free-body diagram for the bucket is

a↑ y, T (the tension in the cord), x, $w = mg$

$$\Sigma F_y = ma_y$$

$$T - mg = ma$$

$$a = \frac{T - mg}{m} = \frac{60.0\,N - (4.40\,kg)(9.80\,m/s^2)}{4.40\,kg}$$

$$a = \frac{60.0\,N - 43.12\,N}{4.40\,kg} = \underline{3.84\,m/s^2}$$

4-21

Take the +y-direction to be downward since that is the direction of the person's acceleration.

The free-body diagram for the person is

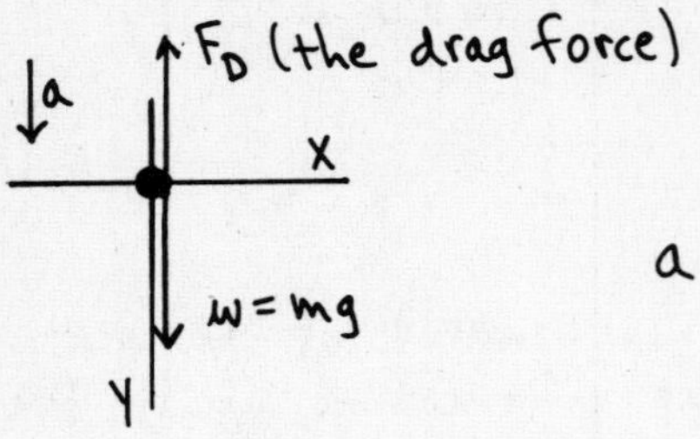

a
F_D (the drag force)
x
w = mg
y

$$\Sigma F_y = ma_y$$

$$mg - F_D = ma$$

$$a = \frac{mg - F_D}{m} = \frac{(50.0\,\text{kg})(9.80\,\text{m/s}^2) - 340\,\text{N}}{50.0\,\text{kg}} = \underline{3.00\,\text{m/s}^2}$$

4-23

a = 2.50 m/s²
T
F
m₁ = 4.00 kg
m₂ = 6.00 kg

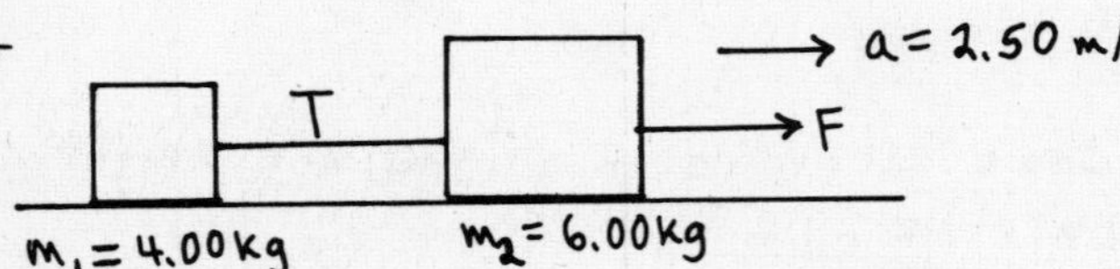

Since the crates are connected by a rope, they both have the same acceleration.

a) Consider the two crates and the rope connecting them as a single object of mass $m = m_1 + m_2 = 10.0$ kg. The free-body diagram is

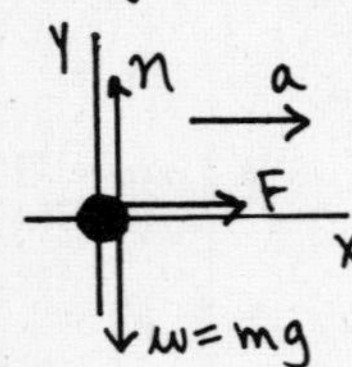

y
n
a
F
x
w = mg

$$\Sigma F_x = ma_x$$

$$F = ma = (10.0\,\text{kg})(2.50\,\text{m/s}^2) = \underline{25.0\,\text{N}}$$

b) Consider the forces on the 4.00 kg crate:

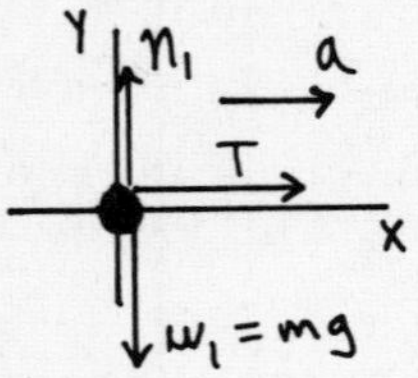

y
n₁
a
T
x
w₁ = mg

$$\Sigma F_x = ma_x$$

$$T = m_1 a = (4.00\,\text{kg})(2.50\,\text{m/s}^2) = \underline{10.0\,\text{N}}$$

As a check, can also consider the forces on the 6.00 kg crate:

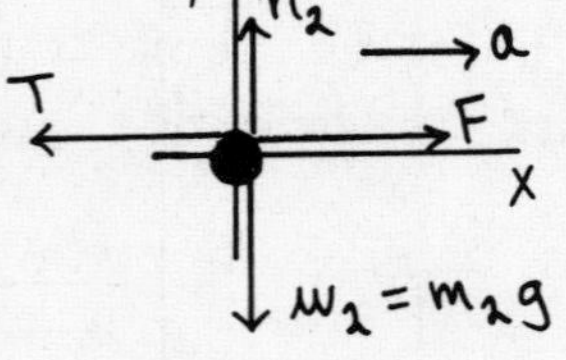

y
n₂
a
T
F
x
w₂ = m₂g

$$\Sigma F_x = ma_x$$

$$F - T = m_2 a$$

$$T = F - m_2 a = 25.0\,\text{N} - (6.00\,\text{kg})(2.50\,\text{m/s}^2)$$

$$T = 25.0\,\text{N} - 15.0\,\text{N} = 10.0\,\text{N} \checkmark$$

Problems

4-27

Use coordinates with the +x-axis along $\vec{F}_1$ and the +y-axis along $\vec{R}$.

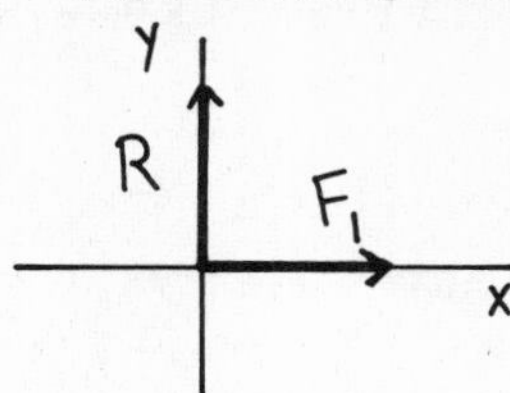

$F_{1x} = +1200\text{N}$ $\quad$ $R_x = 0$

$F_{1y} = 0$ $\quad$ $R_y = +1200\text{N}$

$$\vec{F}_1 + \vec{F}_2 = \vec{R} \Rightarrow \vec{F}_2 = \vec{R} - \vec{F}_1$$

$$F_{2x} = R_x - F_{1x} = 0 - 1200\text{N} = -1200\text{N}$$

$$F_{2y} = R_y - F_{1y} = +1200\text{N} - 0 = +1200\text{N}$$

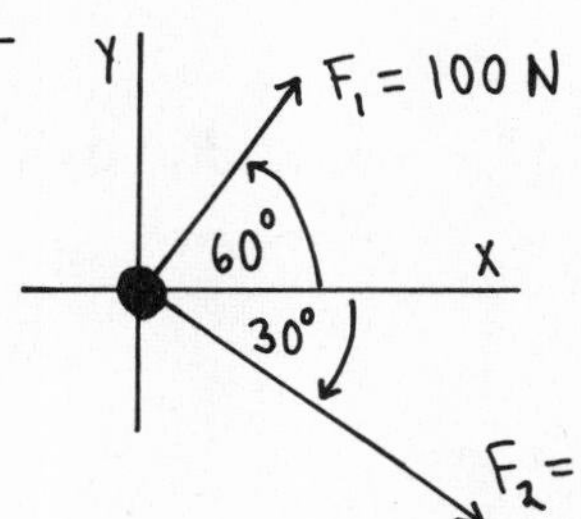

$$F_2 = \sqrt{F_{2x}^2 + F_{2y}^2} = \sqrt{(-1200\text{N})^2 + (1200\text{N})^2} = 1700\text{N}$$

$$\tan\theta = \frac{F_{2y}}{F_{2x}} = \frac{+1200\text{N}}{-1200\text{N}} = -1.00 \Rightarrow \theta = 135°$$

The magnitude of $\vec{F}_2$ is 1700N and its direction is 135° counterclockwise from the direction of $\vec{F}_1$.

4-29

y

$F_1 = 100$ N

60°

x

30°

$F_2 = 140$ N

If the box moves in the +x-direction it must have $a_y = 0$, so $\Sigma F_y = 0$. The smallest force the child can exert and still produce such motion is a force that makes the y-components of all three forces sum to zero, but that doesn't have any x-component.

Let $\vec{F}_3$ be the force exerted by the child.

$$\Sigma F_y = 0 \Rightarrow F_{1y} + F_{2y} + F_{3y} = 0 \Rightarrow F_{3y} = -(F_{1y} + F_{2y})$$

$$F_{1y} = +F_1 \sin 60° = (100\text{N}) \sin 60° = 86.6\text{N}$$

$$F_{2y} = F_2 \sin(-30°) = -F_2 \sin 30° = -(140\text{N}) \sin 30° = -70.0\text{N}$$

Then $F_{3y} = -(F_{1y} + F_{2y}) = -(86.6\text{N} - 70.0\text{N}) = -16.6\text{N}$; $F_{3x} = 0$.

The smallest force the child can exert has magnitude 16.6N and is directed at 90° clockwise from the +x-axis shown in the figure.

4-31

First use the information given about the height of the jump to calculate the speed he has at the instant his feet leave the ground. Use a coordinate system with the +y-axis upward and the origin at the position when his feet leave the ground.

4-31 (cont)

$v_y = 0$ (at the maximum height)
$v_{0y} = ?$
$a_y = -9.80\ m/s^2$
$y - y_0 = +1.2\ m$

$$v_y^2 = v_{0y}^2 + 2a_y(y-y_0)$$
$$v_{0y} = \sqrt{-2a_y(y-y_0)}$$
$$v_{0y} = \sqrt{-2(-9.80\ m/s^2)(1.2\ m)} = 4.85\ m/s$$

Now consider the acceleration phase, from when he starts to jump until when his feet leave the ground. Use a coordinate system where the +y-axis is upward and the origin is at his position when he starts his jump. Calculate the average acceleration:

$$(a_y)_{av} = \frac{v_y - v_{0y}}{t} = \frac{4.85\ m/s - 0}{0.400\ s} = 12.12\ m/s^2$$

Finally, find the average upward force that the ground must exert on him to produce this average upward acceleration. (Don't forget about the downward force of gravity.)

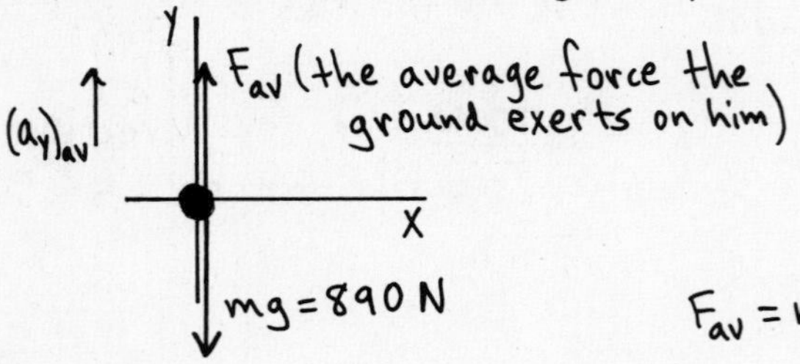

$$m = \frac{w}{g} = \frac{890\ N}{9.80\ m/s^2} = 90.8\ kg$$

$$\Sigma F_y = ma_y$$
$$F_{av} - mg = m(a_y)_{av}$$
$$F_{av} = m(g + (a_y)_{av}) = 90.8\ kg\,(9.80\ m/s^2 + 12.12\ m/s^2) = 1990\ N$$

This is the average force exerted on him by the ground. But by Newton's 3rd Law, the average force he exerts on the ground is equal and opposite, so is <u>1990 N</u>, <u>downward</u>.

<u>4-35</u>

a) Consider the forces on the student. The reading on the scale equals the upward normal force exerted on the student. Also, the student is at rest relative to the elevator, so the acceleration of the student equals that of the elevator.

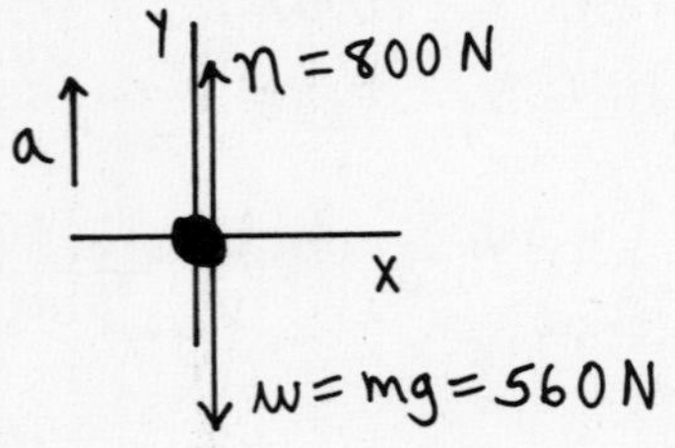

We know that $\vec{a}$ is upward since $n > w$.

$$m = \frac{w}{g} = \frac{560\ N}{9.80\ m/s^2} = 57.1\ kg$$

$$\Sigma F_y = ma_y$$
$$n - w = ma$$
$$a = \frac{n-w}{m} = \frac{800\ N - 560\ N}{57.1\ kg} = \underline{4.20\ m/s^2\ \text{(upward)}}$$

b) Now $n < w$ so the student (and elevator) is accelerating downward. Take the +y-direction to be downward, in the direction of $\vec{a}$.

4-35 (cont)

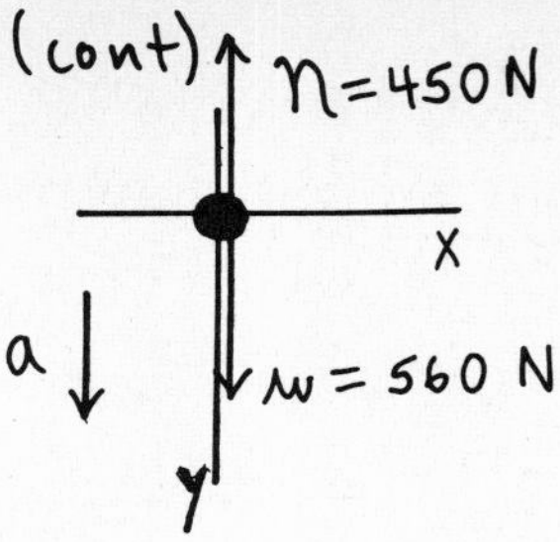

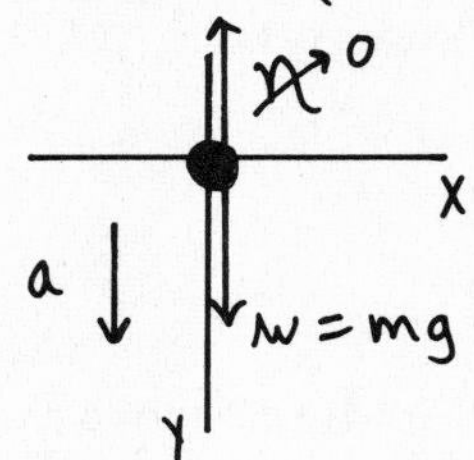

$$\sum F_y = ma_y$$
$$w - n = ma$$
$$a = \frac{w - n}{m} = \frac{560\text{ N} - 450\text{ N}}{57.1\text{ kg}} = \underline{1.93\text{ m/s}^2\text{ (downward)}}$$

c) Now $n = 0$. $n < w$ so the acceleration is downward.

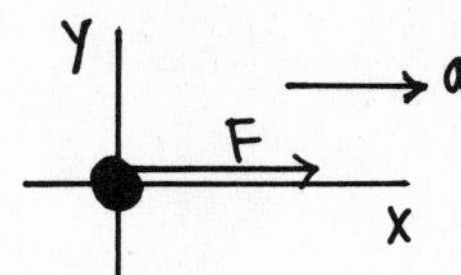

$$\sum F_y = ma_y$$
$$\cancel{m}g = \cancel{m}a$$
$$a = g \text{ (downward)}$$

The elevator and student are in free-fall; the elevator cable may have snapped.

4-37

a) Consider all four cars together as one object. The horizontal force on this combined object is the force of the engine on the first car. The mass of all four cars together is $m = 4(3.90\times10^4\text{ kg}) = 1.56\times10^5\text{ kg}$.

$$\sum F_x = ma_x$$
$$F = ma = (1.56\times10^5\text{ kg})(0.800\text{ m/s}^2) = \underline{1.25\times10^5\text{ N}}$$

b) Treat the last three cars together as one object. The horizontal force on this combined object is the force of the first car on the second car, The mass of these three cars together is $m = 3(3.90\times10^4\text{ kg}) = 1.17\times10^5\text{ kg}$.

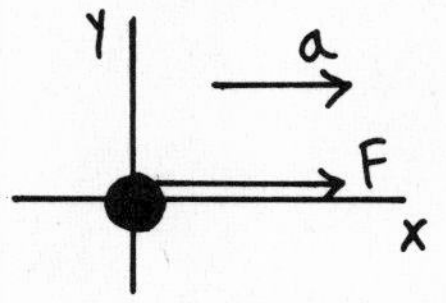

$$\sum F_x = ma_x$$
$$F = ma = (1.17\times10^5\text{ kg})(0.800\text{ m/s}^2) = \underline{9.36\times10^4\text{ N}}$$

c) Treat the last two cars together.

$$m = 2(3.90\times10^4\text{ kg}) = 7.80\times10^4\text{ kg}$$

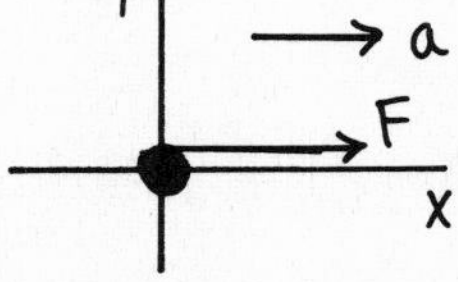

$$\sum F_x = ma_x$$
$$F = ma = (7.80\times10^4\text{ kg})(0.800\text{ m/s}^2) = \underline{6.24\times10^4\text{ N}}$$

4-37 (cont)

d) By Newton's third law the force of the fourth car on the third car is equal and opposite to the force of the third car on the fourth.
Free-body diagram for the fourth car:

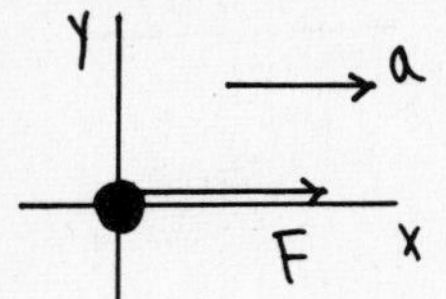

$$\Sigma F_x = ma_x$$
$$F = ma = (3.90 \times 10^4\ \text{kg})(0.800\ \text{m/s}^2) = \underline{3.12 \times 10^4\ \text{N}}$$

e) The forces would all be the same magnitude but would be in the opposite direction.

4-39

a) Take the +y-direction to be upward since that is the direction of the acceleration. The maximum upward acceleration is obtained from the maximum possible tension in the cables.

$$\Sigma F_y = ma_y$$
$$T - mg = ma$$
$$a = \frac{T - mg}{m} = \frac{24{,}000\ \text{N} - (1800\ \text{kg})(9.80\ \text{m/s}^2)}{1800\ \text{kg}}$$
$$a = \frac{6.36 \times 10^3\ \text{N}}{1800\ \text{kg}} = \underline{3.53\ \text{m/s}^2}$$

b) What changes is the weight mg of the elevator.

$$a = \frac{T - mg}{m} = \frac{24{,}000\ \text{N} - (1800\ \text{kg})(1.62\ \text{m/s}^2)}{1800\ \text{kg}} = \frac{2.108 \times 10^4\ \text{N}}{1800\ \text{kg}} = \underline{11.7\ \text{m/s}^2}$$

4-41

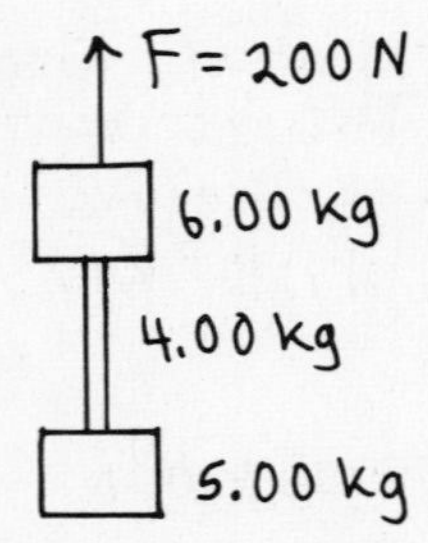

Note that in this problem the mass of the rope is given, and that it is not negligible compared to the other masses.

a) Treat the rope and two blocks together as a single object, with mass $m = 6.00\ \text{kg} + 4.00\ \text{kg} + 5.00\ \text{kg} = 15.0\ \text{kg}$. Take +y to be upward, since the acceleration is upward.

$$\Sigma F_y = ma_y$$
$$F - mg = ma$$
$$a = \frac{F - mg}{m} = \frac{200\ \text{N} - (15.0\ \text{kg})(9.80\ \text{m/s}^2)}{15.0\ \text{kg}} = \frac{200\ \text{N} - 147\ \text{N}}{15.0\ \text{kg}}$$
$$a = \underline{3.53\ \text{m/s}^2}$$

4-41 (cont)

b) Consider the forces on the top block ($m = 6.00$ kg), since the tension at the top of the rope (T_t) will be one of these forces.

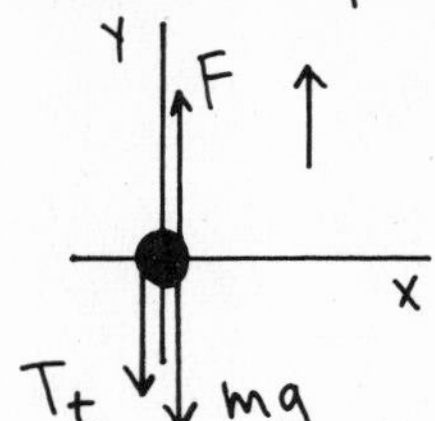

$$\Sigma F_y = ma_y$$
$$F - mg - T_t = ma$$
$$T_t = F - m(g+a)$$
$$T_t = 200\text{N} - (6.00\,\text{kg})(9.80\,\text{m/s}^2 + 3.53\,\text{m/s}^2) = \underline{120\text{ N}}$$

Alternatively, can consider the forces on the combined object rope plus bottom block ($m = 9.00$ kg):

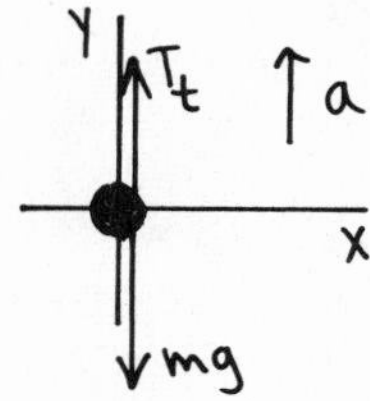

$$\Sigma F_y = ma_y$$
$$T_t - mg = ma$$
$$T_t = m(g+a) = 9.00\,\text{kg}\,(9.80\,\text{m/s}^2 + 3.53\,\text{m/s}^2) = 120\text{ N} \checkmark$$

c) One way to do this is to consider the forces on the top half of the rope ($m = 2.00$ kg). Let T_m be the tension at the midpoint of the rope.

$$\Sigma F_y = ma_y$$
$$T_t - T_m - mg = ma$$
$$T_m = T_t - m(a+g) = 120\text{N} - 2.00\,\text{kg}(3.53\,\text{m/s}^2 + 9.80\,\text{m/s}^2)$$
$$T_m = \underline{93.3\text{ N}}$$

To check this answer we can alternatively consider the forces on the bottom half of the rope plus the lower block taken together as a combined object ($m = 2.00$ kg $+ 5.00$ kg $= 7.00$ kg):

$$\Sigma F_y = ma_y$$
$$T_m - mg = ma$$
$$T_m = m(g+a) = 7.00\,\text{kg}\,(9.80\,\text{m/s}^2 + 3.53\,\text{m/s}^2) = 93.3\text{ N},$$

which checks

CHAPTER 5

Exercises 3, 5, 7, 11, 13, 17, 19, 21, 25, 29, 31, 33, 37, 43, 45

Problems 47, 49, 51, 53, 55, 57, 59, 61, 65, 69, 71, 73, 75, 77

Exercises

5-3

a) Force diagram for the person (T_1 and T_2 are the tension in each half of the rope):

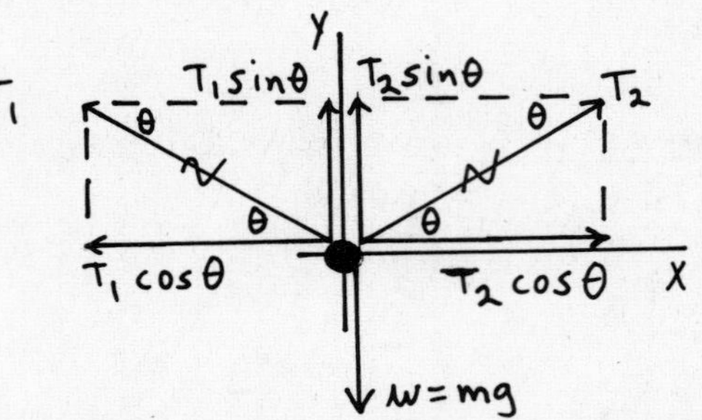

$\Sigma F_x = 0$

$T_2 \cos\theta - T_1 \cos\theta = 0$

$T_1 = T_2 = T$ (The tension is the same on both sides of the person.)

$\Sigma F_y = 0$

$T_1 \sin\theta + T_2 \sin\theta - mg = 0$

But $T_1 = T_2 = T$, so $2T\sin\theta = mg$

$$T = \frac{mg}{2\sin\theta} = \frac{(81.6\text{ kg})(9.80\text{ m/s}^2)}{2\sin(15.0°)}$$

$T = \underline{1540\text{ N}}$

b) The relation $2T\sin\theta = mg$ still applies but now we are given that $T = 24{,}000$ N (the breaking strength) and are asked to find θ.

$$\sin\theta = \frac{mg}{2T} = \frac{(81.6\text{ kg})(9.80\text{ m/s}^2)}{2(24{,}000\text{ N})} = 0.0167 \Rightarrow \theta = \underline{0.955°}$$

5-5

Force diagram for the wrecking ball:

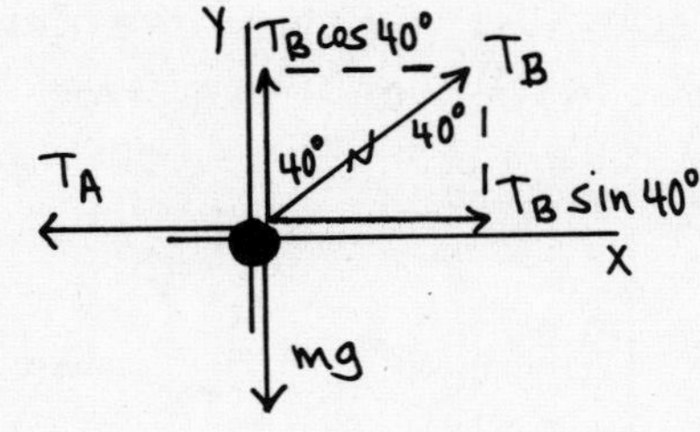

a) $\Sigma F_x = 0$

$T_B \sin 40° - T_A = 0$

$$T_B = \frac{T_A}{\sin 40°} = \frac{460\text{ N}}{\sin 40°} = \underline{716\text{ N}}$$

b) $\Sigma F_y = 0$

$T_B \cos 40° - mg = 0$

$$m = \frac{T_B \cos 40°}{g} = \frac{(716\text{ N})\cos 40°}{9.80\text{ m/s}^2} = \underline{56.0\text{ kg}}$$

5-7

a)

Free-body diagram for the upper knot:

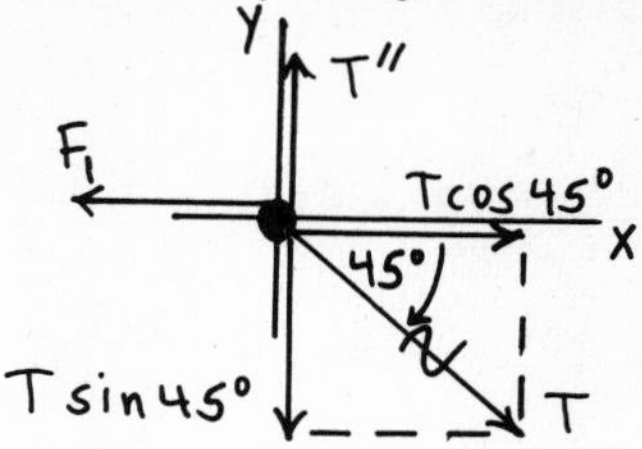

$\Sigma F_x = 0$

$T\cos 45° - F_1 = 0$

$F_1 = T\cos 45°$

$F_1 = (40.0\text{N})\cos 45° = \underline{28.3\text{ N}}$

Free-body diagram for the lower knot:

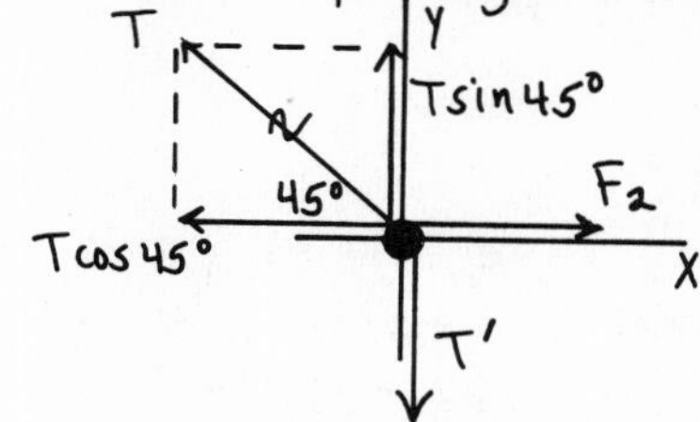

$\Sigma F_x = 0$

$F_2 - T\cos 45° = 0$

$F_2 = T\cos 45° = (40.0\text{ N})\cos 45° = \underline{28.3\text{ N}}$

b) Apply $\Sigma F_y = 0$ to the force diagram for the lower knot:

$\Sigma F_y = 0$

$T\sin 45° - T' = 0$

$T' = T\sin 45° = (40.0\text{N})\cos 45° = 28.3\text{ N}$

Free-body diagram for the block:

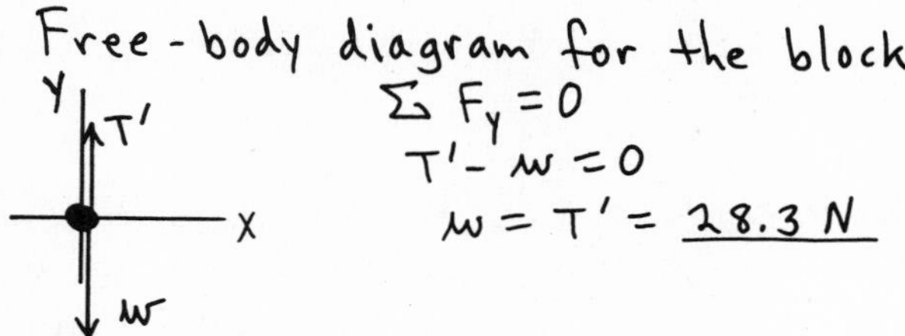

$\Sigma F_y = 0$

$T' - w = 0$

$w = T' = \underline{28.3\text{ N}}$

5-11

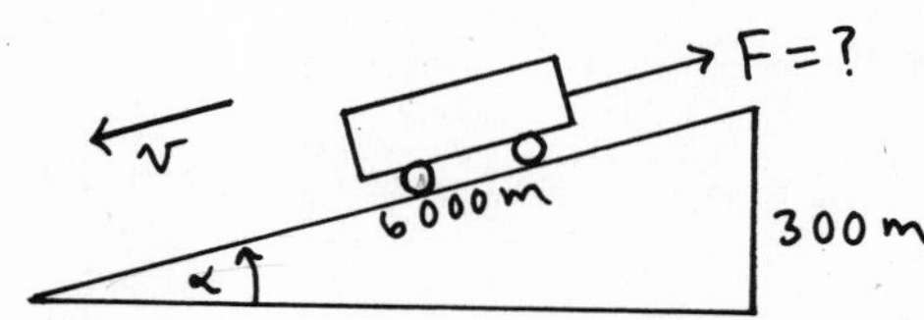

Let $\vec{F}$ be the resistive force.

$\sin\alpha = \dfrac{300\text{ m}}{6000\text{ m}} = 0.0500$

Free-body diagram for the car:

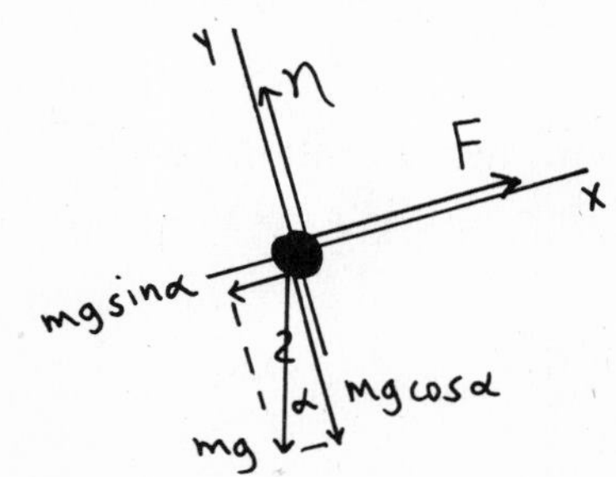

$\Sigma F_x = 0$ (constant speed $\Rightarrow \vec{a} = 0$)

$F - mg\sin\alpha = 0$

$F = mg\sin\alpha = (1200\text{ kg})(9.80\text{m/s}^2)(0.0500)$

$F = \underline{588\text{ N}}$

5-13

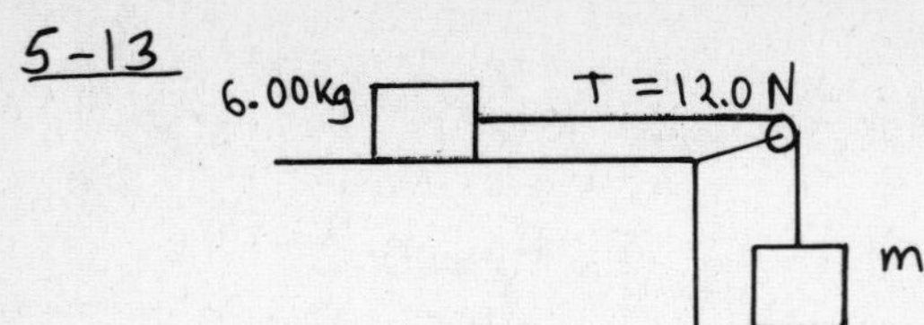

a) Free-body diagram for the 6.00-kg block:

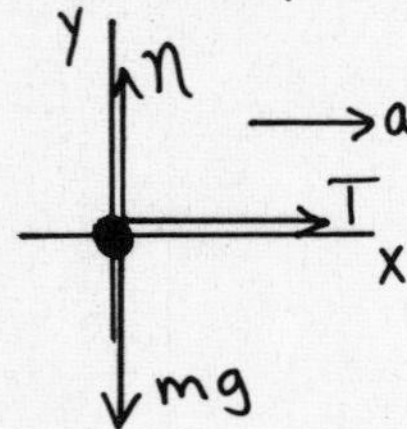

$$\Sigma F_x = ma_x$$
$$T = ma$$
$$a = \frac{T}{m} = \frac{12.0\text{ N}}{6.00\text{ kg}} = \underline{2.00\text{ m/s}^2}$$

b) Free-body diagram for the hanging block: (Take +y to be in the direction of the acceleration (downward). Since the blocks are connected by the rope, they have the same magnitude of acceleration.)

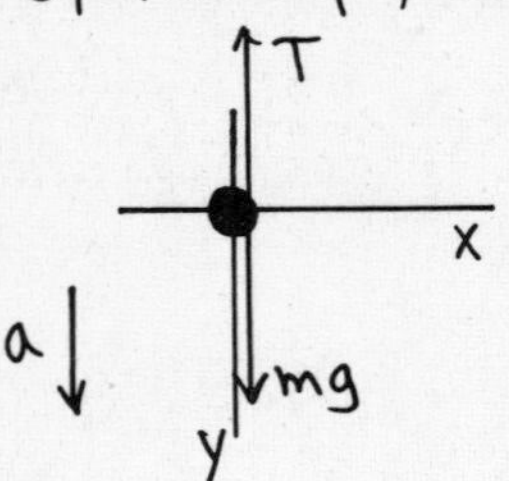

$$\Sigma F_y = ma_y$$
$$mg - T = ma$$
$$m(g-a) = T$$
$$m = \frac{T}{g-a} = \frac{12.0\text{ N}}{9.80\text{ m/s}^2 - 2.00\text{ m/s}^2} = \underline{1.54\text{ kg}}$$

5-17

a)

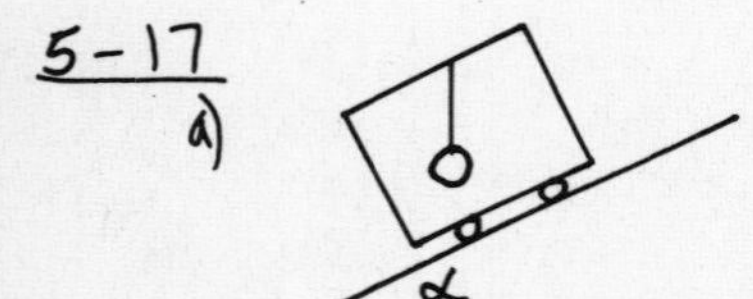

At rest ⇒ $a = 0$, so the forces on the lead sinker must sum to zero. The gravity force on the sinker is vertically downward, toward the center of the earth. The only other force on the sinker is the tension in the string. This force must be vertically upward in order for the two forces to sum to zero.

The sinker hangs such that the string is vertical.

b) car moving up

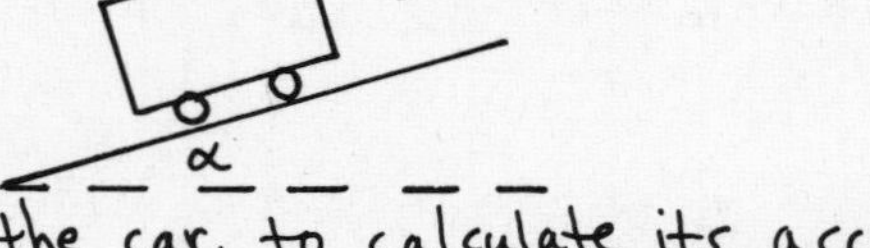

First consider the forces on the car, to calculate its acceleration. Take the x-axis parallel to the incline and the y-axis perpendicular to the incline.

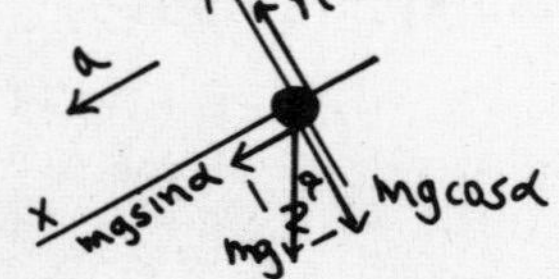

$$\Sigma F_x = ma_x$$
$$\cancel{m}g\sin\alpha = \cancel{m}a$$
$$\boxed{a = g\sin\alpha}$$

5-17 (cont)

If the hanging sinker is at rest relative to the car then it also has acceleration $a = g\sin\alpha$ directed down the incline. Consider the forces on the sinker. Take the same x and y axes as for the car, so that $a_x = g\sin\alpha$ and $a_y = 0$. Assume that the string makes an angle β with the direction perpendicular to the ceiling of the car (the y-axis).

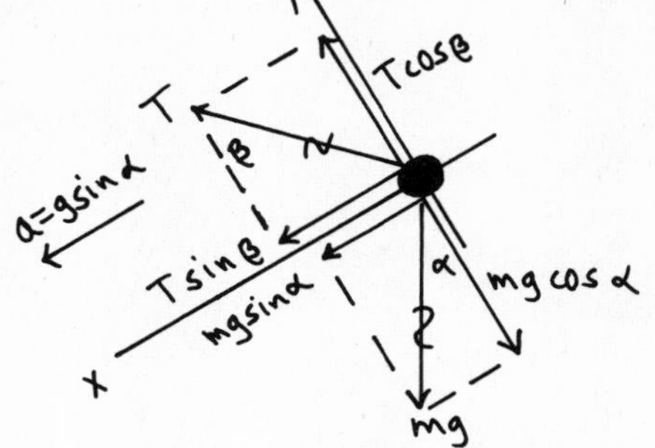

$$\sum F_x = ma_x$$
$$mg\sin\alpha + T\sin\beta = mg\sin\alpha$$

But this says $T\sin\beta = 0$, so $\beta = 0$ and the string is perpendicular to the ceiling of the car.

car moving down incline

The force diagrams for the car and for the sinker are the same as before, so again the sinker hangs such that the string is perpendicular to the ceiling.

There is no deflection during any stage of the motion.

5-19

a) constant speed $\Rightarrow a = 0$

Consider the free-body diagram for the crate. Let $\vec{F}$ be the horizontal force applied by the worker. The friction is kinetic friction since the crate is sliding along the surface.

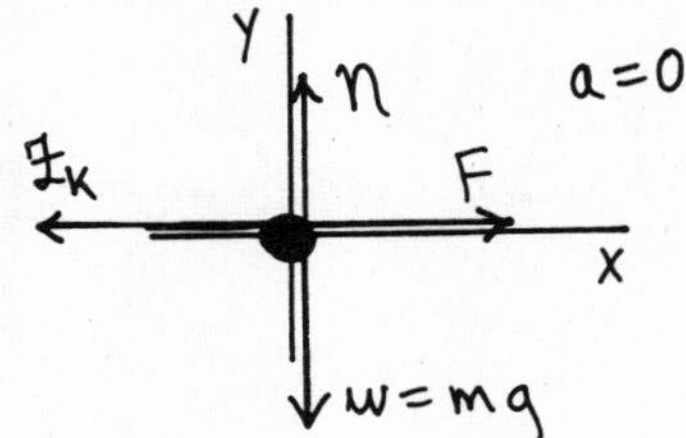

$$\sum F_y = ma_y$$
$$n - mg = 0$$
$$n = mg$$
So $f_k = \mu_k n = \mu_k mg$

$$\sum F_x = ma_x$$
$$F - f_k = 0$$
$$F = f_k = \mu_k mg = (0.20)(8.75\text{ kg})(9.80\text{ m/s}^2)$$
$$F = \underline{17\text{ N}}$$

b) Now the only horizontal force on the crate is the kinetic friction force. Calculate the acceleration it produces. The friction force is $f_k = \mu_k mg$, just as in part (a).

$$\sum F_x = ma_x$$
$$-f_k = ma$$
$$-\mu_k \cancel{m} g = \cancel{m} a$$
$$a = -\mu_k g = -(0.20)(9.80\text{ m/s}^2) = -1.96\text{ m/s}^2$$

5-19 (cont)

Use the constant acceleration equations to find the time it takes this acceleration to reduce the speed from 4.50 m/s to zero:

$v = 0$

$v_0 = 4.50\ \text{m/s}$

$a = -1.96\ \text{m/s}^2$

$t = ?$

$$v = v_0 + at$$

$$t = \frac{v - v_0}{a} = \frac{0 - 4.50\ \text{m/s}}{-1.96\ \text{m/s}^2} = \underline{2.3\ \text{s}}$$

5-21

a) constant speed $\Rightarrow a = 0$

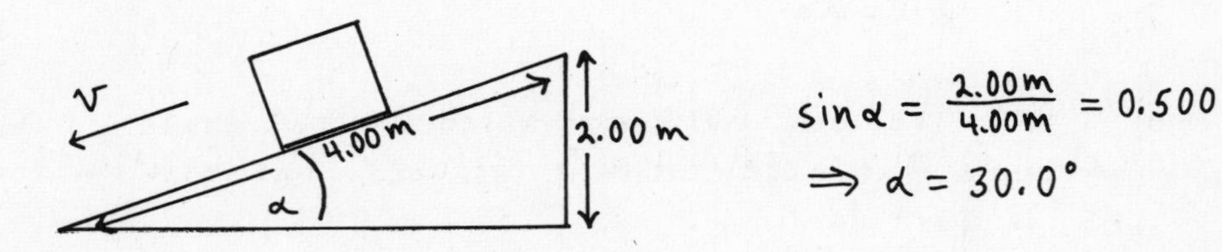

Consider the free-body diagram for the box:

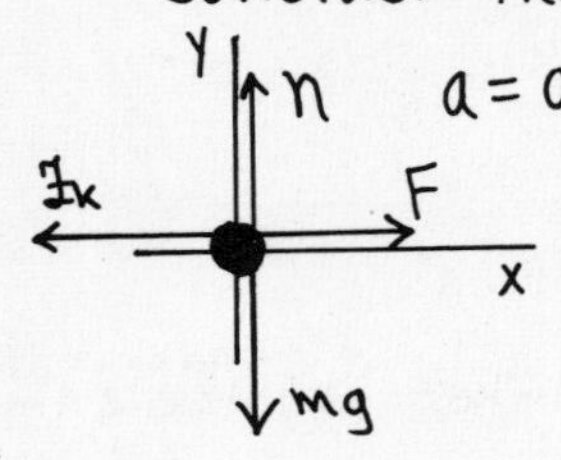

$$\Sigma F_y = ma_y$$
$$n - mg = 0$$
$$n = mg$$
So $f_k = \mu_k n = \mu_k mg$

$$\Sigma F_x = ma_x$$
$$F - f_k = 0$$
$$F = f_k = \mu_k mg$$
$$F = (0.14)(6.00\ \text{kg})(9.80\ \text{m/s}^2) = \underline{8.2\ \text{N}}$$

b)

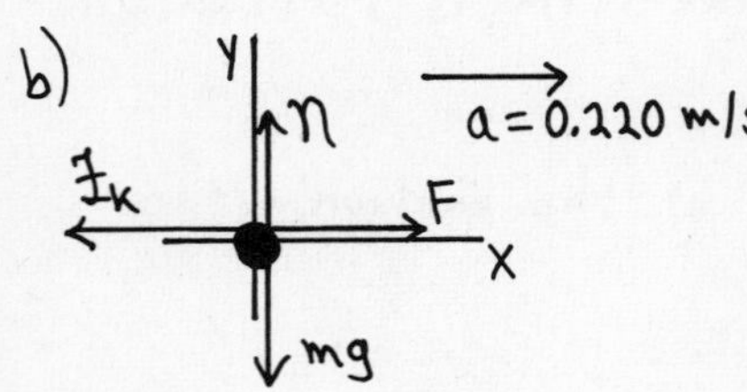

As in part (a), $f_k = \mu_k mg$

$$\Sigma F_x = ma_x$$
$$F - f_k = ma$$
$$F = f_k + ma = \mu_k mg + ma = m(\mu_k g + a)$$
$$F = 6.00\ \text{kg}\left((0.14)(9.80\ \text{m/s}^2) + 0.220\ \text{m/s}^2\right) = \underline{9.6\ \text{N}}$$

c) The normal force $n = mg$ is reduced. This in turn reduces the friction force, so the magnitude of the force $\vec{F}$ required to move the box is less.

part (a): $F = \mu_k mg = (0.14)(6.00\ \text{kg})(1.62\ \text{m/s}^2) = \underline{1.4\ \text{N}}$

part (b): $F = m(\mu_k g + a) = 6.00\ \text{kg}\left((0.14)(1.62\ \text{m/s}^2) + 0.220\ \text{m/s}^2\right) = \underline{2.7\ \text{N}}$

5-25

constant speed

$\Rightarrow a = 0$

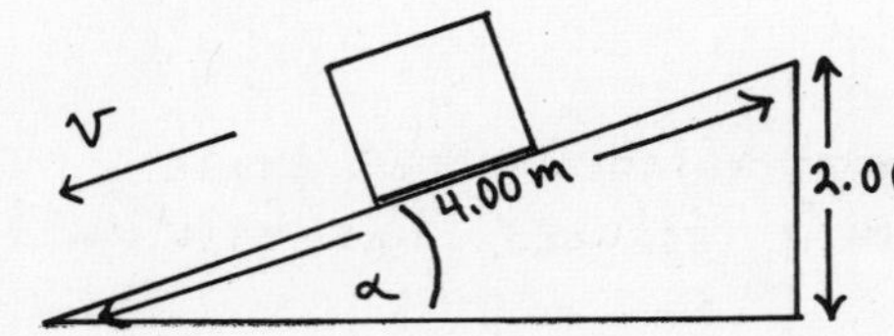

$$\sin\alpha = \frac{2.00\ \text{m}}{4.00\ \text{m}} = 0.500$$
$$\Rightarrow \alpha = 30.0°$$

a) Consider the free-body diagram for the safe, with all the forces except for the applied force we are being asked to calculate. We must decide whether this force must be down the incline or up the incline to make the total resultant force zero. Note that we know that the

5-25 (cont)

Kinetic friction force f_k is directed up the incline, since the friction force opposes the motion. Use coordinates parallel and perpendicular to the incline.

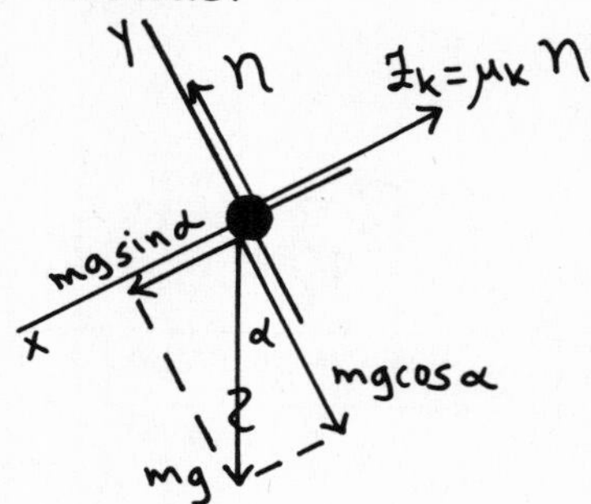

$$mg\sin\alpha = (215\,\text{kg})(9.80\,\text{m/s}^2)\sin 30.0° = 1054\,\text{N}$$

$$\Sigma F_y = ma_y$$

$$n - mg\cos\alpha = 0$$

$$n = mg\cos\alpha = (215\,\text{kg})(9.80\,\text{m/s}^2)\cos 30.0° = 1825\,\text{N}$$

$$f_k = \mu_k n = (0.30)(1825\,\text{N}) = 548\,\text{N}$$

$mg\sin\alpha > f_k$, so for the forces to balance ($a=0$) more force directed up the incline is needed; the safe must be <u>held back</u> if it is to travel at constant speed.

b) Now we can add the applied force $\vec{F}$ to the free-body diagram, since we have determined its direction.

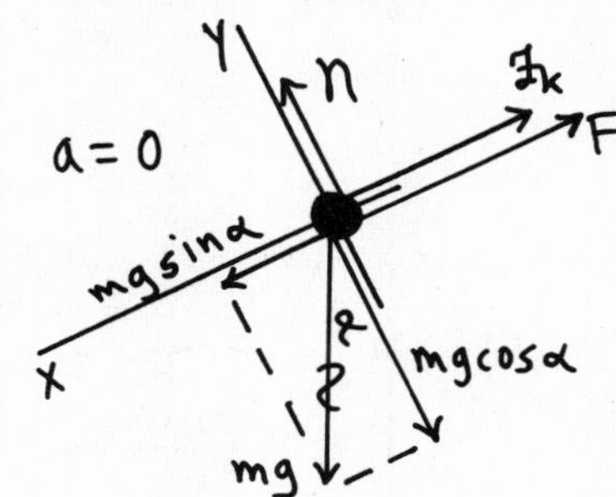

$$\Sigma F_x = ma_x$$

$$mg\sin\alpha - F - f_k = 0$$

$$F = mg\sin\alpha - f_k$$

$$F = 1054\,\text{N} - 548\,\text{N} = \underline{506\,\text{N}}$$

<u>5-29</u> A —T— B → F

constant $v \Rightarrow a = 0$

free-body diagram for A:

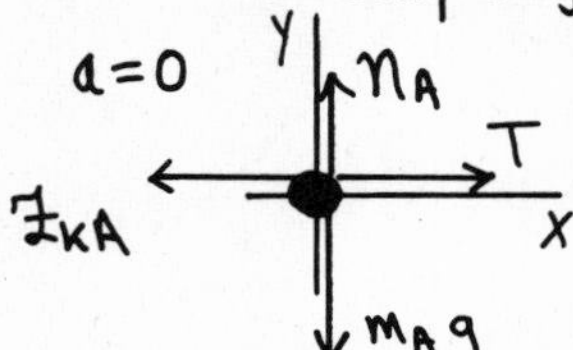

$a=0$; y; n_A; T; f_{kA}; x; $m_A g$

$$\Sigma F_y = ma_y$$

$$n_A - m_A g = 0$$

$$n_A = m_A g$$

$$f_{kA} = \mu_k n_A = \mu_k m_A g$$

$$\Sigma F_x = ma_x$$

$$T - f_{kA} = 0$$

$$\boxed{T = \mu_k m_A g}$$

free-body diagram for B:

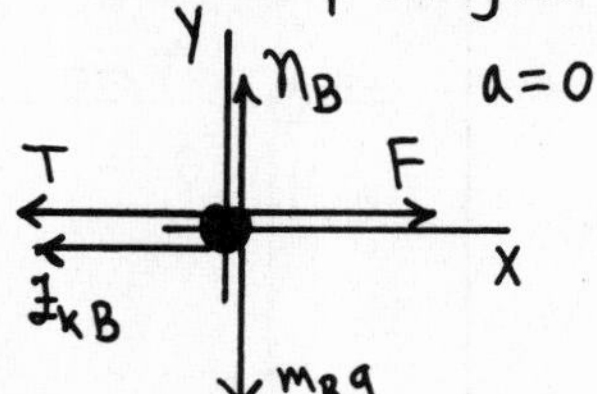

$$\Sigma F_y = ma_y$$

$$n_B - m_B g = 0$$

$$n_B = m_B g$$

$$f_{kB} = \mu_k n_B = \mu_k m_B g$$

$$\Sigma F_x = ma_x$$

$$F - T - f_{kB} = 0$$

$$\boxed{F = T + \mu_k m_B g}$$

Use the first equation to replace T in the second $\Rightarrow F = \mu_k m_A g + \mu_k m_B g = \underline{\mu_k (m_A + m_B) g}$

5-31

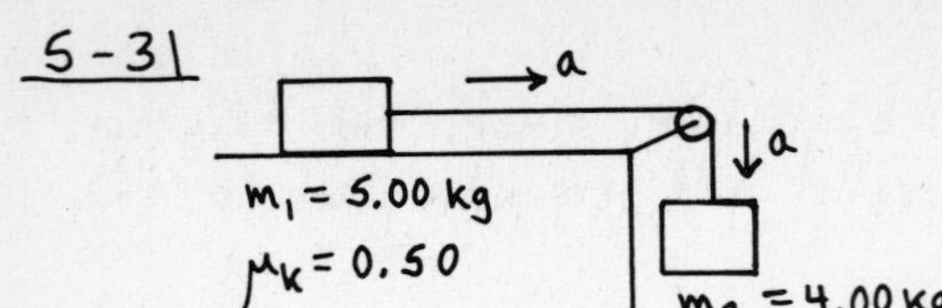

The magnitude of the acceleration is the same for both blocks. For each block take a positive coordinate direction to be the direction of the block's acceleration.

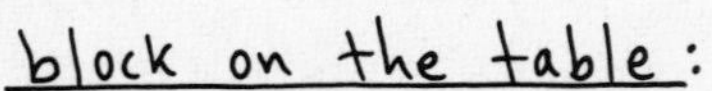

block on the table:

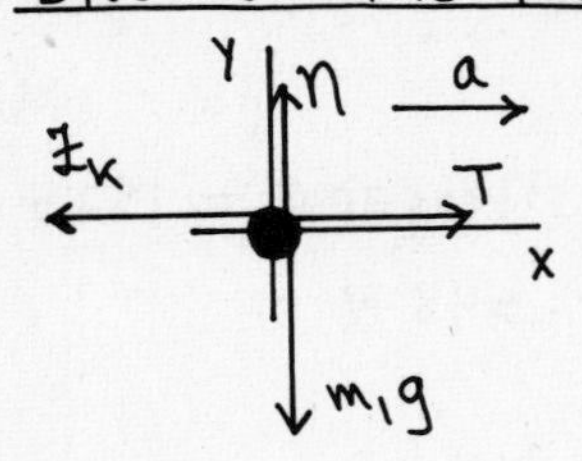

$$\Sigma F_y = ma_y$$
$$n - m_1 g = 0$$
$$n = m_1 g$$

$$f_k = \mu_k n = \mu_k m_1 g$$

$$\Sigma F_x = ma_x$$
$$T - f_k = m_1 a$$
$$\boxed{T - \mu_k m_1 g = m_1 a}$$

hanging block:

T, a, x, m_2g, y

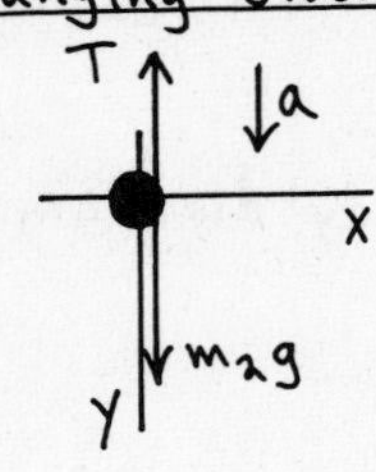

$$\Sigma F_y = ma_y$$
$$m_2 g - T = m_2 a$$
$$\boxed{T = m_2 g - m_2 a}$$

a) Use the second eq. in the first

$$\Rightarrow m_2 g - m_2 a - \mu_k m_1 g = m_1 a$$
$$(m_1 + m_2) a = (m_2 - \mu_k m_1) g$$

$$a = \frac{(m_2 - \mu_k m_1) g}{m_1 + m_2} = \frac{(4.00\text{ kg} - (0.50)(5.00\text{ kg}))(9.80\text{ m/s}^2)}{5.00\text{ kg} + 4.00\text{ kg}} = \underline{1.63\text{ m/s}^2}$$

b) $T = m_2 g - m_2 a = m_2(g - a) = 4.00\text{kg}(9.80\text{m/s}^2 - 1.63\text{ m/s}^2) = \underline{32.7\text{ N}}$

Or, to check

$$T - \mu_k m_1 g = m_1 a \Rightarrow T = m_1(a + \mu_k g) = 5.00\text{kg}(1.63\text{m/s}^2 + (0.50)(9.80\text{m/s}^2))$$
$$T = 32.7\text{ N} \checkmark$$

5-33

a) θ, F

constant $v \Rightarrow a = 0$ and friction is f_k

Free-body diagram for the crate:

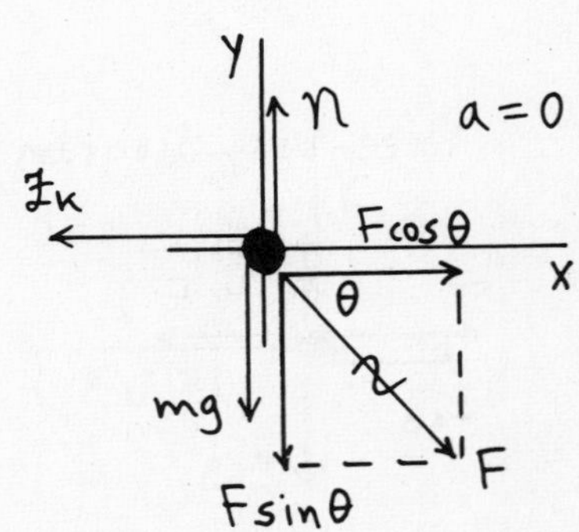

5-33 (cont)

$\Sigma F_y = ma_y$

$n - mg - F\sin\theta = 0$

$n = mg + F\sin\theta$

$f_k = \mu_k n = \mu_k mg + \mu_k F\sin\theta$

$\Sigma F_x = ma_x$

$F\cos\theta - f_k = 0$

$F\cos\theta - \mu_k mg - \mu_k F\sin\theta = 0$

$F(\cos\theta - \mu_k \sin\theta) = \mu_k mg$

$$\boxed{F = \frac{\mu_k mg}{\cos\theta - \mu_k \sin\theta}}$$

b) "start the crate moving" means the force diagram is as in part (a), except that μ_k is replaced by μ_s.

Thus $F = \dfrac{\mu_s mg}{\cos\theta - \mu_s \sin\theta}$

$F \to \infty$ if $\cos\theta - \mu_s \sin\theta = 0 \Rightarrow \mu_s = \dfrac{\cos\theta}{\sin\theta} = \dfrac{1}{\tan\theta}$

5-37

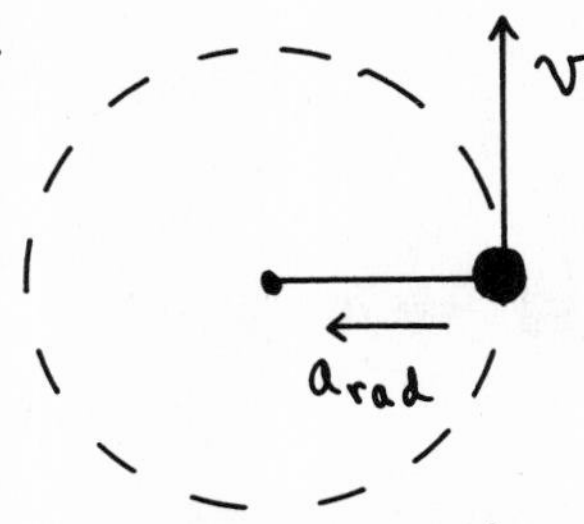

(view from above)

a_{rad} is in toward the center of the circular path.

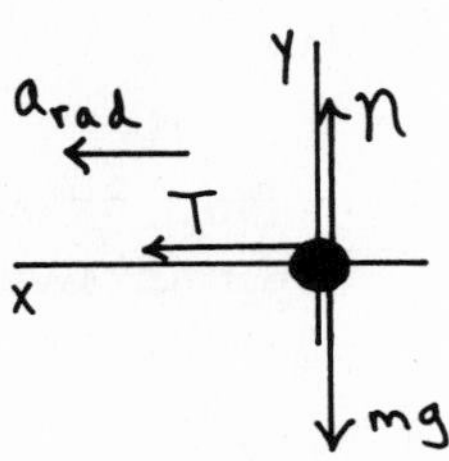

(view from side)

The maximum v produces $T = 500$ N.

$\Sigma F_x = ma_x$

$T = ma_{rad}$

$T = m\dfrac{v^2}{R}$

$v = \sqrt{\dfrac{RT}{m}} = \sqrt{\dfrac{(0.80\text{ m})(500\text{ N})}{1.50\text{ kg}}}$

$v = \underline{16.3\text{ m/s}}$

5-43

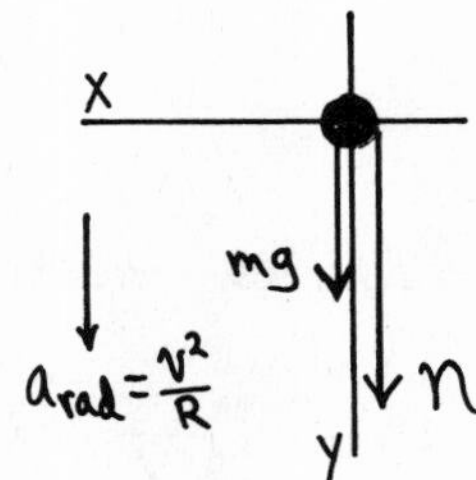

Consider the free-body diagram for the water when the pail is at the top of its circular path. The radial acceleration is in toward the center of the circle so at this point is downward. n is the downward normal force exerted on the water by the bottom of the pail.

x

mg

$a_{rad} = \dfrac{v^2}{R}$

n

y

$\Sigma F_y = ma_y$

$n + mg = m\dfrac{v^2}{R}$

At the minimum speed the water is just ready to lose contact with the bottom of the pail, so at this speed $n \to 0$. (Note that the force n cannot be upward.)

$n \to 0 \Rightarrow \cancel{m}g = \cancel{m}\dfrac{v^2}{R}$

$v = \sqrt{gR} = \sqrt{(9.80\text{ m/s}^2)(0.800\text{ m})} = \underline{2.80\text{ m/s}}$

5-45

a)

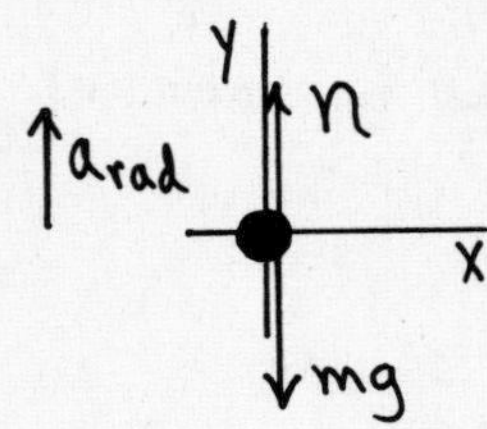

$$a_{rad} = \frac{v^2}{R} \Rightarrow R = \frac{v^2}{a_{rad}}$$

$$a_{rad} = 7g \Rightarrow R = \frac{v^2}{7g} = \frac{(75.0\ \text{m/s})^2}{7(9.80\ \text{m/s}^2)} = \underline{82.0\ \text{m}}$$

b) Free-body diagram for the pilot at the bottom of the path:

a_{rad} ↑, y, n, x, mg

n is the normal force exerted on the pilot by the airplane seat. This force is what is called his apparent weight.

$$\Sigma F_y = ma_y$$

$$n - mg = ma_{rad}$$

$$n = m(g + a_{rad})$$

$$a_{rad} = 7g \Rightarrow n = 8mg$$

$$n = 8(80.0\ \text{kg})(9.80\ \text{m/s}^2) = \underline{6270\ \text{N}}$$

Problems

5-47

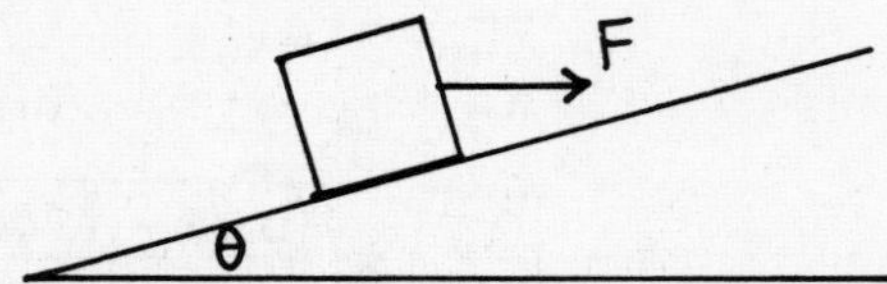

No friction.
Constant speed $\Rightarrow a = 0$.

Since $a = 0$ it is equally convenient to use
(1) coordinate axes parallel and perpendicular to the incline;
(2) coordinate axes that are horizontal and vertical.
We will work the problem both ways.

coordinate axes parallel and perpendicular to the incline.

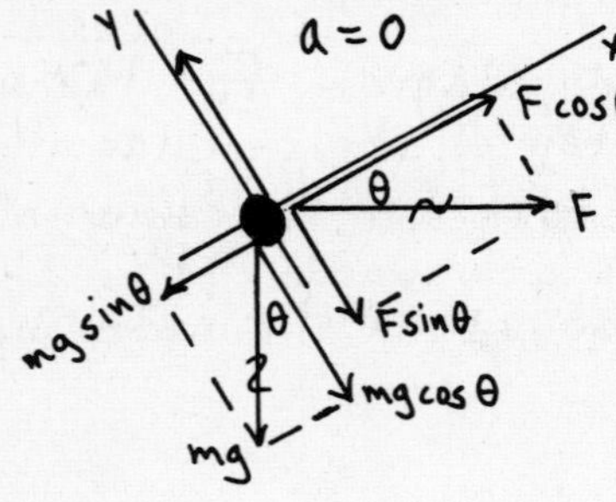

$$\Sigma F_x = ma_x$$

$$F\cos\theta - mg\sin\theta = 0$$

$$F = mg\frac{\sin\theta}{\cos\theta}$$

$$\boxed{F = mg\tan\theta}$$

coordinates that are horizontal and vertical

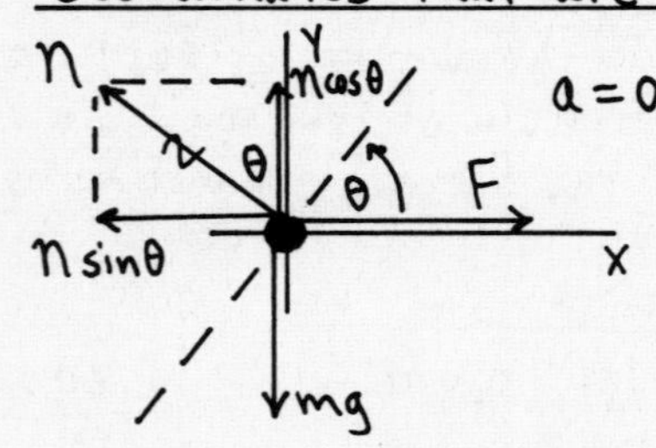

$a = 0$

$$\Sigma F_x = ma_x$$

$$F - n\sin\theta = 0$$

$$\Sigma F_y = ma_y$$

$$n\cos\theta - mg = 0$$

$$n = \frac{mg}{\cos\theta}$$

Combine these two equations to eliminate n:

$$F = n\sin\theta = \frac{mg}{\cos\theta}\sin\theta$$

$F = mg\tan\theta$, the same as before

5-49

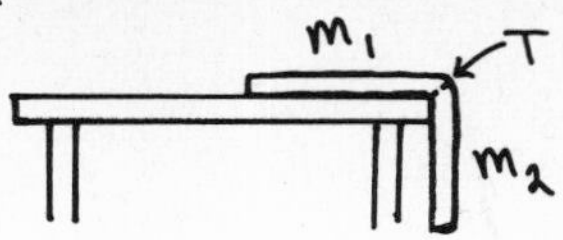

Let m_1 be the mass of that part of the rope that is on the table, and let m_2 be the mass of that part of the rope that is hanging over the edge. ($m_1 + m_2 = m$, the total mass of the rope)

Since the mass of the rope is not being neglected, the tension in the rope varies along the length of the rope. Let T be the tension in the rope at that point that is at the edge of the table.

Free-body diagram for the hanging section of the rope:

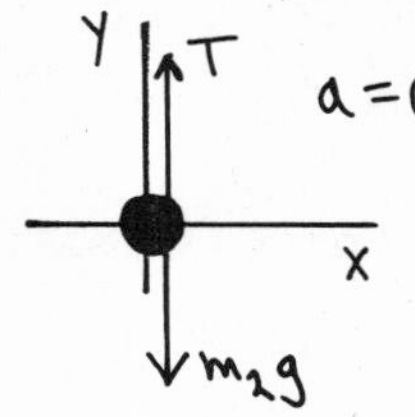

$a = 0$

$$\Sigma F_y = ma_y$$
$$T - m_2 g = 0$$
$$\boxed{T = m_2 g}$$

Free-body diagram for the part of the rope that is on the table:

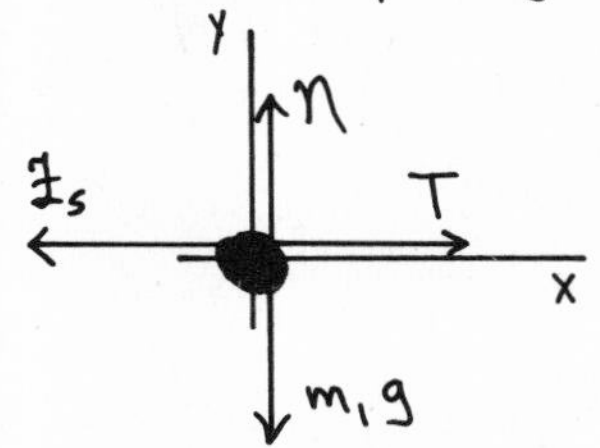

$$\Sigma F_y = ma_y$$
$$n - m_1 g = 0$$
$$n = m_1 g$$

When the maximum amount of rope hangs over the edge the static friction has its maximum value:
$$f_s = \mu_s n = \mu_s m_1 g$$

$$\Sigma F_x = ma_x$$
$$T - f_s = 0$$
$$\boxed{T = \mu_s m_1 g}$$

Use the first equation to replace T:
$$m_2 g = \mu_s m_1 g$$
$$m_2 = \mu_s m_1$$

The fraction that hangs over is $\dfrac{m_2}{m} = \dfrac{m_2}{m_1 + m_2} = \dfrac{\mu_s m_1}{m_1 + \mu_s m_1} = \dfrac{\mu_s}{1 + \mu_s}$.

5-51

a)

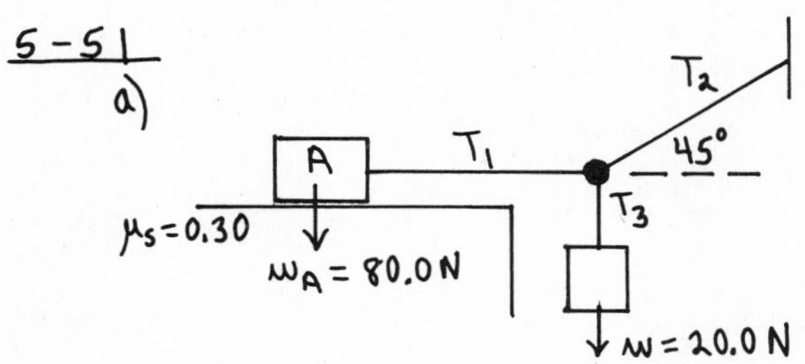

Free-body diagram for the hanging block:

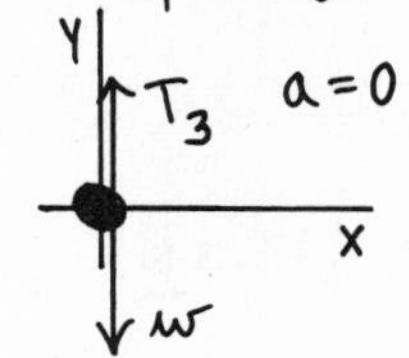

$a = 0$

$$\Sigma F_y = ma_y$$
$$T_3 - w = 0$$
$$T_3 = 20.0 \text{ N}$$

Free-body diagram for the knot:

Y, $T_2 \sin 45°$, T_2, T_1, 45°, $T_2 \cos 45°$, X, T_3

$a = 0$

$$\Sigma F_y = ma_y$$
$$T_2 \sin 45° - T_3 = 0$$
$$T_2 = \frac{T_3}{\sin 45°} = \frac{20.0 \text{ N}}{\sin 45°}$$
$$T_2 = 28.3 \text{ N}$$

$$\Sigma F_x = ma_x$$
$$T_2 \cos 45° - T_1 = 0$$
$$T_1 = T_2 \cos 45°$$
$$T_1 = 20.0 \text{ N}$$

5-51 (cont)

Free-body diagram for block A:

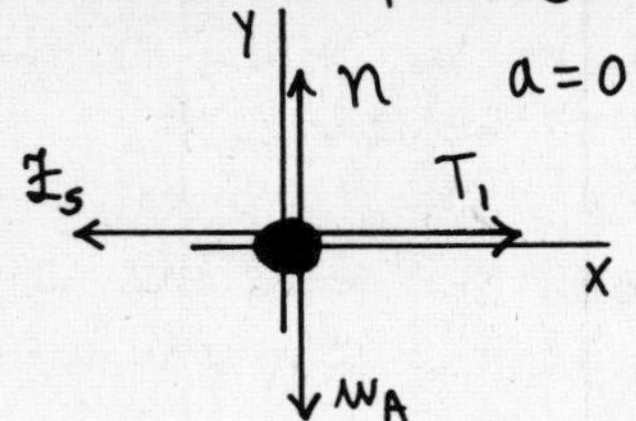

$$\Sigma F_x = ma_x$$
$$T_1 - f_s = 0$$
$$f_s = T_1 = \underline{20.0 \text{ N}}$$

Note: $\Sigma F_y = ma_y$

$$n - w_A = 0$$
$$n = w_A = 80.0 \text{ N}$$

$\mu_s n = (0.30)(80.0 \text{ N}) = 24.0 \text{ N}$

$f_s < \mu_s n$; for this value of w the static friction force can hold the blocks in place.

b) We have all the same free-body diagrams and force equations as in part (a), but now the static friction force has its largest possible value, $f_s = \mu_s n = 24.0$ N. Then $T_1 = f_s = 24.0$ N.

$$T_2 \cos 45° - T_1 = 0 \Rightarrow T_2 = \frac{T_1}{\cos 45°} = \frac{24.0 \text{ N}}{\cos 45°} = 33.9 \text{ N}$$
$$T_2 \sin 45° - T_3 = 0 \Rightarrow T_3 = T_2 \sin 45° = (33.9 \text{ N}) \sin 45° = 24.0 \text{ N}$$
$$T_3 - w = 0 \Rightarrow w = T_3 = \underline{24.0 \text{ N}}$$

5-53

Consider the forces on the scrub brush. Note that the normal force exerted by the wall is horizontal, since it is perpendicular to the wall. The kinetic friction force exerted by the wall is parallel to the wall and opposes the motion, so it is vertically downward.

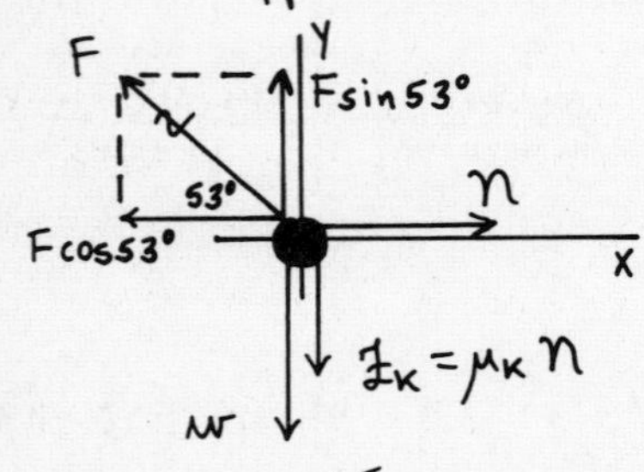

$$\Sigma F_x = ma_x$$
$$n - F\cos 53° = 0$$
$$\boxed{n = F\cos 53°}$$

Then $f_k = \mu_k n$

$$f_k = \mu_k F \cos 53°$$

$$\Sigma F_y = ma_y$$
$$F\sin 53° - w - f_k = 0$$
$$F\sin 53° - w - \mu_k F \cos 53° = 0$$
$$F(\sin 53° - \mu_k \cos 53°) = w$$
$$\boxed{F = \frac{w}{\sin 53° - \mu_k \cos 53°}}$$

a) $F = \dfrac{w}{\sin 53° - \mu_k \cos 53°} = \dfrac{8.00 \text{ N}}{\sin 53° - (0.40)\cos 53°} = \underline{14.3 \text{ N}}$

b) $n = F\cos 53° = (14.3 \text{ N}) \cos 53° = \underline{8.6 \text{ N}}$

5-55

Block B is pulled to the left at constant speed $\Rightarrow$ block A moves to the right at constant speed and $a = 0$ for each block.

5-55 (cont)

Free-body diagram for block A:

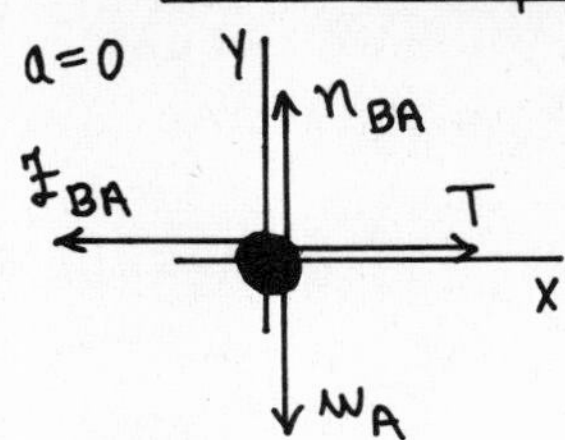

n_{BA} is the normal force that B exerts on A.

$f_{BA} = \mu_k n_{BA}$ is the kinetic friction force that B exerts on A. Block A moves to the right relative to B and f_{BA} opposes this motion $\Rightarrow$ f_{BA} is to the left.

Note also that F acts just on B, not on A.

$\Sigma F_y = ma_y$

$n_{BA} - w_A = 0$

$n_{BA} = 3.60\ N$

$f_{BA} = \mu_k n_{BA} = (0.25)(3.60\ N) = 0.90\ N$

$\Sigma F_x = ma_x$

$T - f_{BA} = 0$

$T = f_{BA} = 0.90\,N$

Free-body diagram for block B:

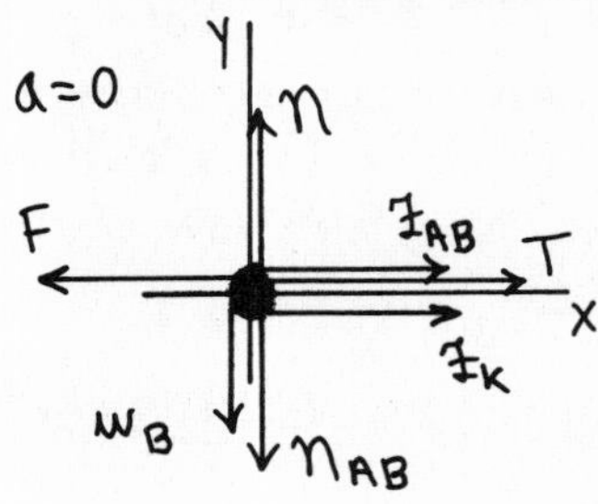

n_{AB} is the normal force that block A exerts on B. By Newton's third law n_{AB} and n_{BA} are equal and opposite $\Rightarrow n_{AB} = 3.60\ N$.

f_{AB} is the kinetic friction force that A exerts on B. Block B moves to the left relative to A and f_{AB} opposes this motion $\Rightarrow$ f_{AB} is to the right.

Also, $f_{AB} = \mu_k n_{AB} = (0.25)(3.60\,N) = 0.90\ N$.

But also note that f_{AB} and f_{BA} are a third law action-reaction pair, so they must be equal and opposite and this is indeed what our calculation gives.

n and f_k are the normal and the friction force exerted by the floor on block B; $f_k = \mu_k n$. Note that block B moves to the left relative to the floor and f_k opposes this motion, so f_k is to the right.

$\Sigma F_y = ma_y$

$n - w_B - n_{AB} = 0$

$n = w_B + n_{AB} = 5.40\,N + 3.60\,N = 9.00\,N$

Then $f_k = \mu_k n = (0.25)(9.00\,N) = 2.25\ N$.

$\Sigma F_x = ma_x$

$f_{AB} + T + f_k - F = 0$

$F = T + f_{AB} + f_k = 0.90\,N + 0.90\,N + 2.25\,N$

$F = \underline{4.05\ N}$

5-57

a) $\rightarrow v_0$

$\longrightarrow a = 2.20\ m/s^2$

First calculate the maximum acceleration that the static friction force can give to the case:

5-57 (cont)

Y, a, n, f_s, X, mg

The static friction force is to the right since it tries to make the case move with the truck. The maximum value it can have is $f_s = \mu_s n$.

$\Sigma F_y = ma_y$
$n - mg = 0$
$n = mg$
$f_s = \mu_s n = \mu_s mg$

$\Sigma F_x = ma_x$
$f_s = ma$
$\mu_s \cancel{m}g = \cancel{m}a$

$a = \mu_s g = (0.30)(9.80\ m/s^2)$
$a = 2.94\ m/s^2$

The truck's acceleration is less than this so the case doesn't slip relative to the truck; the case's acceleration is $a = 2.20\ m/s^2$ (eastward).

Then $f_s = ma = (30.0\ kg)(2.20\ m/s^2) = \underline{66.0\ N}$, $\underline{\text{eastward}}$.

b) Now the acceleration of the truck is greater than the acceleration static friction can give the case. Therefore, the case slips relative to the truck and the friction is kinetic friction. The friction force still tries to keep the case moving with the truck, so the acceleration of the case and the friction force are both <u>westward.</u>

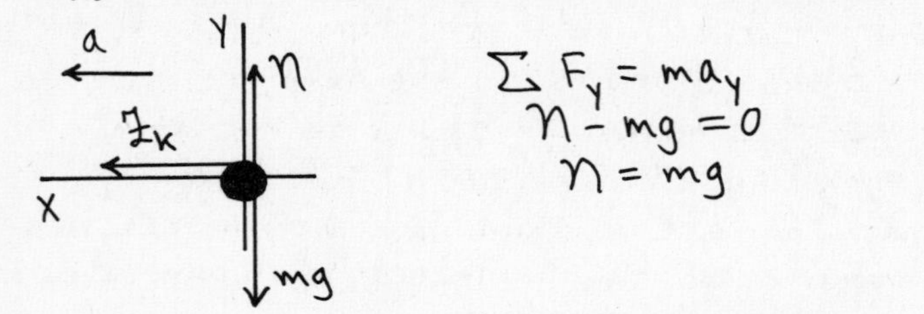

$\Sigma F_y = ma_y$
$n - mg = 0$
$n = mg$

$f_K = \mu_K mg = (0.20)(30.0\ kg)(9.80\ m/s^2)$
$f_K = \underline{58.8\ N}$, $\underline{\text{westward}}$

Note: $f_K = ma \Rightarrow a = \frac{f_K}{m} = \frac{58.8\ N}{30.0\ kg} = 1.96\ m/s^2$; the magnitude of the acceleration of the case is less than that of the truck and the case slides toward the front of the truck.

<u>5-59</u>

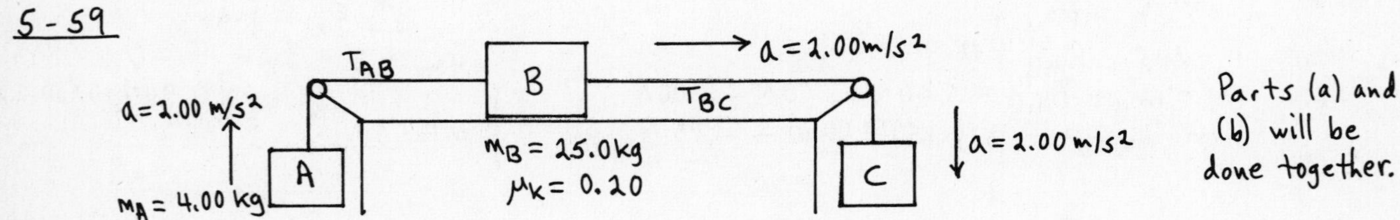

Parts (a) and (b) will be done together.

Consider the forces on each block. Note that each block has the same magnitude of acceleration, but in different directions. For each block let the direction of $\vec{a}$ be a positive coordinate direction.

5-59 (cont)

Block A:

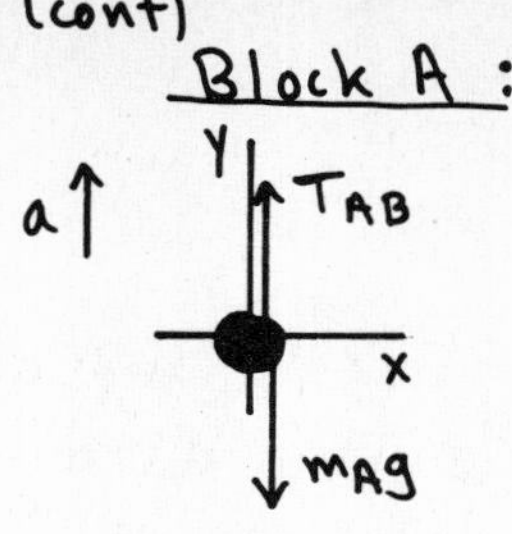

$$\Sigma F_y = ma_y$$

$$T_{AB} - m_A g = m_A a$$

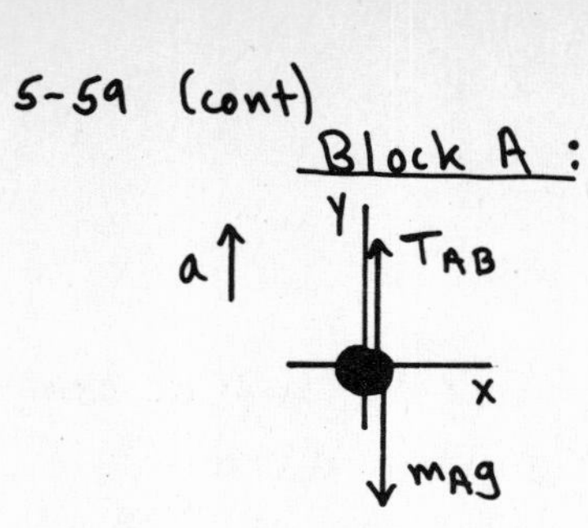

$$T_{AB} = m_A(a+g) = 4.00\text{ kg}(2.00\text{ m/s}^2 + 9.80\text{ m/s}^2) = \underline{47.2\text{ N}}$$

Block B:

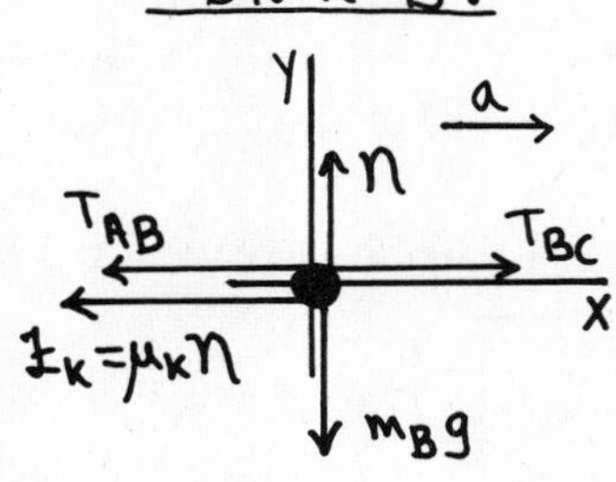

$$\Sigma F_y = ma_y$$

$$n - m_B g = 0$$

$$n = m_B g$$

So $f_k = \mu_k n = \mu_k m_B g$

$$f_k = (0.20)(25.0\text{ kg})(9.80\text{ m/s}^2)$$

$$f_k = 49.0\text{ N}$$

$$\Sigma F_x = ma_x$$

$$T_{BC} - T_{AB} - f_k = m_B a$$

$$T_{BC} = T_{AB} + f_k + m_B a$$

$$T_{BC} = 47.2\text{N} + 49.0\text{N} + (25.0\text{kg})(2.00\text{m/s}^2)$$

$$T_{BC} = \underline{146\text{ N}}$$

Block C:

T_{BC} ↑ ↓ a, x, $m_c g$, y

$$\Sigma F_y = ma_y$$

$$m_c g - T_{BC} = m_c a$$

$$m_c(g-a) = T_{BC}$$

$$m_c = \frac{T_{BC}}{g-a} = \frac{146\text{ N}}{9.80\text{m/s}^2 - 2.00\text{m/s}^2} = \underline{18.7\text{ kg}}$$

5-61

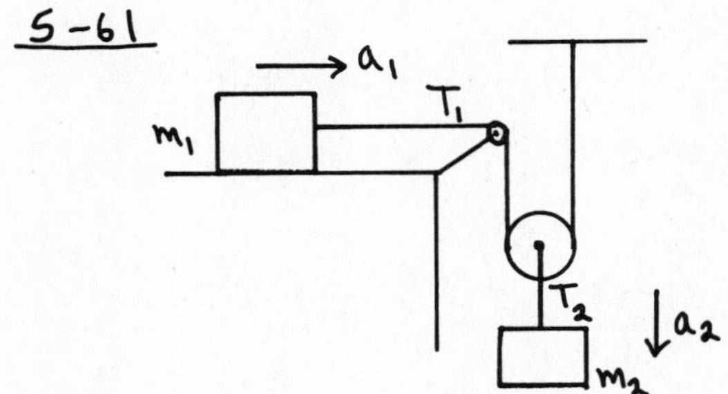

Forces on m_1:

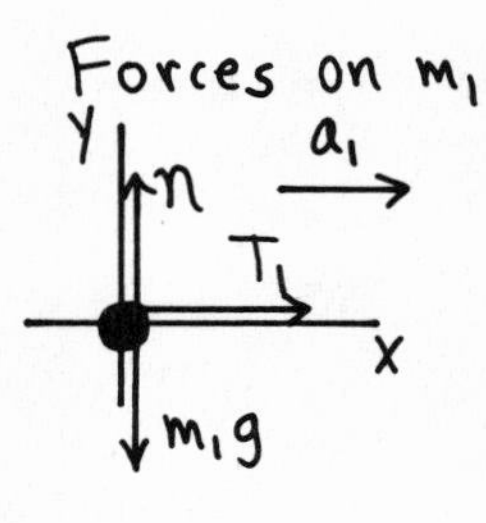

$$\Sigma F_x = ma_x$$

$$\boxed{T_1 = m_1 a_1}$$

Forces on m_2:

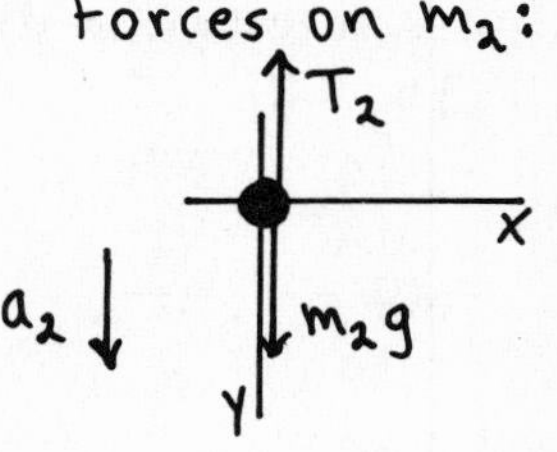

$$\Sigma F_y = ma_y$$

$$\boxed{m_2 g - T_2 = m_2 a_2}$$

This gives us two equations, but there are 4 unknowns (T_1, T_2, a_1, and a_2) so two more equations are required.

Free-body diagram for the moveable pulley (mass m):

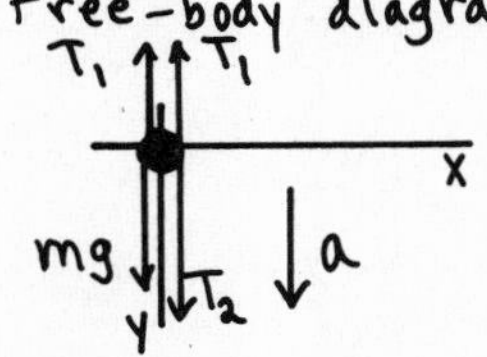

$$\Sigma F_y = ma_y$$

$$mg + T_2 - 2T_1 = ma$$

But our pulleys have negligible mass

$\Rightarrow mg = ma = 0$

$\Rightarrow \boxed{T_2 = 2T_1}$

5-61 (cont)

Combine these three equations to eliminate T_1 and T_2:

$$m_2 g - T_2 = m_2 a_2 \Rightarrow m_2 g - 2T_1 = m_2 a_2$$

$$T_1 = m_1 a_1 \Rightarrow \boxed{m_2 g - 2m_1 a_1 = m_2 a_2}$$

There are still unknowns, a_1 and a_2. But the accelerations a_1 and a_2 are related. In any given time interval if m_1 moves to the right a distance d, then in the same time m_2 moves downward a distance $d/2$. One of the constant acceleration kinematic equations says $x - x_0 = \cancel{v_0 t} + \frac{1}{2}at^2$, so if m_2 moves half the distance it must have half the acceleration of m_1: $a_2 = \frac{1}{2}a_1$, or $\boxed{a_1 = 2a_2}$.

This is the additional equation we need. Use it in the previous boxed equation $\Rightarrow m_2 g - 2m_1(2a_2) = m_2 a_2$

$$a_2(4m_1 + m_2) = m_2 g \Rightarrow \boxed{a_2 = \frac{m_2 g}{4m_1 + m_2}}$$

And then $\boxed{a_1 = 2a_2 = \frac{2m_2 g}{4m_1 + m_2}}$.

5-65

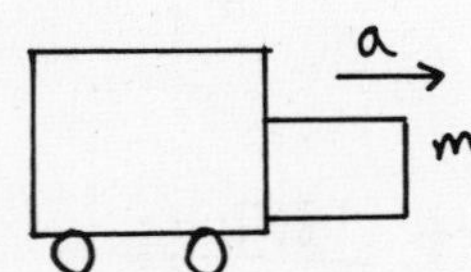

The cart and the block have the same acceleration. The normal force exerted by the cart on the block is perpendicular to the front of the cart, so is horizontal and to the right.

The friction force on the block is directed so as to hold the block up against the downward pull of gravity. We want to calculate the minimum a required, so take static friction to have its maximum value, $f_s = \mu_s n$.

Free-body diagram for the block:

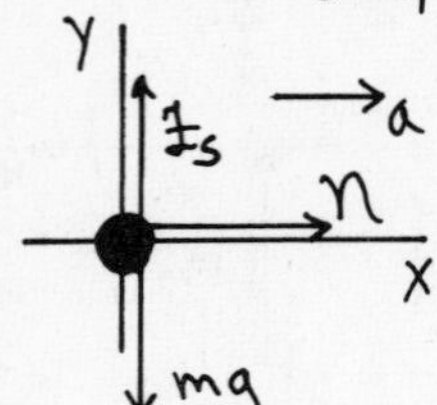

$$\Sigma F_x = ma_x$$
$$n = ma$$
$$f_s = \mu_s n = \mu_s ma$$

$$\Sigma F_y = ma_y$$
$$f_s - mg = 0$$
$$\mu_s \cancel{m} a = \cancel{m} g$$
$$\boxed{a = g/\mu_s}$$

An observer on the cart sees the block pinned there, with no reason for a horizontal force on it because the block is at rest relative to the cart. Therefore, such an observer concludes that $n = 0$ and thus $f_s = 0$, and he doesn't understand what holds the block up against the downward force of gravity. The reason for this difficulty is that $\Sigma \vec{F} = m\vec{a}$ does not apply in a coordinate frame attached to the cart. This reference frame is accelerated, and hence is not inertial.

5-69

a) To keep the car from sliding up the banking the static friction force is directed down the incline. At the maximum speed the static friction force has its maximum value, $f_s = \mu_s n$.

Free-body diagram for the car:

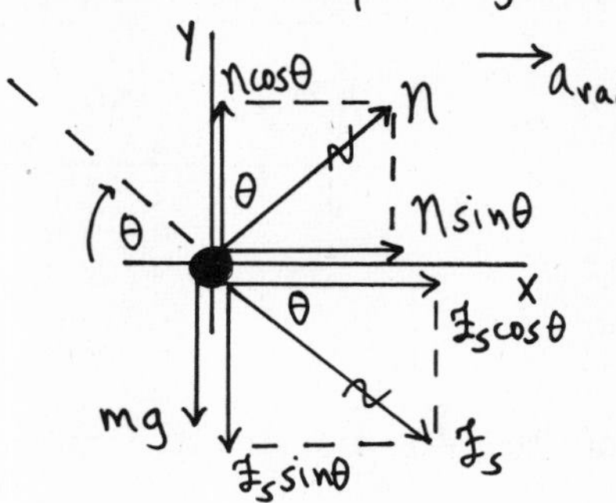

$$\Sigma F_y = ma_y$$
$$n\cos\theta - f_s\sin\theta - mg = 0$$
$$\text{But } f_s = \mu_s n$$
$$\Rightarrow n\cos\theta - \mu_s \sin\theta\, n - mg = 0$$
$$\boxed{n = \frac{mg}{\cos\theta - \mu_s\sin\theta}}$$

$$\Sigma F_x = ma_x$$
$$n\sin\theta + f_s\cos\theta = ma_{rad}$$
$$n\sin\theta + \mu_s\cos\theta\, n = ma_{rad}$$
$$\boxed{n(\sin\theta + \mu_s\cos\theta) = ma}$$

Use the first equation to replace n:

$$\left(\frac{mg}{\cos\theta - \mu_s\sin\theta}\right)(\sin\theta + \mu_s\cos\theta) = ma_{rad}$$

$$a_{rad} = \left(\frac{\sin\theta + \mu_s\cos\theta}{\cos\theta - \mu_s\sin\theta}\right)g = \left(\frac{\sin 25° + (0.35)\cos 25°}{\cos 25° - (0.35)\sin 25°}\right)(9.80\ m/s^2) = 9.56\ m/s^2$$

$$a_{rad} = \frac{v^2}{R} \Rightarrow v = \sqrt{a_{rad} R} = \sqrt{(9.56\ m/s^2)(36\ m)} = \underline{18.6\ m/s}$$

b) To keep the car from sliding <u>down</u> the banking the static friction force is directed up the incline. At the minimum speed the static friction force has its maximum value $f_s = \mu_s n$.

Free-body diagram for the car:

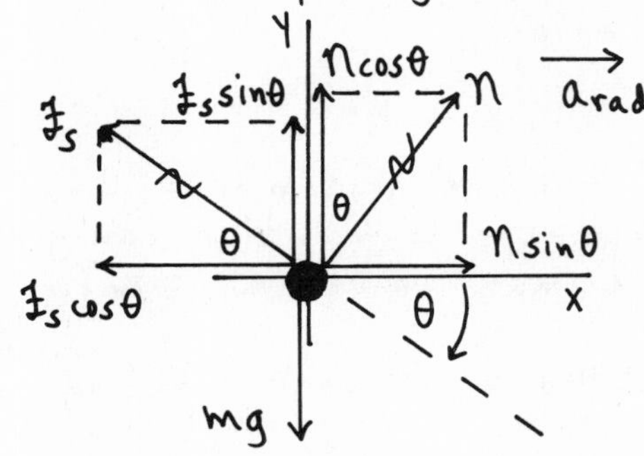

The free-body diagram is identical to that in part (a) except that now the components of f_s have opposite directions. The force equations are all the same except for the opposite sign for terms containing μ_s.

$$\text{Thus, } a_{rad} = \left(\frac{\sin\theta - \mu_s\cos\theta}{\cos\theta + \mu_s\sin\theta}\right)g = \left(\frac{\sin 25° - (0.35)\cos 25°}{\cos 25° + (0.35)\sin 25°}\right)(9.80\ m/s^2)$$

$$a_{rad} = 0.980\ m/s^2$$

$$v = \sqrt{a_{rad} R} = \sqrt{(0.980\ m/s^2)(36\ m)} = \underline{5.9\ m/s}$$

5-71

$$v = (0.60\ \text{rev/s})\left(\frac{2\pi R}{1\ \text{rev}}\right) = (0.60\ \text{rev/s})\left(\frac{2\pi(3.0\ \text{m})}{1\ \text{rev}}\right) = 11.3\ \text{m/s}$$

$$a_{rad} = \frac{v^2}{R} = \frac{(11.3\ \text{m/s})^2}{3.0\ \text{m}} = 42.6\ \text{m/s}^2$$

a)

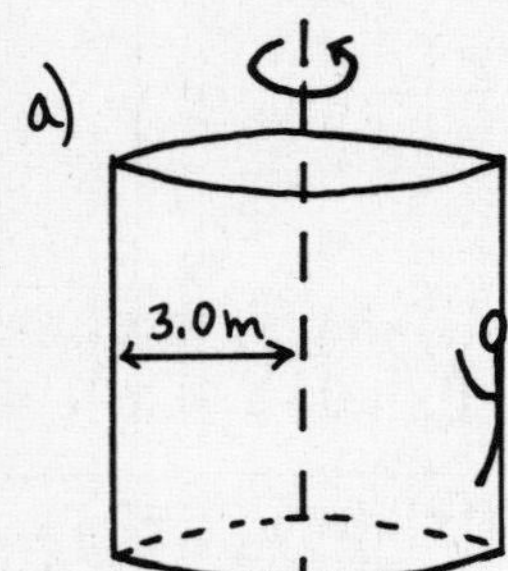

The person is held up against gravity by the static friction force exerted on him by the wall. The acceleration of the person is a_{rad}, directed in toward the axis of rotation.

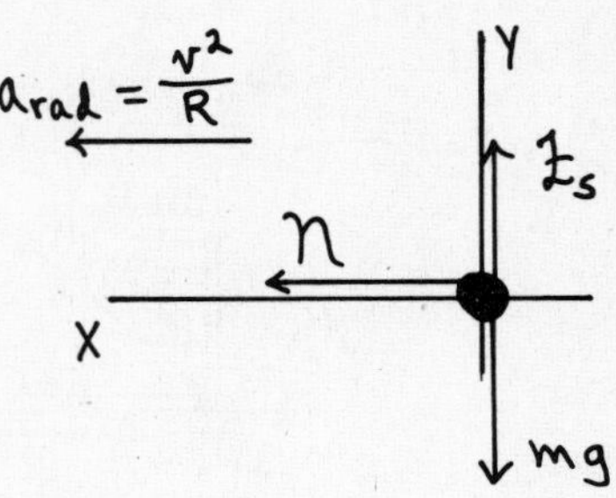

b) To calculate the minimum μ_s required, take f_s to have its maximum value, $f_s = \mu_s n$.

$\sum F_y = ma_y$

$f_s - mg = 0$

$\boxed{\mu_s n = mg}$

$\sum F_x = ma_x$

$\boxed{n = m\frac{v^2}{R}}$

Combine these two equations to eliminate n:

$$\mu_s \left(m \frac{v^2}{R}\right) = mg$$

$$\mu_s = \frac{Rg}{v^2} = \frac{(3.0\ \text{m})(9.80\ \text{m/s}^2)}{(11.3\ \text{m/s})^2} = \underline{0.23}$$

c) No, the mass of the person divided out of the equation for μ_s.

5-73

a)

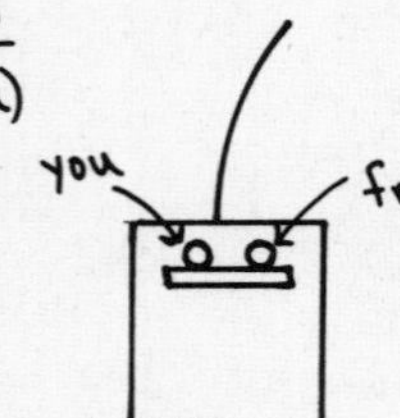

Turn to the right. In the absence of sufficient friction your friend doesn't make the turn completely and you move to the right toward your friend.

b) The maximum radius of the turn is the one that makes a_{rad} just equal to the maximum acceleration that static friction can give to your friend, and for this f_s has its maximum value $f_s = \mu_s n$.

Free-body diagram for your friend, as viewed by someone standing behind the car:

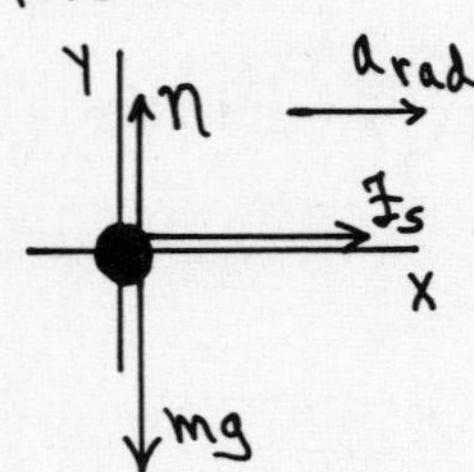

$\sum F_y = ma_y$

$n - mg = 0$

$n = mg$

$\sum F_x = ma_x$

$f_s = ma_{rad}$

$\mu_s n = m\frac{v^2}{R}$

$\mu_s mg = m\frac{v^2}{R}$

$$R = \frac{v^2}{\mu_s g} = \frac{(18\ \text{m/s})^2}{(0.40)(9.80\ \text{m/s}^2)} = \underline{83\ \text{m}}$$

5-75

Use the information given about Kathy to find the time t for one revolution of the merry-go-round. Her acceleration is a_{rad}, directed in toward the axle. Let $\vec{F}_1$ be the horizontal force that keeps her from sliding off. Let her speed be v_1 and let R_1 be her distance from the axis.

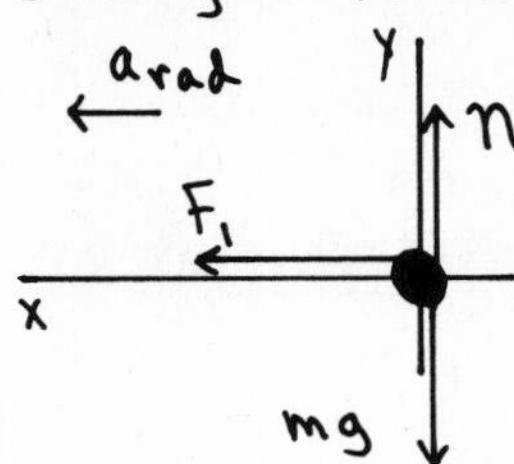

$$\Sigma F_x = ma_x$$
$$F_1 = ma_{rad}$$
$$F_1 = m\frac{v_1^2}{R_1}$$

$$v_1 = \sqrt{\frac{R_1F_1}{m}} = \sqrt{\frac{(2.00\text{ m})(80.0\text{N})}{47.6\text{ kg}}} = 1.83\text{ m/s}$$

The time for one revolution is

$$t = \frac{2\pi R_1}{v_1} = \frac{2\pi(2.00\text{m})}{1.83\text{ m/s}} = 6.87\text{s}$$

Karen goes around once in the same time but her speed (v_2) and the radius of her circular path (R_2) are different.

$$v_2 = \frac{2\pi R_2}{t} = \frac{2\pi(4.00\text{ m})}{6.87\text{s}} = 3.66\text{ m/s}$$

Free-body diagram for Karen:

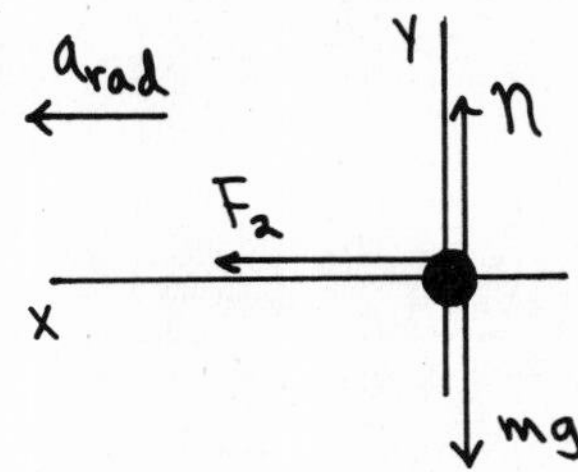

$$\Sigma F_x = ma_x$$
$$F_2 = ma_{rad}$$
$$F_2 = m\frac{v_2^2}{R_2} = 47.6\text{kg}\frac{(3.66\text{ m/s})^2}{4.00\text{m}} = \underline{159\text{ N}}$$

5-77

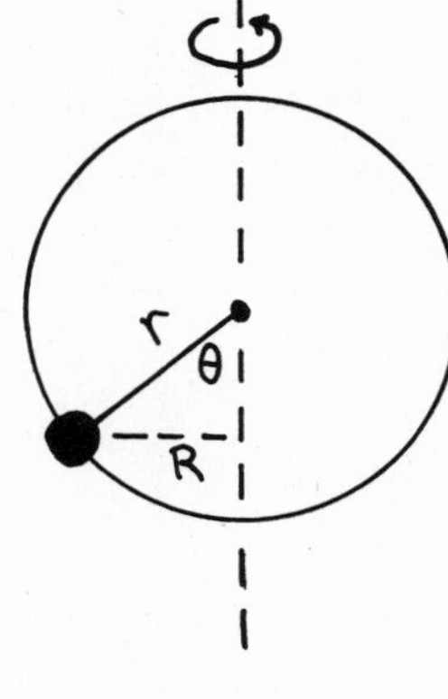

The bead moves in a circle of radius $R = r\sin\theta$. The normal force exerted on the bead by the hoop is radially inward.

Free-body diagram for the bead:

$$\Sigma F_y = ma_y$$
$$n\cos\theta - mg = 0$$
$$n = \frac{mg}{\cos\theta}$$

$$\Sigma F_x = ma_x$$
$$n\sin\theta = ma_{rad}$$

5-77 (cont)

Combine these two equations to eliminate n

$$\left(\frac{\cancel{m}g}{\cos\theta}\right)\sin\theta = \cancel{m}a_{rad}$$

$$\boxed{\frac{\sin\theta}{\cos\theta} = \frac{a_{rad}}{g}}$$

$a_{rad} = \frac{v^2}{R}$ and $v = \frac{2\pi R}{T} \Rightarrow a_{rad} = \frac{4\pi^2 R}{T^2}$, where T is the time for one revolution

And $R = r\sin\theta \Rightarrow a_{rad} = \frac{4\pi^2 r\sin\theta}{T^2}$.

Use this in the above equation:

$$\frac{\sin\theta}{\cos\theta} = \frac{4\pi^2 r\sin\theta}{T^2 g}$$

This equation is satisfied by $\sin\theta = 0$ ($\theta = 0°$)

or $\frac{1}{\cos\theta} = \frac{4\pi^2 r}{T^2 g} \Rightarrow \boxed{\cos\theta = \frac{T^2 g}{4\pi^2 r}}$

a) $4.00 \text{ rev/s} \Rightarrow T = \frac{1}{4.00} \text{ s} = 0.250 \text{ s}$

$$\cos\theta = \frac{(0.250\text{ s})^2(9.80 \text{ m/s}^2)}{4\pi^2(0.100\text{ m})} = 0.1551 \Rightarrow \underline{\theta = 81.1°}$$

b) This would mean $\theta = 90°$. But $\cos 90° = 0$ so this requires $T \to 0$. θ <u>approaches</u> 90° as the hoop rotates very fast, but $\theta = 90°$ is not possible.

c) $1.00 \text{ rev/s} \Rightarrow T = 1.00\text{ s}$

The $\cos\theta = \frac{T^2 g}{4\pi^2 r}$ equation then says $\cos\theta = \frac{(1.00\text{s})^2(9.80\text{ m/s}^2)}{4\pi^2(0.100\text{ m})} = 2.48$, which is not possible.

The only way to have the $\sum\vec{F} = m\vec{a}$ equations satisfied is for $\sin\theta = 0$. This means $\theta = 0°$; the bead sits at the bottom of the hoop.

CHAPTER 6

Exercises 1, 5, 7, 9, 11, 17, 19, 23, 25, 27, 31, 33, 37

Problems 39, 41, 43, 47, 51, 55, 57

Exercises

6-1

a) $W_F = (F\cos\phi)s = (2.00\text{ N}\cos 0°)(1.50\text{ m}) = \underline{3.00\text{ J}}$
(The force and displacement vectors are in the same direction.)

b) $W_f = (f\cos\phi)s = (0.400\text{ N}\cos 180°)(1.50\text{ m}) = \underline{-0.600\text{ J}}$
(The force and displacement vectors are in opposite directions.)

6-5

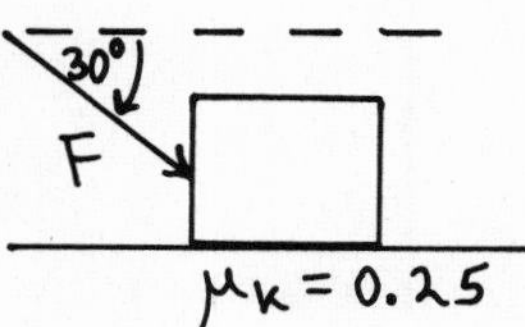

constant speed
$\Rightarrow a = 0$

a) Free-body diagram for the crate:

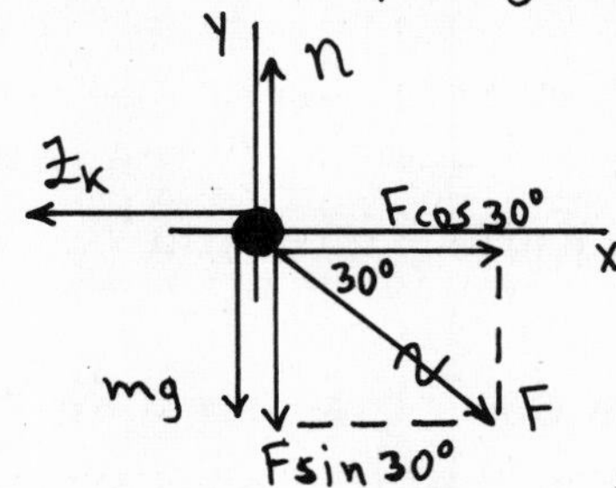

$\sum F_y = ma_y$
$n - mg - F\sin 30° = 0$
$n = mg + F\sin 30°$

$f_k = \mu_k n = \mu_k mg + F\mu_k \sin 30°$

$\sum F_x = ma_x$
$F\cos 30° - f_k = 0$
$F\cos 30° - \mu_k mg - \mu_k \sin 30° F = 0$
$F = \dfrac{\mu_k mg}{\cos 30° - \mu_k \sin 30°}$

$$F = \frac{0.25(35.0\text{ kg})(9.80\text{ m/s}^2)}{\cos 30° - (0.25)\sin 30°} = \underline{116\text{ N}}$$

b) $W = (F\cos\phi)s = (116\text{ N}\cos 30°)(5.0\text{ m}) = \underline{500\text{ J}}$
($F\cos 30°$ is the horizontal component of $\vec{F}$; the work done by $\vec{F}$ is the displacement times the component of $\vec{F}$ in the direction of the displacement.)

c) We have an expression for f_k from part (a):
$f_k = \mu_k(mg + F\sin 30°) = 0.25((35.0\text{ kg})(9.80\text{ m/s}^2) + (116\text{ N})(\sin 30°)) = 100\text{ N}$

$\phi = 180°$ since f_k is opposite to the displacement.
$\Rightarrow W_f = (f_k\cos\phi)s = (100\text{ N})(\cos 180°)(5.0\text{ m}) = \underline{-500\text{ J}}$.

d) The normal force is perpendicular to the displacement so $\phi = 90°$ and $W_n = 0$.
The gravity force (the weight) is perpendicular to the displacement so $\phi = 90°$ and $W_w = 0$.

6-7

Use the information given to calculate the force constant k of the spring:

$F = kx \Rightarrow k = \frac{F}{x} = \frac{90.0\ N}{0.400\ m} = 225\ N/m.$

a) $F = kx = (225\ N/m)(0.100\ m) = \underline{22.5\ N}$

$F = kx = (225\ N/m)(-0.200\ m) = -45.0\ N$ (magnitude $\underline{45.0\ N}$)

b) $W = \frac{1}{2}kx^2 = \frac{1}{2}(225\ N/m)(0.100\ m)^2 = \underline{1.12\ J}$

$W = \frac{1}{2}kx^2 = \frac{1}{2}(225\ N/m)(-0.200\ m)^2 = \underline{4.50\ J}$

6-9

$W = \frac{1}{2}kx^2 \Rightarrow k = \frac{2W}{x^2} = \frac{2(20.0\ J)}{(0.150\ m)^2} = 1780\ N/m$

Then $F = kx = (1780\ N/m)(0.150\ m) = \underline{267\ N}$.

6-11

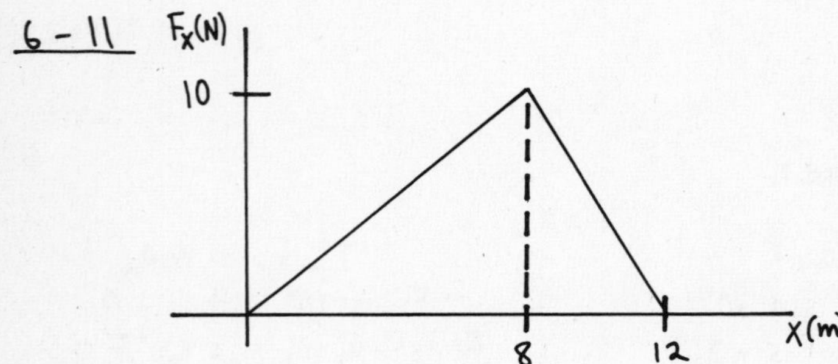

a) The displacement is in the +x-direction and the x-component of this force is positive, so the work done is positive. The magnitude of the work done is the area under the curve on the graph between x=0 and x= 12 m.

$\Rightarrow W = \frac{1}{2}(8m)(10\ N) + \frac{1}{2}(4m)(10\ N) = \underline{60\ J}$

b) The displacement is in the -x-direction so now the work done is negative. The magnitude of the work is the area under the curve between x=8 m and x=12 m.

$\Rightarrow W = -\frac{1}{2}(4m)(10\ N) = \underline{-20\ J}$

6-17

a) $W_{tot} = K_2 - K_1 \Rightarrow K_2 = W_{tot} + K_1$

$K_1 = \frac{1}{2}mv_1^2 = \frac{1}{2}(2.50\ kg)(3.00\ m/s)^2 = 11.25\ J$

The only force that does work on the wagon is the 2.50 N force. This force is in the direction of the displacement so $\phi = 0°$ and the force does positive work:

$W = (F\cos\phi)s = (2.50\ N)(\cos 0°)(4.0\ m) = 10.0\ J$

$K_2 = W_{tot} + K_1 = 10.0\ J + 11.25\ J = 21.25\ J$

6-17 (cont)

$$K_2 = \frac{1}{2}mv_2^2 \Rightarrow v_2 = \sqrt{\frac{2K_2}{m}} = \sqrt{\frac{2(21.25\text{ J})}{2.50\text{ kg}}} = \underline{4.12\text{ m/s}}$$

b)

Y n a F x mg

$$\sum F_x = ma_x$$

$$F = ma$$

$$a = \frac{F}{m} = \frac{2.50\text{ N}}{2.50\text{ kg}} = 1.00\text{ m/s}^2$$

$$v_2^2 = v_1^2 + 2a(x - x_0)$$

$$v_2 = \sqrt{v_1^2 + 2a(x-x_0)} = \sqrt{(3.00\text{ m/s})^2 + 2(1.00\text{ m/s}^2)(4.0\text{ m})} = \underline{4.12\text{ m/s}}$$

This agrees with the result calculated in part (a).

6-19

$$W_{tot} = K_2 - K_1$$

$$K_1 = \frac{1}{2}mv_1^2 = \frac{1}{2}(0.42\text{ kg})(4.00\text{ m/s})^2 = 3.36\text{ J}$$

$$K_2 = \frac{1}{2}mv_2^2 = \frac{1}{2}(0.42\text{ kg})(6.00\text{ m/s})^2 = 7.56\text{ J}$$

$$W_{tot} = K_2 - K_1 = 7.56\text{ J} - 3.36\text{ J} = 4.20\text{ J}$$

The 30.0 N force is the only force doing work on the ball, so it must do 4.20 J of work.

$$W = (F\cos\phi)s \Rightarrow s = \frac{W}{F\cos\phi} = \frac{4.20\text{ J}}{30.0\text{N}\cos 0°} = \underline{0.140\text{ m}}$$

6-23

a) $W_{tot} = K_2 - K_1$

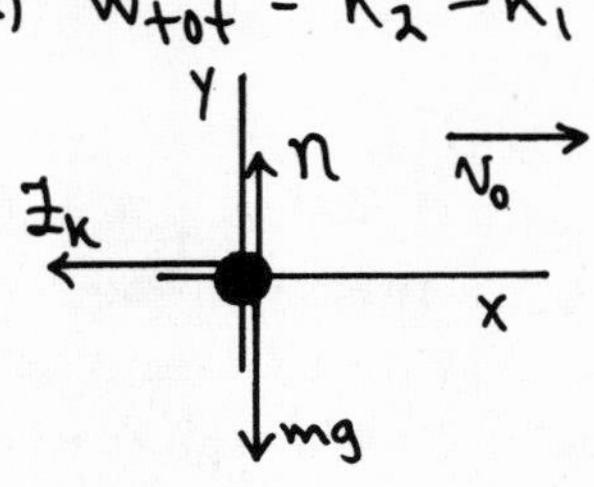

$$f_k = \mu_k n = \mu_k mg$$

f_k is the only force that does work, so $W_{tot} = W_{f_k}$.

f_k is opposite to the displacement so $\phi = 180°$ and $W_{f_k} = f_k(\cos\phi)s = \mu_k mg(\cos 180°)s = -\mu_k mgs$

$K_2 = 0$ (car stops)

$K_1 = \frac{1}{2}mv_0^2$

$$W_{tot} = K_2 - K_1 \Rightarrow -\mu_k \cancel{m} gs = -\frac{1}{2}\cancel{m}v_0^2 \Rightarrow s = \frac{v_0^2}{2\mu_k g}$$

b) $s = \frac{v_0^2}{2\mu_k g} \Rightarrow 2\mu_k g = \frac{v_0^2}{s} = \frac{(17.9\text{ m/s})^2}{42.7\text{ m}} = 7.50\text{ m/s}^2$

Then for $v_0 = 26.8\text{ m/s}$, $s = \frac{v_0^2}{2\mu_k g} = \frac{(26.8\text{ m/s})^2}{7.50\text{ m/s}^2} = \underline{95.8\text{ m}}$

6-25

a) $W_{tot} = K_2 - K_1 \Rightarrow K_2 = K_1 + W_{tot}$

$K_1 = 0$ (released with no initial velocity)

The only force doing work is the spring force. Eq. (6-7) gives the work done <u>on</u> the spring to move its end from x_1 to x_2. The force the spring exerts on an object attached to it is $F = -kx_1$ so the work the spring force does is $W_{spr} = -(\frac{1}{2}kx_2^2 - \frac{1}{2}kx_1^2) = \frac{1}{2}kx_1^2 - \frac{1}{2}kx_2^2$. Here $x_1 = -0.100\text{ m}$ and $x_2 = 0$. Thus $W_{spr} = \frac{1}{2}(300\text{ N/m})(-0.100\text{ m})^2 - 0 = 1.50\text{ J}$.

$$K_2 = K_1 + W_{tot} = 0 + 1.50\text{ J} = 1.50\text{ J}$$

$$K_2 = \frac{1}{2}mv_2^2 \Rightarrow v_2 = \sqrt{\frac{2K_2}{m}} = \sqrt{\frac{2(1.50\text{ J})}{1.60\text{ kg}}} = \underline{1.37\text{ m/s}}$$

b) $K_2 = K_1 + W_{tot}$

$K_1 = 0$

$W_{tot} = W_{spr} = \frac{1}{2}kx_1^2 - \frac{1}{2}kx_2^2$

Now $x_2 = -0.060\text{ m}$

$\Rightarrow W_{spr} = \frac{1}{2}(300\text{ N/m})(-0.100\text{ m})^2 - \frac{1}{2}(300\text{ N/m})(-0.060\text{ m})^2 = 1.50\text{ J} - 0.54\text{ J} = 0.96\text{ J}$

$$K_2 = 0 + 0.96\text{ J} = 0.96\text{ J}$$

$$K_2 = \frac{1}{2}mv_2^2 \Rightarrow v_2 = \sqrt{\frac{2K_2}{m}} = \sqrt{\frac{2(0.96\text{ J})}{1.60\text{ kg}}} = \underline{1.10\text{ m/s}}$$

6-27

a)

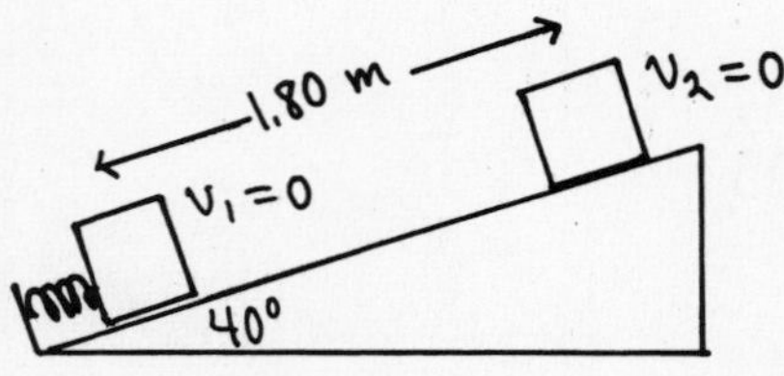

$W_{tot} = K_2 - K_1 = 0$

$W_{tot} = W_{spr} + W_w = 0 \Rightarrow W_{spr} = -W_w$

(The spring does positive work on the glider since the spring force is directed up the incline, the same as the direction of the displacement.)

s
40°
130°
w = mg

$W_w = (w\cos\phi)s = (mg\cos 130°)s = (0.0500\text{ kg})(9.80\text{ m/s}^2)(\cos 130°)(1.80\text{ m})$

$W_w = -0.567\text{ J}$

(The component of w parallel to the incline is directed down the incline, opposite to the displacement, so gravity does negative work.)

$W_{spr} = -W_w = +0.567\text{ J}$

$W_{spr} = \frac{1}{2}kx^2$

$$x = \sqrt{\frac{2W_{spr}}{k}} = \sqrt{\frac{2(0.567\text{ J})}{150\text{ N/m}}} = \underline{0.0869\text{ m}}$$

6-27 (cont)

b)

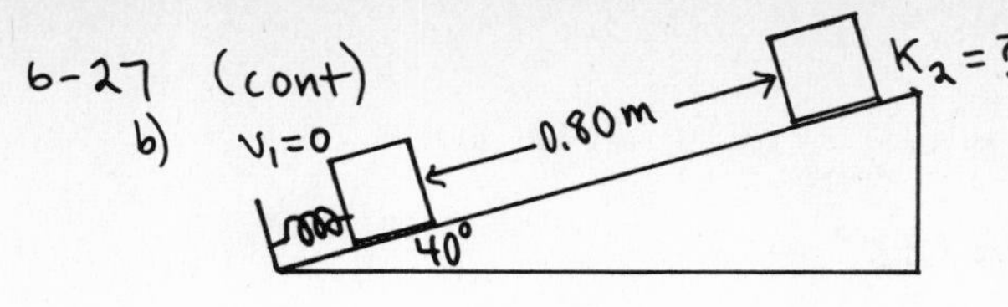

$$W_{tot} = K_2 - K_1$$
$$K_2 = K_1 + W_{tot}$$
$$K_1 = 0$$

$W_{tot} = W_{spr} + W_w$

From part (a), $W_{spr} = 0.567\ \text{J}$ and $W_w = (mg\cos 130°)s$

$W_w = (0.0500\ \text{kg})(9.80\ \text{m/s}^2)(\cos 130°)(0.80\ \text{m}) = -0.252\ \text{J}$

Then $K_2 = W_{spr} + W_w = +0.567\ \text{J} - 0.252\ \text{J} = \underline{+0.315\ \text{J}}$.
(Kinetic energy must always be positive.)

6-31

The man's total work is his weight mg multiplied by the height h he climbs: $W = mgh = (80.0\ \text{kg})(9.80\ \text{m/s}^2)(12.0\ \text{m}) = 9408\ \text{J}$.

His average power is $P_{av} = \frac{W}{t} = \frac{9408\ \text{J}}{20\ \text{s}} = \underline{470\ \text{W}}$

6-33

$$P = \vec{F}\cdot\vec{v} = Fv \Rightarrow F = \frac{P}{v} = \frac{30.0\times10^3\ \text{W}}{12.0\ \text{m/s}} = \underline{2500\ \text{N}}$$

This is the resisting force against which the motor pushes the boat, so it is the tension in the towline if the boat is being pulled through the water at this same speed.

6-37

a) $P = \vec{F}\cdot\vec{v} = Fv \Rightarrow F = \frac{P}{v}$

$P = (6.00\ \text{hp})\left(\frac{746\ \text{W}}{1\ \text{hp}}\right) = 4476\ \text{W}$

$v = (50.0\ \text{km/h})\left(\frac{1\times10^3\ \text{m}}{1\ \text{km}}\right)\left(\frac{1\ \text{h}}{3600\ \text{s}}\right) = 13.89\ \text{m/s}$

$F = \frac{P}{v} = \frac{4476\ \text{W}}{13.89\ \text{m/s}} = \underline{322\ \text{N}}$

b) The power required is the 6.00 hp of part (a) plus the power P_g required to lift the car against gravity.

$\tan\alpha = \frac{10\ \text{m}}{100\ \text{m}} = 0.10 \Rightarrow \alpha = 5.71°$

The vertical component of the velocity of the car is $v\sin\alpha = (13.89\ \text{m/s})\sin 5.71° = 1.382\ \text{m/s}$.

$P_g = F(v\sin\alpha) = mgv\sin\alpha = (1200\ \text{kg})(9.80\ \text{m/s}^2)(1.382\ \text{m/s}) = 1.63\times10^4\ \text{W}$

$P_g = 1.63\times10^4\ \text{W}\left(\frac{1\ \text{hp}}{746\ \text{W}}\right) = 21.8\ \text{hp}$

The total power required is $6.00\ \text{hp} + 21.8\ \text{hp} = \underline{27.8\ \text{hp}}$.

6-37 (cont)

c) The power is reduced by the rate at which gravity does work:

$\tan\alpha = 0.02 \Rightarrow \alpha = 1.15°$

$$P_g = mg(v \sin\alpha) = (1200\ \text{kg})(9.80\ \text{m/s}^2)(13.89\ \text{m/s}) \sin 1.15° = 3.278 \times 10^3\ \text{W}\left(\frac{1\ \text{hp}}{746\ \text{W}}\right) = 4.39\ \text{hp}$$

The power required from the engine is then $6.00\ \text{hp} - 4.39\ \text{hp} = \underline{1.61\ \text{hp}}$.

d) No power is needed from the engine if gravity does work at the rate of $P_g = 6.00\ \text{hp} = 4476\ \text{W}$.

$$P_g = mg\, v \sin\alpha \Rightarrow \sin\alpha = \frac{P_g}{mg\,v} = \frac{4476\ \text{W}}{(1200\ \text{kg})(9.80\ \text{m/s}^2)(13.89\ \text{m/s})} = 0.0274$$

$\alpha = 1.57° \Rightarrow \tan\alpha = 0.0274$; a 2.74% grade.

Problems

6-39

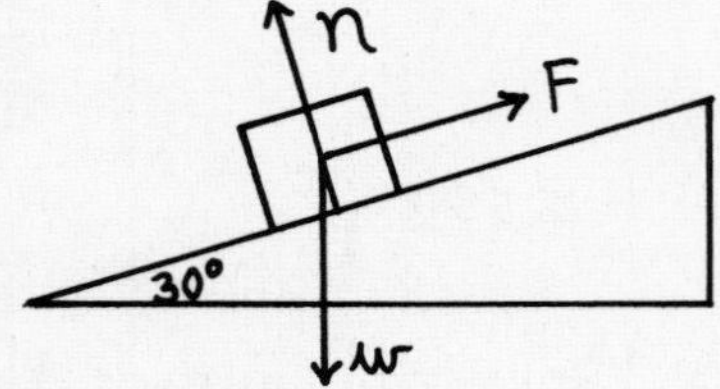

a) $W_F = (F\cos\phi)s$

Both $\vec{F}$ and $\vec{s}$ are parallel to the incline, so $\phi = 0°$ and

$W_F = Fs = (80.0\ \text{N})(4.60\ \text{m}) = \underline{368\ \text{J}}$

b) $W_w = (w \cos\phi)\, s$

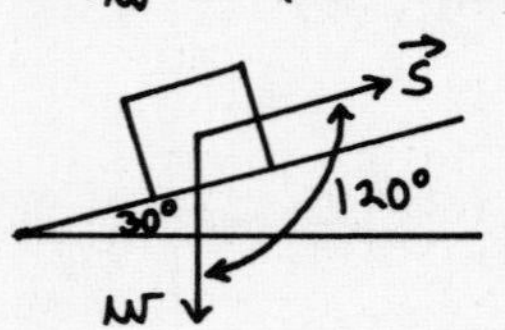

$\phi = 120°$, so $W_w = (90.0\ \text{N})(\cos 120°)(4.60\ \text{m}) = \underline{-207\ \text{J}}$

Alternatively, the component of w parallel to the incline is $w \sin 30°$. This component is down the incline so its angle with $\vec{s}$ is $\phi = 180°$.

$$W_{w \sin 30°} = (90.0\ \text{N} \sin 30°)(\cos 180°)(4.60\ \text{m}) = -207\ \text{J}.$$

The other component of w, $w \cos 30°$, is perpendicular to $\vec{s}$ and hence does no work. Thus $W_w = W_{w \sin 30°} = -207\ \text{J}$, which agrees with the above.

c) The normal force is perpendicular to the displacement ($\phi = 90°$), so $W_n = 0$.

d) $W_{tot} = W_F + W_w + W_n = 368\ \text{J} - 207\ \text{J} + 0 = \underline{161\ \text{J}}$

6-41

$F = \alpha x^3$, with $\alpha = 8.00\ N/m^3$.

a) $x = 1.00\ m \Rightarrow F = (8.00\ N/m^3)(1.00\ m)^3 = 8.00\ N$
Force directed toward origin $\Rightarrow \vec{F}$ is in $-x$-direction; $\vec{F} = -(8.00\ N)\hat{i}$.

b) $x = 2.00\ m \Rightarrow F = (8.00\ N/m^3)(2.00\ m)^3 = 64.0\ N$; $\vec{F} = -(64.0\ N)\hat{i}$.

c) $W = \int_{x_1}^{x_2} F_x\,dx$

$F_x = -\alpha x^3$, minus since $\vec{F}$ is toward the origin.

$$W = \int_{x_1}^{x_2} (-\alpha x^3)\,dx = -\frac{\alpha}{4}\left(x^4\Big|_{x_1}^{x_2}\right) = -\frac{\alpha}{4}(x_2^4 - x_1^4)$$

$$W = -\tfrac{1}{4}(8.00\ N/m^3)\left((2.00\ m)^4 - (1.00\ m)^4\right) = \underline{-30.0\ J}$$

The work done is negative since the object moves away from the origin and the force is in the opposite direction, toward the origin.

6-43

a) Free-body diagram for the block:

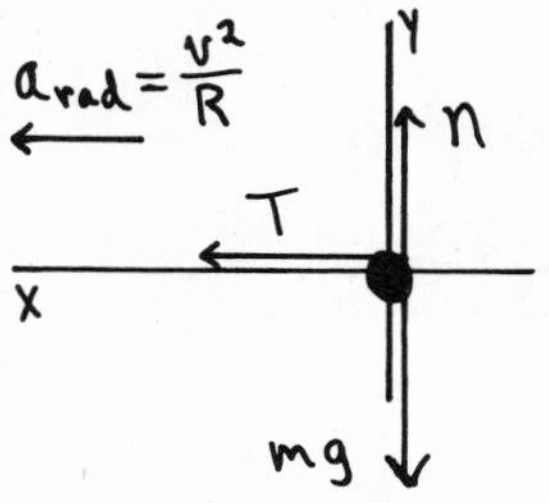

$\Sigma F_x = ma_x$

$$T = m\frac{v^2}{R} = (0.0500\ kg)\frac{(0.80\ m/s)^2}{0.40\ m} = \underline{0.080\ N}$$

b) $$T = m\frac{v^2}{R} = (0.0500\ kg)\frac{(1.60\ m/s)^2}{0.20\ m} = \underline{0.64\ N}$$

c) The tension changes as the distance of the block from the hole changes. We could use $W = \int_{x_1}^{x_2} F_x\,dx$ to calculate the work. But a much simpler approach is to use $W_{tot} = K_2 - K_1$.

The only force doing work on the block is the tension in the cord, so $W_{tot} = W_T$.

$K_1 = \frac{1}{2}mv_1^2 = \frac{1}{2}(0.0500\ kg)(0.80\ m/s)^2 = 0.016\ J$

$K_2 = \frac{1}{2}mv_2^2 = \frac{1}{2}(0.0500\ kg)(1.60\ m/s)^2 = 0.064\ J$

$W_{tot} = K_2 - K_1 \Rightarrow W_T = 0.064\ J - 0.016\ J = \underline{0.048\ J}$

This is the amount of work done by the person who pulled on the cord.

6-47

a) $W_{tot} = K_2 - K_1$

$K_1 = \frac{1}{2}mv_1^2 = \frac{1}{2}(80.0\ kg)(3.00\ m/s)^2 = 360\ J$
$K_2 = \frac{1}{2}mv_2^2 = \frac{1}{2}(80.0\ kg)(1.50\ m/s)^2 = 90\ J$
$W_{tot} = 90\ J - 360\ J = \underline{-270\ J}$

b) Neglecting friction, work is done by you (with the force you apply to the pedals) and by gravity: $W_{tot} = W_{you} + W_{gravity}$.

The gravity force is $w = mg = (80.0\ kg)(9.80\ m/s^2) = 784\ N$, downward. The displacement is 5.20 m, upward. Thus $\phi = 180°$ and
$W_{gravity} = (F\cos\phi)s = (784\ N)(5.20\ m)\cos 180° = -4077\ J$.

$W_{tot} = W_{you} + W_{gravity} \Rightarrow W_{you} = W_{tot} - W_{gravity} = -270\ J - (-4077\ J) = \underline{+3807\ J}$

6-51

a) As in Example 6-9, $W = mgh$.

We need the mass of blood lifted; we are given the volume
$V = (7500\ \cancel{L})\left(\frac{1\times10^{-3}\ m^3}{1\ \cancel{L}}\right) = 7.50\ m^3$.
$m = \text{density} \times \text{volume} = (1.00\times10^3\ kg/m^3)(7.50\ m^3) = 7.50\times10^3\ kg$
Then
$W = mgh = (7.50\times10^3\ kg)(9.80\ m/s^2)(1.63\ m) = \underline{1.20\times10^5\ J}$

b) $P_{av} = \frac{\Delta W}{\Delta t} = \frac{1.20\times10^5\ J}{(24\ \cancel{h})(3600\ s/\cancel{h})} = \underline{1.39\ W}$

6-55

$P_{av} = 2000\ MW = 2.00\times10^9\ W$.

$P_{av} = \frac{\Delta W}{\Delta t}$, so in 1.00 s the amount of work done on the water by gravity is
$W = P_{av}\Delta t = (2.00\times10^9\ W)(1.00\ s) = 2.00\times10^9\ J$

$W = mgh$, so the mass of water flowing over the dam in 1.00 s must be
$m = \frac{W}{gh} = \frac{2.00\times10^9\ J}{(9.80\ m/s^2)(170\ m)} = 1.20\times10^6\ kg$

$\text{density} = \frac{m}{V} \Rightarrow V = \frac{m}{\text{density}} = \frac{1.20\times10^6\ kg}{1.00\times10^3\ kg/m^3} = \underline{1.20\times10^3\ m^3}$

6-57

a) $P = F_{tot} v$, with $F_{tot} = F_{roll} + F_{air}$

$F_{air} = \frac{1}{2} C A \rho v^2 = \frac{1}{2}(1.0)(0.463\,m^2)(1.2\,kg/m^3)(14.0\,m/s)^2 = 54.4\,N$

$F_{roll} = \mu_r n = \mu_r w = (0.0045)(540\,N + 111\,N) = 2.93\,N$

$P = (F_{roll} + F_{air})v = (2.93\,N + 54.4\,N)(14.0\,m/s) = \underline{803\,W}$

b) $F_{air} = \frac{1}{2} C A \rho v^2 = \frac{1}{2}(0.88)(0.366\,m^2)(1.2\,kg/m^3)(14.0\,m/s)^2 = 37.9\,N$

$F_{roll} = \mu_r n = \mu_r w = (0.0030)(540\,N + 89\,N) = 1.89\,N$

$P = (F_{roll} + F_{air})v = (1.89\,N + 37.9\,N)(14.0\,m/s) = \underline{557\,W}$

c) $F_{air} = \frac{1}{2} C A \rho v^2 = \frac{1}{2}(0.88)(0.366\,m^2)(1.20\,kg/m^3)(7.0\,m/s)^2 = 9.47\,N$

$F_{roll} = \mu_r w = 1.89\,N$ (unchanged)

$P = (F_{roll} + F_{air})v = (1.89\,N + 9.47\,N)(7.0\,m/s) = \underline{79.5\,W}$

CHAPTER 7

Exercises 3, 5, 7, 13, 15, 17

Problems 21, 23, 25, 27, 29, 33, 35

Exercises

7-3

a)

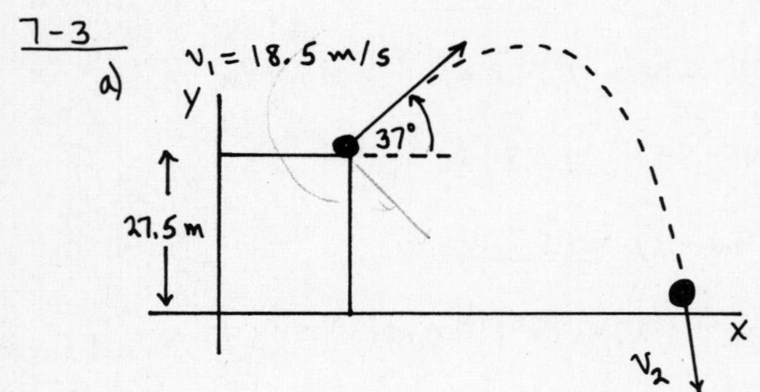

$K_1 + U_1 + W_{other} = K_2 + U_2$

$W_{other} = 0$ (The only force on the ball while it is in the air is gravity.)

$K_1 = \frac{1}{2}mv_1^2$

$K_2 = \frac{1}{2}mv_2^2$

$U_1 = mgy_1,\ y_1 = 27.5\text{ m}$

$U_2 = mgy_2 = 0$, since $y_2 = 0$ for our choice of coordinates

$$\frac{1}{2}mv_1^2 + mgy_1 = \frac{1}{2}mv_2^2$$

$$v_2 = \sqrt{v_1^2 + 2gy_1} = \sqrt{(18.5\text{ m/s})^2 + 2(9.80\text{ m/s}^2)(27.5\text{ m})} = \underline{29.7\text{ m/s}}$$

Note that the projection angle of 37° doesn't enter into the calculation. The kinetic energy depends only on the magnitude of the velocity; it is independent of the direction of the velocity.

b) Nothing changes in the calculation. The expression derived in part (a) for v_2 is independent of the angle, so $v_2 = 29.7$ m/s, the same as before.

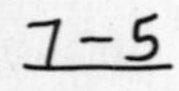

7-5

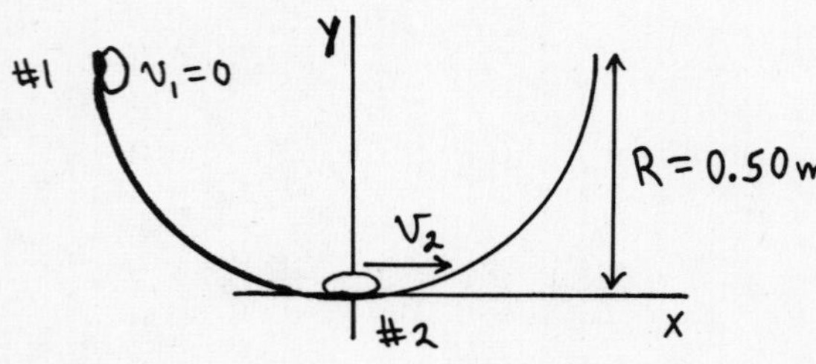

$y_1 = R$; $y_2 = 0$

The forces on the object are gravity, the normal force, and friction. The normal force is at all points in the motion perpendicular to the displacement $\Rightarrow W_n = 0$. Hence $W_{other} = W_f$, the work done by friction.

$K_1 + U_1 + W_{other} = K_2 + U_2$

$K_1 = \frac{1}{2}mv_1^2 = 0$

$K_2 = \frac{1}{2}mv_2^2 = \frac{1}{2}(0.10\text{ kg})(1.8\text{ m/s})^2 = 0.162\text{ J}$

$U_1 = mgy_1 = (0.10\text{kg})(9.80\text{ m/s}^2)(0.50\text{ m}) = 0.490\text{ J}$

$U_2 = mgy_2 = 0$

$W_{other} = W_f$

7-5 (cont)

Thus $0 + U_1 + W_f = K_2 + 0$

$$W_f = K_2 - U_1 = 0.162\ \text{J} - 0.490\ \text{J} = \underline{-0.33\ \text{J}}$$

(The friction work is negative as expected, since the friction force is directed opposite to the displacement.)

7-7

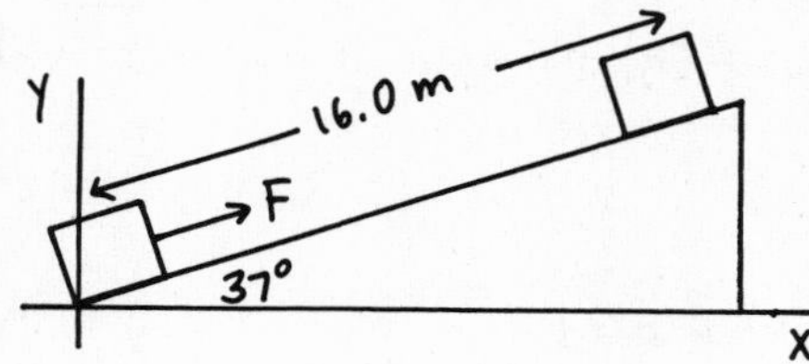

$y_1 = 0$

$y_2 = (16.0\text{m})\sin 37° = 9.63\text{m}$

a) $W_F = (F\cos\phi)s = (120\text{N})(\cos 0°)(16.0\text{m}) = \underline{1920\ \text{J}}$

b) Use the free-body diagram for the oven to calculate the normal force n; then the friction force can be calculated from $f_k = \mu_k n$. For this calculation use coordinates parallel and perpendicular to the incline.

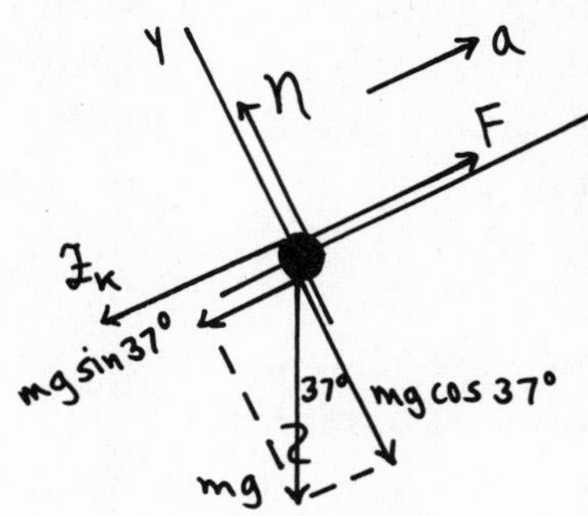

$\Sigma F_y = ma_y$

$n - mg\cos 37° = 0$

$n = mg\cos 37°$

$f_k = \mu_k n = \mu_k mg\cos 37° = (0.25)(12.0\text{kg})(9.80\text{m/s}^2)\cos 37°$

$f_k = 23.5\ \text{N}$

$W_{f_k} = (f_k\cos\phi)s = (23.5\text{N})(\cos 180°)(16.0\text{m}) = \underline{-376\ \text{J}}$

c) $\Delta U = U_2 - U_1 = mg(y_2 - y_1) = (12.0\text{kg})(9.80\text{m/s}^2)(9.63\text{m} - 0) = \underline{1132\ \text{J}}$

d) $K_1 + U_1 + W_{other} = K_2 + U_2$

$\Delta K = K_2 - K_1 = U_1 - U_2 + W_{other}$

$\Delta K = W_{other} - \Delta U$

$W_{other} = W_F + W_{f_k} = 1920\text{J} - 376\text{J} = 1544\ \text{J}$

$\Delta U = 1132\text{J}$

$\Rightarrow \Delta K = 1544\text{J} - 1132\text{J} = \underline{412\ \text{J}}$

e) We can use the free-body diagram that is in part (b):

$\Sigma F_x = ma_x$

$F - f_k - mg\sin 37° = ma$

$$a = \frac{F - f_k - mg\sin 37°}{m} = \frac{120\text{N} - 23.5\text{N} - (12.0\text{kg})(9.80\text{m/s}^2)\sin 37°}{12.0\text{ kg}}$$

$a = 2.144\ \text{m/s}^2$

7-7 (cont)

$v_1 = 0$
$a = 2.144\ m/s^2$
$x - x_0 = 16.0\ m$
$v_2 = ?$

$$v_2^2 = v_1^{2\nearrow 0} + 2a(x - x_0)$$

$$v_2 = \sqrt{2a(x-x_0)} = \sqrt{2(2.144\ m/s^2)(16.0\ m)} = 8.283\ m/s$$

Then $\Delta K = K_2 - K_1^{\nearrow 0} = \frac{1}{2}mv_2^2 = \frac{1}{2}(12.0\ kg)(8.283\ m/s)^2 = \underline{412\ J}$;
this agrees with the result calculated in part (d) using energy methods.

7-13

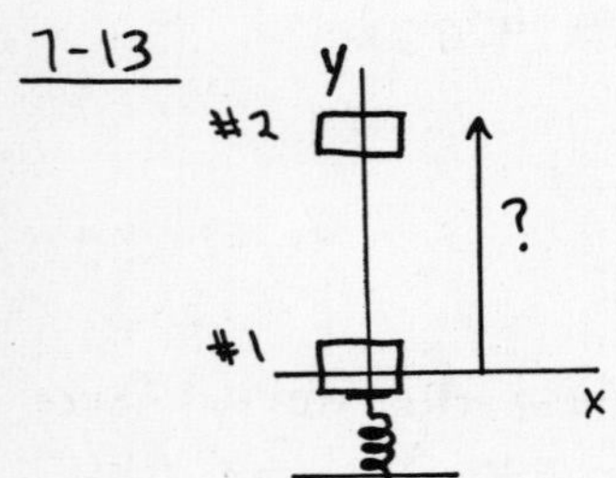

$$K_1 + U_1 + W_{other} = K_2 + U_2$$

The spring force and gravity are the only forces doing work on the brick so $W_{other} = 0$ and $U = U_{grav} + U_{el}$.
Brick released from rest $\Rightarrow K_1 = 0$.
At the maximum height $v_2 = 0$ so $K_2 = 0$.
$U_1 = U_{1,grav} + U_{1,el}$; $y_1 = 0 \Rightarrow U_{1,grav} = 0$
$U_{1,el} = \frac{1}{2}kx_1^2 = \frac{1}{2}(500\ N/m)(0.20\ m)^2 = 10.0\ J$

(Here x_1 refers to the amount the spring is stretched or compressed when the brick is at position #1; it is not the x-coordinate of the brick in the coordinate system shown in the sketch.)

$U_2 = U_{2,grav} + U_{2,el}$
$U_{2,grav} = mgy_2$, where y_2 is the height we are solving for
$U_{2,el} = 0$ since now the spring is no longer compressed

Putting all this into $K_1 + U_1 + W_{other} = K_2 + U_2$ gives
$U_{1,el} = U_{2,grav}$

The description in terms of energy is very simple: the elastic energy originally stored in the spring is converted to gravitational potential energy of the brick.

$$10.0\ J = mgy_2$$

$$y_2 = \frac{10.0\ J}{mg} = \frac{10.0\ J}{(0.600\ kg)(9.80\ m/s^2)} = \underline{1.7\ m}$$

7-15

Use coordinates where the origin is at one atom. The other atom then has coordinate x.

$$F_x = -\frac{dU}{dx} = -\frac{d}{dx}\left(-\frac{C_6}{x^6}\right) = +C_6\frac{d}{dx}\left(\frac{1}{x^6}\right) = -\frac{6C_6}{x^7}$$

The minus sign means that F_x is directed in the $-x$-direction, toward the origin.

The force has magnitude $6C_6/x^7$ and is attractive.

7-17

$$\mathcal{U}(x,y) = k(x^2+y^2) + k'xy$$

$$F_x = -\frac{\partial \mathcal{U}}{\partial x} = -(2kx + k'y)$$

$$F_y = -\frac{\partial \mathcal{U}}{\partial y} = -(2ky + k'x)$$

$$F_z = -\frac{\partial \mathcal{U}}{\partial z} = 0$$

$$\vec{F} = -(2kx + k'y)\vec{i} - (2ky + k'x)\vec{j}$$

Problems

7-21

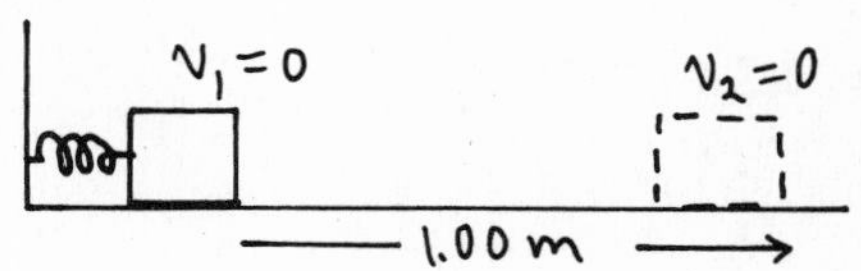

Work is done on the block by the spring force and by friction, so $W_{other} = W_{\mathcal{F}}$ and $\mathcal{U} = \mathcal{U}_{el}$.

$$K_1 + \mathcal{U}_1 + W_{other} = K_2 + \mathcal{U}_2$$

$K_1 = K_2 = 0$

$\mathcal{U}_1 = \mathcal{U}_{1,el} = \frac{1}{2}kx_1^2 = \frac{1}{2}(100\ \text{N/m})(0.20\ \text{m})^2 = 2.00\ \text{J}$

$\mathcal{U}_2 = \mathcal{U}_{2,el} = 0$, since after the block leaves the spring has given up all its stored energy.

$W_{other} = W_{\mathcal{F}} = (\mathcal{F}_k \cos\phi)s = \mu_k mg(\cos\phi)s = -\mu_k mgs$ since $\phi = 180°$ (The friction force is directed opposite to the displacement.)

Putting all this into $K_1 + \mathcal{U}_1 + W_{other} = K_2 + \mathcal{U}_2$ gives

$$\mathcal{U}_{1,el} + W_{\mathcal{F}} = 0$$

(The potential energy originally stored in the spring is taken out of the system by the negative work done by friction.)

$$\mu_k mgs = \mathcal{U}_{1,el}$$

$$\mu_k = \frac{\mathcal{U}_{1,el}}{mgs} = \frac{2.00\ \text{J}}{(0.50\ \text{kg})(9.80\ \text{m/s}^2)(1.00\ \text{m})} = \underline{0.408}$$

7-23

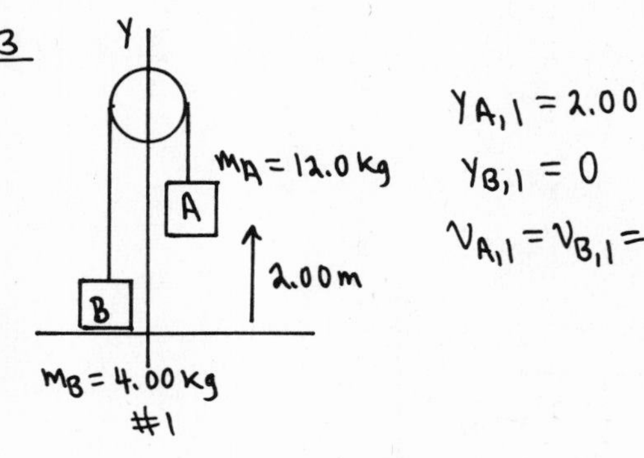

$y_{A,1} = 2.00$ m

$y_{B,1} = 0$

$v_{A,1} = v_{B,1} = 0$

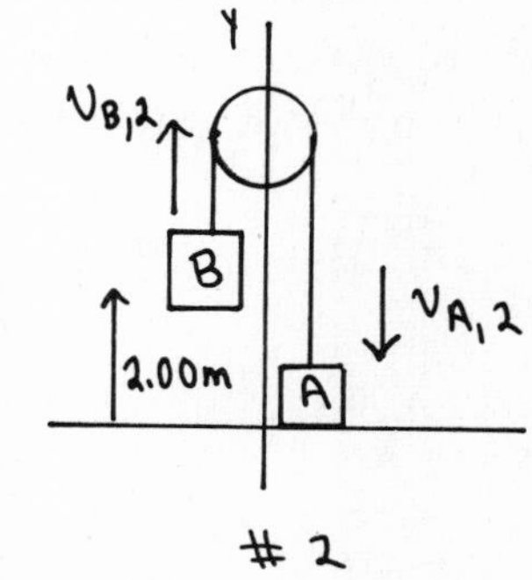

$y_{A,2} = 0$

$y_{B,2} = 2.00$ m

7-23 (cont)

The tension force does positive work on the 4.0 kg block and an equal amount of negative work on the 12.0 kg block, so the net work done by the tension is zero.

Work is done on the system only by gravity, so $W_{other} = 0$ and $U = U_{gravity}$.

$$K_1 + U_1 + W_{other} = K_2 + U_2$$

$K_1 = 0$

$K_2 = \frac{1}{2} m_A v_{A,2}^2 + \frac{1}{2} m_B v_{B,2}^2$

But since the two blocks are connected by a rope they move together and have the same speed: $v_{A,2} = v_{B,2} = v_2$.

$K_2 = \frac{1}{2}(m_A + m_B)\, v_2^2 = (8.00\ \text{kg})\, v_2^2$

$U_1 = m_A g y_{A,1} = (12.0\ \text{kg})(9.80\ \text{m/s}^2)(2.00\ \text{m}) = 235.2\ \text{J}$

$U_2 = m_B g y_{B,2} = (4.0\ \text{kg})(9.80\ \text{m/s}^2)(2.00\ \text{m}) = 78.4\ \text{J}$

Put all this into $K_1 + U_1 + W_{other} = K_2 + U_2$

$$\Rightarrow \quad U_1 = K_2 + U_2$$

$$235.2\ \text{J} = (8.00\ \text{kg})\, v_2^2 + 78.4\ \text{J}$$

$$v_2 = \sqrt{\frac{235.2\ \text{J} - 78.4\ \text{J}}{8.00\ \text{kg}}} = \underline{4.43\ \text{m/s}}$$

7-25

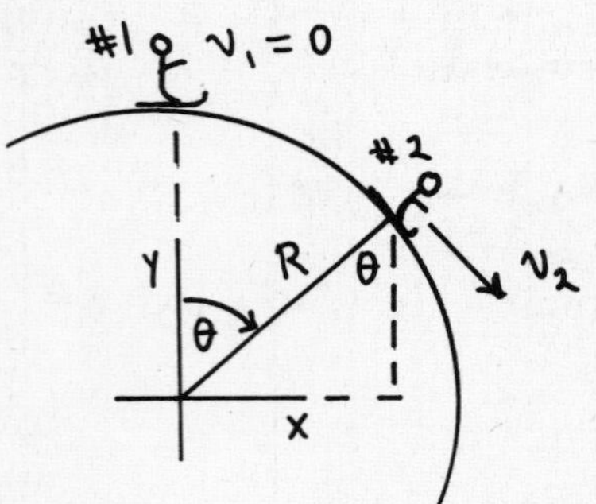

Let point #2 be where the skier loses contact with the snowball.

Loses contact $\Rightarrow n \to 0$

$y_1 = R, \quad y_2 = R\cos\theta$

First, analyze the forces on the skier when she is at point #2. For this use coordinates that are in the tangential and radial directions. The skier moves in an arc of a circle, so her acceleration is $a_{rad} = \frac{v_2^2}{R}$, directed in towards the center of the snowball.

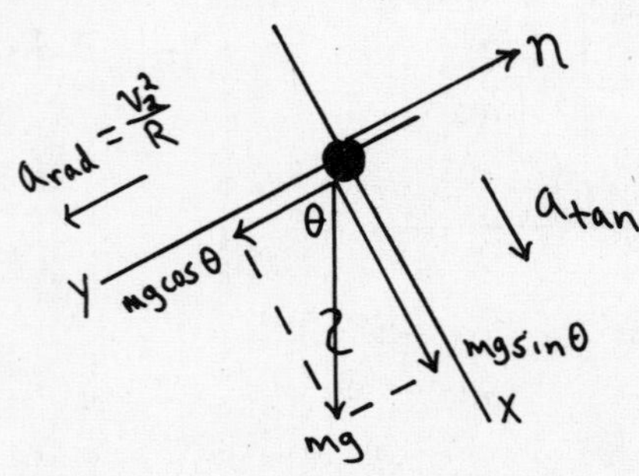

$$\Sigma F_y = m a_y$$

$$mg\cos\theta - \cancel{n}^{\,0} = m\frac{v_2^2}{R}$$

$$\cancel{m} g\cos\theta = \cancel{m}\frac{v_2^2}{R}$$

$$\boxed{v_2^2 = Rg\cos\theta}$$

7-25 (cont)

Now use conservation of energy to get another equation relating v_2 to θ:

$$K_1 + U_1 + W_{other} = K_2 + U_2$$

The only force that does work on the skier is gravity $\Rightarrow W_{other} = 0$.

$K_1 = 0$

$U_1 = mgy_1 = mgR$

$K_2 = \frac{1}{2}mv_2^2$

$U_2 = mgy_2 = mgR\cos\theta$

$$\Rightarrow mgR = \frac{1}{2}mv_2^2 + mgR\cos\theta$$

$$\boxed{v_2^2 = 2gR(1-\cos\theta)}$$

Combine this with the $\Sigma F_y = ma_y$ eq. $\Rightarrow Rg\cos\theta = 2gR(1-\cos\theta)$

$$\cos\theta = 2 - 2\cos\theta$$

$$3\cos\theta = 2 \Rightarrow \cos\theta = \tfrac{2}{3} \Rightarrow \theta = \underline{48.2°}$$

7-27

a)

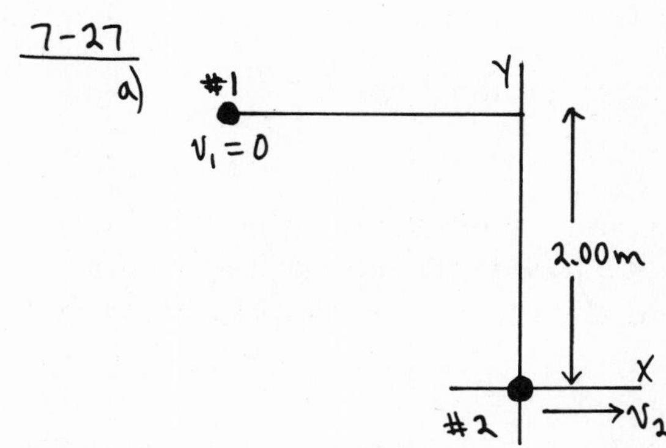

$y_1 = 2.00\text{ m}$

$y_2 = 0$

The tension in the string is at all points in the motion perpendicular to the displacement, so $W_T = 0$.

The only force that does work on the ball is gravity, so $W_{other} = 0$.

$$K_1 + U_1 + W_{other}^{\,0} = K_2 + U_2$$

$K_1 = 0$

$U_2 = 0$

$U_1 = mgy_1$

$K_2 = \frac{1}{2}mv_2^2$

$\Rightarrow U_1 = K_2$

$$mgy_1 = \tfrac{1}{2}mv_2^2 \Rightarrow v_2 = \sqrt{2gy_1} = \sqrt{2(9.80\text{ m/s}^2)(2.00\text{ m})} = 6.26\text{ m/s}$$

b) Free-body diagram for the ball as it swings through its lowest point:

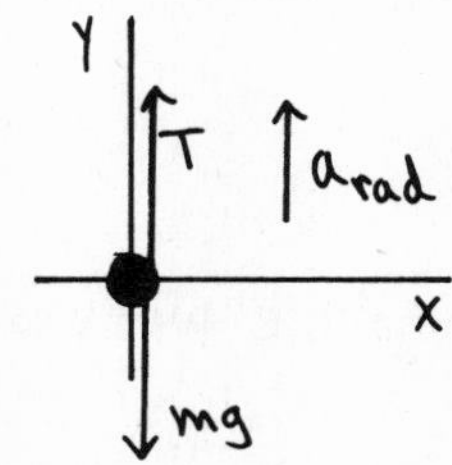

The acceleration $\vec{a}_{rad}$ is directed in toward the center of the circular path, so at this point it is upward.

$\Sigma F_y = ma_y$

$T - mg = ma_{rad}$

$T = m(g + a_{rad}) = m\left(g + \frac{v_2^2}{R}\right)$, where the radius R for the circular motion is the length L of the string.

7-27 (cont)

It is instructive to use the algebraic expression for v_2 from part (a) rather than just putting in the numerical value:

$$v_2 = \sqrt{2gy_1} = \sqrt{2gL} \Rightarrow v_2^2 = 2gL$$

$T = m\left(g + \frac{v_2^2}{L}\right) = m\left(g + \frac{2gL}{L}\right) = 3mg$; the tension at this point is three times the weight of the ball.

$$T = 3mg = 3(0.500\ \text{kg})(9.80\ \text{m/s}^2) = \underline{14.7\ \text{N}}$$

7-29

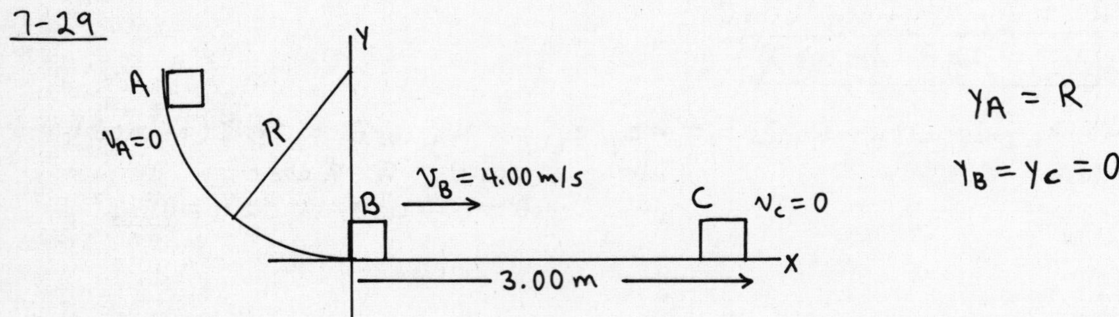

a) Apply conservation of energy to the motion from B to C:

$$K_B + U_B + W_{other} = K_C + U_C$$

The only force that does work on the package during this part of the motion is friction $\Rightarrow W_{other} = W_f = f_k(\cos\phi)s = \mu_k mg(\cos 180°)s = -\mu_k mgs$

$K_B = \frac{1}{2}mv_B^2$

$U_B = U_C = 0$

$K_C = 0$

$\Rightarrow \quad K_B + W_f = 0$ (The negative friction work takes away all the kinetic energy.)

$\frac{1}{2}\cancel{m}v_B^2 - \mu_k \cancel{m}gs = 0$

$$\mu_k = \frac{v_B^2}{2gs} = \frac{(4.00\ \text{m/s})^2}{2(9.80\ \text{m/s}^2)(3.00\ \text{m})} = \underline{0.272}$$

b) Apply conservation of energy to the motion from A to B:

$$K_A + U_A + W_{other} = K_B + U_B$$

Work is done by gravity and by friction, so $W_{other} = W_f$.

$K_A = 0$

$U_A = mgy_A = mgR = (2.00\ \text{kg})(9.80\ \text{m/s}^2)(1.60\ \text{m}) = 31.36\ \text{J}$

$K_B = \frac{1}{2}mv_B^2 = \frac{1}{2}(2.00\ \text{kg})(4.00\ \text{m/s})^2 = 16.00\ \text{J}$

$U_B = 0$

$\Rightarrow \quad U_A + W_f = K_B$

$$W_f = K_B - U_A = 16.00\ \text{J} - 31.36\ \text{J} = \underline{-15.4\ \text{J}}$$

(W_f is negative as expected; the friction force does negative work since it is directed opposite to the displacement.)

7-33

$F_x = -\alpha x - \beta x^2$; $\alpha = 70.0\ N/m$ and $\beta = 12.0\ N/m^2$

a) $W_{F_x} = U_1 - U_2 = \int_{x_1}^{x_2} F_x(x)\,dx$

Let $x_1 = 0$ and $U_1 = 0$. Let x_2 be some arbitrary point $x \Rightarrow U_2 = U(x)$.
Then

$$U(x) = -\int_0^x F_x(x)\,dx = -\int_0^x (-\alpha x - \beta x^2)\,dx = \int_0^x (\alpha x + \beta x^2)\,dx = \tfrac{1}{2}\alpha x^2 + \tfrac{1}{3}\beta x^3$$

$$\boxed{U(x) = \tfrac{1}{2}\alpha x^2 + \tfrac{1}{3}\beta x^3}$$

b)

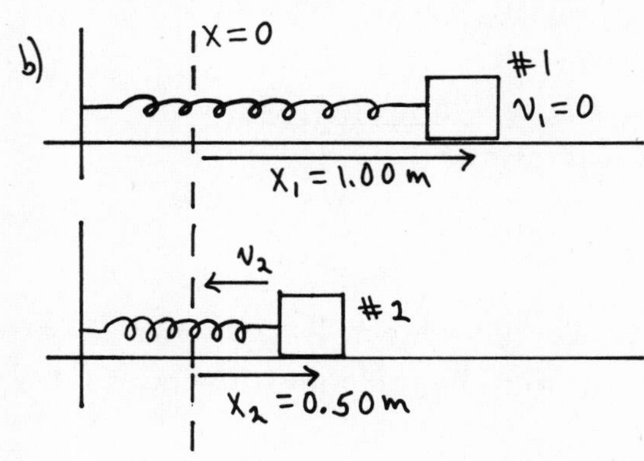

$K_1 + U_1 + W_{other} = K_2 + U_2$

The only force that does work on the object is the spring force $\Rightarrow W_{other} = 0$.

$K_1 = 0$

$K_2 = \frac{1}{2} m v_2^2$

$U_1 = U(x_1) = \frac{1}{2}\alpha x_1^2 + \frac{1}{3}\beta x_1^3$

$U_1 = \frac{1}{2}(70.0\ N/m)(1.00\ m)^2 + \frac{1}{3}(12.0\ N/m^2)(1.00\ m)^3 = 39.0\ J$

$U_2 = U(x_2) = \frac{1}{2}\alpha x_2^2 + \frac{1}{3}\beta x_2^3$

$U_2 = \frac{1}{2}(70.0\ N/m)(0.50\ m)^2 + \frac{1}{3}(12.0\ N/m^2)(0.50\ m)^3 = 9.25\ J$

So, $39.0\ J = \frac{1}{2} m v_2^2 + 9.25\ J$

$$v_2 = \sqrt{\frac{2(39.0\ J - 9.25\ J)}{2.00\ kg}} = \underline{5.45\ m/s}$$

7-35

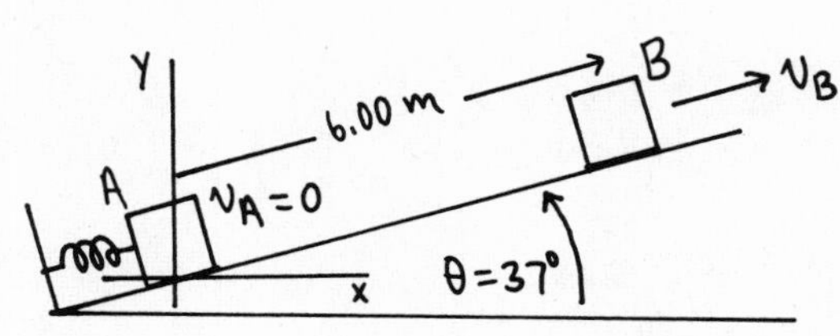

$\rightarrow v_B = 4.00\ m/s$

The normal force is $n = mg\cos\theta$, so $f_k = \mu_k n = \mu_k mg\cos\theta$.

$y_A = 0$; $y_B = (6.00\ m)\sin 37° = 3.61\ m$

Apply conservation of energy to the motion of the block from point A to point B:

$K_A + U_A + W_{other} = K_B + U_B$

Work is done by gravity, by the spring force, and by friction, so $W_{other} = W_f$ and $U = U_{el} + U_{gravity}$.

$K_A = 0$

$K_B = \frac{1}{2} m v_B^2 = \frac{1}{2}(2.00\ kg)(4.00\ m/s)^2 = 16.0\ J$

$U_A = U_{A,el} + \cancel{U_{A,gravity}}^{0} = U_{A,el}$

$U_B = \cancel{U_{B,el}} + U_{B,gravity} = mgy_B = (2.00\ kg)(9.80\ m/s^2)(3.61\ m) = 70.76\ J$

$W_{other} = W_f = (f_k \cos\phi)s = \mu_k mg\cos\theta(\cos 180°)s = -\mu_k mg\cos\theta\, s$

$W_{other} = -(0.50)(2.00\ kg)(9.80\ m/s^2)(\cos 37°)(6.00\ m) = -46.96\ J$

So, $U_{A,el} - 46.96\ J = 16.0\ J + 70.76\ J$

$U_{A,el} = 46.96\ J + 16.0\ J + 70.76\ J = \underline{134\ J}$

CHAPTER 8

Exercises 3, 5, 11, 13, 15, 19, 21, 23, 25, 29, 31, 33, 37, 41

Problems 45, 47, 49, 51, 55, 57, 59, 63, 65, 69, 71

Exercises

8-3

$p_{x1} = m v_{x1} = (0.145\ \text{kg})(1.20\ \text{m/s}) = 0.174\ \text{kg}\cdot\text{m/s}$
$p_{x2} = m v_{x2} = (0.0550\ \text{kg})(-6.20\ \text{m/s}) = -0.341\ \text{kg}\cdot\text{m/s}$

$P_x = p_{x1} + p_{x2} = 0.174\ \text{kg}\cdot\text{m/s} - 0.341\ \text{kg}\cdot\text{m/s} = \underline{-0.167\ \text{kg}\cdot\text{m/s}}$
(minus sign means that is in the $-x$-direction)

8-5

a) Let Gretzky be object A and the defender be object B. Let the $+x$-direction be the direction in which Gretzky is moving initially. No horizontal external forces, so P_x is constant.

before: A, $v_{A1} = 13.0\ \text{m/s}$ → ; ← $v_{B1} = 5.00\ \text{m/s}$, B; x

after: A, $v_{A2} = 2.50\ \text{m/s}$ → ; B, $v_{B2} = ?$; x

P_x constant $\Rightarrow m_A v_{A1x} + m_B v_{B1x} = m_A v_{A2x} + m_B v_{B2x}$
If we multiply through by g we get an equation that uses the object's weight rather than its mass:

$$w_A v_{A1x} + w_B v_{B1x} = w_A v_{A2x} + w_B v_{B2x}$$

The x-components can be positive or negative depending on the directions of the velocities relative to our coordinate system.

Putting in the numbers,
$(756\ \text{N})(+13.0\ \text{m/s}) + (900\ \text{N})(-5.00\ \text{m/s}) = (756\ \text{N})(+2.50\ \text{m/s}) + (900\ \text{N})\, v_{B2x}$
$9828\ \text{N}\cdot\text{m/s} - 4500\ \text{N}\cdot\text{m/s} = 1890\ \text{N}\cdot\text{m/s} + (900\ \text{N})\, v_{B2x}$

$$v_{B2x} = \frac{9828\ \text{N}\cdot\text{m/s} - 4500\ \text{N}\cdot\text{m/s} - 1890\ \text{N}\cdot\text{m/s}}{900\ \text{N}} = 3.82\ \text{m/s}$$

After the collision the defender is moving at $\underline{3.82\ \text{m/s}}$ in the same direction as Gretzky.

b) $K_1 = \frac{1}{2} m_A v_{A1}^2 + \frac{1}{2} m_B v_{B1}^2 = \frac{1}{2}\left(\frac{756\ \text{N}}{9.80\ \text{m/s}^2}\right)(13.0\ \text{m/s})^2 + \frac{1}{2}\left(\frac{900\ \text{N}}{9.80\ \text{m/s}^2}\right)(5.00\ \text{m/s})^2$
$K_1 = 6519\ \text{J} + 1148\ \text{J} = 7667\ \text{J}$
(Note that the kinetic energy of an object is always positive, and that it does not depend on the direction of the object's velocity.)

8-5 (cont)

$$K_2 = \tfrac{1}{2} m_A v_{A2}^2 + \tfrac{1}{2} m_B v_{B2}^2 = \tfrac{1}{2}\left(\frac{756\ \text{N}}{9.80\ \text{m/s}^2}\right)(2.50\ \text{m/s})^2 + \tfrac{1}{2}\left(\frac{900\ \text{N}}{9.80\ \text{m/s}^2}\right)(3.82\ \text{m/s})^2$$

$$K_2 = 241\ \text{J} + 670\ \text{J} = 911\ \text{J}$$

$$\Delta K = K_2 - K_1 = 911\ \text{J} - 7667\ \text{J} = -6756\ \text{J}$$

The kinetic energy decreases by 6760 J.

8-11

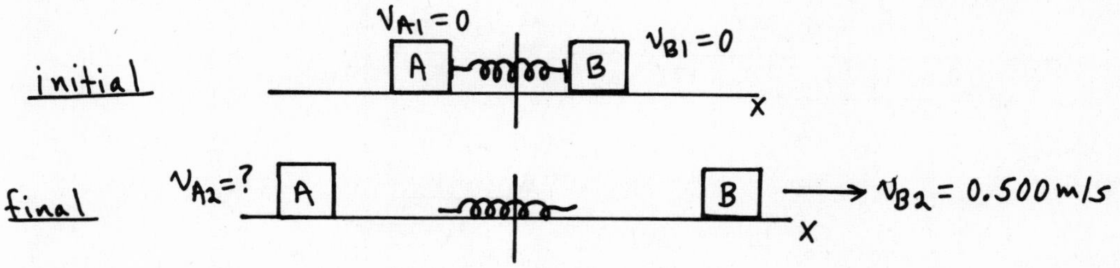

initial: $v_{A1} = 0$, $v_{B1} = 0$ — A, B, x

final: $v_{A2} = ?$ — A, B — $v_{B2} = 0.500\ \text{m/s}$ — x

a) No horizontal force $\Rightarrow P_x$ is constant.

$$m_A v_{A1x} + m_B v_{B1x} = m_A v_{A2x} + m_B v_{B2x} \quad (v_{A1x} = 0,\ v_{B1x} = 0)$$

$$0 = m_A v_{A2x} + m_B v_{B2x}$$

$$v_{A2x} = -\left(\frac{m_B}{m_A}\right) v_{B2x} = -\left(\frac{3.00\ \text{kg}}{1.00\ \text{kg}}\right)(+0.500\ \text{m/s}) = -1.50\ \text{m/s}$$

Block A has a final speed of 1.50 m/s, and moves off in the opposite direction to B.

b) Use energy conservation

$$K_1 + U_1 + W_{other} = K_2 + U_2$$

Only the spring force does work $\Rightarrow W_{other} = 0$ and $U = U_{el}$

$K_1 = 0$ (the blocks initially are at rest)

$U_2 = 0$ (no potential energy is left stored in the spring)

$K_2 = \tfrac{1}{2} m_A v_{A2}^2 + \tfrac{1}{2} m_B v_{B2}^2 = \tfrac{1}{2}(1.00\ \text{kg})(1.50\ \text{m/s})^2 + \tfrac{1}{2}(3.00\ \text{kg})(0.500\ \text{m/s})^2 = 1.50\ \text{J}$

$U_1 = U_{1,el}$, the potential energy stored in the compressed spring

$$\Rightarrow U_{1,el} = K_2 = 1.50\ \text{J}$$

8-13

Take the x-axis to lie along the initial velocity of A.

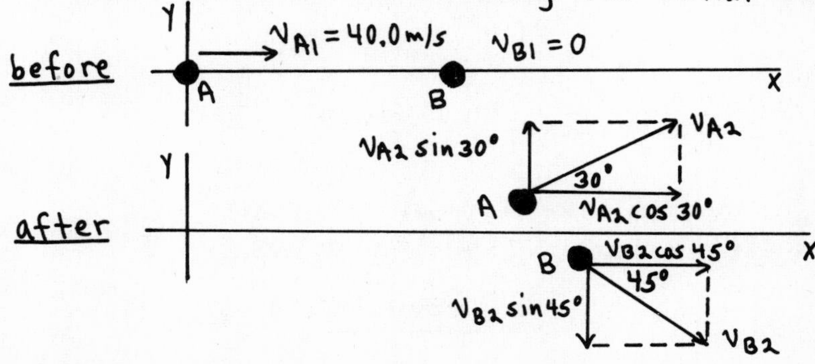

8-13 (cont)

a) P_x is constant

$$\Rightarrow m_A v_{A1x} + m_B v_{B1x} = m_A v_{A2x} + m_B v_{B2x}$$

$$\cancel{m} v_{A1} = \cancel{m} v_{A2} \cos 30° + \cancel{m} v_{B2} \cos 45°$$

$$\boxed{40.0 \text{ m/s} = 0.8660\, v_{A2} + 0.7071\, v_{B2}}$$

P_y is constant

$$\Rightarrow m_A v_{A1y} + m_B v_{B1y} = m_A v_{A2y} + m_B v_{B2y}$$

$$0 = \cancel{m} v_{A2} \sin 30° - \cancel{m} v_{B2} \sin 45°$$

$$\boxed{0 = 0.5000\, v_{A2} - 0.7071\, v_{B2}}$$

Add these two eqs. $\Rightarrow 1.366\, v_{A2} = 40.0$ m/s

$$v_{A2} = \underline{29.3 \text{ m/s}}$$

Then $v_{B2} = \frac{0.5000}{0.7071} v_{A2} = \left(\frac{0.5000}{0.7071}\right)(29.3 \text{ m/s}) = \underline{20.7 \text{ m/s}}$.

b) $K_1 = K_{A1} = \frac{1}{2} m v_{A1}^2$

$K_2 = K_{A2} + K_{B2} = \frac{1}{2} m v_{A2}^2 + \frac{1}{2} m v_{B2}^2$

$\Delta K = K_2 - K_1$

The fraction of the original kinetic energy of puck A dissipated during the collision is

$$-\frac{\Delta K}{K_1} = -\frac{K_2 - K_1}{K_1} = -\left(\frac{\frac{1}{2} m v_{A2}^2 + \frac{1}{2} m v_{B2}^2 - \frac{1}{2} m v_{A1}^2}{\frac{1}{2} m v_{A1}^2}\right) = 1 - \left(\frac{v_{A2}}{v_{A1}}\right)^2 - \left(\frac{v_{B2}}{v_{A1}}\right)^2$$

$$-\frac{\Delta K}{K_1} = 1 - \left(\frac{29.3 \text{ m/s}}{40.0 \text{ m/s}}\right)^2 - \left(\frac{20.7 \text{ m/s}}{40.0 \text{ m/s}}\right)^2 = 1 - 0.5366 - 0.2679 = 0.1956; \quad \underline{19.6\%}$$

8-15

a)

before: A, $v_{A1} = 4.00$ m/s →; B, $v_{B1} = 0$; x

after: A—B, $v_2 = ?$; x

P_x is constant

$$\Rightarrow m_A v_{A1x} + m_B v_{B1x} = (m_A + m_B) v_{2x}$$

$$m_A v_{A1} = (m_A + m_B) v_2$$

$$v_2 = \left(\frac{m_A}{m_A + m_B}\right) v_{A1} = \left(\frac{10{,}000 \text{ kg}}{10{,}000 \text{ kg} + 20{,}000 \text{ kg}}\right) 4.00 \text{ m/s} = \underline{1.33 \text{ m/s}}$$

b) $K_1 = \frac{1}{2} m_A v_{A1}^2 + \frac{1}{2} m_B v_{B1}^2 \rightarrow 0 = \frac{1}{2}(10{,}000 \text{ kg})(4.00 \text{ m/s})^2 = 8.00 \times 10^4$ J

$K_2 = \frac{1}{2}(m_A + m_B) v_2^2 = \frac{1}{2}(30{,}000 \text{ kg})(1.33 \text{ m/s})^2 = 2.653 \times 10^4$ J

$\Delta K = K_2 - K_1 = 2.653 \times 10^4 \text{ J} - 8.00 \times 10^4 \text{ J} = \underline{-5.35 \times 10^4 \text{ J}}$ (decreases)

8-15 (cont)

c)

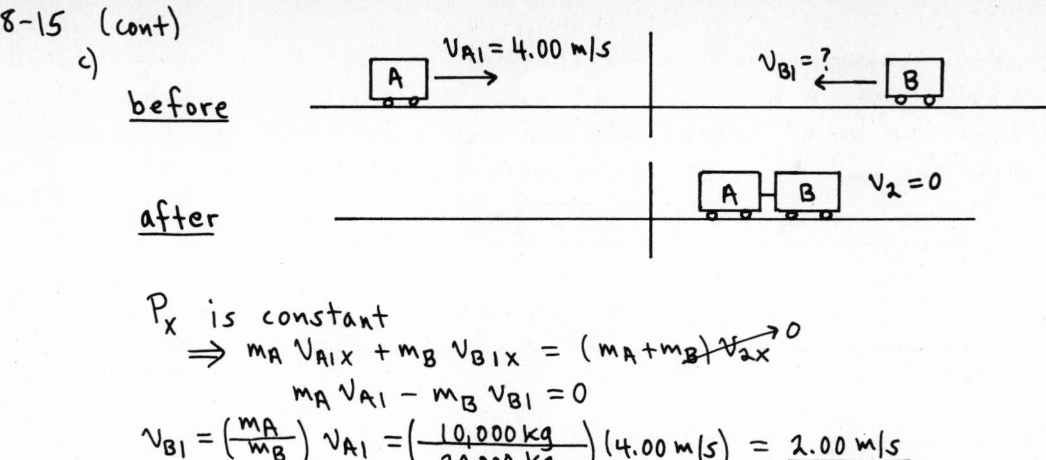

P_x is constant

$$\Rightarrow m_A v_{A1x} + m_B v_{B1x} = (m_A + m_B) v_{2x} \rightarrow 0$$

$$m_A v_{A1} - m_B v_{B1} = 0$$

$$v_{B1} = \left(\frac{m_A}{m_B}\right) v_{A1} = \left(\frac{10{,}000 \text{ kg}}{20{,}000 \text{ kg}}\right)(4.00 \text{ m/s}) = \underline{2.00 \text{ m/s}}$$

8-19

W, N, E, S

Let the automobile be object A and the truck be object B.

before: $v_{A1} = 40.0$ km/h (A moving along +x); $v_{B1} = 20.0$ km/h (B moving along −y)

after: A+B, $v_{2x} = v_2 \cos\theta$, $v_{2y} = v_2 \sin\theta$, θ, v_2

P_x is constant

$$\Rightarrow m_A v_{A1x} + m_B v_{B1x} = (m_A + m_B) v_{2x}$$

$$m_A v_{A1} = (m_A + m_B) v_{2x}$$

$$(2000 \text{ kg})(40.0 \text{ km/h}) = (2000 \text{ kg} + 4000 \text{ kg}) v_{2x}$$

$$\boxed{v_{2x} = 13.33 \text{ km/h}}$$

P_y is constant

$$\Rightarrow m_A v_{A1y} + m_B v_{B1y} = (m_A + m_B) v_{2y}$$

$$-m_B v_{B1} = (m_A + m_B) v_{2y}$$

$$-(4000 \text{ kg})(20.0 \text{ km/h}) = (2000 \text{ kg} + 4000 \text{ kg}) v_{2y}$$

$$\boxed{v_{2y} = -13.33 \text{ km/h}}$$

$$v_2 = \sqrt{v_{2x}^2 + v_{2y}^2} = \sqrt{(13.33 \text{ km/h})^2 + (-13.33 \text{ km/h})^2} = \underline{18.9 \text{ km/h}}$$

$$\tan\theta = \frac{v_{2y}}{v_{2x}} = \frac{-13.33 \text{ km/h}}{13.33 \text{ km/h}} = -1.00 \Rightarrow \theta = -45° \text{ (minus means clockwise from +x-axis)}$$

The wreckage is moving in the direction <u>45.0° S of E</u>.

8-21

Apply conservation of momentum to the collision between the bullet and the block. Let object A be the bullet and object B be the block. Let v_A be the speed of the bullet before the collision and let V be the speed of the block with the bullet inside just after the collision.

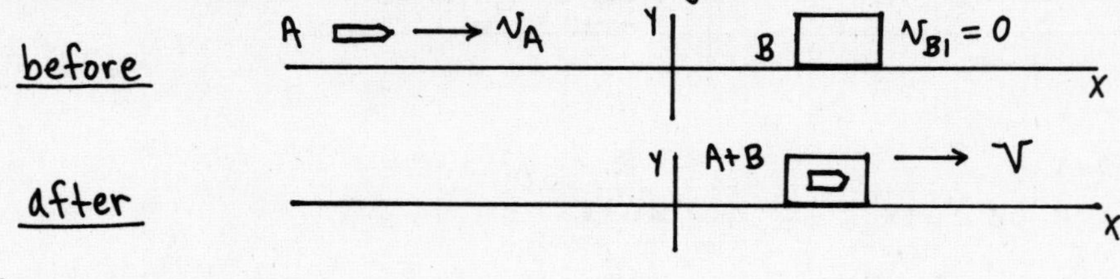

P_x is constant

$$\Rightarrow \boxed{m_A v_A = (m_A + m_B) V}$$

Apply conservation of energy to the motion of the block after the collision

Y

m → V #2 v = 0

#1 ——— 0.250 m ——→ X

$$K_1 + U_1 + W_{other} = K_2 + U_2$$

Work is done by friction $\Rightarrow W_{other} = W_f = (f_k \cos\phi)s = -f_k s = -\mu_k mgs$

$U_1 = U_2 = 0$ (no work done by gravity)

$K_1 = \frac{1}{2}mV^2$

$K_2 = 0$ (block has come to rest)

$$\Rightarrow \frac{1}{2}mV^2 - \mu_k mgs = 0$$

$$V = \sqrt{2\mu_k gs} = \sqrt{2(0.20)(9.80\,\text{m/s}^2)(0.250\,\text{m})} = 0.9899\,\text{m/s}$$

Use this in the conservation of momentum equation

$$\Rightarrow v_A = \left(\frac{m_A + m_B}{m_A}\right)V = \left(\frac{5.00\times10^{-3}\,\text{kg} + 1.50\,\text{kg}}{5.00\times10^{-3}\,\text{kg}}\right)(0.9899\,\text{m/s}) = \underline{496\,\text{m/s}}$$

8-23

before: Y; A, $v_{A1} = 0.60\,\text{m/s}$ →; $m_A = 0.300\,\text{kg}$; ← $v_{B1} = 1.50\,\text{m/s}$, B; $m_B = 0.200\,\text{kg}$; X

after: Y; A, $v_{A2} = ?$; B, $v_{B2} = ?$; X

From conservation of x-component of momentum:

$$m_A v_{A1x} + m_B v_{B1x} = m_A v_{A2x} + m_B v_{B2x}$$

8-23 (cont)

$$m_A v_{A1} - m_B v_{B1} = m_A v_{A2x} + m_B v_{B2x}$$
$$(0.300\text{kg})(0.60\text{m/s}) - (0.200\text{ kg})(1.50\text{m/s}) = (0.300\text{kg})\, v_{A2x} + (0.200\text{ kg})\, v_{B2x}$$
$$0.180\text{ kg}\cdot\text{m/s} - 0.300\text{ kg}\cdot\text{m/s} = (0.300\text{ kg})\, v_{A2x} + (0.200\text{kg})\, v_{B2x}$$
$$\boxed{-1.20\text{ m/s} = 3\,v_{A2x} + 2\,v_{B2x}}$$

From the relative velocity relation for an elastic collision (Eq. (8-19)):
$$v_{B2x} - v_{A2x} = -(v_{B1x} - v_{A1x})$$
$$v_{B2x} - v_{A2x} = -(-1.50\text{ m/s} - 0.60\text{ m/s}) = +2.10\text{ m/s}$$
Multiply this eq. by three $\Rightarrow$ $\boxed{6.30\text{ m/s} = 3v_{B2x} - 3v_{A2x}}$

Add the two eqs. $\Rightarrow 5.10\text{ m/s} = 5\,v_{B2x}$
$$v_{B2x} = +1.02\text{m/s}$$
and then $v_{A2x} = v_{B2x} - 2.10\text{ m/s} = 1.02\text{ m/s} - 2.10\text{ m/s} = -1.08\text{ m/s}$

The 0.300 kg block (block A) is moving to the left at 1.08 m/s and the 0.200 kg block (block B) is moving to the right at 1.02 m/s.

8-25

Take the x-axis to be toward the right, so $v_{x1} = +5.00\text{ m/s}$.

a) $J_x = p_{x2} - p_{x1}$

$J_x = F_x(t_2 - t_1) = (+5.00\text{N})(5.00\text{s}) = +25.0\text{ kg}\cdot\text{m/s}$

$\Rightarrow\ p_{x2} = J_x + p_{x1} = +25.0\text{ kg}\cdot\text{m/s} + (1.50\text{kg})(+5.00\text{ m/s}) = +32.5\text{ kg}\cdot\text{m/s}$

$v_{x2} = \dfrac{p_{x2}}{m} = \dfrac{32.5\text{ kg}\cdot\text{m/s}}{1.50\text{ kg}} = \underline{21.7\text{ m/s}}$ (to the right)

b) $J_x = F_x(t_2 - t_1) = (-7.00\text{N})(5.00\text{ s}) = -35.0\text{ kg}\cdot\text{m/s}$ (minus since force is to left)

$p_{x2} = J_x + p_{x1} = -35.0\text{kg}\cdot\text{m/s} + (1.50\text{kg})(+5.00\text{m/s}) = -27.5\text{ kg}\cdot\text{m/s}$

$v_{x2} = \dfrac{p_{x2}}{m} = \dfrac{-27.5\text{ kg}\cdot\text{m/s}}{1.50\text{ kg}} = \underline{-18.3\text{ m/s}}$ (to the left)

8-29

Take the x-axis to be toward the right, so $F_x = +(A + Bt^2)$

a) $J_x = \int_{t_1}^{t_2} F_x\,dt = \int_0^{t_2}(A + Bt^2)\,dt = (At + \frac{1}{3}Bt^3)\Big|_0^{t_2} = At_2 + \frac{1}{3}Bt_2^3$

The impulse has magnitude $J = At_2 + \frac{1}{3}Bt_2^3$ and is directed to the right.

b) $J_x = p_{x2} - p_{x1} \Rightarrow p_{x2} = J_x + p_{x1}$

Initially at rest $\Rightarrow p_{x1} = 0$.

$v_{x2} = \dfrac{p_{x2}}{m} = \dfrac{J_x}{m} = \dfrac{A}{m}t_2 + \dfrac{B}{3m}t_2^3$

Her speed is $v = \frac{A}{m}t_2 + \frac{B}{3m}t_2^3$ and she is moving to the right.

8-31

Apply Eq. (8-25) with the earth as mass #1 and the moon as mass #2. Take the origin at the earth and let the moon lie on the positive x-axis.

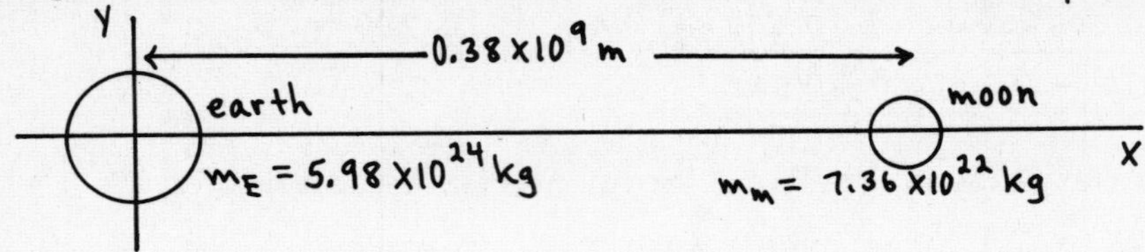

$$x_{cm} = \frac{m_1 x_1 + m_2 x_2}{m_1 + m_2}$$

$x_1 = 0$ and $x_2 = 0.38 \times 10^9$ m

$$x_{cm} = \frac{(7.36 \times 10^{22}\ \text{kg})(0.38 \times 10^9\ \text{m})}{5.98 \times 10^{24}\ \text{kg} + 7.36 \times 10^{22}\ \text{kg}} = 4.6 \times 10^6\ \text{m}$$

The center of mass is 4.6×10^6 m from the center of the earth and is on the line connecting the centers of the earth and of the moon.

8-33

Apply Eq. (8-34): $a = -\frac{v_{ex}}{m}\frac{dm}{dt}$

$$\frac{dm}{dt} = -\frac{ma}{v_{ex}} = -\frac{(5000\ \text{kg})(20.0\ \text{m/s}^2)}{2000\ \text{m/s}} = -50.0\ \text{kg/s}$$

So in 1s the rocket must eject 50.0 kg of gas.

8-37

Use Eq. (8-35): $v - v_0 = v_{ex} \ln\left(\frac{m_0}{m}\right)$

$v_0 = 0$ ("fired from rest"), so $\frac{v}{v_{ex}} = \ln\left(\frac{m_0}{m}\right)$

$\Rightarrow \frac{m_0}{m} = e^{v/v_{ex}}$, or $\frac{m}{m_0} = e^{-v/v_{ex}}$

If v is the final speed then m is the mass left when all the fuel has been expended; $\frac{m}{m_0}$ is the fraction of the initial mass that is not fuel.

a) $v = 1.00 \times 10^{-3} c = 2.998 \times 10^5\ \text{m/s} \Rightarrow \frac{m}{m_0} = e^{-(2.998 \times 10^5\ \text{m/s})/(2000\ \text{m/s})} = 8 \times 10^{-66}$

This is clearly not feasible, for so little of the initial mass to not be fuel.

b) $v = 3000\ \text{m/s} \Rightarrow \frac{m}{m_0} = e^{-(3000\ \text{m/s})/(2000\ \text{m/s})} = \underline{0.223}$

(22.3% of the total initial mass not fuel, so 77.7% fuel; this possible.)

8-41

Let $\vec{p}_e$, $\vec{p}_{\bar{\nu}}$, and $\vec{p}_N$ be the final momenta of the electron, the antineutrino, and the recoiling ^{210}Po nucleus.

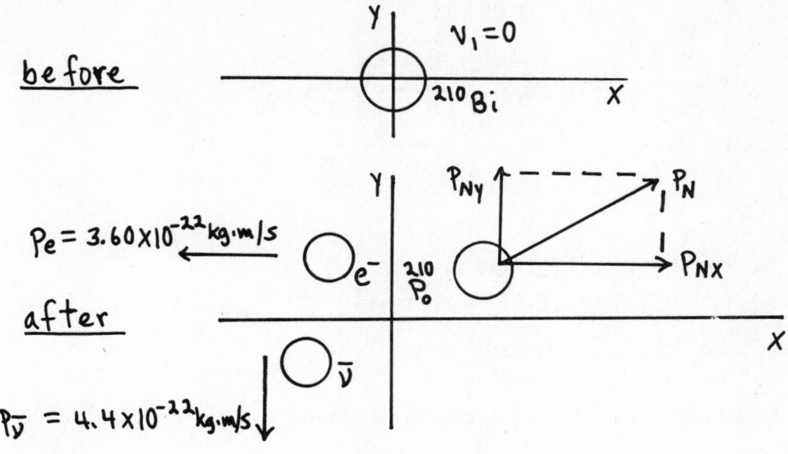

P_x is conserved and initially $P_x = 0$

$\Rightarrow \quad 0 = p_{Nx} - p_e$

$$p_{Nx} = p_e = 3.60\times10^{-22}\ \text{kg·m/s}$$

P_y is conserved and initially $P_y = 0$

$\Rightarrow \quad 0 = p_{Ny} - p_{\bar{\nu}}$

$$p_{Ny} = p_{\bar{\nu}} = 4.40\times10^{-22}\ \text{kg·m/s}$$

Then $p_N = \sqrt{p_{Nx}^2 + p_{Ny}^2} = \sqrt{(3.60\times10^{-22})^2 + (4.40\times10^{-22})^2}\ \text{kg·m/s} = \underline{5.69\times10^{-22}\ \text{kg·m/s}}$

b) $K_N = \frac{1}{2} m_N v_N^2$ and $p_N = m_N v_N \Rightarrow K_N = \frac{p_N^2}{2m_N}$

$$K_N = \frac{(5.69\times10^{-22}\ \text{kg·m/s})^2}{2(3.50\times10^{-25}\ \text{kg})} = \underline{4.63\times10^{-19}\ \text{J}}$$

Problems

8-45

Use a coordinate system attached to the ground. Take the +x-axis to be east (along the tracks) and the y-axis to be north (parallel to the ground and perpendicular to the tracks). Then P_x is conserved but P_y is not conserved due to the sideways force exerted by the tracks, the force that keeps the hand car on the tracks.

a) Let A be the 20.0 kg mass and B be the car (mass 180 kg). After the mass is thrown sideways relative to the car, it still has the same eastward component of velocity, 5.00 m/s, as it had before it was thrown.

8-45 (cont)

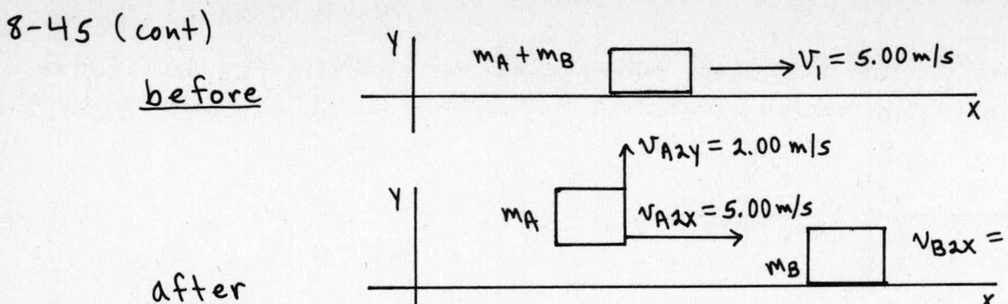

P_x is conserved

$$\Rightarrow (m_A + m_B) v_1 = m_A v_{A2x} + m_B v_{B2x}$$

$$(200 \text{ kg})(5.00 \text{ m/s}) = (20.0 \text{ kg})(5.00 \text{ m/s}) + (180 \text{ kg})\, v_{B2x}$$

$$v_{B2x} = \frac{1000 \text{ kg}\cdot\text{m/s} - 100 \text{ kg}\cdot\text{m/s}}{180 \text{ kg}} = 5.00 \text{ m/s}$$

The final velocity of the car is 5.00 m/s, east (unchanged).

b) We need the final velocity of A relative to the ground.

$$\vec{v}_{A/E} = \vec{v}_{A/B} + \vec{v}_{B/E}$$

$v_{B/E} = +5.00$ m/s

$v_{A/B} = -5.00$ m/s (minus since the mass is moving west relative to the car)

This gives $v_{A/E} = 0$; the mass is at rest relative to the earth after it is thrown backwards from the car.

As in part (a), $(m_A + m_B) v_1 = m_A v_{A2x} + m_B v_{B2x}$.

Now $v_{A2x} = 0$, so $(m_A + m_B) v_1 = m_B v_{B2x}$

$$v_{B2x} = \left(\frac{m_A + m_B}{m_B}\right) v_1 = \left(\frac{200 \text{ kg}}{180 \text{ kg}}\right)(5.00 \text{ m/s}) = 5.56 \text{ m/s}$$

The final velocity of the car is <u>5.56 m/s, east</u>.

c) Let A be the 20.0 kg mass and B be the car (mass $m_B = 200$ kg).

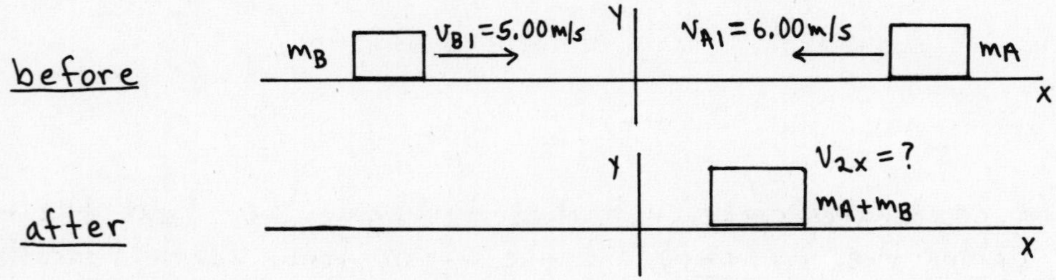

P_x is conserved

$$\Rightarrow m_A v_{A1x} + m_B v_{B1x} = (m_A + m_B) v_{2x}$$

$$-m_A v_{A1} + m_B v_{B1} = (m_A + m_B) v_{2x}$$

$$v_{2x} = \frac{m_B v_{B1} - m_A v_{A1}}{m_A + m_B} = \frac{(200 \text{ kg})(5.00 \text{ m/s}) - (20.0 \text{ kg})(6.00 \text{ m/s})}{200 \text{ kg} + 20.0 \text{ kg}} = 4.00 \text{ m/s}$$

The final velocity of the car is <u>4.00 m/s, east</u>.

8-47

Apply conservation of momentum to the collision between the bullet and the block and apply conservation of energy to the motion of the block after the collision.

Collision between the bullet and the block: Let object A be the bullet and object B be the block. Apply momentum conservation to find the speed v_{B2} of the block just after the collision.

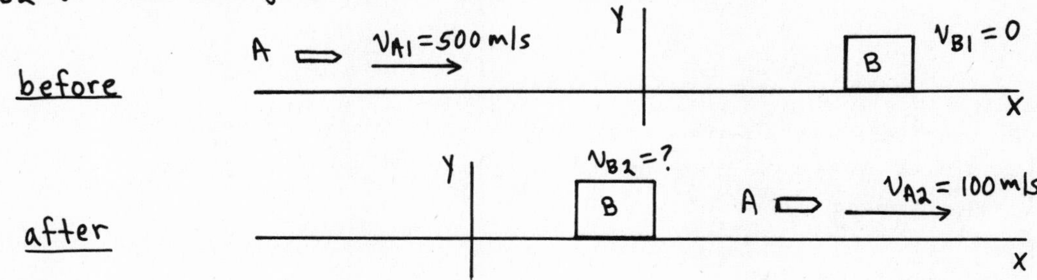

P_x is conserved, so

$$m_A v_{A1x} + m_B v_{B1x}^{\;0} = m_A v_{A2x} + m_B v_{B2x}$$

$$m_A v_{A1} = m_A v_{A2} + m_B v_{B2x}$$

$$v_{B2x} = \frac{m_A(v_{A1} - v_{A2})}{m_B} = \frac{(2.00\times10^{-3}\text{ kg})(500\text{ m/s} - 100\text{ m/s})}{1.00\text{ kg}} = 0.800\text{ m/s}$$

Motion of the block after the collision:
Let point 1 in the motion be just after the collision, where the block has the speed 0.800 m/s calculated above, and let point 2 be where the block has come to rest.

Y; $v_1 = 0.800$ m/s; $v_2 = 0$; #1 —— 0.30 m ——→ #2; X

$$K_1 + U_1 + W_{other} = K_2 + U_2$$

Work is done on the block by friction, so $W_{other} = W_f$ and $U_1 = U_2 = 0$.

$W_{other} = W_f = (f_k \cos\phi)s = -f_k s = -\mu_k mgs$, where $s = 0.30$ m

$K_1 = \frac{1}{2}mv_1^2$

$K_2 = 0$ (block has come to rest)

$$\Rightarrow \tfrac{1}{2}\cancel{m}v_1^2 - \mu_k \cancel{m}gs = 0$$

$$\mu_k = \frac{v_1^2}{2gs} = \frac{(0.800\text{ m/s})^2}{2(9.80\text{ m/s}^2)(0.30\text{ m})} = \underline{0.109}$$

b) For the bullet,

$K_1 = \frac{1}{2}mv_1^2 = \frac{1}{2}(2.00\times10^{-3}\text{ kg})(500\text{ m/s})^2 = 250$ J

$K_2 = \frac{1}{2}mv_2^2 = \frac{1}{2}(2.00\times10^{-3}\text{ kg})(100\text{ m/s})^2 = 10$ J

$\Delta K = K_2 - K_1 = 10\text{ J} - 250\text{ J} = -240$ J

The kinetic energy of the bullet decreases by <u>240 J.</u>

8-47 (cont)

c) Immediately after the collision the speed of the block is 0.800 m/s so its kinetic energy is $K = \frac{1}{2}mv^2 = \frac{1}{2}(1.00\text{ kg})(0.800\text{ m/s})^2 = \underline{0.320\text{ J}}$.

Note that the collision is highly inelastic. The bullet loses 240 J of kinetic energy, but only 0.320 J is gained by the block. But momentum is conserved in the collision. All the momentum lost by the bullet is gained by the block.

8-49

Use the free-body diagram for the frame when it hangs at rest on the end of the spring to find the force constant k of the spring. Let s be the amount the spring is stretched.

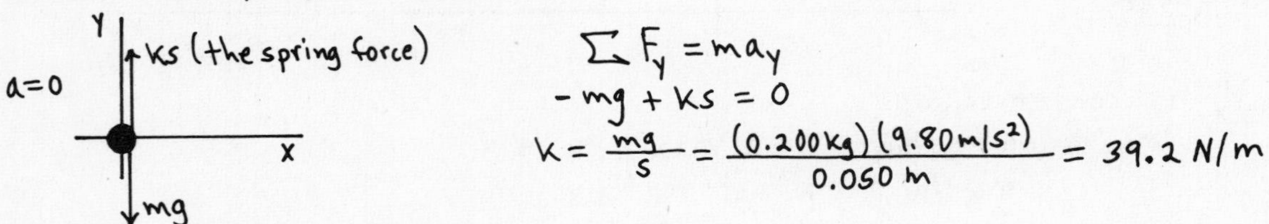

$$\Sigma F_y = ma_y$$
$$-mg + ks = 0$$
$$k = \frac{mg}{s} = \frac{(0.200\text{ kg})(9.80\text{ m/s}^2)}{0.050\text{ m}} = 39.2\text{ N/m}$$

Next find the speed of the putty when it reaches the frame. The putty falls with acceleration $a = g$, downward.

$v_0 = 0$

$y - y_0 = 0.300\text{ m}$

$v = ?$

$a = +9.80\text{ m/s}^2$

$$v^2 = v_0^2 + 2a(y - y_0), \quad v_0 = 0$$
$$v = \sqrt{2a(y-y_0)}$$
$$v = \sqrt{2(9.80\text{ m/s}^2)(0.300\text{ m})} = 2.425\text{ m/s}$$

Apply conservation of momentum to the collision between the putty (A) and the frame (B):

P_y is conserved, so $-m_A v_{A1} = -(m_A + m_B)v_2$

$$v_2 = \left(\frac{m_A}{m_A + m_B}\right)v_{A1} = \left(\frac{0.200\text{ kg}}{0.400\text{ kg}}\right)(2.425\text{ m/s}) = 1.212\text{ m/s}$$

Apply conservation of energy to the motion of the frame on the end of the spring after the collision. Let point #1 be just after the putty strikes and point #2 be when the frame has its maximum downward displacement. Let d be the amount the frame moves downward.

8-49 (cont)

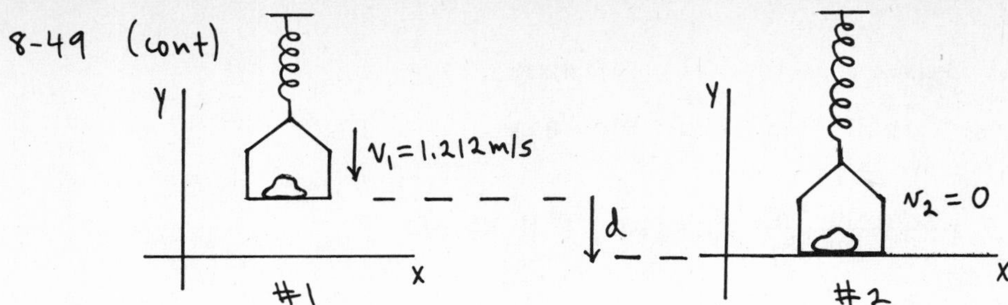

When the frame is at position #1 the spring is stretched a distance $x_1 = 0.050$ m. When the frame is at position #2 the spring is stretched a distance $x_2 = 0.050\text{ m} + d$. Use coordinates with the +y-direction upward and $y = 0$ at the lowest point reached by the frame, so that $y_1 = d$ and $y_2 = 0$. Work is done on the frame by gravity and by the spring force, so $W_{other} = 0$ and $U = U_{el} + U_{gravity}$.

$$K_1 + U_1 + W_{other} = K_2 + U_2$$

$W_{other} = 0$

$K_1 = \frac{1}{2}mv_1^2 = \frac{1}{2}(0.400\text{ kg})(1.212\text{ m/s})^2 = 0.2938\text{ J}$

$K_2 = 0$

$U_1 = U_{1,el} + U_{1,gravity} = \frac{1}{2}kx_1^2 + mgy_1 = \frac{1}{2}(39.2\text{ N/m})(0.050\text{ m})^2 + (0.400\text{kg})(9.80\text{m/s}^2)\,d$

$U_1 = 0.0490\text{ J} + (3.92\text{ N})\,d$

$U_2 = U_{2,el} + U_{2,gravity} = \frac{1}{2}kx_2^2 + mgy_2 \rightarrow 0 = \frac{1}{2}(39.2\text{ N/m})(0.050\text{ m} + d)^2$

$U_2 = 0.0490\text{ J} + (1.96\text{ N})d + (19.6\text{ N/m})\,d^2$

$\Rightarrow 0.2938\text{ J} + 0.0490\text{ J} + (3.92\text{ N})d = 0.0490\text{ J} + (1.96\text{ N})d + (19.6\text{ N/m})\,d^2$

$(19.6\text{ N/m})d^2 - (1.96\text{ N})d - 0.2938\text{ J} = 0$

$d = \frac{1}{2(19.6)}\left[1.96 \pm \sqrt{(1.96)^2 - 4(19.6)(-0.2938)}\,\right]\text{ m}$

$d = 0.050\text{ m} \pm 0.1322\text{ m}$

The solution we want is a positive (downward) distance, so

$d = 0.050\text{ m} + 0.1322\text{ m} = \underline{0.182\text{ m}}$

8-51

Let A be the bullet and B be the stone.

a)

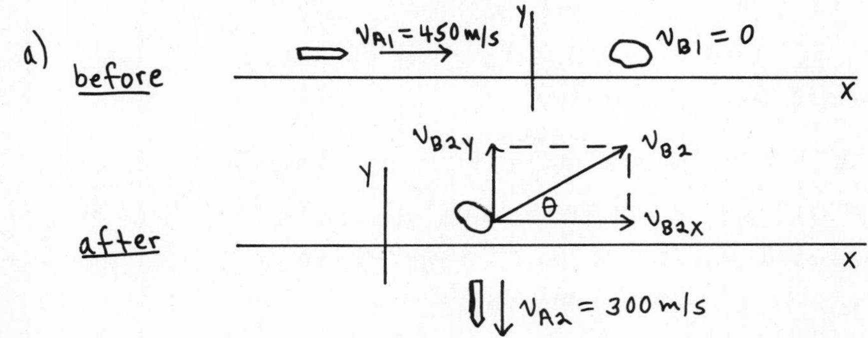

8-51 (cont)

P_x is conserved, so

$$m_A v_{A1x} + m_B v_{B1x}^{\nearrow 0} = m_A v_{A2x}^{\nearrow 0} + m_B v_{B2x}$$

$$m_A v_{A1} = m_B v_{B2x}$$

$$v_{B2x} = \left(\frac{m_A}{m_B}\right) v_{A1} = \left(\frac{2.50\times10^{-3}\,\text{kg}}{0.100\,\text{kg}}\right)(450\,\text{m/s}) = 11.25\,\text{m/s}$$

P_y is conserved, so

$$m_A v_{A1y}^{\nearrow 0} + m_B v_{B1y}^{\nearrow 0} = m_A v_{A2y} + m_B v_{B2y}$$

$$0 = -m_A v_{A2} + m_B v_{B2y}$$

$$v_{B2y} = \left(\frac{m_A}{m_B}\right) v_{A2} = \left(\frac{2.50\times10^{-3}\,\text{kg}}{0.100\,\text{kg}}\right)(300\,\text{m/s}) = 7.50\,\text{m/s}$$

$$v_{B2} = \sqrt{v_{B2x}^2 + v_{B2y}^2} = \sqrt{(11.25\,\text{m/s})^2 + (7.50\,\text{m/s})^2} = \underline{13.5\,\text{m/s}}$$

$$\tan\theta = \frac{v_{B2y}}{v_{B2x}} = \frac{7.50\,\text{m/s}}{11.25\,\text{m/s}} = 0.667 \Rightarrow \theta = 33.7^\circ \text{ (defined in the sketch)}$$

b) To answer this question compare K_1 and K_2 for the system:

$$K_1 = \tfrac{1}{2} m_A v_{A1}^2 + \tfrac{1}{2} m_B v_{B1}^{2\,\nearrow 0} = \tfrac{1}{2}(2.50\times10^{-3}\,\text{kg})(450\,\text{m/s})^2 = 253\,\text{J}$$

$$K_2 = \tfrac{1}{2} m_A v_{A2}^2 + \tfrac{1}{2} m_B v_{B2}^2 = \tfrac{1}{2}(2.50\times10^{-3}\,\text{kg})(300\,\text{m/s})^2 + \tfrac{1}{2}(0.100\,\text{kg})(13.5\,\text{m/s})^2 = 122\,\text{J}$$

$$\Delta K = K_2 - K_1 = 122\,\text{J} - 253\,\text{J} = -131\,\text{J}$$

The kinetic energy of the system decreases by 131 J as a result of the collision; the collision is <u>not</u> elastic.

8-55

a) $K = \tfrac{1}{2} m_A v_A^2 + \tfrac{1}{2} m_B v_B^2$

Note that $\vec{v}_A'$ and $\vec{v}_B'$ as defined in the problem are the velocities of A and B in coordinates moving with the center of mass. Note also that $m_A \vec{v}_A' + m_B \vec{v}_B' = M \vec{v}_{cm}'$ where $\vec{v}_{cm}'$ is the velocity of the cm in these coordinates. But that's zero, so $m_A \vec{v}_A' + m_B \vec{v}_B' = 0$; we use this in the proof.

$$\vec{v}_A = \vec{v}_A' + \vec{v}_{cm} \Rightarrow v_A^2 = v_A'^2 + v_{cm}^2 + 2\,\vec{v}_A' \cdot \vec{v}_{cm}$$

$$\vec{v}_B = \vec{v}_B' + \vec{v}_{cm} \Rightarrow v_B^2 = v_B'^2 + v_{cm}^2 + 2\,\vec{v}_B' \cdot \vec{v}_{cm}$$

(This uses that for a vector $\vec{A}$, $A^2 = \vec{A}\cdot\vec{A}$)

Thus $K = \tfrac{1}{2} m_A v_A'^2 + \tfrac{1}{2} m_A v_{cm}^2 + m_A \vec{v}_A' \cdot \vec{v}_{cm} + \tfrac{1}{2} m_B v_B'^2 + \tfrac{1}{2} m_B v_{cm}^2 + m_B \vec{v}_B' \cdot \vec{v}_{cm}$

$$K = \tfrac{1}{2}(m_A + m_B) v_{cm}^2 + \tfrac{1}{2}(m_A v_A'^2 + m_B v_B'^2) + (m_A \vec{v}_A' + m_B \vec{v}_B') \cdot \vec{v}_{cm}$$

But $m_A + m_B = M$ and as noted earlier $m_A \vec{v}_A' + m_B \vec{v}_B' = 0$, so

$$K = \tfrac{1}{2} M v_{cm}^2 + \tfrac{1}{2}(m_A v_A'^2 + m_B v_B'^2)$$

This is the result the problem asked us to derive.

8-55 (cont)

In the collision $\vec{P} = M\vec{v}_{cm}$ is constant, so $\frac{1}{2}Mv_{cm}^2$ stays constant. The asteroids can lose all their relative kinetic energy but $\frac{1}{2}Mv_{cm}^2$ must remain.

8-57

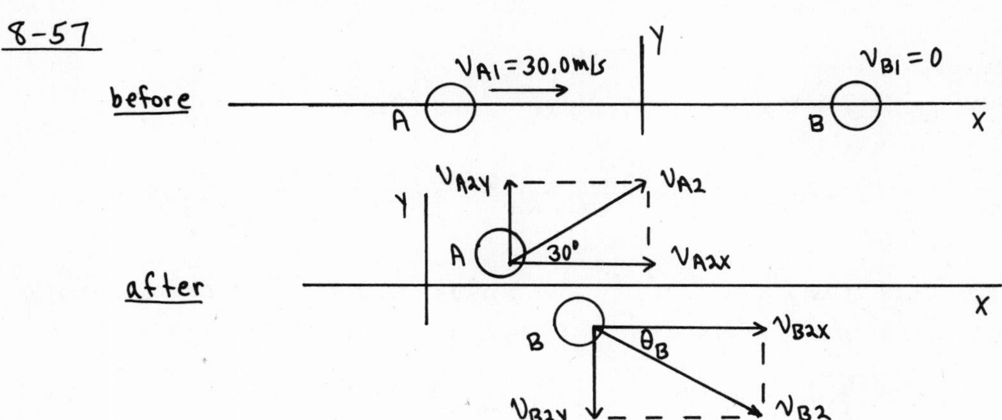

The result of Problem 8-56 part (d) applies here and says that $30.0° + \theta_B = 90.0°$, so that $\theta_B = 60.0°$. (A and B move off in perpendicular directions.)

P_x is conserved

$$\Rightarrow m_A v_{A1x} + m_B \cancel{v_{B1x}}^{0} = m_A v_{A2x} + m_B v_{B2x}$$

But $m_A = m_B$, so

$$\boxed{v_{A1} = v_{A2}\cos 30° + v_{B2}\cos 60°}$$

P_y is conserved

$$\Rightarrow m_A \cancel{v_{A1y}}^{0} + m_B \cancel{v_{B1y}}^{0} = m_A v_{A2y} + m_B v_{B2y}$$

$$0 = \cancel{m_A} v_{A2y} + \cancel{m_B} v_{B2y}$$

$$0 = v_{A2}\sin 30° - v_{B2}\sin 60°$$

$$\boxed{v_{B2} = \left(\frac{\sin 30°}{\sin 60°}\right) v_{A2}}$$

Use this result in the first eq.

$$\Rightarrow v_{A1} = v_{A2}\cos 30° + \left(\frac{\sin 30° \cos 60°}{\sin 60°}\right) v_{A2}$$

$$v_{A1} = 1.155\, v_{A2}$$

$$v_{A2} = \frac{v_{A1}}{1.155} = \frac{30.0 \text{ m/s}}{1.155} = \underline{26.0 \text{ m/s}}$$

And then $v_{B2} = \left(\frac{\sin 30°}{\sin 60°}\right)(26.0 \text{ m/s}) = \underline{15.0 \text{ m/s}}$

8-59

a) Use coordinates fixed to the earth and take the +x-axis to be in the direction that they jump. The crate is at rest when they jump, so each person has a speed of 5.00 m/s relative to the earth after they jump. Let A be Jack, B be Jill, and C be the crate.

8-59 (cont)

$P_x = 0$ before they jump since everything is at rest.

after

$v_{C2} = ?$ ← C |y   A → $v_{A2} = 5.00$ m/s

x

B → $v_{B2} = 5.00$ m/s

P_x is conserved

$$\Rightarrow 0 = m_A v_{A2x} + m_B v_{B2x} + m_C v_{C2x}$$

$$0 = m_A v_{A2} + m_B v_{B2} - m_C v_{C2}$$

$$v_{C2} = \frac{m_A v_{A2} + m_B v_{B2}}{m_C} = \frac{(80.0\,\text{kg})(5.00\,\text{m/s}) + (50.0\,\text{kg})(5.00\,\text{m/s})}{20.0\,\text{kg}} = \underline{32.5\,\text{m/s}}$$

b) First consider Jack jumping. Before he jumps $P_x = 0$. The crate initially is at rest so he jumps with a speed of 5.00 m/s relative to the ground.

after

$v_{B+C} = ?$ ← B+C   y| A → $v_{A2} = 5.00$ m/s

x

P_x is conserved $\Rightarrow 0 = m_A v_{A2} - (m_B + m_C) v_{B+C}$

$$v_{B+C} = \left(\frac{m_A}{m_B + m_C}\right) v_{A2} = \left(\frac{80.0\,\text{kg}}{50.0\,\text{kg} + 20.0\,\text{kg}}\right)(5.00\,\text{m/s}) = 5.71\,\text{m/s};$$

Jill and the crate are moving in the $-x$- direction at 5.71 m/s after Jack jumps.

Jill then jumps at 5.00 m/s, relative to the crate, in the $+x$-direction. We need her velocity relative to the earth:

$$\vec{v}_{J/E} = \vec{v}_{J/C} + \vec{v}_{C/E} \Rightarrow v_{J/E} = +5.00\,\text{m/s} - 5.71\,\text{m/s} = -0.71\,\text{m/s}$$

Now apply conservation of momentum to Jill's jumping off:

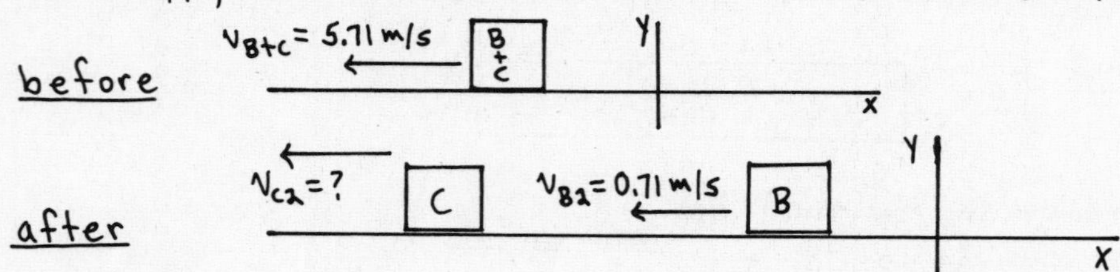

P_x is conserved $\Rightarrow -(m_B + m_C) v_{B+C} = -m_C v_{C2} - m_B v_{B2}$

$$v_{C2} = \frac{(m_B + m_C) v_{B+C} - m_B v_{B2}}{m_C} = \frac{(70.0\,\text{kg})(5.71\,\text{m/s}) - (50.0\,\text{kg})(0.71\,\text{m/s})}{20.0\,\text{kg}} = \underline{18.2\,\text{m/s}}$$

c) Jill jumps: Before she jumps $P_x = 0$. The crate is initially at rest so she jumps with a speed of 5.00 m/s relative to the ground.

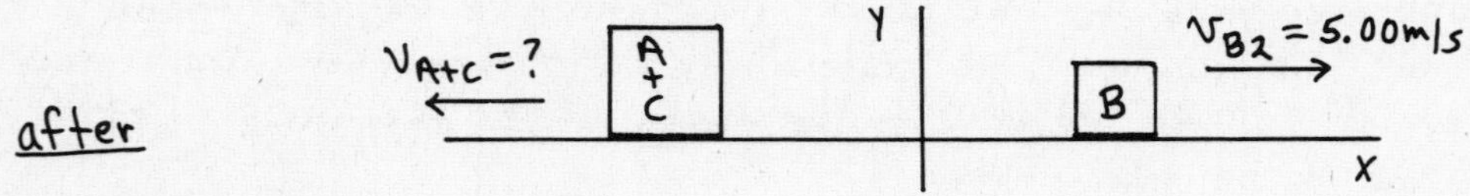

8-59 (cont)

P_x is conserved $\Rightarrow 0 = m_B v_{B2} - (m_A + m_c) v_{A+c}$

$$v_{A+c} = \frac{m_B v_{B2}}{m_A + m_c} = \frac{(50.0\,\text{kg})(5.00\,\text{m/s})}{80.0\,\text{kg} + 20.0\,\text{kg}} = 2.50\,\text{m/s}$$

Jack and the crate are moving at 2.50 m/s in the −x-direction after Jill jumps.

Jack jumps at 5.00 m/s in the +x-direction relative to the crate, so his velocity relative to the ground is

$$\vec{v}_{J/E} = \vec{v}_{J/c} + \vec{v}_{c/E} \Rightarrow v_{J/E} = +5.00\,\text{m/s} - 2.50\,\text{m/s} = +2.50\,\text{m/s}$$

Jack jumps:

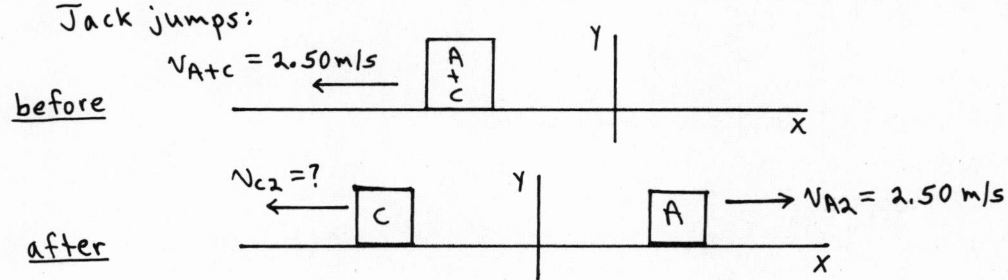

P_x is conserved $\Rightarrow -(m_A + m_c) v_{A+c} = m_A v_{A2} - m_c v_{c2}$

$$v_{c2} = \frac{(m_A + m_c) v_{A+c} + m_A v_{A2}}{m_c} = \frac{(80.0\,\text{kg} + 20.0\,\text{kg})(2.50\,\text{m/s}) + (80.0\,\text{kg})(2.50\,\text{m/s})}{20.0\,\text{kg}} = \underline{22.5\,\text{m/s}}$$

8-63

Use coordinates fixed to the ice, with the direction you walk the +x-direction. $\vec{v}_{cm}$ is constant and initially $v_{cm} = 0$.

Y

v_y

v_s ← concrete slab

X

$$v_{cm} = \frac{m_y \vec{v}_y + m_s \vec{v}_s}{m_y + m_s} = 0$$

$$\Rightarrow m_y \vec{v}_y + m_s \vec{v}_s = 0$$

$$m_y v_{yx} + m_s v_{sx} = 0$$

$$v_{sx} = -\left(\frac{m_y}{m_s}\right) v_{yx} = -\left(\frac{\cancel{m_y}}{5\cancel{m_y}}\right) 3.00\,\text{m/s} = -0.600\,\text{m/s}$$

The slab moves at $\underline{0.600\,\text{m/s}}$, in the direction opposite to the direction you are walking.

8-65

Apply momentum conservation in the x and y direction:

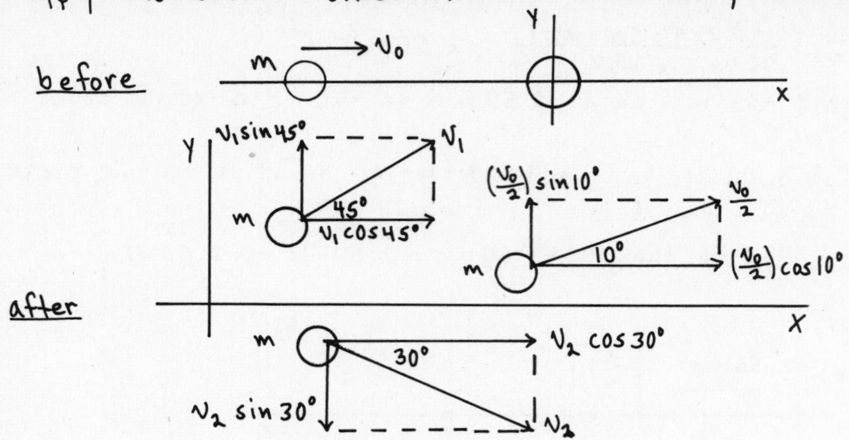

P_x is conserved

$\Rightarrow m v_0 = m\left(\frac{v_0}{2}\right)\cos 10° + m v_1 \cos 45° + m v_2 \cos 30°$

$v_0 = 0.4924\, v_0 + 0.7071\, v_1 + 0.8660\, v_2$

$\boxed{0.5076\, v_0 = 0.7071\, v_1 + 0.8660\, v_2}$

P_y is conserved

$\Rightarrow 0 = m\left(\frac{v_0}{2}\right)\sin 10° + m v_1 \sin 45° - m v_2 \sin 30°$

$\boxed{0.0868\, v_0 = -\,0.7071\, v_1 + 0.5000\, v_2}$

Add these two equations $\Rightarrow 0.5944\, v_0 = 1.366\, v_2$

$v_2 = 0.4351\, v_0 = 0.4351\,(4.0 \times 10^6\ \text{m/s}) = 1.74 \times 10^6\ \text{m/s}$

Then $v_1 = \dfrac{0.5076\, v_0 - 0.8660\, v_2}{0.7071} = \dfrac{(0.5076)(4.0 \times 10^6\ \text{m/s}) - (0.8660)(1.74 \times 10^6\ \text{m/s})}{0.7071} = 0.74 \times 10^5\ \text{m/s}$

The two emitted neutrons have speeds of $\underline{1.7 \times 10^6\ \text{m/s}}$ and $\underline{0.74 \times 10^5\ \text{m/s}}$.

The speeds of the Ba and Kr nuclei are related by P_z conservation.

P_z is conserved $\Rightarrow 0 = m_{Ba} v_{Ba} - m_{Kr} v_{Kr}$

$v_{Kr} = \left(\frac{m_{Ba}}{m_{Kr}}\right) v_{Ba} = \left(\frac{2.3 \times 10^{-25}\ \text{kg}}{1.5 \times 10^{-25}\ \text{kg}}\right) v_{Ba} = 1.5\, v_{Ba}\ ;\ \boxed{v_{Kr} = 1.5\, v_{Ba}}$

(We can't say what these speeds are, but they must satisfy this relation. The value of v_{Ba} depends on energy considerations.)

8-69

$\vec{F} = (\alpha t^2)\vec{i} - (\beta - \gamma t)\vec{j}\ ;\ \alpha = 15.0\ \text{N/s}^2,\ \beta = 12.0\ \text{N},\ \gamma = 20.0\ \text{N/s}$

$J_x = \int_{t_1}^{t_2} F_x(t)\,dt = \int_0^{t_2} (\alpha t^2)\,dt = \frac{1}{3}\alpha t_2^3 = \frac{1}{3}(15.0\ \text{N/s}^2)(0.500\ \text{s})^3 = 0.625\ \text{N}\cdot\text{s}$

$J_y = \int_{t_1}^{t_2} F_y(t)\,dt = \int_0^{t_2} -(\beta - \gamma t)\,dt = -(\beta t_2 - \frac{1}{2}\gamma t_2^2)$

$J_y = -[(12.0\ \text{N})(0.500\ \text{s}) - \frac{1}{2}(20.0\ \text{N/s})(0.500\ \text{s})^2] = -3.50\ \text{N}\cdot\text{s}$

8-69 (cont)

$$J_x = p_{x2} - p_{x1} = m(v_{x2} - \cancel{v_{x1}}^{0})$$

$$v_{x2} = \frac{J_x}{m} = \frac{0.625\ \text{N}\cdot\text{s}}{2.00\ \text{kg}} = 0.312\ \text{m/s}$$

$$J_y = p_{y2} - p_{y1} = m(v_{y2} - \cancel{v_{y1}}^{0})$$

$$v_{y2} = \frac{J_y}{m} = \frac{-3.50\ \text{N}\cdot\text{s}}{2.00\ \text{kg}} = -1.75\ \text{m/s}$$

Thus $\vec{v} = (0.312\ \text{m/s})\vec{i} - (1.75\ \text{m/s})\vec{j}$.

8-71

a) Eq. (8-35) $\Rightarrow v - \cancel{v_0}^{0} = v_{ex} \ln\left(\frac{m_0}{m}\right)$

$$v = v_{ex} \ln\left(\frac{m_0}{m}\right)$$

The total initial mass of the rocket is $m_0 = 12{,}000\ \text{kg} + 1000\ \text{kg} = 13{,}000\ \text{kg}$. Of this, 9000 kg + 600 kg = 9600 kg is fuel, so the mass m left after all the fuel is burned is 13,000 kg − 9600 kg = 3400 kg.

$$\Rightarrow v = v_{ex} \ln\left(\frac{13{,}000\ \text{kg}}{3400\ \text{kg}}\right) = \underline{1.34\ v_{ex}}$$

b) First stage: $v = v_{ex} \ln\left(\frac{m_0}{m}\right)$

$m_0 = 13{,}000\ \text{kg}$

The first stage has 9000 kg of fuel, so the mass left after the first-stage fuel has burned is 13,000 kg − 9000 kg = 4000 kg.

$$v = v_{ex} \ln\left(\frac{13{,}000\ \text{kg}}{4000\ \text{kg}}\right) = \underline{1.18\ v_{ex}}$$

c) Second stage:

$m_0 = 1000\ \text{kg}$

$m = 1000\ \text{kg} - 600\ \text{kg} = 400\ \text{kg}$

$$v = v_0 + v_{ex} \ln\left(\frac{m_0}{m}\right) = 1.18\ v_{ex} + v_{ex} \ln\left(\frac{1000\ \text{kg}}{400\ \text{kg}}\right) = \underline{2.10\ v_{ex}}$$

d) $v = 8.00\ \text{km/s}$

$$\Rightarrow v_{ex} = \frac{v}{2.10} = \frac{8.00\ \text{km/s}}{2.10} = \underline{3.81\ \text{km/s}}$$

CHAPTER 9

Exercises 3, 5, 7, 9, 13, 15, 19, 21, 23, 27, 29, 33, 35

Problems 39, 43, 45, 47, 49, 53

Exercises

9-3

$\theta = \gamma t + \beta t^3$; $\gamma = 2.50$ rad/s , $\beta = 0.0400$ rad/s^3

a) $\omega = \frac{d\theta}{dt} = \gamma + 3\beta t^2$

b) $t = 0 \Rightarrow \omega = \gamma = 2.50$ rad/s

c) $t = 5.00\text{ s} \Rightarrow \omega = 2.50\text{ rad/s} + 3(0.0400\text{ rad/s}^3)(5.00\text{ s})^2 = \underline{5.50\text{ rad/s}}$

$\omega_{av} = \frac{\Delta\theta}{\Delta t} = \frac{\theta_2 - \theta_1}{t_2 - t_1}$

$t_1 = 0 \Rightarrow \theta_1 = 0$

$t_2 = 5.00\text{ s} \Rightarrow \theta_2 = (2.50\text{ rad/s})(5.00\text{ s}) + (0.0400\text{ rad/s}^3)(5.00\text{ s})^3 = 17.5\text{ rad}$

So $\omega_{av} = \frac{17.5\text{ rad} - 0}{5.00\text{ s} - 0} = \underline{3.50\text{ rad/s}}$

ω at 5.00 s is larger than ω_{av} for the time interval 0 to 5.00 s. The angular velocity is increasing in time so its value at the end of the interval is larger than its average value during the interval.

9-5

$\theta = a + bt^2 + ct^3$

$\omega = \frac{d\theta}{dt} = 2bt + 3ct^2$

$\alpha = \frac{d\omega}{dt} = 2b + 6ct$

9-7

a) $\alpha = 0.520$ rad/s^2

$\omega_0 = 0$ (starts from rest)

$\omega = 8.00$ rad/s

$t = ?$

$\omega = \omega_0 + \alpha t$

$t = \frac{\omega - \omega_0}{\alpha} = \frac{8.00\text{ rad/s} - 0}{0.520\text{ rad/s}^2} = \underline{15.4\text{ s}}$

b) $\theta - \theta_0 = ?$

$\theta - \theta_0 = \omega_0 t + \frac{1}{2}\alpha t^2 = 0 + \frac{1}{2}(0.520\text{ rad/s}^2)(15.4\text{ s})^2 = 61.7\text{ rad}$

$\theta - \theta_0 = 61.7\text{ rad}\left(\frac{1\text{ rev}}{2\pi\text{ rad}}\right) = \underline{9.81\text{ rev/s}}$

9-9

a) $\omega_0 = (900\ \text{rev/min})\left(\frac{1\ \text{min}}{60\ \text{s}}\right) = 15.0\ \text{rev/s}$

$\omega = (400\ \text{rev/min})\left(\frac{1\ \text{min}}{60\ \text{s}}\right) = 6.67\ \text{rev/s}$

$t = 5.00\ \text{s}$

$\alpha = ?$

$\omega = \omega_0 + \alpha t$

$\alpha = \frac{\omega - \omega_0}{t} = \frac{6.67\ \text{rev/s} - 15.0\ \text{rev/s}}{5.00\ \text{s}}$

$\alpha = \underline{-1.67\ \text{rev/s}^2}$

$\theta - \theta_0 = ?$

$\theta - \theta_0 = \omega_0 t + \frac{1}{2}\alpha t^2 = (15.0\ \text{rev/s})(5.00\ \text{s}) + \frac{1}{2}(-1.67\ \text{rev/s}^2)(5.00\ \text{s})^2 = \underline{54.1\ \text{rev}}$

b) $\omega = 0$ (comes to rest)

$\omega_0 = 6.67\ \text{rev/s}$

$\alpha = -1.67\ \text{rev/s}^2$

$t = ?$

$\omega = \omega_0 + \alpha t$

$t = \frac{\omega - \omega_0}{\alpha} = \frac{0 - 6.67\ \text{rev/s}}{-1.67\ \text{rev/s}^2} = \underline{3.99\ \text{s}}$

9-13

a) $v = r\omega$, but in this equation ω must be in rad/s

$\omega = (700\ \text{rev/min})\left(\frac{2\pi\ \text{rad}}{1\ \text{rev}}\right)\left(\frac{1\ \text{min}}{60\ \text{s}}\right) = 73.3\ \text{rad/s}$

$r = 0.075\ \text{m}$

$v = r\omega = (0.075\ \text{m})(73.3\ \text{rad/s}) = \underline{5.50\ \text{m/s}}$

b) $v = r\omega \Rightarrow \omega = \frac{v}{r} = \frac{0.60\ \text{m/s}}{0.045\ \text{m}} = 13.33\ \text{rad/s}$

$\omega = 13.33\ \text{rad/s}\left(\frac{1\ \text{rev}}{2\pi\ \text{rad}}\right)\left(\frac{60\ \text{s}}{1\ \text{min}}\right) = \underline{127\ \text{rev/min}}$

9-15

The tangential speed of a blade tip is $v = r\omega$.

$\omega = (400\ \text{rev/min})\left(\frac{2\pi\ \text{rad}}{1\ \text{rev}}\right)\left(\frac{1\ \text{min}}{60\ \text{s}}\right) = 41.9\ \text{rad/s}$

$v = r\omega = (5.0\ \text{m})(41.9\ \text{rad/s}) = 209.5\ \text{m/s}$

The upward velocity of the entire blade has magnitude 20.0 m/s.

The tangential velocity and the upward velocity of a blade tip are perpendicular, so their resultant has magnitude

$v_{res} = \sqrt{(209.5\ \text{m/s})^2 + (20.0\ \text{m/s})^2} = \underline{210\ \text{m/s}}$

9-19

a) at the start

$t = 0$

flywheel starts from rest $\Rightarrow \omega = \omega_0 = 0$

$a_{tan} = r\alpha = (0.300\ \text{m})(0.600\ \text{rad/s}^2) = \underline{0.180\ \text{m/s}^2}$

9-19 (cont)

$$a_{rad} = r\omega^2 = \underline{0}$$

$$a = \sqrt{a_{rad}^2 + a_{tan}^2} = a_{tan} = \underline{0.180\ m/s^2}$$

b) $\underline{\theta - \theta_0 = 120°}$

$$a_{tan} = r\alpha = \underline{0.180\ m/s^2}$$

Calculate ω:

$\theta - \theta_0 = 120°\left(\frac{\pi\ rad}{180°}\right) = 2.094\ rad$

$\omega_0 = 0$

$\alpha = 0.600\ rad/s^2$

$\omega = ?$

$\omega^2 = \cancel{\omega_0^2}^{\,0} + 2\alpha(\theta - \theta_0)$

$\omega = \sqrt{2\alpha(\theta-\theta_0)} = \sqrt{2(0.600\ rad/s^2)(2.094\ rad)}$

$\omega = 1.585\ rad/s$

Then $a_{rad} = r\omega^2 = (0.300\ m)(1.585\ rad/s)^2 = \underline{0.754\ m/s^2}$

$$a = \sqrt{a_{rad}^2 + a_{tan}^2} = \sqrt{(0.754\ m/s^2)^2 + (0.180\ m/s^2)^2} = \underline{0.775\ m/s^2}$$

c) $\underline{\theta - \theta_0 = 240°}$

$$a_{tan} = r\alpha = \underline{0.180\ m/s^2}$$

Calculate ω:

$\theta - \theta_0 = 240°\left(\frac{\pi\ rad}{180°}\right) = 4.189\ rad$

$\omega_0 = 0$

$\alpha = 0.600\ rad/s^2$

$\omega = ?$

$\omega^2 = \cancel{\omega_0^2}^{\,0} + 2\alpha(\theta - \theta_0)$

$\omega = \sqrt{2\alpha(\theta-\theta_0)} = \sqrt{2(0.600\ rad/s^2)(4.189\ rad)}$

$\omega = 2.242\ rad/s$

$$a_{rad} = r\omega^2 = (0.300\ m)(2.242\ rad/s)^2 = \underline{1.51\ m/s^2}$$

$$a = \sqrt{a_{rad}^2 + a_{tan}^2} = \sqrt{(1.51\ m/s^2)^2 + (0.180\ m/s^2)^2} = \underline{1.52\ m/s^2}$$

9-21

a)

0.400 m

0.200 kg

0.400 m

r r r r

0.200 m

0.200 m

$$r = \sqrt{(0.200\ m)^2 + (0.200\ m)^2} = 0.2828\ m$$

$$I = \sum_i m_i r_i^2 = 4(0.200\ kg)(0.2828\ m)^2$$

$$I = \underline{0.0640\ kg \cdot m^2}$$

9-21 (cont)

b)

0.200 m, 0.200 m, 0.200 kg, axis

$r = 0.200\text{ m}$

$I = \sum_i m_i r_i^2 = 4(0.200\text{ kg})(0.200\text{ m})^2$

$I = \underline{0.0320\text{ kg}\cdot\text{m}^2}$

9-23

$I = \sum_i m_i r_i^2 \Rightarrow I = I_{rim} + I_{spokes}$

$I_{rim} = MR^2 = (1.20\text{ kg})(0.300\text{ m})^2 = 0.108\text{ kg}\cdot\text{m}^2$

Each spoke can be treated as a slender rod with the axis through one end $\Rightarrow I_{spokes} = 8\left(\frac{1}{3}ML^2\right) = \frac{8}{3}(0.375\text{ kg})(0.300\text{ m})^2 = 0.090\text{ kg}\cdot\text{m}^2$

$I = I_{rim} + I_{spokes} = 0.108\text{ kg}\cdot\text{m}^2 + 0.090\text{ kg}\cdot\text{m}^2 = \underline{0.198\text{ kg}\cdot\text{m}^2}$

9-27

$K = \frac{1}{2}I\omega^2$

$a_{rad} = R\omega^2 \Rightarrow \omega = \sqrt{\frac{a_{rad}}{R}} = \sqrt{\frac{5000\text{ m/s}^2}{1.20\text{ m}}} = 64.55\text{ rad/s}$

Disk $\Rightarrow I = \frac{1}{2}MR^2 = \frac{1}{2}(80.0\text{ kg})(1.20\text{ m})^2 = 57.6\text{ kg}\cdot\text{m}^2$

Thus $K = \frac{1}{2}I\omega^2 = \frac{1}{2}(57.6\text{ kg}\cdot\text{m}^2)(64.55\text{ rad/s})^2 = \underline{1.20\times10^5\text{ J}}$

9-29

Eq. (9-18) says $U = Mg\,y_{cm}$

The positive work done by the wrestler must equal in magnitude the negative work done by gravity

$\Rightarrow W = -W_{grav} = U_2 - U_1 = Mg(\Delta y_{cm})$

$W = Mg(\Delta y_{cm}) = (120\text{ kg})(9.80\text{ m/s}^2)(0.400\text{ m}) = \underline{470\text{ J}}$

9-33

a)

Y, L, dx, X

Let the rod lie along the x-axis, with one end at the origin. Let A be the cross-section area of the rod, and slice the rod up into narrow sections of width dx. Let dm be the mass of each slice.

Then $dm = \gamma x\,dx$ and $M = \int dm$.

$M = \int dm = \gamma\int_0^L x\,dx = \gamma\left(\frac{1}{2}x^2\Big|_0^L\right) = \frac{1}{2}\gamma L^2$

9-33 (cont)

b) $I = \int r^2\,dm = \int x^2\,dm = \gamma \int_0^L x^3\,dx = \gamma\left(\frac{1}{4}x^4\Big|_0^L\right) = \frac{1}{4}\gamma L^4$

But $M = \frac{1}{2}\gamma L^2 \Rightarrow \gamma = \frac{2M}{L^2}$, so $I = \frac{1}{4}\left(\frac{2M}{L^2}\right)L^4 = \frac{1}{2}ML^2$.

For a uniform rod with the axis about one end $I = \frac{1}{3}ML^2$. The present result is larger than this. The mass per unit length increases along the rod, so more of the mass is farther from the other end than for a uniform rod.

9-35

From part (c) of Fig. 9-2, $I_{cm} = \frac{1}{12}M(a^2+a^2) = \frac{1}{6}Ma^2$.

$$I_p = I_{cm} + Md^2$$

The distance d of P from the cm is $d = \sqrt{\left(\frac{a}{2}\right)^2 + \left(\frac{a}{2}\right)^2} = \frac{a}{\sqrt{2}}$.

$$I_p = I_{cm} + Md^2 = \frac{1}{6}Ma^2 + M\left(\frac{a}{\sqrt{2}}\right)^2 = Ma^2\left(\frac{1}{6}+\frac{1}{2}\right) = 2Ma^2/3$$

Problems

9-39

$\theta(t) = \gamma t^2 - \beta t^3$; $\gamma = 2.50\ \text{rad/s}^2$, $\beta = 0.400\ \text{rad/s}^3$

a) $\omega(t) = \frac{d\theta}{dt} = \frac{d}{dt}(\gamma t^2 - \beta t^3) = 2\gamma t - 3\beta t^2$

b) $\alpha(t) = \frac{d\omega}{dt} = \frac{d}{dt}(2\gamma t - 3\beta t^2) = 2\gamma - 6\beta t$

c) The maximum angular velocity occurs when $\alpha = 0$.

$$2\gamma - 6\beta t = 0 \Rightarrow t = \frac{2\gamma}{6\beta} = \frac{\gamma}{3\beta} = \frac{2.50\ \text{rad/s}^2}{3(0.400\ \text{rad/s}^3)} = 2.083\ \text{s}$$

At this t, $\omega = 2\gamma t - 3\beta t^2 = 2(2.50\ \text{rad/s}^2)(2.083\ \text{s}) - 3(0.400\ \text{rad/s}^3)(2.083\ \text{s})^2 = 5.21\ \text{rad/s}$

The maximum positive angular velocity is 5.21 rad/s and it occurs at 2.08 s.

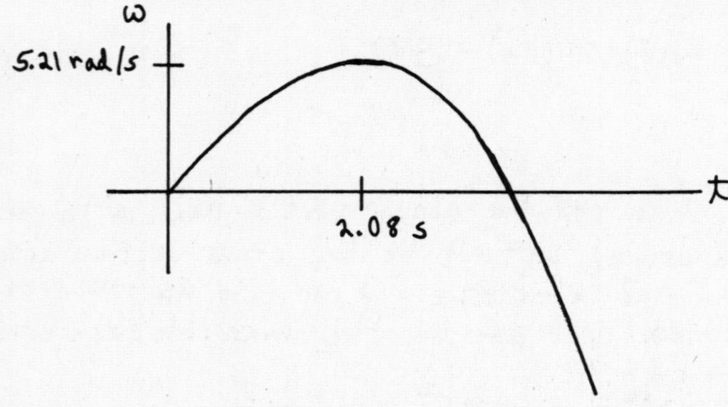

9-43

a) $\frac{v_{toy}}{v_{scale}} = \frac{L_{toy}}{L_{real}} \Rightarrow v_{toy} = v_{scale}\left(\frac{L_{toy}}{L_{real}}\right) = (800\ km/h)\left(\frac{0.100\ m}{3.0\ m}\right) = 26.67\ km/h$

$v_{toy} = (26.67\ km/h)\left(\frac{1000\ m}{1\ km}\right)\left(\frac{1\ h}{3600\ s}\right) = \underline{7.41\ m/s}$

b) $K = \frac{1}{2}mv^2 = \frac{1}{2}(0.120\ kg)(7.41\ m/s)^2 = \underline{3.29\ J}$

c) $K = \frac{1}{2}I\omega^2 \Rightarrow \omega = \sqrt{\frac{2K}{I}} = \sqrt{\frac{2(3.29\ J)}{400\times10^{-5}\ kg\cdot m^2}} = \underline{406\ rad/s}$

9-45

a) $W_{tot} = K_2 - K_1 \Rightarrow K_2 = K_1 + W_{tot}$

$W_{tot} = -4000\ J$ (the amount of energy given up by the fly wheel)

$K_1 = \frac{1}{2}I\omega_1^2$, but ω_1 must be in rad/s.

$\omega_1 = (300\ rev/min)\left(\frac{2\pi\ rad}{1\ rev}\right)\left(\frac{1\ min}{60\ s}\right) = 31.42\ rad/s$

$K_1 = \frac{1}{2}(16.0\ kg\cdot m^2)(31.42\ rad/s)^2 = 7898\ J$

$K_2 = K_1 + W_{tot} = 7898\ J - 4000\ J = 3898\ J$

$K_2 = \frac{1}{2}I\omega_2^2 \Rightarrow \omega_2 = \sqrt{\frac{2K_2}{I}} = \sqrt{\frac{2(3898\ J)}{16.0\ kg\cdot m^2}} = 22.07\ rad/s$

$\omega_2 = 22.07\ rad/s\left(\frac{1\ rev}{2\pi\ rad}\right)\left(\frac{60\ s}{1\ min}\right) = \underline{211\ rev/min}$

b) The 4000 J of energy must be restored to the flywheel, so

$P_{av} = \frac{\Delta W}{\Delta t} = \frac{4000\ J}{5.00\ s} = \underline{800\ W.}$

9-47

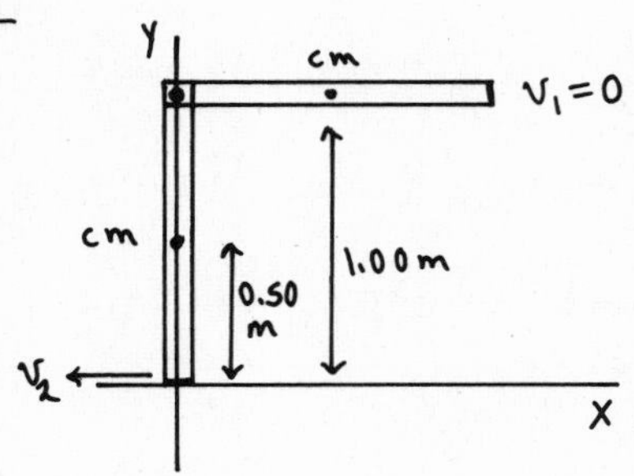

Take the origin of coordinates at the lowest point reached by the stick and take the positive y-direction to be upward.

a) Use Eq. (9-18): $U = Mgy_{cm}$

$\Delta U = U_2 - U_1 = Mg(y_{cm2} - y_{cm1})$

The center of mass of the meter stick is at its geometrical center, so $y_{cm1} = 1.00\ m$ and $y_{cm2} = 0.50\ m$.

Then $\Delta U = (0.300\ kg)(9.80\ m/s^2)(0.50\ m - 1.00\ m) = \underline{-1.47\ J}$

9-47 (cont)

b) Use conservation of energy: $K_1 + U_1 + W_{other} = K_2 + U_2$

Gravity is the only force that does work on the meter stick, so $W_{other} = 0$.
$K_1 = 0$

Thus $K_2 = U_1 - U_2 = -\Delta U$, where ΔU was calculated in part (a).
$K_2 = \frac{1}{2} I \omega_2^2 \Rightarrow \frac{1}{2} I \omega_2^2 = -\Delta U$ and $\omega_2 = \sqrt{\frac{2(-\Delta U)}{I}}$

Stick pivoted about one end $\Rightarrow I = \frac{1}{3} ML^2$ where $L = 1.00\text{ m}$, so

$$\omega_2 = \sqrt{\frac{6(-\Delta U)}{ML^2}} = \sqrt{\frac{6(1.47\text{ J})}{(0.300\text{ kg})(1.00\text{ m})^2}} = \underline{5.42\text{ rad/s}}$$

c) $v = r\omega = (1.00\text{ m})(5.42\text{ rad/s}) = \underline{5.42\text{ m/s}}$

d) $v_{0y} = 0$
$y - y_0 = -1.00\text{ m}$
$a_y = -9.80\text{ m/s}^2$
$v = ?$

$$v^2 = v_{0y}^2 + 2a_y(y - y_0)$$

$$v = -\sqrt{2a_y(y - y_0)} = -\sqrt{2(-9.80\text{ m/s}^2)(-1.00\text{ m})} = \underline{-4.43\text{ m/s}}$$

The answer in part (c) is larger.

9-49

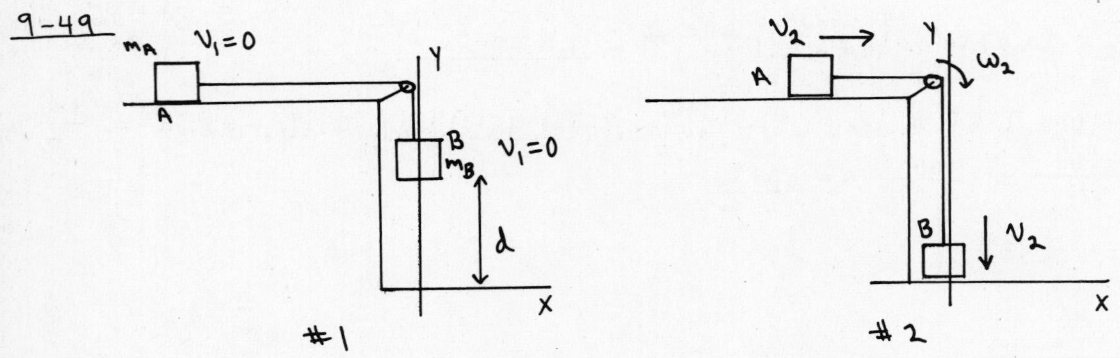

Use the work-energy relation $K_1 + U_1 + W_{other} = K_2 + U_2$.

Use coordinates where $+y$ is upward and where the origin is at the position of block B after it has descended.

The tension in the rope does positive work on block A and negative work of the same magnitude on block B, so the net work done by the tension in the rope is zero.

Gravity does work on block B and kinetic friction does work on block A.
Thus $W_{other} = W_f = -\mu_k m_A g d$.

9-49 (cont)

$K_1 = 0$ (system is released from rest)

$U_1 = m_B g y_{B1} = m_B g d$

$U_2 = m_B g y_{B2} = 0$

$K_2 = \frac{1}{2} m_A v_2^2 + \frac{1}{2} m_B v_2^2 + \frac{1}{2} I \omega_2^2$. But $v(\text{blocks}) = R\omega\,(\text{pulley})$, so $\omega_2 = \frac{v_2}{R}$

and $K_2 = \frac{1}{2}(m_A + m_B) v_2^2 + \frac{1}{2} I \left(\frac{v_2}{R}\right)^2 = \frac{1}{2}\left(m_A + m_B + \frac{I}{R^2}\right) v_2^2$

Putting all this into the work-energy relation gives

$$m_B g d - \mu_K m_A g d = \frac{1}{2}\left(m_A + m_B + \frac{I}{R^2}\right) v_2^2$$

$$\frac{1}{2}\left(m_A + m_B + \frac{I}{R^2}\right) v_2^2 = g d (m_B - \mu_K m_A)$$

$$v_2 = \sqrt{\frac{2gd(m_B - \mu_K m_A)}{m_A + m_B + I/R^2}}$$

9-53

Let L be the length of the cylinder. Divide the cylinder into thin cylindrical shells of inner radius r and outer radius r+dr. An end view is

dr
r
R

$\rho = \alpha r$

The mass of the thin cylindrical shell is

$dm = \rho\, dV = \rho(2\pi r\, dr) L = 2\pi\alpha L r^2 dr.$

$$I = \int r^2 dm = 2\pi\alpha L \int_0^R r^4 dr = 2\pi\alpha L\left(\tfrac{1}{5}R^5\right) = \tfrac{2}{5}\pi\alpha L R^5$$

Relate M to α:

$$M = \int dm = 2\pi\alpha L \int_0^R r^2 dr = 2\pi\alpha L\left(\tfrac{1}{3}R^3\right) = \tfrac{2}{3}\pi\alpha L R^3, \text{ so } \pi\alpha L R^3 = \frac{3M}{2}.$$

Use this in the above result for I

$$\Rightarrow I = \tfrac{2}{5}\left(\tfrac{3M}{2}\right)R^2 = \underline{\tfrac{3}{5} M R^2}$$

CHAPTER 10

Exercises 1, 3, 5, 9, 11, 13, 17, 19, 21, 23, 25, 27, 31

Problems 35, 39, 41, 43, 47, 49

Exercises

10-1

a)

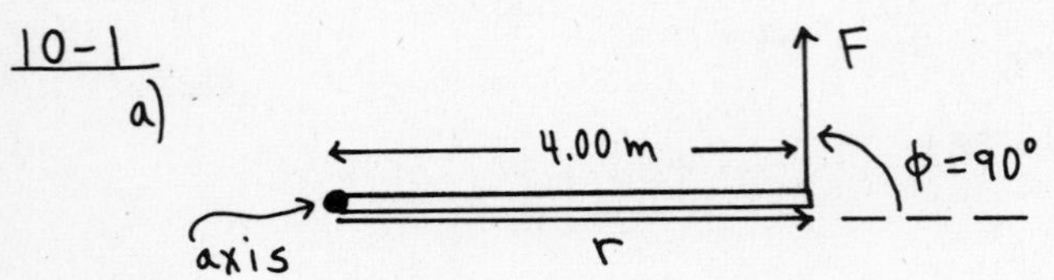

$\tau = Fl$

$l = r\sin\phi = (4.00\ \text{m})\sin 90° = 4.00\ \text{m}$

$\tau = (20.0\ \text{N})(4.00\ \text{m}) = \underline{80.0\ \text{N}\cdot\text{m}}$

The force tends to produce a counterclockwise (↺) rotation about the axis; by the right-hand rule the vector $\vec{\tau}$ is directed out of the plane of the figure.

b)

$\tau = Fl$

$l = r\sin\phi = (4.00\ \text{m})\sin 120° = 3.464\ \text{m}$

$\tau = (20.0\ \text{N})(3.464\ \text{m}) = \underline{69.3\ \text{N}\cdot\text{m}}$

The force tends to produce a counterclockwise (↺) rotation about the axis; by the right-hand rule the vector $\vec{\tau}$ is directed out of the plane of the figure.

c)

$\tau = Fl$

$l = r\sin\phi = (4.00\ \text{m})\sin 30° = 2.00\ \text{m}$

$\tau = (20.0\ \text{N})(2.00\ \text{m}) = \underline{40.0\ \text{N}\cdot\text{m}}$

The force tends to produce a counterclockwise (↺) rotation about the axis; by the right-hand rule the vector $\vec{\tau}$ is directed out of the plane of the figure.

d)

$\tau = Fl$

$l = r\sin\phi = (2.00\ \text{m})\sin 60° = 1.732\ \text{m}$

$\tau = (20.0\ \text{N})(1.732\ \text{m}) = \underline{34.6\ \text{N}\cdot\text{m}}$

The force tends to produce a clockwise (↻) rotation about the axis; by the right-hand rule the vector $\vec{\tau}$ is directed into the plane of the figure.

e)

$\tau = Fl$

$r = 0$ so $l = 0 \Rightarrow \underline{\tau = 0}$

f)

$\tau = Fl$

$l = r\sin\phi;\ \phi = 180° \Rightarrow l = 0 \Rightarrow \underline{\tau = 0}$

10-3

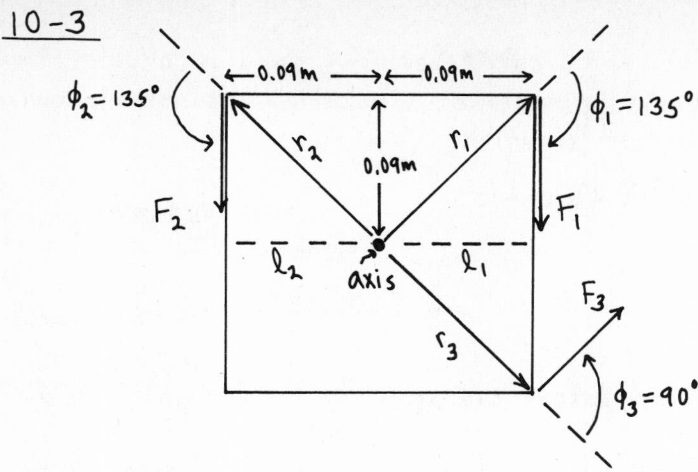

Let (+↺) be the positive sense of rotation.

$$r_1 = r_2 = r_3 = \sqrt{(0.090\ \text{m})^2 + (0.090\ \text{m})^2} = 0.1273\ \text{m}$$

$$\tau_1 = -F_1 l_1$$
$$l_1 = r_1 \sin\phi_1 = (0.1273\ \text{m}) \sin 135°$$
$$l_1 = 0.0900\ \text{m}$$
$$\tau_1 = -(24.0\ \text{N})(0.0900\ \text{m}) = -2.16\ \text{N}\cdot\text{m}$$
($\vec{\tau}_1$ is directed into paper)

$$\tau_2 = +F_2 l_2$$
$$l_2 = r_2 \sin\phi_2 = (0.1273\ \text{m}) \sin 135° = 0.0900\ \text{m}$$
$$\tau_2 = +(16.0\ \text{N})(0.0900\ \text{m}) = +1.44\ \text{N}\cdot\text{m}$$
($\vec{\tau}_2$ is directed out of paper)

$$\tau_3 = +F_3 l_3$$
$$l_3 = r_3 \sin\phi_3 = (0.1273\ \text{m}) \sin 90° = 0.1273\ \text{m}$$
$$\tau_3 = +(18.0\ \text{N})(0.1273\ \text{m}) = +2.29\ \text{N}\cdot\text{m}$$ ($\vec{\tau}_3$ is directed out of paper)

$$\Sigma\tau = \tau_1 + \tau_2 + \tau_3 = -2.16\ \text{N}\cdot\text{m} + 1.44\ \text{N}\cdot\text{m} + 2.29\ \text{N}\cdot\text{m} = \underline{+1.57\ \text{N}\cdot\text{m}}$$

$+ \Rightarrow$ the net torque tends to produce a counterclockwise (↺) rotation; the net vector torque is directed out of the plane of the paper.

10-5

$$\vec{r} = (-0.300\ \text{m})\vec{i} + (0.500\ \text{m})\vec{j}$$
$$\vec{F} = (4.00\ \text{N})\vec{i} - (5.00\ \text{N})\vec{j}$$

$$\vec{\tau} = \vec{r} \times \vec{F} = [(-0.300\ \text{m})\vec{i} + (0.500\ \text{m})\vec{j}] \times [(4.00\ \text{N})\vec{i} - (5.00\ \text{N})\vec{j}]$$
$$\vec{\tau} = -(1.20\ \text{N}\cdot\text{m})(\vec{i}\times\vec{i}) + (1.50\ \text{N}\cdot\text{m})\,\vec{i}\times\vec{j} + (2.00\ \text{N}\cdot\text{m})\,\vec{j}\times\vec{i} - (2.50\ \text{N}\cdot\text{m})\,\vec{j}\times\vec{j}$$
$$\vec{i}\times\vec{i} = \vec{j}\times\vec{j} = 0$$
$$\vec{i}\times\vec{j} = \vec{k}, \quad \vec{j}\times\vec{i} = -\vec{k}$$
Thus $\vec{\tau} = (1.50\ \text{N}\cdot\text{m})\vec{k} + (2.00\ \text{N}\cdot\text{m})(-\vec{k}) = \underline{(-0.50\ \text{N}\cdot\text{m})\vec{k}}$

Note: The calculation gives that $\vec{\tau}$ is in the -z-direction. This agrees with what one gets from the right-hand rule.

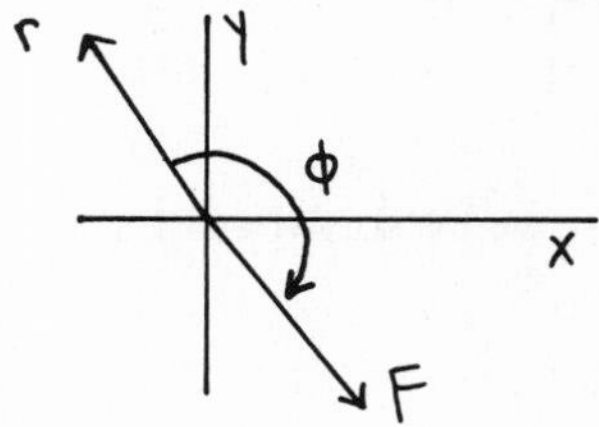

When the fingers of your right hand curl from the direction of $\vec{r}$ into the direction of $\vec{F}$ (through the smaller of the two angles, angle ϕ) your thumb points into the paper (the direction of $\vec{\tau}$, the -z-direction).

10-9

Use the kinematic information to solve for the angular acceleration of the grindstone. Assume that the grindstone is rotating counterclockwise and let that be the positive sense of rotation, (+↺).

$\omega_0 = 900 \text{ rev/min} \left(\frac{2\pi \text{ rad}}{1 \text{ rev}}\right)\left(\frac{1 \text{ min}}{60 \text{ s}}\right) = 94.25 \text{ rad/s}$

$t = 10.0 \text{ s}$

$\omega = 0$ (comes to rest)

$\alpha = ?$

$$\omega = \omega_0 + \alpha t$$

$$\alpha = \frac{\omega - \omega_0}{t}$$

$$\alpha = \frac{0 - 94.25 \text{ rad/s}}{10.0 \text{ s}} = -9.42 \text{ rad/s}^2$$

Now consider the torques on the grindstone and apply $\Sigma \tau = I\alpha$, with (+↺)

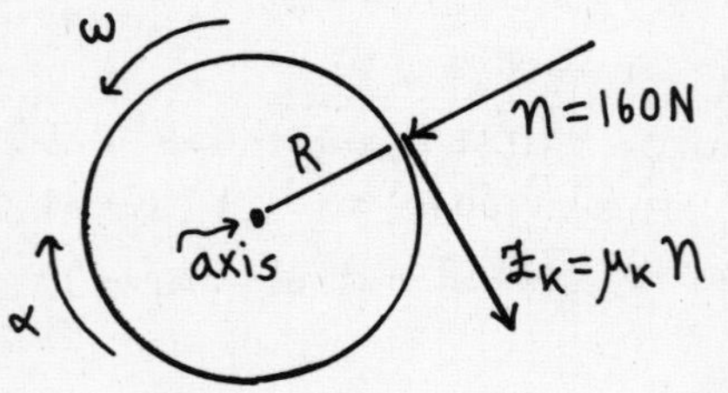

The normal force has zero moment arm for rotation about an axis at the center of the grindstone, and therefore zero torque. The only torque on the grindstone is that due to the friction force f_K exerted by the ax; for this force the moment arm is $l = R$ and the torque is negative.

$$\Sigma \tau = -f_K R = -\mu_K n R$$

$$I = \tfrac{1}{2} M R^2 \quad \text{(solid disk, axis through center)}$$

$$\Sigma \tau = I\alpha \Rightarrow -\mu_K n R = \left(\tfrac{1}{2} M R^2\right)\alpha$$

$$\mu_K = -\frac{MR\alpha}{2n} = -\frac{(50.0 \text{ kg})(0.300 \text{ m})(-9.42 \text{ rad/s}^2)}{2(160 \text{ N})} = \underline{+0.442}$$

10-11

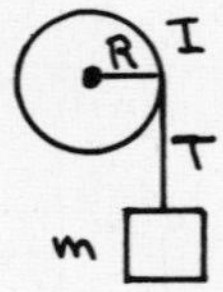

Apply $\Sigma \tau = I\alpha$ to the rotating cylinder (take the direction of α to be the positive sense of rotation):

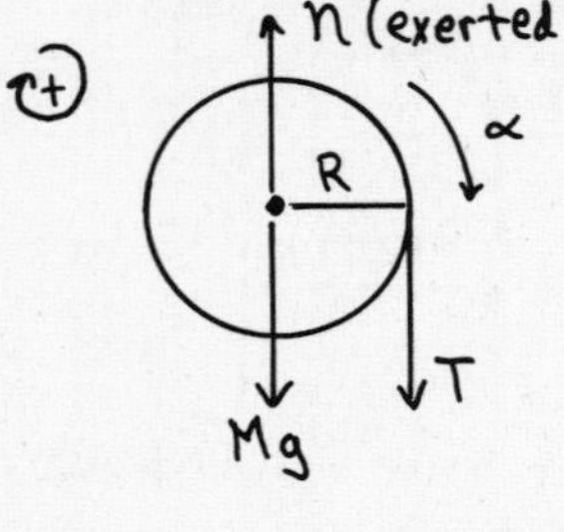

$$\Sigma \tau = TR$$

cylinder, axis through center $\Rightarrow I = \tfrac{1}{2} M R^2$

$$\Sigma \tau = I\alpha \Rightarrow TR = \left(\tfrac{1}{2} M R^2\right)\alpha$$

$$\boxed{T = \tfrac{1}{2} M R \alpha}$$

Apply $\Sigma \vec{F} = m\vec{a}$ to the descending bucket (take +y to be downward since that is the direction of $\vec{a}$)

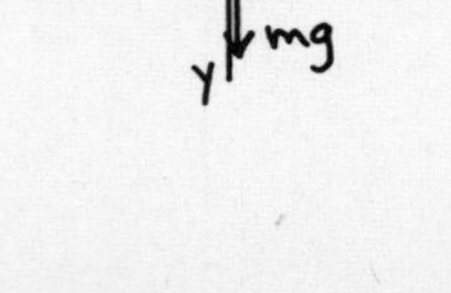

$$\Sigma F_y = m a_y$$

$$\boxed{mg - T = ma}$$

10-11 (cont)

The angular acceleration α of the cylinder is related to the linear acceleration a of the bucket by $a = R\alpha \Rightarrow \alpha = \frac{a}{R}$

Use this in the first eq. $\Rightarrow T = \frac{1}{2}MR\left(\frac{a}{R}\right)$

$$T = \tfrac{1}{2}Ma$$

Use this in the second eq. $\Rightarrow mg - \frac{1}{2}Ma = ma$

$$a = \frac{mg}{m + M/2} = \frac{(20.0\,\text{kg})(9.80\,\text{m/s}^2)}{20.0\,\text{kg} + 20.0\,\text{kg}/2} = 6.533\,\text{m/s}^2$$

Then $T = \frac{1}{2}Ma = \frac{1}{2}(20.0\,\text{kg})(6.533\,\text{m/s}^2) = \underline{65.3\,\text{N}}$

b) Use the constant acceleration equations, with positive y downward:

$v_{oy} = 0$

$a_y = 6.53\,\text{m/s}^2$

$y - y_o = +12.0\,\text{m}$

$v_y = ?$

$$v_y^2 = v_{oy}^2 + 2a_y(y - y_o) \quad (v_{oy}^2 \to 0)$$

$$v_y = \sqrt{2a_y(y-y_o)} = \sqrt{2(6.53\,\text{m/s}^2)(12.0\,\text{m})} = 12.5\,\text{m/s}$$

c) $v_y = 12.5\,\text{m/s}$

$v_{oy} = 0$

$a_y = 6.53\,\text{m/s}^2$

$t = ?$

$$v_y = v_{oy} + a_y t$$

$$t = \frac{v_y - v_{oy}}{a_y} = \frac{12.5\,\text{m/s} - 0}{6.53\,\text{m/s}^2} = \underline{1.91\,\text{s}}$$

d) Apply $\Sigma F_y = ma_y$ to the free-body diagram for the cylinder, with the origin of coordinates at the axle and positive y upward. The linear (translational) acceleration a of the cylinder is zero.

$a = 0$

$$\Sigma F_y = ma_y$$

$$n - Mg - T = 0$$

$$n = Mg + T$$

$$n = (20.0\,\text{kg})(9.80\,\text{m/s}^2) + 65.3\,\text{N} = \underline{261\,\text{N}}$$

10-13

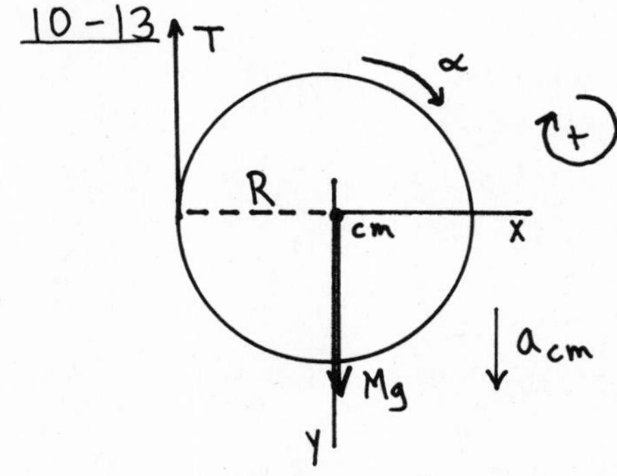

For translational motion of the center of mass of the hoop take the origin of coordinates at the cm and take +y to be downward. Then $\Sigma F_y = ma_y$ gives

$$\boxed{Mg - T = Ma_{cm}}$$

Apply $\Sigma\tau = I\alpha$ for rotation about the center of mass with the clockwise sense of rotation taken to be positive. The weight Mg has zero moment arm and therefore zero torque.

Thus $TR = I_{cm}\alpha$

For a hoop $I_{cm} = MR^2 \Rightarrow TR = MR^2\alpha$

$$T = MR\alpha$$

10-13 (cont)

$$a_{cm} = R\alpha \Rightarrow \boxed{T = Ma_{cm}}$$

Combine these two equations to eliminate T $\Rightarrow \cancel{M}g - \cancel{M}a_{cm} = \cancel{M}a_{cm}$

$$2a_{cm} = g \Rightarrow a_{cm} = g/2$$

Then $T = M(g/2) = \frac{1}{2}Mg = \frac{1}{2}(0.140\text{ kg})(9.80\text{ m/s}^2) = \underline{0.686\text{ N}}$

b) Apply the constant acceleration kinematic equations to the motion of the center of mass:

$v_{0y} = 0$

$y - y_0 = 0.600\text{ m}$

$a_y = \frac{1}{2}g = 4.90\text{ m/s}^2$

$t = ?$

$$y - y_0 = \cancel{v_{0y}t}^{\,0} + \frac{1}{2}a_y t^2$$

$$t = \sqrt{\frac{2(y-y_0)}{a_y}} = \sqrt{\frac{2(0.600\text{ m})}{4.90\text{ m/s}^2}} = \underline{0.495\text{ s}}$$

c) We can use the constant angular accelerations for rotational motion:

$t = 0.495\text{ s}$

$\omega_0 = 0$

$\alpha = \frac{a_{cm}}{R} = \frac{\frac{1}{2}(9.80\text{ m/s}^2)}{0.0800\text{ m}} = 61.25\text{ rad/s}^2$

$\omega = ?$

$$\omega = \omega_0 + \alpha t$$

$$\omega = 0 + (61.25\text{ rad/s}^2)(0.495\text{ s})$$

$$\omega = \underline{30.3\text{ rad/s}}$$

Alternatively, we can find v_{cm} at this point and then use $\omega = \frac{v_{cm}}{R}$:

$v_{0y} = 0$

$v_y = ?$

$a_y = 4.90\text{ m/s}^2$

$y - y_0 = 0.600\text{ m}$

$$v_y^2 = \cancel{v_{0y}^2} + 2a_y(y-y_0)$$

$$v_y = \sqrt{2a_y(y-y_0)} = \sqrt{2(4.90\text{ m/s}^2)(0.600\text{ m})} = 2.42\text{ m/s}$$

Then $\omega = \frac{v_{cm}}{R} = \frac{2.42\text{ m/s}}{0.0800\text{ m}} = 30.3\text{ rad/s}$, the same as before.

<u>10-17</u>

a) α, F, R, axis, (+)

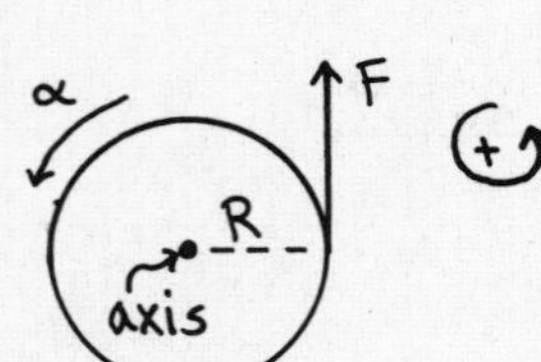

Apply $\Sigma\tau = I\alpha$ to find the angular acceleration α:

$$FR = I\alpha$$

$$\alpha = \frac{FR}{I} = \frac{(25.0\text{ N})(4.40\text{ m})}{160\text{ kg}\cdot\text{m}^2} = 0.6875\text{ rad/s}^2$$

Use the constant α kinematic equations to find ω:

$\omega = ?$

$\omega_0 = 0$ (initially at rest)

$\alpha = 0.6875\text{ rad/s}^2$

$t = 20.0\text{ s}$

$$\omega = \omega_0 + \alpha t$$

$$\omega = 0 + (0.6875\text{ rad/s}^2)(20.0\text{ s}) = 13.75\text{ rad/s}$$

So $\omega = \underline{13.8\text{ rad/s}}$.

b) This question can be answered either of two ways:

(1) $W = \tau\Delta\theta$ (Eq. (10-23))

$\Delta\theta = \theta - \theta_0 = \omega_0 t + \frac{1}{2}\alpha t^2 = 0 + \frac{1}{2}(0.6875\text{ rad/s}^2)(20.0\text{ s})^2 = 137.5\text{ rad}$

$\tau = FR = (25.0\text{ N})(4.40\text{ m}) = 110\text{ N}\cdot\text{m}$

Then $W = \tau\Delta\theta = (110\text{ N}\cdot\text{m})(137.5\text{ rad}) = \underline{1.51\times 10^4\text{ J}}$.

<u>or</u>

10-17 (cont)

(2) $W_{tot} = K_2 - K_1$ (the work-energy relation from Chapter 6)

$W_{tot} = W$, the work done by the child

$K_1 = 0$

$K_2 = \frac{1}{2} I \omega^2 = \frac{1}{2}(160 \text{ kg}\cdot\text{m}^2)(13.75 \text{ rad/s})^2 = 1.51 \times 10^4 \text{ J}$

$\Rightarrow$ $W = 1.51 \times 10^4$ J, the same as before

c) $P_{av} = \dfrac{\Delta W}{\Delta t} = \dfrac{1.51 \times 10^4 \text{ J}}{20.0 \text{ s}} = \underline{755 \text{ W}}$

10-19

a) Use Eq. (10-25): $P = \tau \omega$, where ω must be in rad/s

$\omega = (2400 \text{ rev/min})\left(\frac{2\pi \text{ rad}}{1 \text{ rev}}\right)\left(\frac{1 \text{ min}}{60 \text{ s}}\right) = 251.3 \text{ rad/s}$

$\tau = \dfrac{P}{\omega} = \dfrac{1.80 \times 10^6 \text{ W}}{251.3 \text{ rad/s}} = \underline{7.16 \times 10^3 \text{ N}\cdot\text{m}}$

b) [diagram: drum of radius R, rope tension T, weight w moving up with speed v]

v constant $\Rightarrow a = 0 \Rightarrow T = w$

$\tau = TR \Rightarrow T = \dfrac{\tau}{R} = \dfrac{7.16 \times 10^3 \text{ N}\cdot\text{m}}{0.250 \text{ m}} = 2.86 \times 10^4 \text{ N}$

Thus, a weight $w = \underline{2.86 \times 10^4 \text{ N}}$ can be lifted.

c) $v = R\omega$ and the drum has $\omega = 251.3$ rad/s

$\Rightarrow v = (0.250 \text{ m})(251.3 \text{ rad/s}) = \underline{62.8 \text{ m/s}}$

10-21

Use $L = I\omega$.

The second hand makes 1 revolution in 1 min

$\Rightarrow \omega = (1.00 \text{ rev/min})\left(\frac{2\pi \text{ rad}}{1 \text{ rev}}\right)\left(\frac{1 \text{ min}}{60 \text{ s}}\right) = 0.1047 \text{ rad/s}$

Slender rod, axis about one end $\Rightarrow I = \frac{1}{3}ML^2 = \frac{1}{3}(20.0 \times 10^{-3} \text{ kg})(0.250 \text{ m})^2$

$I = 4.167 \times 10^{-4} \text{ kg}\cdot\text{m}^2$

Then $L = I\omega = (4.167 \times 10^{-4} \text{ kg}\cdot\text{m}^2)(0.1047 \text{ rad/s}) = \underline{4.36 \times 10^{-5} \text{ kg}\cdot\text{m}^2/\text{s}}$

10-23

Use $L = mvr\sin\phi$ (Eq. (10-28)):

[diagram: mass m moving with velocity $\vec{v}$; position vector $\vec{r}$ from axis; $\phi = 143°$, 37°; $l = r\sin 37° = r\sin\phi$; axis]

$L = mvr\sin\phi$

$L = (0.600 \text{ kg})(12.0 \text{ m/s})(8.00 \text{ m}) \sin 143°$

$L = \underline{34.7 \text{ kg}\cdot\text{m}^2/\text{s}}$

10-25

a) $L_1 = L_2 \Rightarrow I_1\omega_1 = I_2\omega_2$

Block treated as a point mass $\Rightarrow I = mr^2$ where r is the distance of the block from the hole.

$$\cancel{m}r_1^2\omega_1 = \cancel{m}r_2^2\omega_2$$

$$\omega_2 = \left(\frac{r_1}{r_2}\right)^2\omega_1 = \left(\frac{0.200\text{ m}}{0.100\text{ m}}\right)^2(1.50\text{ rad/s}) = \underline{6.00\text{ rad/s}}$$

b) $K_1 = \frac{1}{2}I_1\omega_1^2 = \frac{1}{2}mr_1^2\omega_1^2 = \frac{1}{2}mv_1^2$

$v_1 = r_1\omega_1 = (0.200\text{ m})(1.50\text{ rad/s}) = 0.300\text{ m/s}$

$K_1 = \frac{1}{2}mv_1^2 = \frac{1}{2}(0.0500\text{ kg})(0.300\text{ m/s})^2 = 2.25\times10^{-3}\text{ J}$

$K_2 = \frac{1}{2}mv_2^2$

$v_2 = r_2\omega_2 = (0.100\text{ m})(6.00\text{ rad/s}) = 0.600\text{ m/s}$

$K_2 = \frac{1}{2}mv_2^2 = \frac{1}{2}(0.0500\text{ kg})(0.600\text{ m/s})^2 = 9.00\times10^{-3}\text{ J}$

$\Delta K = K_2 - K_1 = 9.00\times10^{-3}\text{ J} - 2.25\times10^{-3}\text{ J} = \underline{6.75\times10^{-3}\text{ J}}$

c) $W_{tot} = \Delta K$

But $W_{tot} = W$, the work done by the tension in the cord $\Rightarrow W = \underline{6.75\times10^{-3}\text{ J}}$.

10-27

$L_1 = L_2$, with the axis at the hinge

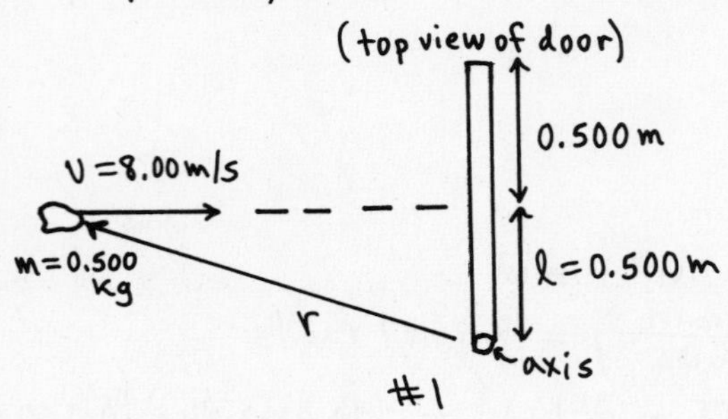

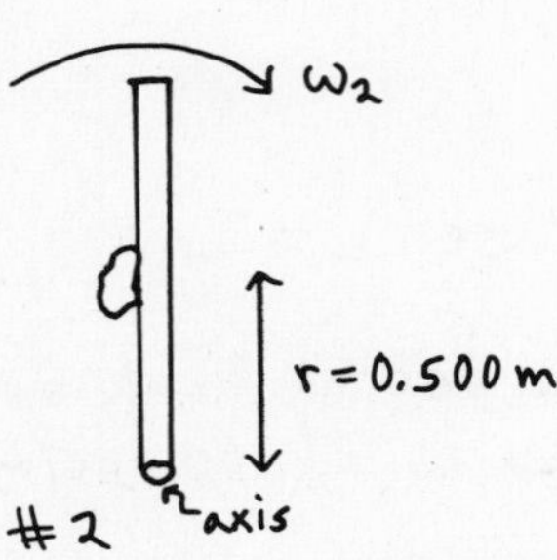

Before impact, $L_1 = L_{mud} + L_{door}$, and $L_{door} = 0$.

$L_{mud} = mvr\sin\phi = mvl$ (Eq. 10-28)

$L_{mud} = (0.500\text{ kg})(8.00\text{ m/s})(0.500\text{ m}) = 2.00\text{ kg}\cdot\text{m}^2/\text{s}$

$\Rightarrow L_1 = 2.00\text{ kg}\cdot\text{m}^2/\text{s}$

After impact the mud sticks to the door and both objects rotate with the same angular velocity ω_2.

$I_2 = I_{door} + I_{mud}$

$I_{door} = \frac{1}{3}ML^2 = \frac{1}{3}(50.0\text{ kg})(1.00\text{ m})^2 = 16.67\text{ kg}\cdot\text{m}^2$

Treat the mud as a point mass, so $I_{mud} = mr^2 = (0.500\text{ kg})(0.500\text{ m})^2 = 0.125\text{ kg}\cdot\text{m}^2$

Thus $I_2 = 16.67\text{ kg}\cdot\text{m}^2 + 0.125\text{ kg}\cdot\text{m}^2 = 16.80\text{ kg}\cdot\text{m}^2$

10-27 (cont)

Then $L_1 = L_2 \Rightarrow I_2 \omega_2 = 2.00\,kg\cdot m^2/s$

and $\omega_2 = \frac{2.00\,kg\cdot m^2/s}{16.80\,kg\cdot m^2} = \underline{0.119\,rad/s}$

<u>10-31</u>

a) By the work-energy relation $W = \Delta K$ the work done on the gyroscope equals its increase in kinetic energy.

Thus $P = \frac{W}{t} = \frac{\Delta K}{t} \Rightarrow t = \frac{\Delta K}{P}$

$\Delta K = K_2 - \cancel{K_1}^{\,0}$, the amount of energy stored in the gyroscope when it is up to speed

$\Delta K = K_2 = \frac{1}{2} I \omega^2$

Solid disk $\Rightarrow I = \frac{1}{2}MR^2 = \frac{1}{2}(50{,}000\,kg)(2.00\,m)^2 = 1.00 \times 10^5\,kg\cdot m^2$

$\omega = 800\,rev/min\left(\frac{2\pi\,rad}{1\,rev}\right)\left(\frac{1\,min}{60\,s}\right) = 83.78\,rad/s$

$\Rightarrow \Delta K = \frac{1}{2}I\omega^2 = \frac{1}{2}(1.00 \times 10^5\,kg\cdot m^2)(83.78\,rad/s)^2 = 3.51 \times 10^8\,J$

$t = \frac{\Delta K}{P} = \frac{3.51 \times 10^8\,J}{7.46 \times 10^4\,W} = 4705\,s = \underline{78.4\,min}$

b) Eq. (10-35) says $\Omega = \frac{\tau}{L} = \frac{\tau}{I\omega}$.

Thus $\tau = I\omega\Omega$

$\Omega = 1.00°/s\left(\frac{\pi\,rad}{180°}\right) = 0.01745\,rad/s$

$\tau = I\omega\Omega = (1.00 \times 10^5\,kg\cdot m^2)(83.78\,rad/s)(0.01745\,rad/s) = \underline{1.46 \times 10^5\,N\cdot m}$

Problems

<u>10-35</u>

Use $\Sigma\tau = I\alpha$ to find α, and then use the constant α kinematic equations to solve for t.

F=180N

l=1.00m

axis at hinge

$\Sigma\tau = Fl = (180\,N)(1.00\,m) = 180\,N\cdot m$

From Table 9-2(d), $I = \frac{1}{3}Ml^2 = \frac{1}{3}\left(\frac{700\,N}{9.80\,m/s^2}\right)(1.00\,m)^2 = 23.81\,kg\cdot m^2$

$\Sigma\tau = I\alpha \Rightarrow \alpha = \frac{\Sigma\tau}{I} = \frac{180\,N\cdot m}{23.81\,kg\cdot m^2} = 7.56\,rad/s^2$

$\alpha = 7.56\,rad/s^2$

$\theta - \theta_0 = 90°\left(\frac{\pi\,rad}{180°}\right) = \frac{\pi}{2}\,rad$

$\omega_0 = 0$ (door initially at rest)

$t = ?$

$\theta - \theta_0 = \cancel{\omega_0 t} + \frac{1}{2}\alpha t^2$

$t = \sqrt{\frac{2(\theta-\theta_0)}{\alpha}} = \sqrt{\frac{2(\frac{\pi}{2}\,rad)}{7.56\,rad/s^2}} = \underline{0.645\,s}$

10-39

a)

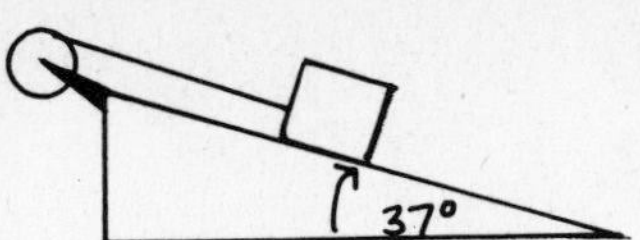

Apply $\Sigma \tau = I\alpha$ to the rotation of the flywheel about the axis:

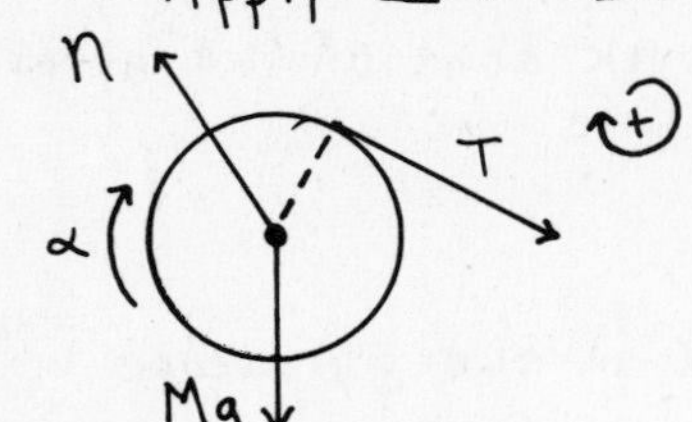

The forces n and Mg act at the axis so have zero torque.

Thus $\Sigma \tau = TR$

then $\Sigma \tau = TR \Rightarrow$ $\boxed{TR = I\alpha}$

Apply $\Sigma \vec{F} = m\vec{a}$ to the translational motion of the block:

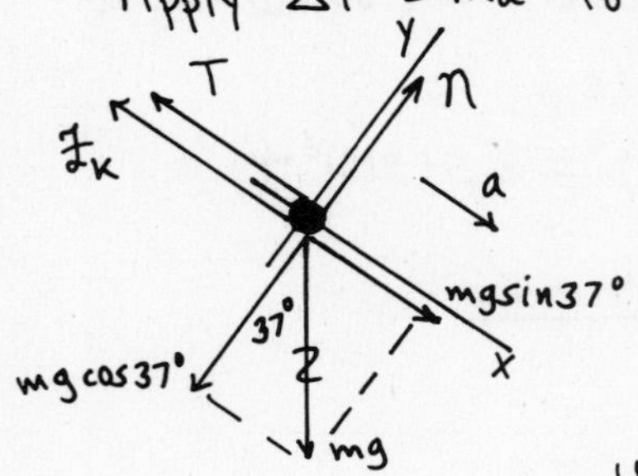

$\Sigma F_y = ma_y$

$n - mg\cos 37° = 0$

$n = mg\cos 37°$

$f_K = \mu_K n = \mu_K mg\cos 37°$

$\Sigma F_x = ma_x$

$mg\sin 37° - T - f_K = ma$

$mg\sin 37° - T - \mu_K mg\cos 37° = ma$

$\boxed{mg(\sin 37° - \mu_K \cos 37°) - T = ma}$

But we also know that $a_{block} = R\alpha_{wheel} \Rightarrow \alpha = \frac{a}{R}$

Use this in the first eq. $\Rightarrow TR = I\frac{a}{R}$

$T = \left(\frac{I}{R^2}\right)a$

Use this to replace T in the second eq.

$\Rightarrow mg(\sin 37° - \mu_K \cos 37°) - \left(\frac{I}{R^2}\right)a = ma$

$$a = \frac{mg(\sin 37° - \mu_K \cos 37°)}{m + I/R^2} = \frac{(5.00\,kg)(9.80\,m/s^2)(\sin 37° - (0.20)\cos 37°)}{5.00\,kg + 0.300\,kg\cdot m^2/(0.200\,m)^2} = \underline{1.73\,m/s^2}$$

b) $T = \frac{0.300\,kg\cdot m^2}{(0.200\,m)^2}(1.73\,m/s^2) = \underline{13.0\,N}$

10-41

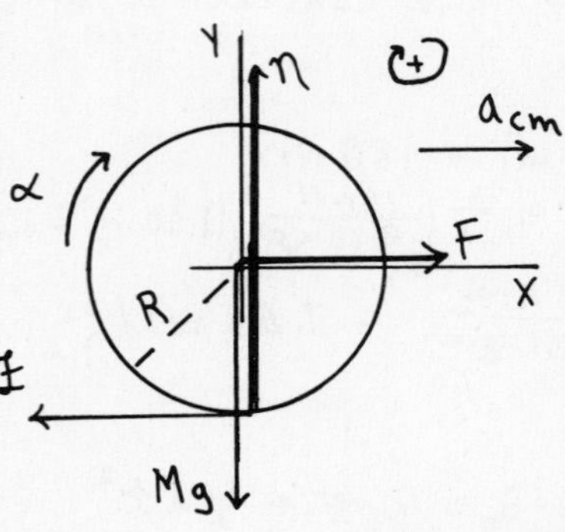

Apply $\Sigma \vec{F} = m\vec{a}$ to the translational motion of the center of mass:

$\Sigma F_x = ma_x$

$\boxed{F - f = Ma_{cm}}$

Apply $\Sigma \tau = I\alpha$ to rotation about the center of mass:

$\Sigma \tau = fR$

thin-walled hollow cylinder $\Rightarrow I = MR^2$

Then $\Sigma \tau = I\alpha \Rightarrow fR = MR^2\alpha$

But $a_{cm} = R\alpha \Rightarrow$ $\boxed{f = Ma_{cm}}$

Use this in the first eq. $\Rightarrow F - Ma_{cm} = Ma_{cm}$

$a_{cm} = \frac{F}{2M}$

And then $f = Ma_{cm} = M\left(\frac{F}{2M}\right) = \frac{F}{2}$.

10-43

This problem can be done either of two ways. We will do it both ways.

(1) Conservation of energy: $K_1 + U_1 + W_{other} = K_2 + U_2$

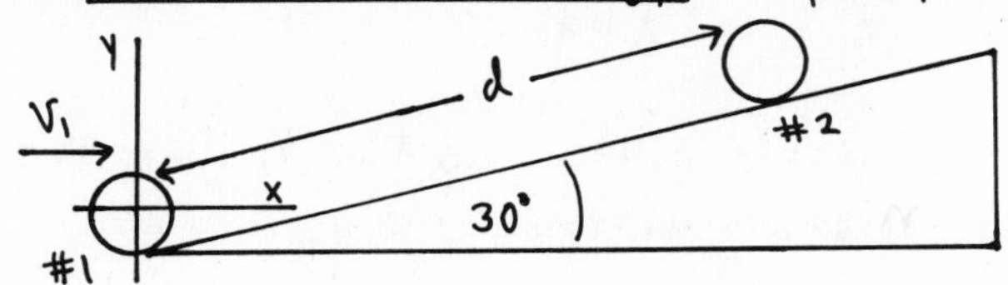

Take position #1 to be the location of the disk at the bare of the ramp and #2 to be where the disk momentarily stops before rolling back down.

Take the origin of coordinates at the center of the disk at position #1 and take +y to be upward. Then $y_1 = 0$ and $y_2 = d \sin 30°$, where d is the distance that the disk rolls up the ramp.

"rolls without slipping" and neglect rolling friction $\Rightarrow W_f = 0$ and only gravity does work on the disk, so $W_{other} = 0$.

$U_1 = Mgy_1 = 0$

$K_1 = \frac{1}{2}Mv_1^2 + \frac{1}{2}I_{cm}\omega_1^2$ (Eq. 10-20)

But $\omega_1 = v_1/R$ and $I_{cm} = \frac{1}{2}MR^2 \Rightarrow \frac{1}{2}I_{cm}\omega_1^2 = \frac{1}{2}\left(\frac{1}{2}MR^2\right)\left(\frac{v_1}{R}\right)^2 = \frac{1}{4}Mv_1^2$.

Thus $K_1 = \frac{1}{2}Mv_1^2 + \frac{1}{4}Mv_1^2 = \frac{3}{4}Mv_1^2$.

$U_2 = Mgy_2 = Mgd \sin 30°$

$K_2 = 0$ (disk is at rest here)

Thus $\frac{3}{4}Mv_1^2 = Mgd \sin 30°$

$$d = \frac{3v_1^2}{4g \sin 30°} = \frac{3(4.00 \text{ m/s})^2}{4(9.80 \text{ m/s}^2)\sin 30°} = \underline{2.49 \text{ m}}$$

(2) Force and acceleration:

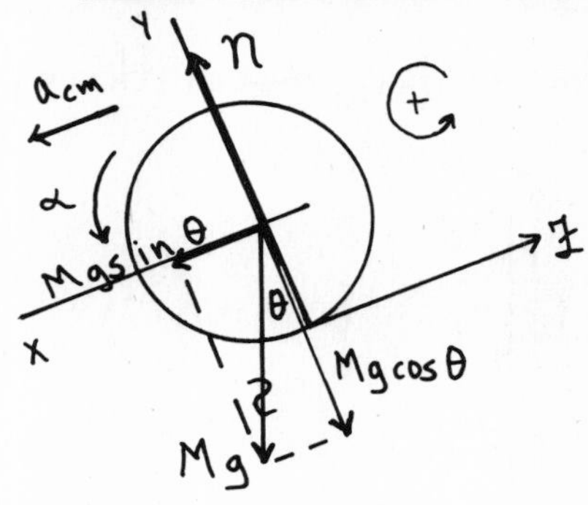

Apply $\sum F_x = ma_x$ to the translational motion of the center of mass:

$$\boxed{Mg \sin\theta - f = Ma_{cm}}$$

Apply $\sum \tau = I\alpha$ to the rotation about the center of mass:

$$fR = \left(\tfrac{1}{2}MR^2\right)\alpha$$

$$\boxed{f = \tfrac{1}{2}MR\alpha}$$

But $a_{cm} = R\alpha \Rightarrow f = \frac{1}{2}Ma_{cm}$

Use this in the first equation to eliminate f:

$$Mg\sin\theta - \tfrac{1}{2}Ma_{cm} = Ma_{cm}$$

$$\tfrac{3}{2}a_{cm} = g\sin\theta$$

$$a_{cm} = \tfrac{2}{3}g\sin\theta = \tfrac{2}{3}(9.80 \text{ m/s}^2)\sin 30° = 3.267 \text{ m/s}^2$$

Apply the constant acceleration equations to the motion of the center of mass. Note that in our coordinates the positive x-direction is down the incline.

10-43 (cont)

$v_{0x} = -4.00\,\text{m/s}$ (directed up incline)
$a_x = +3.267\,\text{m/s}^2$
$v_x = 0$ (momentarily comes to rest)
$x - x_0 = ?$

$$v_x^2 = v_{0x}^2 + 2a(x - x_0)$$
$$x - x_0 = -\frac{v_{0x}^2}{2a} = -\frac{(-4.00\,\text{m/s})^2}{2(3.267\,\text{m/s}^2)}$$
$$x - x_0 = -2.49\,\text{m}$$

The calculation says that the disk travels 2.49 m up the incline, the same result that we obtained from conservation of energy.

10-47

$$I_A = \tfrac{1}{2} I_B \Rightarrow I_B = 2 I_A$$

There is no external torque on the system consisting of the two disks, so we can apply conservation of angular momentum, $L_1 = L_2$.
L_1 is the initial angular momentum of disk A, $L_1 = I_A \omega_0$.
L_2 is the final angular momentum of the two disks after they are connected and reach a common angular velocity ω, $L_2 = (I_A + I_B)\omega = 3 I_A \omega$

Then $L_1 = L_2 \Rightarrow I_A \omega_0 = 3 I_A \omega \Rightarrow \omega = \omega_0/3$.

$$W_{tot} = K_2 - K_1$$
$$W_{tot} = W_f = -4000\,\text{J}$$
$$K_1 = \tfrac{1}{2} I_A \omega_0^2$$
$$K_2 = \tfrac{1}{2}(I_A + I_B)\omega^2 = \tfrac{1}{2}(3I_A)\left(\tfrac{\omega_0}{3}\right)^2 = \tfrac{1}{3}\left(\tfrac{1}{2} I_A \omega_0^2\right) = \tfrac{1}{3} K_1$$

Thus $-4000\,\text{J} = \tfrac{1}{3} K_1 - K_1$

$$\tfrac{2}{3} K_1 = 4000\,\text{J} \Rightarrow K_1 = \tfrac{3}{2}(4000\,\text{J}) = \underline{6000\,\text{J}}$$

10-49

a) Apply conservation of angular momentum to the collision between the bullet and the board:

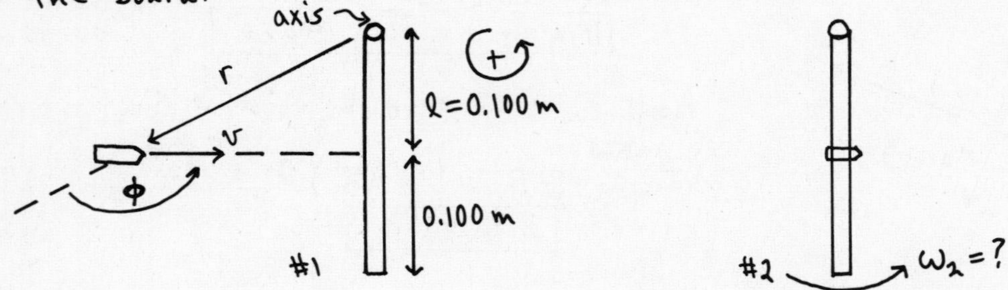

$$L_1 = L_2$$
$$L_1 = mvr\sin\phi = mv\ell = (5.00\times10^{-3}\,\text{kg})(300\,\text{m/s})(0.100\,\text{m}) = 0.150\,\text{kg}\cdot\text{m}^2/\text{s}$$
$$L_2 = I_2 \omega_2$$
$$I_2 = I_{board} + I_{bullet} = \tfrac{1}{3}ML^2 + mr^2 = \tfrac{1}{3}(2.20\,\text{kg})(0.200\,\text{m})^2 + (5.00\times10^{-3}\,\text{kg})(0.100\,\text{m})^2$$
$$I_2 = 0.02938\,\text{kg}\cdot\text{m}^2$$

Then $L_1 = L_2 \Rightarrow \omega_2 = \dfrac{L_1}{I_2} = \dfrac{0.150\,\text{kg}\cdot\text{m}^2/\text{s}}{0.02938\,\text{kg}\cdot\text{m}^2} = \underline{5.10\,\text{rad/s}}$.

10-49 (cont)

b) Apply conservation of energy to the motion of the board after the collision. Take the origin of coordinates at the center of the board and +y to be upward, so $y_{cm1} = 0$ and $y_{cm2} = h$, the height being asked for.

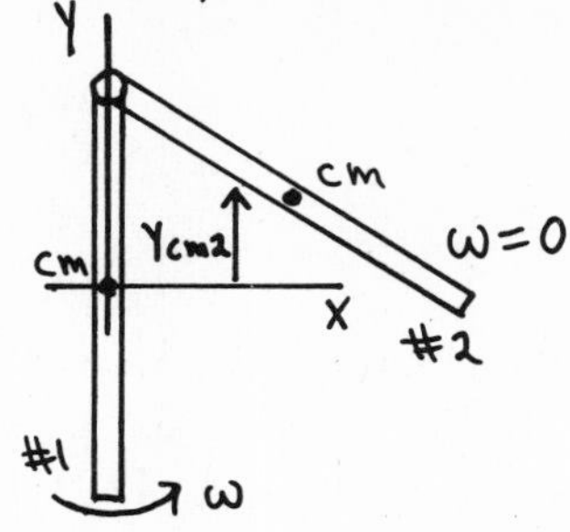

$$K_1 + U_1 + W_{other} = K_2 + U_2$$

Only gravity does work, so $W_{other} = 0$.

$K_1 = \frac{1}{2} I \omega^2$

$U_1 = mg\, y_{cm1} = 0$

$K_2 = 0$

$U_2 = mg\, y_{cm2} = mgh$

Thus $\frac{1}{2} I \omega^2 = mgh$

$$h = \frac{I\omega^2}{2mg} = \frac{(0.02938\ \text{kg}\cdot\text{m}^2)(5.10\ \text{rad/s})^2}{2(2.20\ \text{kg} + 5.00\times10^{-3}\ \text{kg})(9.80\ \text{m/s}^2)} = 0.01768\ \text{m} = \underline{1.77\ \text{cm}}$$

c)

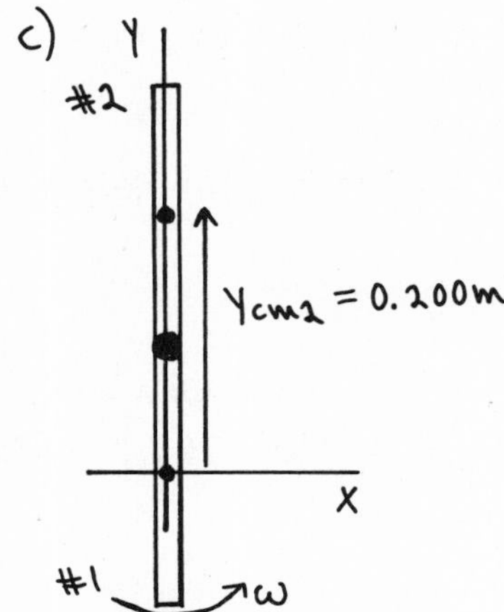

Apply conservation of energy as in part (b), except now we want $y_{cm2} = h = 0.200\ \text{m}$. Solve for the ω after the collision required for this to happen.

$$\frac{1}{2} I \omega^2 = mgh$$

$$\omega = \sqrt{\frac{2mgh}{I}} = \sqrt{\frac{2(2.20\ \text{kg} + 5.00\times10^{-3}\ \text{kg})(9.80\ \text{m/s}^2)(0.200\ \text{m})}{0.02938\ \text{kg}\cdot\text{m}^2}}$$

$$\omega = 17.15\ \text{rad/s}$$

Now go back to the equation that results from applying conservation of angular momentum to the collision and solve for the initial speed of the bullet.

$$L_1 = L_2 \Rightarrow m_{bullet}\, v\, l = I_2\, \omega_2$$

$$v = \frac{I_2 \omega_2}{m_{bullet}\, l} = \frac{(0.02938\ \text{kg}\cdot\text{m}^2)(17.15\ \text{rad/s})}{(5.00\times10^{-3}\ \text{kg})(0.100\ \text{m})} = \underline{1.01\times10^3\ \text{m/s}}$$

CHAPTER 11

Exercises 1, 3, 7, 11, 13, 15, 19, 23, 25, 27

Problems 31, 37, 43, 45, 47, 49, 51, 53, 55, 57

Exercises

11-1

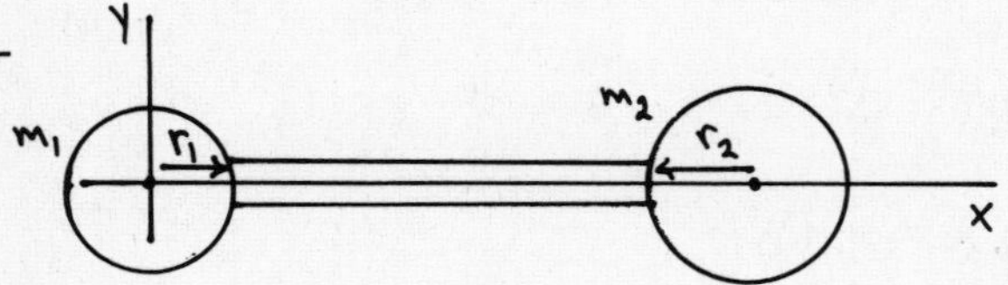

Use coordinates with the origin at the center of the 1.00 kg ball and the +x-axis along the rod.

The x-coordinate of the center of gravity is given by

$$x_{cm} = \frac{m_1x_1 + m_2x_2}{m_1 + m_2}$$

The center of gravity of ball #1 is at $x_1 = 0$ and the center of gravity of ball #2 is at $x_2 = 0.400\text{ m} + r_1 + r_2 = 0.400\text{ m} + 0.060\text{ m} + 0.080\text{ m} = 0.540\text{ m}$.

Then $x_{cm} = \dfrac{0 + (4.00\text{ kg})(0.540\text{ m})}{1.00\text{ kg} + 4.00\text{ kg}} = 0.432\text{ m}$.

The center of gravity is 0.432 m to the right of the center of the 1.00 kg ball.

11-3

Free-body diagram for the board: Let w_m be the weight of the motor. Take the origin of coordinates at the end where the 700 N force is applied (point A).

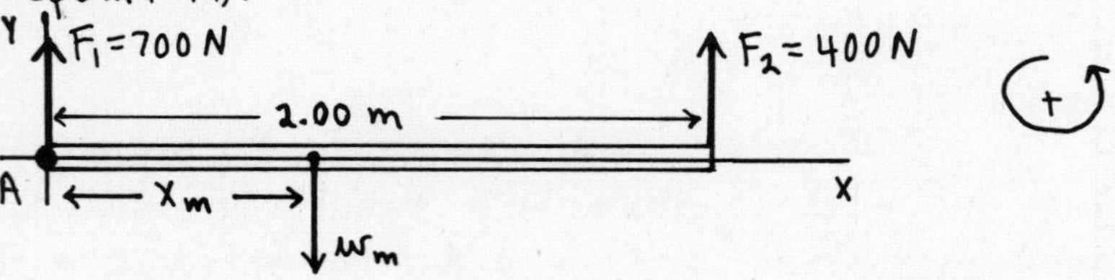

$$\Sigma F_y = ma_y \Rightarrow F_1 + F_2 - w_m = 0$$

$$w_m = F_1 + F_2 = 700\text{ N} + 400\text{ N} = 1100\text{ N}$$

$$\Sigma \tau_A = 0 \Rightarrow +F_2(2.00\text{ m}) - w_m x_m = 0$$

$$x_m = 2.00\text{ m}\left(\frac{F_2}{w_m}\right) = 2.00\text{ m}\left(\frac{400\text{ N}}{1100\text{ N}}\right) = 0.727\text{ m}$$

The center of gravity of the motor is 0.727 m from the end where the 700 N force is applied.

11-7

a) Free-body diagram for the diving board:

Take the origin of coordinates at the left-hand end of the board (point A).

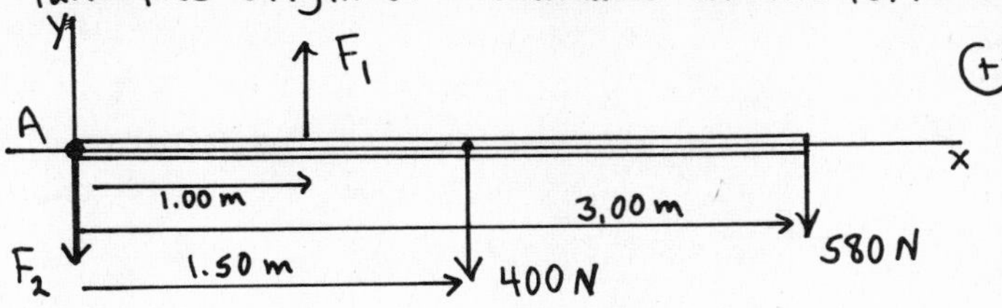

$\vec{F}_1$ is the force applied at the support point and $\vec{F}_2$ is the force at the end that is held down.

$$\sum \tau_A = 0 \Rightarrow +F_1(1.00\text{ m}) - (580\text{ N})(3.00\text{ m}) - (400\text{ N})(1.50\text{ m}) = 0$$

$$F_1 = \frac{(580\text{N})(3.00\text{m}) + (400\text{N})(1.50\text{m})}{1.00\text{ m}} = \underline{2340\text{ N}}$$

b) $\sum F_y = ma_y$

$F_1 - F_2 - 400\text{N} - 580\text{ N} = 0$

$F_2 = F_1 - 400\text{ N} - 580\text{ N} = 2340\text{N} - 400\text{N} - 580\text{N} = \underline{1360\text{ N}}$

11-11

(a) Free-body diagram for the strut. Take the origin of coordinates at the hinge (point A) and +y upward. Let F_h and F_v be the horizontal and vertical components of the force $\vec{F}$ exerted on the strut by the pivot. The tension in the vertical cable is the weight w of the suspended object. The weight w of the strut can be taken to act at the center of the strut. Let L be the length of the strut.

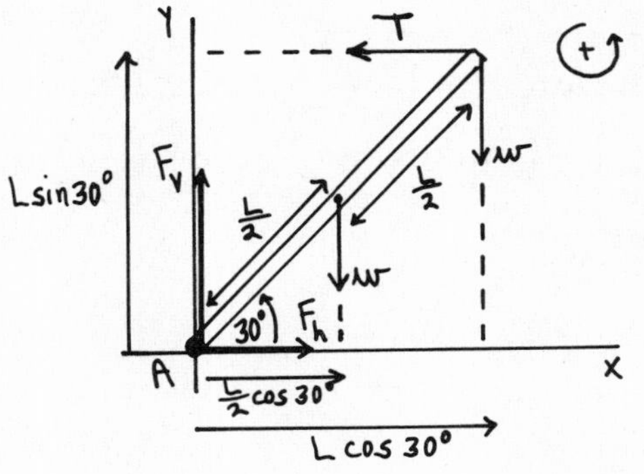

$\sum F_y = ma_y \Rightarrow F_v - w - w = 0$

$F_v = 2w$

Sum torques about point A. The pivot force has zero moment arm for this axis and so doesn't enter into the torque equation.

$\sum \tau_A = 0 \Rightarrow +TL\sin 30° - w(\frac{L}{2}\cos 30°) - w(L\cos 30°) = 0$

$T\sin 30° - \frac{3w}{2}\cos 30° = 0$

$T = \frac{3w\cos 30°}{2\sin 30°} = \underline{2.60\,w}$

Then $\sum F_x = ma_x \Rightarrow T - F_h = 0$

$F_h = 2.60\,w$

We have the components of $\vec{F}$ so can find its magnitude and direction:

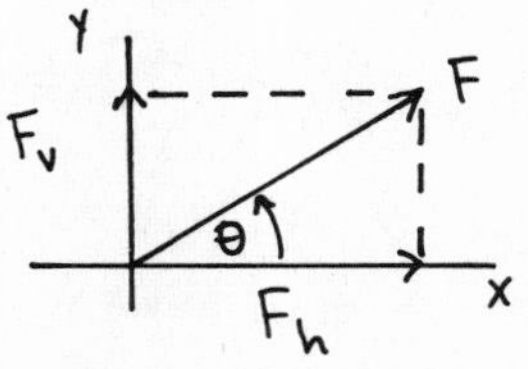

$F = \sqrt{F_h^2 + F_v^2} = \sqrt{(2.60w)^2 + (2.00w)^2} = \underline{3.28\,w}$

$\tan\theta = \frac{F_v}{F_h} = \frac{2.00\,w}{2.60\,w} = 0.7692 \Rightarrow \theta = \underline{37.6°}$

11-11 (cont)

(b) Free-body diagram for the strut:

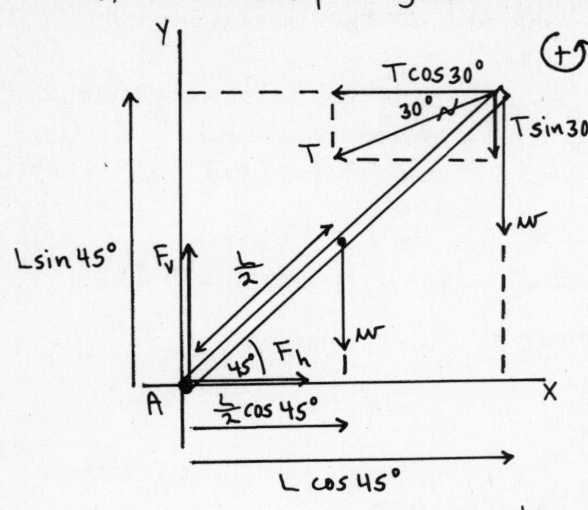

The tension T has been replaced by its x and y components. The torque due to T equals the sum of the torques of its components, and the latter are easier to calculate.

$$\Sigma \tau_A = 0$$

$$+(T\cos 30°)(L\sin 45°) - (T\sin 30°)(L\cos 45°) - w\left(\tfrac{L}{2}\cos 45°\right) - w(L\cos 45°) = 0$$

The length L divides out of the equation. The equation can also be simplified by noting that $\sin 45° = \cos 45°$.

Then $T(\cos 30° - \sin 30°) = \frac{3w}{2}$.

$$T = \frac{3w}{2(\cos 30° - \sin 30°)} = \underline{4.10\,w}$$

$\Sigma F_x = ma_x$

$F_h - T\cos 30° = 0$

$F_h = T\cos 30° = (4.10\,w)(\cos 30°) = 3.55\,w$

$\Sigma F_y = ma_y$

$F_v - w - w - T\sin 30° = 0$

$F_v = 2w + (4.10\,w)\sin 30° = 4.05\,w$

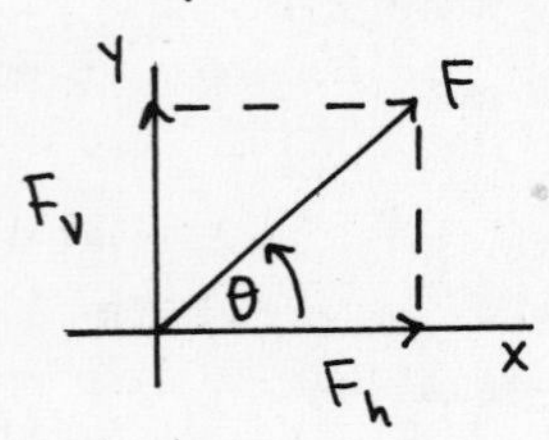

$$F = \sqrt{F_h^2 + F_v^2} = \sqrt{(3.55w)^2 + (4.05w)^2} = \underline{5.39\,w}$$

$$\tan\theta = \frac{F_v}{F_h} = \frac{4.05\,w}{3.55\,w} = 1.14 \Rightarrow \theta = \underline{48.8°}$$

11-13

Free-body diagram for the door:

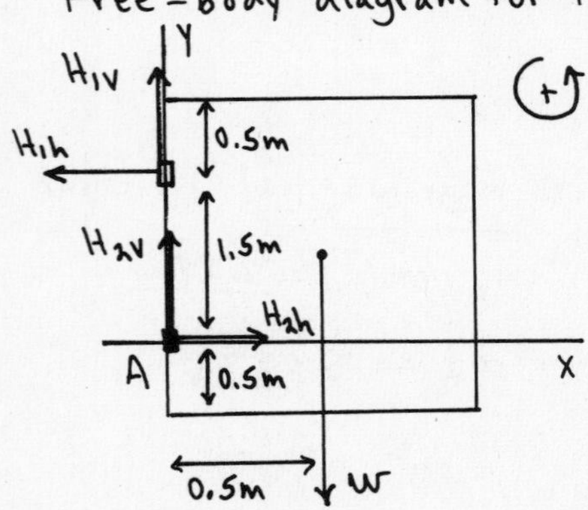

Let $\vec{H}_1$ and $\vec{H}_2$ be the forces exerted by the upper and lower hinges. Take the origin of coordinates at the bottom hinge (point A) and +y upward.

We are given that $H_{1v} = H_{2v} = \frac{w}{2} = 125$ N.

$$\Sigma F_x = ma \Rightarrow H_{2h} - H_{1h} = 0 \Rightarrow H_{1h} = H_{2h}$$

The horizontal components of the hinge forces are equal in magnitude and opposite in direction.

11-13 (cont)

Sum torques about point A. H_{1v}, H_{2v}, and H_{2h} all have zero moment arm and hence zero torque about an axis at this point.

Thus $\Sigma \tau_A = 0 \Rightarrow H_{1h}(1.50\text{ m}) - w(0.50\text{ m}) = 0$

$$H_{1h} = w\left(\frac{0.50\text{ m}}{1.50\text{ m}}\right) = (250\text{ N})\left(\frac{0.50\text{ m}}{1.50\text{ m}}\right) = 83.3\text{ N}$$

The horizontal component of each hinge force is <u>83.3 N.</u>

<u>11-15</u>

a)

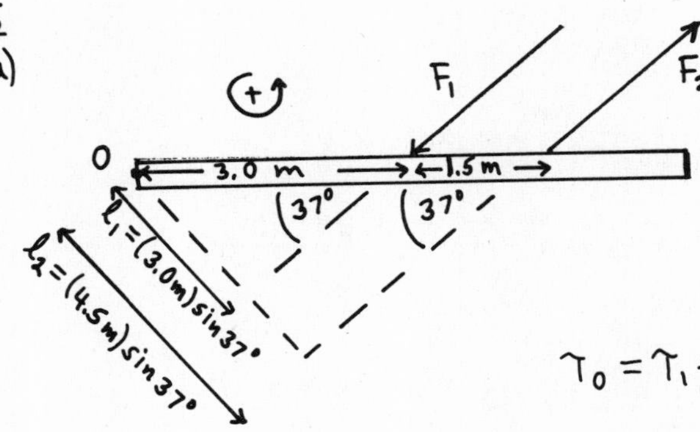

$\tau_1 = -F_1 l_1 = -(6.00\text{ N})(3.00\text{ m})\sin 37°$

$\tau_1 = -10.83\text{ N}\cdot\text{m}$

$\tau_2 = +F_2 l_2 = +(6.00\text{ N})(4.50\text{ m})\sin 37°$

$\tau_2 = +16.25\text{ N}\cdot\text{m}$

$\tau_O = \tau_1 + \tau_2 = -10.83\text{ N} + 16.25\text{ N}\cdot\text{m} = +5.42\text{ N}\cdot\text{m}$

<u>5.42 N·m, counterclockwise</u>

b)

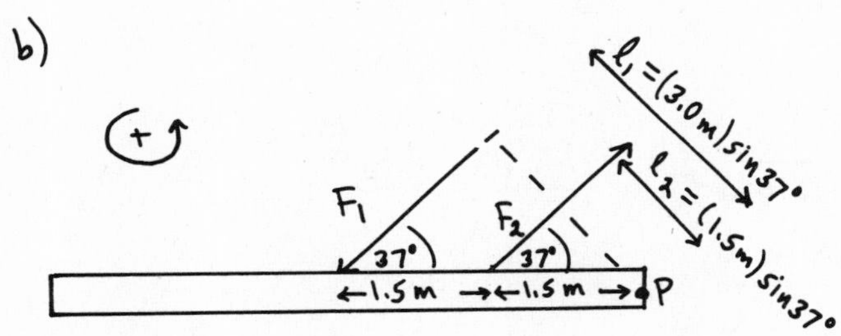

$\tau_1 = +F_1 l_1 = +(6.00\text{ N})(3.00\text{ m})\sin 37°$

$\tau_1 = +10.833\text{ N}\cdot\text{m}$

$\tau_2 = -F_2 l_2 = -(6.00\text{ N})(1.50\text{ m})\sin 37°$

$\tau_2 = -5.416\text{ N}\cdot\text{m}$

$\tau_P = \tau_1 + \tau_2 = +10.833\text{ N}\cdot\text{m} - 5.416\text{ N}\cdot\text{m} = +5.42\text{ N}\cdot\text{m}$

<u>5.42 N·m, counterclockwise</u>

c) Eq. (11-8) says $\Sigma \tau = lF$, for an axis through any point.
l is the perpendicular distance between the lines of action of the two forces:

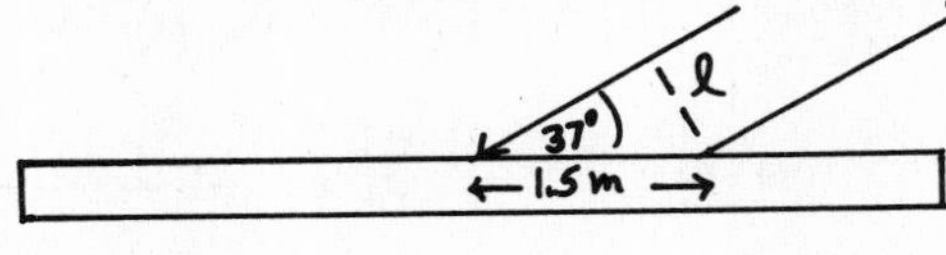

$l = (1.5\text{ m})\sin 37° = 0.9027\text{ m}$

$\Rightarrow \Sigma\tau = lF = (0.9027\text{ m})(6.00\text{ N}) = 5.42\text{ N}\cdot\text{m}$, which agrees with the results of parts (b) and (a).

<u>11-19</u>

$Y = \frac{l_0 F_\perp}{A\Delta l} \Rightarrow A = \frac{l_0 F_\perp}{Y\Delta l}$ (A is the cross-section area of the wire)

steel $\Rightarrow Y = 2.0\times10^{11}$ Pa (Table 11-1)

Thus $A = \frac{(3.00\text{ m})(300\text{ N})}{(2.0\times10^{11}\text{ Pa})(0.20\times10^{-2}\text{ m})} = 2.25\times10^{-6}\text{ m}^2$

$A = \pi r^2$, so $r = \sqrt{A/\pi} = \sqrt{2.25\times10^{-6}\text{ m}^2/\pi} = 8.463\times10^{-4}\text{ m}$

$d = 2r = 1.69\times10^{-3}\text{ m} = $ <u>1.69 mm</u>

11-23

$0.50\,m$ T_1 $m_1 = 5.0\,kg$

$0.50\,m$ T_2 $m_2 = 10.0\,kg$

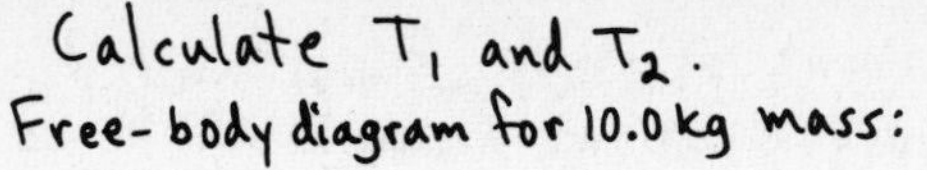

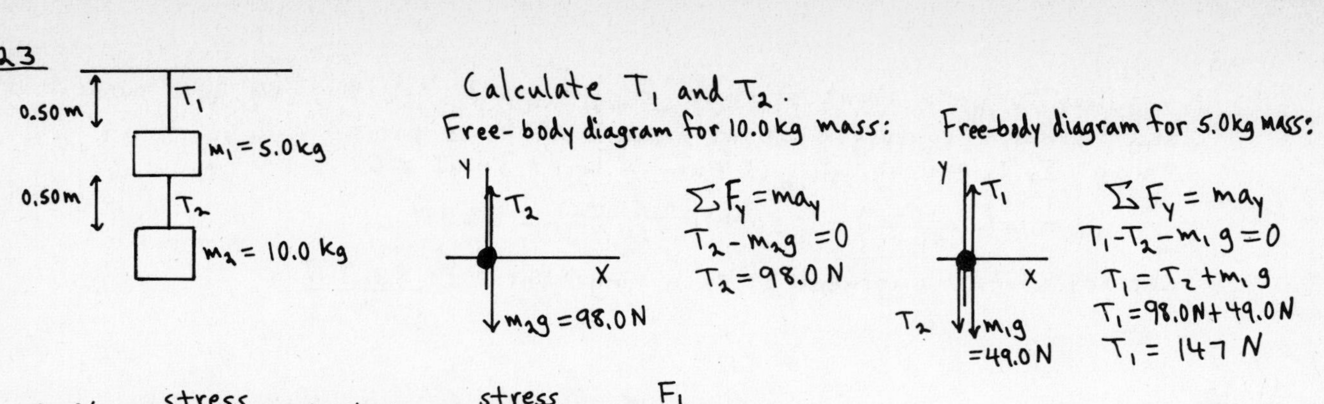

Calculate T_1 and T_2.

Free-body diagram for 10.0 kg mass:

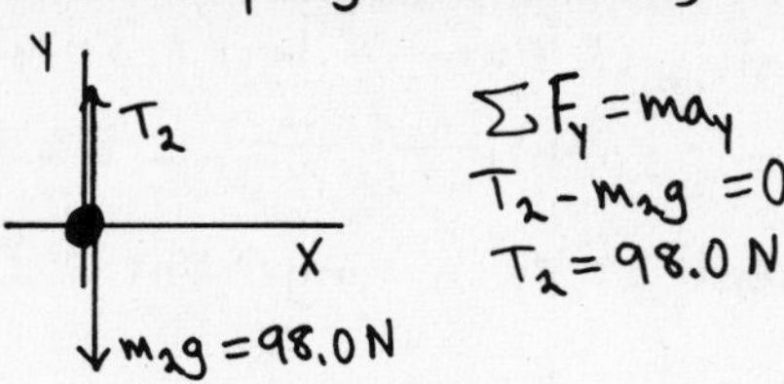

Y, T_2, X, $m_2 g = 98.0\,N$

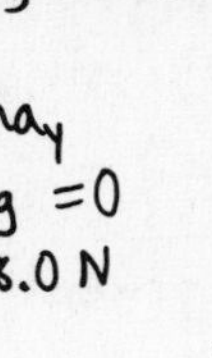

$\Sigma F_y = m a_y$

$T_2 - m_2 g = 0$

$T_2 = 98.0\,N$

Free-body diagram for 5.0 kg MASS:

Y, T_1, X, T_2, $m_1 g = 49.0\,N$

$\Sigma F_y = m a_y$

$T_1 - T_2 - m_1 g = 0$

$T_1 = T_2 + m_1 g$

$T_1 = 98.0\,N + 49.0\,N$

$T_1 = 147\,N$

a) $Y = \dfrac{\text{stress}}{\text{strain}} \Rightarrow \text{strain} = \dfrac{\text{stress}}{Y} = \dfrac{F_\perp}{AY}$

upper wire: $\text{strain} = \dfrac{T_1}{AY} = \dfrac{147\,N}{(3.0\times10^{-7}\,m^2)(2.0\times10^{11}\,Pa)} = \underline{2.45\times10^{-3}}$

lower wire: $\text{strain} = \dfrac{T_2}{AY} = \dfrac{98\,N}{(3.0\times10^{-7}\,m^2)(2.0\times10^{11}\,Pa)} = \underline{1.63\times10^{-3}}$

b) $\text{strain} = \dfrac{\Delta l}{l_0} \Rightarrow \Delta l = (l_0)(\text{strain})$

upper wire: $\Delta l = (0.50\,m)(2.45\times10^{-3}) = 1.22\times10^{-3}\,m = \underline{1.22\,mm}$

lower wire: $\Delta l = (0.50)(1.63\times10^{-3}) = 8.15\times10^{-4}\,m = \underline{0.82\,mm}$

11-25

$B = -\dfrac{\Delta p}{\Delta V/V_0} = -\dfrac{1.8\times10^6\,Pa}{(-0.30\,cm^3)/(1000\,cm^3)} = \underline{+6.0\times10^9\,Pa}$

$k = \dfrac{1}{B} = \dfrac{1}{6.0\times10^9\,Pa} = \underline{1.7\times10^{-10}\,Pa^{-1}}$

11-27

$\text{shear stress} = \dfrac{F_\parallel}{A}$

If the shear stress on one rivet is $6.00\times10^8\,Pa$ then the force $F_\parallel$ on each rivet is $F_\parallel = A(6.00\times10^8\,Pa) = \pi r^2(6.00\times10^8\,Pa) = \pi(0.150\times10^{-2}\,m)^2(6.00\times10^8\,Pa)$

$F_\parallel = 4.24\times10^3\,N$

Each rivet carries one quarter of the load $\Rightarrow$ total load of $4F_\parallel = \underline{1.70\times10^4\,N}$

Problems

11-31

a) Free-body diagram for the car:

$n_r = (1-f)w$ $\quad n_f = fw$

d, A, cg, x, w, (+)

Let the cg be a distance x from the rear axle. Sum torques about the rear axle (point A).

$\Sigma \tau_A = 0 \Rightarrow n_f d - wx = 0$

$fwd - wx = 0$

$x = fd$, as was to be shown.

11-31 (cont)

b) In Example 11-2, $n_f = 0.53\,w \Rightarrow f = 0.53$ and $d = 2.46$ m.
The general result derived in part (a) gives $x = fd = (0.53)(2.46\text{ m}) = 1.30$ m, the same as calculated in the example.

11-37

Free-body diagram for the bar:

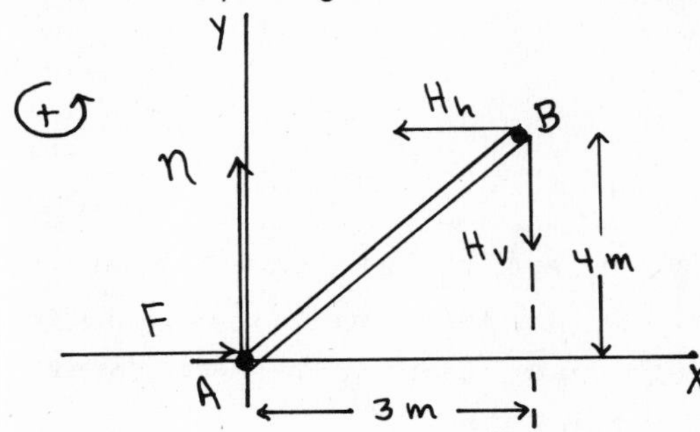

n is the normal force exerted on the bar by the surface. There is no friction force here.
H_h and H_v are the components of the force exerted on the bar by the hinge. The components of the force of the bar on the hinge will be equal in magnitude and opposite in direction.

$\Sigma F_x = ma_x \Rightarrow F = H_h = 75.0$ N

$\Sigma F_y = ma_y \Rightarrow n - H_v = 0$

$H_v = n$, but we don't know either of these forces

$\Sigma \tau_B = 0 \Rightarrow F(4.00\text{m}) - n(3.00\text{ m}) = 0$

$n = \left(\frac{4.00\text{ m}}{3.00\text{ m}}\right) F = \frac{4}{3}(75.0\text{N}) = 100.0\text{ N} \Rightarrow H_v = 100.0\text{ N}$

Force of bar on hinge:

horizontal component 75.0 N, to right
vertical component 100.0 N, upward

11-43

Free-body diagram for the gate:

Y, H_{Av}, T, $T\sin 30°$, A, 30°, $T\cos 30°$, (+↺), 2.00 m, 2.00 m, 2.00 m, H_{Bv}, B, H_{Bh}, w, X

Use coordinates with the origin at B. Let $\vec{H}_A$ and $\vec{H}_B$ be the forces exerted by the hinges A and B. The problem states that $\vec{H}_A$ has no horizontal component.
Replace the tension $\vec{T}$ by its horizontal and vertical components.

a) $\Sigma \tau_B = 0 \Rightarrow +(T\sin 30°)(4.00\text{ m}) + (T\cos 30°)(2.00\text{ m}) - w(2.00\text{ m}) = 0$

$T(2\sin 30° + \cos 30°) = w$

11-43 (cont) $T = \dfrac{w}{2\sin 30^\circ + \cos 30^\circ} = \dfrac{500\ \text{N}}{2\sin 30^\circ + \cos 30^\circ} = \underline{268\ \text{N}}$

b) $\sum F_x = ma_x \Rightarrow H_{Bh} - T\cos 30^\circ = 0$

$H_{Bh} = T\cos 30^\circ = (268\ \text{N})\cos 30^\circ = \underline{232\ \text{N}}$

c) $\sum F_y = ma_y \Rightarrow H_{Av} + H_{Bv} + T\sin 30^\circ - w = 0$

$H_{Av} + H_{Bv} = w - T\sin 30^\circ = 500\ \text{N} - (268\ \text{N})\sin 30^\circ = \underline{366\ \text{N}}$

11-45

Free-body diagram for the crate:

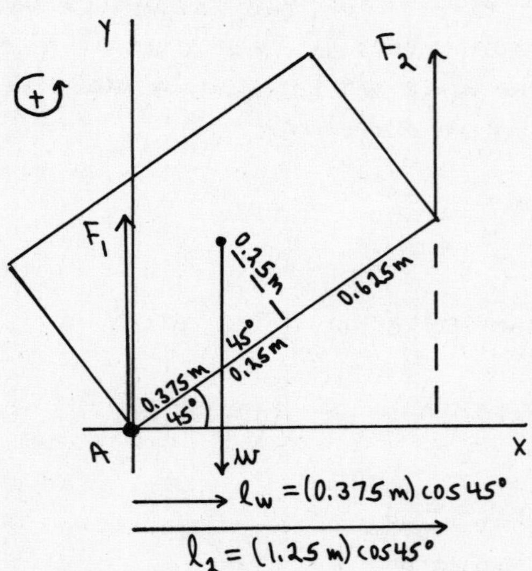

Let $\vec{F}_1$ and $\vec{F}_2$ be the vertical forces exerted by you and your friend. Take the origin at the lower left-hand corner of the crate (point A).

$\sum F_y = ma_y \Rightarrow F_1 + F_2 - w = 0$

$F_1 + F_2 = w = (200\ \text{kg})(9.80\ \text{m/s}^2) = 1960\ \text{N}$

$\sum \tau_A = 0 \Rightarrow F_2 l_2 - w l_w = 0$

$F_2 = w\left(\dfrac{l_w}{l_2}\right) = 1960\ \text{N}\left(\dfrac{(0.375\ \text{m})\cos 45^\circ}{(1.25\ \text{m})\cos 45^\circ}\right) = 588\ \text{N}$

Then $F_1 = w - F_2 = 1960\ \text{N} - 588\ \text{N} = 1372\ \text{N}$.

The person below (you) applies a force of $\underline{1372\ \text{N}}$. The person above (your friend) applies a force of $\underline{588\ \text{N}}$. It is better to be the person above.

11-47

a) Find the angle where the bale starts to tip:

Starts to tip $\Rightarrow$ only lower left-hand corner of the bale makes contact with the conveyor belt $\Rightarrow$ the line of action of the normal force n passes through the left-hand edge of the bale. Consider $\sum \tau_A = 0$ with point A at the lower left-hand corner. Then $\tau_n = 0$ and $\tau_{f_s} = 0$, so it must be that $\tau_{mg} = 0$ also. This means that the line of action of the gravity force must pass through point A. Thus the free-body diagram must be as follows:

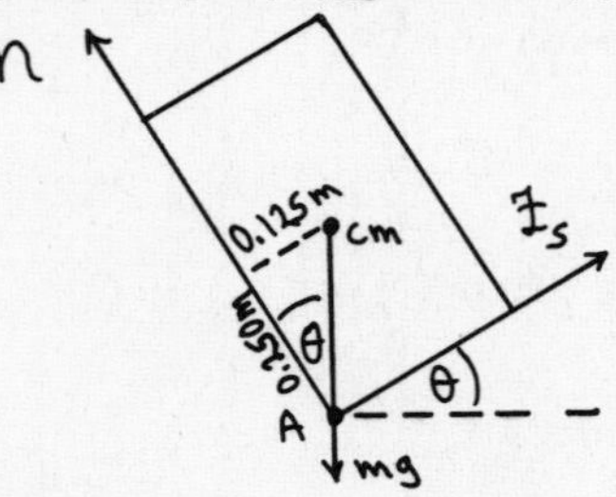

$\tan\theta = \dfrac{0.125\ \text{m}}{0.250\ \text{m}} = 0.500 \Rightarrow \theta = \underline{26.6^\circ}$, angle where tips

11-47 (cont)

At the angle where the bale is ready to slip down the incline f_s has its maximum possible value, $f_s = \mu_s n$. Free-body diagram for the bale, with the origin of coordinates at the cm.

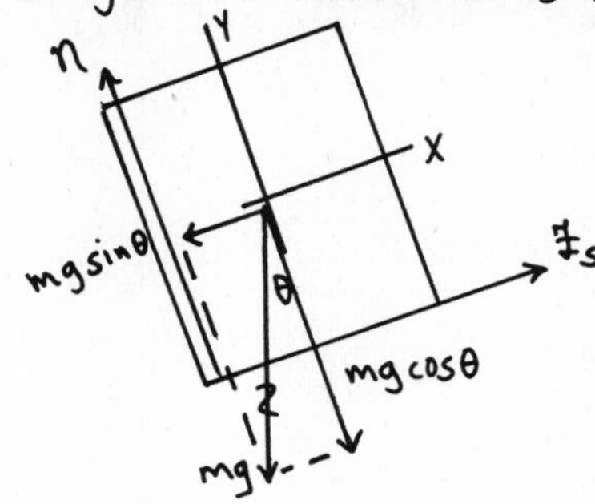

$$\Sigma F_y = ma_y$$
$$n - mg\cos\theta = 0$$
$$n = mg\cos\theta \Rightarrow f_s = \mu_s mg\cos\theta$$

$$\Sigma F_x = ma_x$$
$$f_s - mg\sin\theta = 0$$
$$\mu_s mg\cos\theta - mg\sin\theta = 0$$
$$\boxed{\tan\theta = \mu_s} \qquad \mu_s = 0.30 \Rightarrow \theta = 16.7°$$

$\theta = 26.6°$ to tip ; $\theta = 16.7°$ to slip $\Rightarrow$ slips first

b) The magnitude of the friction force didn't enter into the calculation of the tipping angle $\Rightarrow$ still tips at $\theta = 26.6°$.

$\mu_s = 0.75 \Rightarrow$ slips at $\theta = \arctan(0.75) = 36.9°$

Now the bale will tip over before it will slide down the incline.

11-49

a) Free-body diagram for the door:

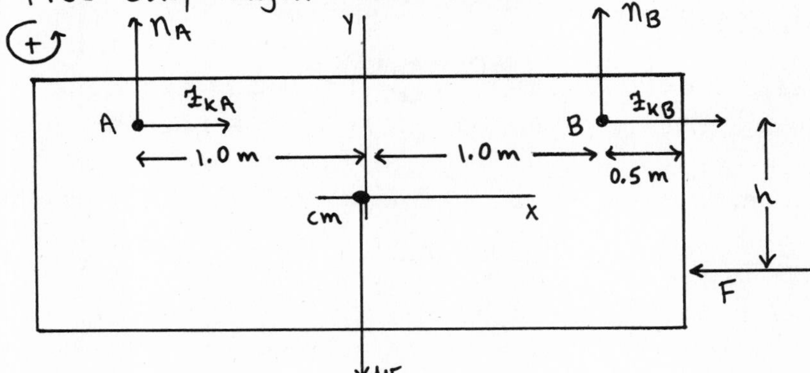

Take the origin of coordinates at the center of the door (at the cm). Let n_A, f_{KA}, and n_B, f_{KB} be the normal and friction forces exerted on the door at each wheel.

$$\Sigma F_y = ma_y$$
$$n_A + n_B - w = 0$$
$$n_A + n_B = w = 800\text{ N}$$

$$\Sigma F_x = ma_x$$
$$f_{KA} + f_{KB} - F = 0$$
$$F = f_{KA} + f_{KB}$$

But $f_K = \mu_K n \Rightarrow F = \mu_K (n_A + n_B) = (0.55)(800\text{ N}) = 440\text{ N}$

$$\Sigma \tau_B = 0$$

n_B, f_{KB}, and f_{KA} all have zero moment arm and hence zero torque about this point.

Thus $+w(1.00\text{ m}) - n_A(2.00\text{ m}) - F(h) = 0$

$$n_A = \frac{w(1.00\text{ m}) - F(h)}{2.00\text{ m}} = \frac{(800\text{ N})(1.00\text{ m}) - (440\text{ N})(1.50\text{ m})}{2.00\text{ m}} = 70\text{ N}$$

11-49 (cont)

And then $n_B = 800\text{ N} - n_A = 800\text{ N} - 70\text{ N} = \underline{730\text{ N}}$

b) If h is too large the torque of F will cause wheel A to leave the track. When wheel A just starts to lift off the track n_A and $f_{kA} \rightarrow 0$.

The equations in part (a) still apply:

$$n_A^{\,0} + n_B - w = 0 \Rightarrow n_B = w = 800\text{ N}$$

$$f_{kB} = \mu_k n_B = (0.55)(800\text{ N}) = 440\text{ N}$$

$$F = f_{kA}^{\,0} + f_{kB} = 440\text{N}$$

$$w(1.00\text{ m}) - n_A^{\,0}(2.00\text{m}) - F(h) = 0$$

$$h = \frac{w(1.00\text{m})}{F} = \frac{(800\text{N})(1.00\text{m})}{440\text{ N}} = \underline{1.8\text{ m}}$$

11-51

Calculate the tension in the wire as the mass passes through the lowest point. Free-body diagram for the mass:

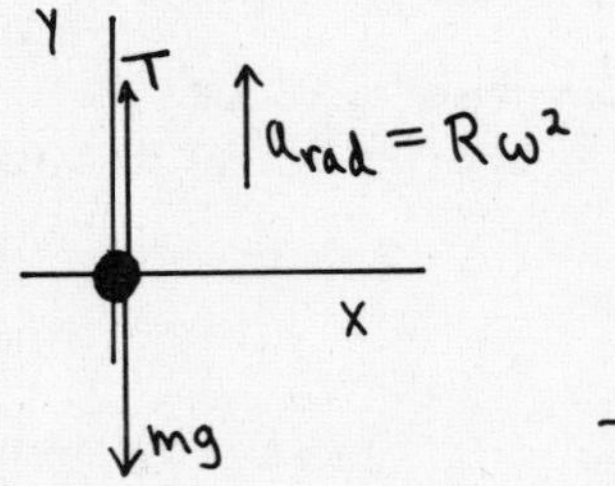

The mass moves in an arc of a circle with radius R = 0.50 m. It has acceleration $\vec{a}_{rad}$ directed in toward the center of the circle so at this point $\vec{a}_{rad}$ is upward.

$$\Sigma F_y = ma_y$$

$$T - mg = mR\omega^2 \Rightarrow T = m(g + R\omega^2)$$

But ω must be in rad/s: $\omega = (2.00\text{ rev/s})\left(\frac{2\pi\text{ rad}}{1\text{ rev}}\right) = 12.57\text{ rad/s}$.

Then $T = (15.0\text{ kg})(9.80\text{ m/s}^2 + (0.50\text{m})(12.57\text{ rad/s})^2) = 1332\text{ N}$.

Now calculate the elongation Δl of the wire that this tensile force produces:

$$Y = \frac{F_\perp l_0}{A\,\Delta l} \Rightarrow \Delta l = \frac{F_\perp l_0}{YA} = \frac{(1332\text{ N})(0.50\text{ m})}{(2.0\times10^{11}\text{Pa})(0.014\times10^{-4}\text{m}^2)} = 2.4\times10^{-3}\text{ m} = \underline{2.4\text{ mm}}$$

11-53

a) stress $= \frac{F_\perp}{A}$, so equal stresses implies $\frac{T}{A}$ same for each wire.

$$\frac{T_A}{1.00\text{ mm}^2} = \frac{T_B}{4.00\text{mm}^2} \Rightarrow T_B = 4.00T_A$$

The question is where along the rod to hang the weight in order to produce this relation between the tensions in the two wires. Let the weight be suspended at point C, a distance x to the right of wire A. The free-body diagram for the rod is then

11-53 (cont)

T_A ↑ (+↺) T_B ↑

C

← x →|← 1.05 m − x →

↓ w

$\Sigma \tau_c = 0$

$+T_B(1.05\,m - x) - T_A x = 0$

But $T_B = 4.00 T_A \Rightarrow 4.00 \not{T_A}(1.05\,m - x) - \not{T_A} x = 0$

$4.20\,m - 4.00x = x$

$x = \frac{4.20\,m}{5.00} = \underline{0.84\,m}$ (measured from A)

b) $Y = \frac{stress}{strain} \Rightarrow strain = \frac{stress}{Y} = \frac{F_\perp}{AY}$

Equal strains thus means $\frac{T_A}{(1.00\,mm^2)(2.40\times10^{11}\,Pa)} = \frac{T_B}{(4.00\,mm^2)(1.60\times10^{11}\,Pa)}$

$T_B = \left(\frac{4.00}{1.00}\right)\left(\frac{1.60}{2.40}\right) T_A = 2.667\,T_A$

The $\Sigma \tau_c = 0$ equation still gives $T_B(1.05\,m - x) - T_A x = 0$.
But now $T_B = 2.667\,T_A$ so $(2.667\,\not{T_A})(1.05\,m - x) - \not{T_A} x = 0$

$2.80\,m = 3.667x \Rightarrow x = \frac{2.80\,m}{3.667} = \underline{0.76\,m}$ (measured from A)

11-55

Each piece of the composite rod is subjected to a tensile force of 4.00×10^4 N.

a) $Y = \frac{F_\perp l_o}{A\Delta l} \Rightarrow \Delta l = \frac{F_\perp l_o}{YA}$

$\Delta l_c = \Delta l_s \Rightarrow \frac{F_\perp l_{o,c}}{Y_c A_c} = \frac{F_\perp l_{o,s}}{Y_s A_s}$ ($c \Rightarrow$ copper, $s \Rightarrow$ steel)

But the $F_\perp$ is the same for both

$\Rightarrow l_{o,s} = \frac{Y_s}{Y_c}\frac{A_s}{A_c} l_{o,c}$

$L = \left(\frac{2.0\times10^{11}\,Pa}{1.1\times10^{11}\,Pa}\right)\left(\frac{1.00\,cm^2}{2.00\,cm^2}\right)(1.40\,m) = \underline{1.27\,m}$

b) $stress = \frac{F_\perp}{A} = \frac{T}{A}$

copper: $stress = \frac{T}{A} = \frac{4.00\times10^4\,N}{2.00\times10^{-4}\,m^2} = \underline{2.00\times10^8\,Pa}$

steel: $stress = \frac{T}{A} = \frac{4.00\times10^4\,N}{1.00\times10^{-4}\,m^2} = \underline{4.00\times10^8\,Pa}$

c) $Y = \frac{stress}{strain} \Rightarrow strain = \frac{stress}{Y}$

copper: $strain = \frac{2.00\times10^8\,Pa}{1.1\times10^{11}\,Pa} = \underline{1.8\times10^{-3}}$

steel: $strain = \frac{4.00\times10^8\,Pa}{2.0\times10^{11}\,Pa} = \underline{2.0\times10^{-3}}$

11-57

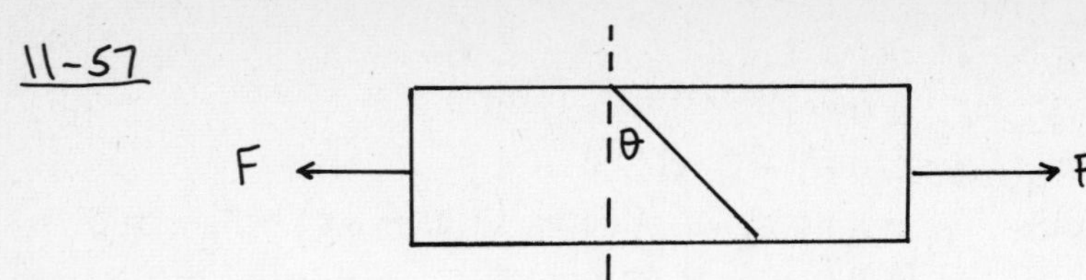

a)

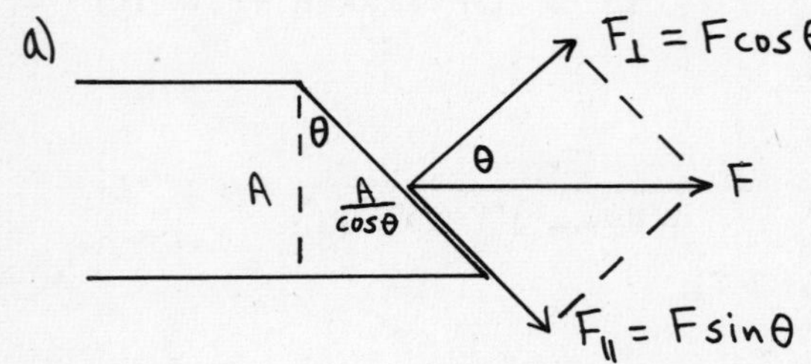

The area of the diagonal face is $\frac{A}{\cos\theta}$.

$$\text{tensile stress} = \frac{F_\perp}{\left(\frac{A}{\cos\theta}\right)} = \frac{F\cos\theta}{\left(\frac{A}{\cos\theta}\right)} = \frac{F\cos^2\theta}{A}$$

b) $$\text{shear stress} = \frac{F_\parallel}{\left(\frac{A}{\cos\theta}\right)} = \frac{F\sin\theta}{\left(\frac{A}{\cos\theta}\right)} = \frac{F\sin\theta\cos\theta}{A} = \frac{F\sin 2\theta}{2A}$$ (using a trig identity)

c) From the result of part (a) the tensile stress is a maximum for $\cos\theta = 1 \Rightarrow \underline{\theta = 0°}$.

d) From the result of part (b) the shear stress is a maximum for $\sin 2\theta = 1 \Rightarrow 2\theta = 90°$ and $\underline{\theta = 45°}$.

CHAPTER 12

Exercises 3, 5, 7, 9, 11, 17, 19, 21, 25, 29

Problems 33, 35, 39, 43, 45, 47, 49

Exercises

12-3

earth m_E — F_E ← m → F_s — sun m_s; ← x → ← r−x →; ← r →

Let $\vec{F}_E$ and $\vec{F}_s$ be the gravitational forces exerted on the spaceship by the earth and by the sun. The distance from the earth to the sun is $r = 1.49\times10^{11}$ m.

Let the ship be a distance x from the earth; it is then a distance r−x from the sun.

$$F_E = F_S \Rightarrow G\frac{m m_E}{x^2} = G\frac{m m_s}{(r-x)^2}$$

$$\frac{m_E}{x^2} = \frac{m_s}{(r-x)^2} \Rightarrow (r-x)^2 = x^2\frac{m_s}{m_E}$$

$$r-x = x\sqrt{\frac{m_s}{m_E}} \Rightarrow r = x\left(1+\sqrt{\frac{m_s}{m_E}}\right)$$

$$x = \frac{r}{1+\sqrt{\frac{m_s}{m_E}}} = \frac{1.49\times10^{11}\text{ m}}{1+\sqrt{1.99\times10^{30}\text{ kg}/5.98\times10^{24}}} = \underline{2.58\times10^{8}\text{ m}}$$ (from center of the earth)

12-5

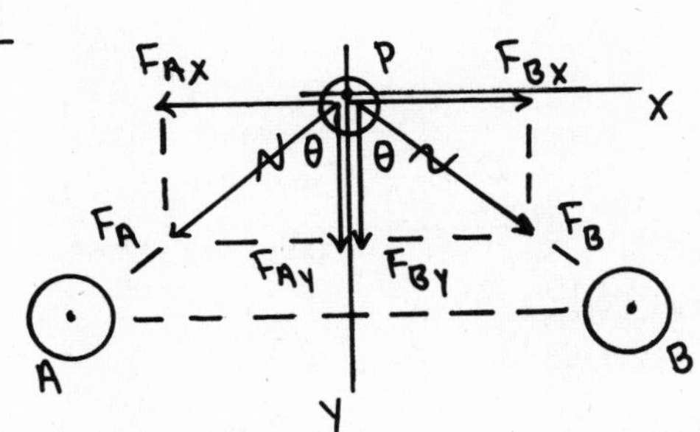

$\sin\theta = 0.80$
$\cos\theta = 0.60$

Take the origin of coordinates at point P.

$$F_A = G\frac{m_A m}{r^2} = (6.673\times10^{-11}\text{ N}\cdot\text{m}^2/\text{kg}^2)\frac{(6.40\text{ kg})(0.010\text{ kg})}{(0.100\text{ m})^2} = 4.271\times10^{-10}\text{ N}$$

$$F_B = G\frac{m_B m}{r^2} = 4.271\times10^{-10}\text{ N}$$

$$F_{Ax} = -F_A\sin\theta = -(4.271\times10^{-10}\text{ N})(0.80) = -3.42\times10^{-10}\text{ N}$$

$$F_{Ay} = +F_A\cos\theta = +(4.271\times10^{-10}\text{ N})(0.60) = 2.56\times10^{-10}\text{ N}$$

$$F_{Bx} = +F_B\sin\theta = +3.42\times10^{-10}\text{ N}$$

$$F_{By} = +F_B\cos\theta = +2.56\times10^{-10}\text{ N}$$

$$\Sigma F_x = ma_x \Rightarrow F_{Ax} + F_{Bx} = ma_x$$

$$0 = ma_x \Rightarrow a_x = 0$$

12-5 (cont)

$$\Sigma F_y = ma_y \Rightarrow F_{Ay} + F_{By} = ma_y$$
$$2(2.56 \times 10^{-10}\,N) = (0.010\,kg)\,a_y$$
$a_y = \underline{5.1 \times 10^{-8}\,m/s^2}$, directed downward midway between A and B

12-7

Use the measured gravitational force to calculate the gravitational constant G:

$$F_g = G\frac{m_1 m_2}{r^2} \Rightarrow G = \frac{F_g r^2}{m_1 m_2} = \frac{(1.30 \times 10^{-10}\,N)(0.0400\,m)^2}{(0.800\,kg)(4.00 \times 10^{-3}\,kg)} = 6.50 \times 10^{-11}\,N\cdot m^2/kg^2$$

Then use $g = \frac{Gm_E}{R_E^2}$ (Eq. 12-4) to calculate the mass of the earth:

$$g = \frac{Gm_E}{R_E^2} \Rightarrow m_E = \frac{R_E^2 g}{G} = \frac{(6.38 \times 10^6\,m)^2(9.80\,m/s^2)}{6.50 \times 10^{-11}\,N\cdot m^2/kg^2} = \underline{6.14 \times 10^{24}\,kg}$$

12-9

$$g = \frac{GM}{r^2} = \frac{(6.673 \times 10^{-11}\,N\cdot m^2/kg^2)(5.00\,kg)}{(2.00\,m)^2} = \underline{8.34 \times 10^{-11}\,m/s^2}$$

12-11

$$\vec{F} = m\vec{g} \Rightarrow g_x = \frac{F_x}{m} \text{ and } g_y = \frac{F_y}{m}$$

$$g_x = \frac{-0.225\,N}{0.0100\,kg} = \underline{-22.5\,m/s^2}$$

$$g_y = \frac{+0.540\,N}{0.0100\,kg} = \underline{+54.0\,m/s^2}$$

12-17

Example 12-7 gives the escape speed as $v_1 = \sqrt{\frac{2GM}{R}}$, where M and R are the mass and radius of the astronomical object.

$$\Rightarrow v_1 = \sqrt{2(6.673 \times 10^{-11}\,N\cdot m^2/kg^2)(2.0 \times 10^{15}\,kg)/5.0 \times 10^3\,m} = \underline{7.3\,m/s}$$

At this speed a person can run 100 m in 14 s; barely possible for the average person.

12-19

m, h, R_E

The radius of the orbit is $r = h + R_E$.

$r = 1.20 \times 10^6\,m + 6.38 \times 10^6\,m = 7.58 \times 10^6\,m$.

12-19 (cont.)

Free-body diagram for the satellite:

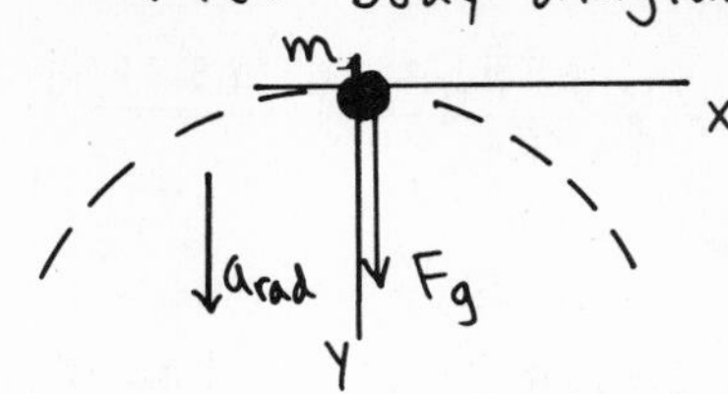

$$\Sigma F_y = ma_y$$
$$F_g = ma_{rad}$$
$$G\frac{mm_E}{r^2} = m\frac{v^2}{r}$$
$$v = \sqrt{\frac{Gm_E}{r}} = \sqrt{\frac{(6.673\times10^{-11}\ \text{N}\cdot\text{m}^2/\text{kg}^2)(5.98\times10^{24}\ \text{kg})}{7.58\times10^{6}\ \text{m}}}$$
$$v = \underline{7.26\times10^{7}\ \text{m/s}}$$

12-21

a) Find the orbit radius r:

$$\frac{Gmm_E}{r^2} = \frac{mv^2}{r}$$
$$\Rightarrow r = \frac{Gm_E}{v^2} = \frac{(6.673\times10^{-11}\ \text{N}\cdot\text{m}^2/\text{kg}^2)(5.98\times10^{24}\ \text{kg})}{(6400\ \text{m/s})^2} = 9.742\times10^{6}\ \text{m}$$

The period (time for one revolution) is then given by

$$T = \frac{2\pi r}{v} = \frac{2\pi\,(9.742\times10^{6}\ \text{m})}{6400\ \text{m/s}} = 9.56\times10^{3}\ \text{s} = \underline{159\ \text{min.}}$$

b) $$a_{rad} = \frac{v^2}{r} = \frac{(6400\ \text{m/s})^2}{9.742\times10^{6}\ \text{m}} = \underline{4.20\ \text{m/s}^2}$$

12-25

a) $F = w_0 = mg_0 = (5.0\ \text{kg})(25\ \text{m/s}^2) = \underline{125\ \text{N}}$; this is the true weight of the object.

b) From Eq. (12-33), $w = w_0 - \frac{mv^2}{R}$

$$T = \frac{2\pi r}{v} \Rightarrow v = \frac{2\pi r}{T} = \frac{2\pi\,(7.1\times10^{7}\ \text{m})}{(10\ \text{h})(3600\ \text{s}/1\ \text{h})} = 1.24\times10^{4}\ \text{m/s}$$
$$\frac{v^2}{R} = \frac{(1.24\times10^{4}\ \text{m/s})^2}{7.1\times10^{7}\ \text{m}} = 2.17\ \text{m/s}^2$$

Then $w = 125\ \text{N} - (5.0\ \text{kg})(2.17\ \text{m/s}^2) = \underline{114\ \text{N}}$

12-29

A black hole with the sun's mass M has the Schwarzschild radius given by Eq. (12-37):

$$R_s = \frac{2GM}{c^2} = \frac{2(6.673\times10^{-11}\ \text{N}\cdot\text{m}^2/\text{kg}^2)(1.99\times10^{30}\ \text{kg})}{(2.998\times10^{8}\ \text{m/s})^2} = 2.95\times10^{3}\ \text{m}$$

The ratio of R_s to the current radius R is

$$\frac{R_s}{R} = \frac{2.95\times10^{3}\ \text{m}}{6.96\times10^{8}\ \text{m}} = \underline{4.24\times10^{-6}}$$

Problems

12-33

a)

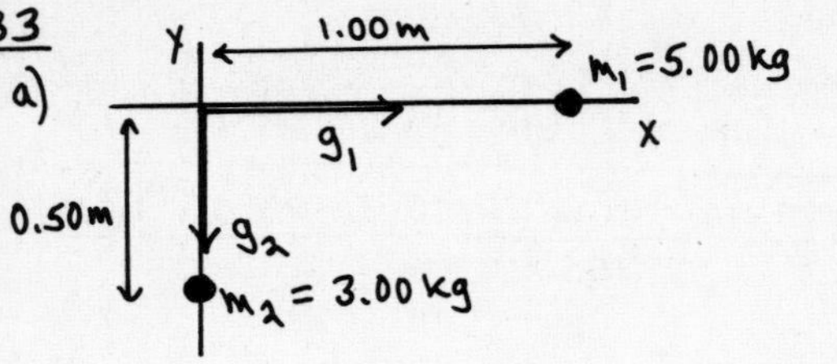

$$g_1 = G\frac{m_1}{r_1^2} = \frac{(6.673\times10^{-11}\ \text{N}\cdot\text{m}^2/\text{kg}^2)(5.00\ \text{kg})}{(1.00\ \text{m})^2}$$

$g_1 = 3.34\times10^{-10}\ \text{m/s}^2$

$$g_2 = G\frac{m_2}{r_2^2} = \frac{(6.673\times10^{-11}\ \text{N}\cdot\text{m}^2/\text{kg}^2)(3.00\ \text{kg})}{(0.50\ \text{m})^2}$$

$g_2 = 8.01\times10^{-10}\ \text{m/s}^2$

$g_x = g_{1x} + g_{2x} = \underline{+3.34\times10^{-10}\ \text{m/s}^2}$

$g_y = g_{1y} + g_{2y} = \underline{-8.01\times10^{-10}\ \text{m/s}^2}$

b) $F_x = mg_x = (0.0100\ \text{kg})(3.34\times10^{-10}\ \text{m/s}^2) = 3.34\times10^{-12}\ \text{N}$

$F_y = mg_y = (0.0100\ \text{kg})(-8.01\times10^{-10}\ \text{m/s}^2) = -8.01\times10^{-12}\ \text{N}$

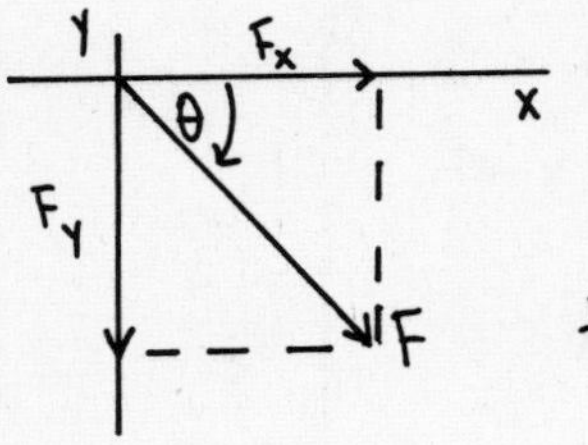

$$F = \sqrt{F_x^2 + F_y^2} = \sqrt{(3.34\times10^{-12}\ \text{N})^2 + (-8.01\times10^{-12}\ \text{N})^2}$$

$F = \underline{8.68\times10^{-12}\ \text{N}}$

$$\tan\theta = \frac{F_y}{F_x} = \frac{-8.01\times10^{-12}\ \text{N}}{3.34\times10^{-12}\ \text{N}} = -2.40 \Rightarrow \theta = \underline{-67.4^\circ}$$

12-35

To stay above the same point on the surface of the earth the orbital period of the satellite must equal the orbital period of the earth:

$$T = 1\ \text{d}\left(\frac{24\ \text{h}}{1\ \text{d}}\right)\left(\frac{3600\ \text{s}}{1\ \text{h}}\right) = 8.64\times10^4\ \text{s}$$

Eq. (12-20) gives the relation between the orbit radius and the period

$$T = \frac{2\pi r^{3/2}}{\sqrt{Gm_E}} \Rightarrow T^2 = \frac{4\pi^2 r^3}{Gm_E}$$

$$r = \left(\frac{T^2 G m_E}{4\pi^2}\right)^{1/3} = \left(\frac{(8.64\times10^4\ \text{s})^2(6.673\times10^{-11}\ \text{N}\cdot\text{m}^2/\text{kg}^2)(5.98\times10^{24}\ \text{kg})}{4\pi^2}\right)^{1/3}$$

$r = 4.23\times10^7\ \text{m}$

This is the radius of the orbit; it is related to the height h above the earth's surface and the radius R_E of the earth by $r = h + R_E$.
Thus $h = r - R_E = 4.23\times10^7\ \text{m} - 6.38\times10^6\ \text{m} = \underline{3.59\times10^7\ \text{m}}$.

12-39

Section 12-7 proves that any two spherically symmetric point masses interact as though they were point masses with all the mass concentrated at their centers.

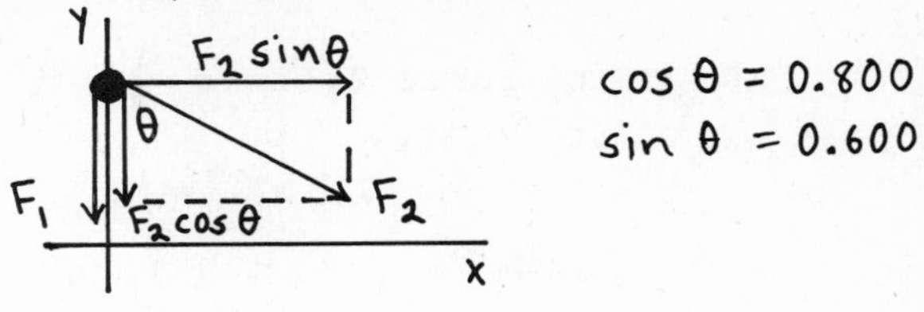

12-39 (cont)

$m_3 = 0.050$ kg
$m_2 = 0.800$ kg
$m_1 = 0.600$ kg
0.400 m
0.500 m
0.300 m

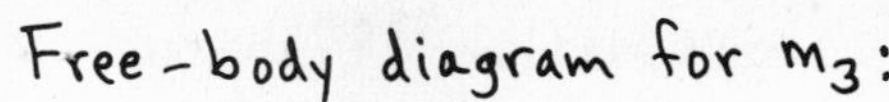

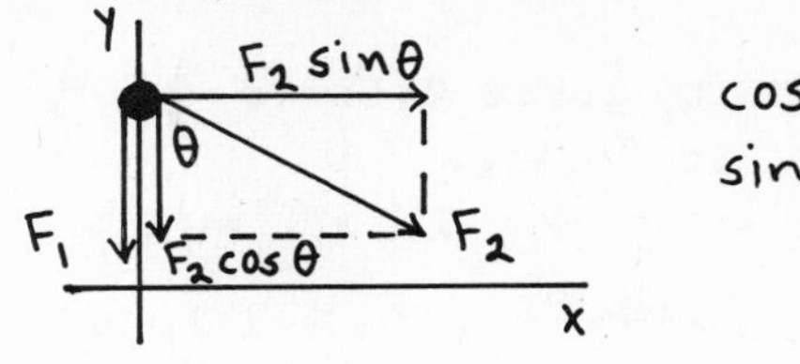

$\cos\theta = 0.800$
$\sin\theta = 0.600$

$$F_1 = G\frac{m_1 m_3}{r_{13}^2} = \frac{(6.673\times10^{-11}\ \text{N}\cdot\text{m}^2/\text{kg}^2)(0.600\ \text{kg})(0.050\ \text{kg})}{(0.400\ \text{m})^2} = 1.251\times10^{-11}\ \text{N}$$

$$F_2 = G\frac{m_2 m_3}{r_{23}^2} = \frac{(6.673\times10^{-11}\ \text{N}\cdot\text{m}^2/\text{kg}^2)(0.800\ \text{kg})(0.050\ \text{kg})}{(0.500\ \text{m})^2} = 1.068\times10^{-11}\ \text{N}$$

$F_{1x} = 0$; $F_{1y} = -1.251\times10^{-11}$ N

$F_{2x} = +F_2\sin\theta = +(1.068\times10^{-11}\ \text{N})(0.600) = 0.641\times10^{-11}$ N

$F_{2y} = -F_2\cos\theta = -(1.068\times10^{-11}\ \text{N})(0.800) = -0.854\times10^{-11}$ N

$F_x = F_{1x} + F_{2x} = 0 + 0.641\times10^{-11}\ \text{N} = 0.641\times10^{-11}$ N

$F_y = F_{1y} + F_{2y} = -1.251\times10^{-11}\ \text{N} - 0.854\times10^{-11}\ \text{N} = -2.105\times10^{-11}$ N

y, F_x, x, θ, F_y, F

$$F = \sqrt{F_x^2 + F_y^2} = \sqrt{(0.641\times10^{-11}\ \text{N})^2 + (-2.105\times10^{-11}\ \text{N})^2} = \underline{2.20\times10^{-11}\ \text{N}}$$

$$\tan\theta = \frac{F_y}{F_x} = \frac{-2.105\times10^{-11}\ \text{N}}{0.641\times10^{-11}\ \text{N}} = -3.284 \Rightarrow \theta = \underline{-73.1^\circ}$$

12-43

#1 $v_1 = 0$
h
#2 $v_2 = ?$
R_E

Take point #1 to be where the hammer is released and point #2 to be just above the surface of the earth, so $r_1 = R_E + h$ and $r_2 = R_E$.

$$K_1 + U_1 + W_{other} = K_2 + U_2$$

Only gravity does work so $W_{other} = 0$.

$K_1 = 0$

$K_2 = \frac{1}{2}mv_2^2$

$U_1 = -\frac{Gmm_E}{r_1} = -\frac{Gmm_E}{h+R_E}$

$U_2 = -G\frac{mm_E}{r_2} = -\frac{Gmm_E}{R_E}$

$$\Rightarrow -G\frac{\cancel{m}m_E}{h+R_E} = \frac{1}{2}\cancel{m}v_2^2 - G\frac{\cancel{m}m_E}{R_E}$$

$$v_2^2 = 2Gm_E\left(\frac{1}{R_E} - \frac{1}{R_E+h}\right) = \frac{2Gm_E}{R_E(R_E+h)}(\cancel{R_E} + h - \cancel{R_E}) = \frac{2Gm_E h}{R_E(R_E+h)}$$

$$v_2 = \sqrt{\frac{2Gm_E h}{R_E(R_E+h)}}$$

12-45

Use conservation of energy: $K_1 + U_1 + W_{other} = K_2 + U_2$.

The gravity force exerted by the sun is the only force that does work on the comet, so $W_{other} = 0$.

$K_1 = \frac{1}{2} m v_1^2$, $v_1 = 2.0 \times 10^4$ m/s

$U_1 = -G\frac{m_S m}{r_1}$, $r_1 = 2.0 \times 10^{11}$ m

$K_2 = \frac{1}{2} m v_2^2$

$U_2 = -G\frac{m_S m}{r_2}$, $r_2 = 4.0 \times 10^{10}$ m

$$\Rightarrow \frac{1}{2} \cancel{m} v_1^2 - G\frac{m_S \cancel{m}}{r_1} = \frac{1}{2}\cancel{m} v_2^2 - G\frac{m_S \cancel{m}}{r_2}$$

$$v_2^2 = v_1^2 + 2Gm_S\left(\frac{1}{r_2} - \frac{1}{r_1}\right) = v_1^2 + 2Gm_S\left(\frac{r_1 - r_2}{r_1 r_2}\right)$$

$$v_2 = \sqrt{(2.0\times10^4\ \text{m/s})^2 + 2(6.673\times10^{-11}\ \text{N}\cdot\text{m}^2/\text{kg}^2)(1.99\times10^{30}\ \text{kg})\ \frac{2.0\times10^{11}\ \text{m} - 4.0\times10^{10}\ \text{m}}{(2.0\times10^{11}\ \text{m})(4.0\times10^{10}\ \text{m})}}$$

$v_2 =$ <u>7.56×10^4 m/s</u> (The comet has greater speed when it is closer to the earth.)

12-47

Apply conservation of linear momentum and conservation of energy.

y
$m_A = 50.0$ kg
$r_1 = 40.0$ m
$v_{B1} = 0$
A
B
x
$v_{A1} = 0$
$m_B = 100.0$ kg
#1

y
v_{A2}
v_{B2}
A
B
m_A
m_B
x
$r_2 = 20.0$ m
#2

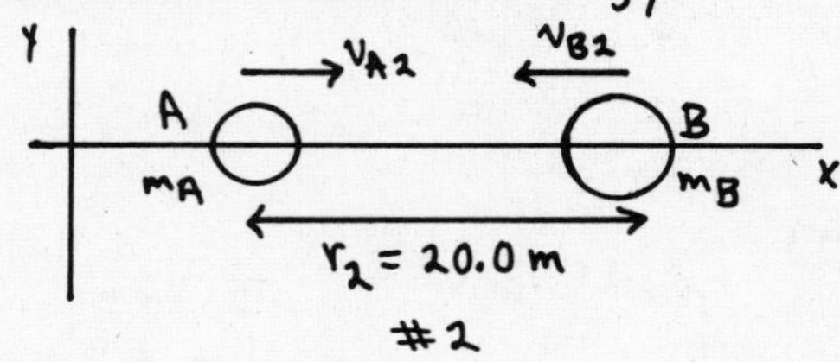

P_x is conserved $\Rightarrow 0 = m_A v_{A2x} + m_B v_{B2x}$

$0 = m_A v_{A2} - m_B v_{B2}$

$0 = (50.0\ \text{kg})v_{A2} - (100.0\ \text{kg})v_{B2}$

$\boxed{v_{A2} = 2v_{B2}}$

Conservation of energy $\Rightarrow K_1 + U_1 + W_{other} = K_2 + U_2$

$W_{other} = 0$ (only the gravity force does work)

$K_1 = 0$

$U_1 = -G\frac{m_A m_B}{r_1} = -(6.673\times10^{-11}\ \text{N}\cdot\text{m}^2/\text{kg}^2)\frac{(50.0\ \text{kg})(100.0\ \text{kg})}{40.0\ \text{m}} = -8.341\times10^{-9}$ J

$U_2 = -G\frac{m_A m_B}{r_2} = -(6.673\times10^{-11}\ \text{N}\cdot\text{m}^2/\text{kg}^2)\frac{(50.0\ \text{kg})(100.0\ \text{kg})}{20.0\ \text{m}} = -1.668\times10^{-8}$ J

$K_2 = \frac{1}{2} m_A v_{A2}^2 + \frac{1}{2} m_B v_{B2}^2$

$$\Rightarrow -8.341\times10^{-9}\ \text{J} = -1.668\times10^{-8}\ \text{J} + \frac{1}{2} m_A v_{A2}^2 + \frac{1}{2} m_B v_{B2}^2$$

$$\frac{1}{2} m_A v_{A2}^2 + \frac{1}{2} m_B v_{B2}^2 = 8.33\times10^{-9}\ \text{J}$$

12-47 (cont)

Use $v_{A2} = 2v_{B2}$ to eliminate v_{A2}

$$\Rightarrow \tfrac{1}{2} m_A (2v_{B2})^2 + \tfrac{1}{2} m_B v_{B2}^2 = 8.33\times10^{-9}\,\text{J}$$

$$v_{B2}^2 \left(2m_A + \tfrac{1}{2} m_B\right) = 8.33\times10^{-9}\,\text{J}$$

$$v_{B2} = \sqrt{\frac{8.33\times10^{-9}\,\text{J}}{2(50.0\,\text{kg}) + \frac{1}{2}(100.0\,\text{kg})}} = \underline{7.45\times10^{-6}\,\text{m/s}}$$

$$v_{A2} = 2v_{B2} = \underline{1.49\times10^{-5}\,\text{m/s}}$$

b) $v_{rel} = v_{A2} + v_{B2}$ (Since the spheres are moving toward each other.)

$$v_{rel} = 1.49\times10^{-5}\,\text{m/s} + 7.45\times10^{-6}\,\text{m/s} = \underline{2.24\times10^{-5}\,\text{m/s}}$$

12-49

Use a coordinate system with the origin at the left-hand end of the rod and the x-axis along the rod. Divide the rod into small segments of length dx.

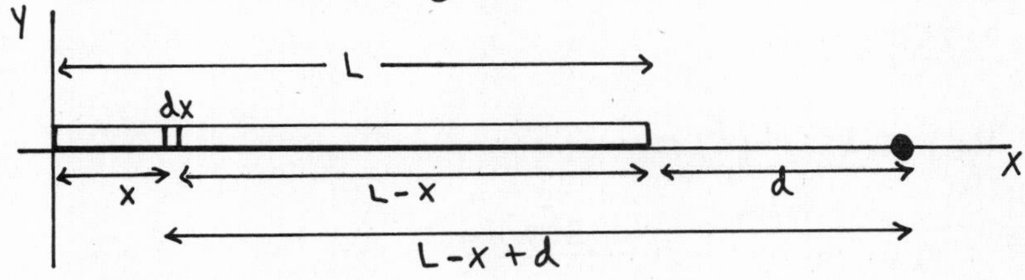

The mass of each segment is $dM = dx\,\frac{M}{L}$. Each segment is a distance $L-x+d$ from the point, so $g = \frac{GM}{r^2}$

$$\Rightarrow dg = \frac{G\,dM}{(L-x+d)^2} = \frac{GM}{L}\,\frac{dx}{(L-x+d)^2}.$$

$$g = \int_0^L dg = \frac{GM}{L}\int_0^L \frac{dx}{(L-x+d)^2} = \frac{GM}{L}\left(+\left.\frac{1}{(L-x+d)}\right|_0^L\right) = \frac{GM}{L}\left(\frac{1}{d} - \frac{1}{L+d}\right)$$

$$g = \frac{GM}{L}\,\frac{L+d-d}{d(L+d)} = \underline{\frac{GM}{d(L+d)}}$$

$d \gg L \Rightarrow g \to \frac{GM}{d^2}$, the same as for a point mass.

CHAPTER 13

Exercises 5, 9, 13, 15, 17, 19, 21, 25, 27, 29, 33

Problems 37, 39, 41, 43, 45, 49, 51, 55

Exercises

13-5

a) $E = \frac{1}{2}mv^2 + \frac{1}{2}kx^2$

$E = \frac{1}{2}(0.500\text{ kg})(0.300\text{ m/s})^2 + \frac{1}{2}(300\text{ N/m})(0.012\text{m})^2 = 0.0225\text{J} + 0.0216\text{J} = \underline{0.0441\text{ J}}$

b) $E = \frac{1}{2}kA^2 \Rightarrow A = \sqrt{\frac{2E}{k}} = \sqrt{\frac{2(0.0441\text{ J})}{300\text{ N/m}}} = \underline{0.017\text{ m}}$

c) $E = \frac{1}{2}m\,v_{max}^2 \Rightarrow v_{max} = \sqrt{\frac{2E}{m}} = \sqrt{\frac{2(0.0441\text{ J})}{0.500\text{ kg}}} = \underline{0.420\text{ m/s}}$

13-9

a) $T = \frac{1}{f} = \frac{1}{4.00\text{ Hz}} = \underline{0.250\text{ s}}$

b) $\omega = 2\pi f = 2\pi\,(4.00\text{ Hz}) = \underline{25.1\text{ rad/s}}$

c) $\omega = \sqrt{\frac{k}{m}} \Rightarrow m = \frac{k}{\omega^2} = \frac{200\text{ N/m}}{(25.1\text{ rad/s})^2} = \underline{0.317\text{ kg}}$

13-13

a) Eq. (13-25): $A = \sqrt{x_0^2 + \frac{v_0^2}{\omega^2}} = \sqrt{x_0^2 + \frac{m v_0^2}{k}} = \sqrt{(0.200\text{ m})^2 + \frac{(3.00\text{kg})(-6.00\text{ m/s})^2}{150\text{ N/m}}}$

$A = \underline{0.872\text{ m}}$

b) Eq. (13-24): $\phi = \arctan\left(-\frac{v_0}{\omega x_0}\right)$

$\omega = \sqrt{\frac{k}{m}} = \sqrt{\frac{150\text{ N/m}}{3.00\text{ kg}}} = 7.071\text{ rad/s}$

$\phi = \arctan\left(-\frac{(-6.00\text{ m/s})}{(7.071\text{ rad/s})(0.200\text{m})}\right) = \arctan(+4.243) = \underline{76.7°}$ (or $\underline{1.34\text{rad}}$)

c) $E = \frac{1}{2}mv_0^2 + \frac{1}{2}kx_0^2 = \frac{1}{2}(3.00\text{kg})(-6.00\text{ m/s})^2 + \frac{1}{2}(150\text{N/m})(0.200\text{ m})^2 = 54.0\text{J} + 3.0\text{J} = \underline{57.0\text{ J}}$

d) $x = A\cos(\omega t + \phi) \Rightarrow x = (0.872\text{ m})\cos[(7.07\text{ rad/s})t - 1.34\text{ rad}]$

13-15

a) $-kx = ma \Rightarrow a = -\frac{k}{m}x$ (Eq. 13-2)

But the maximum $|x|$ is A, so $a_{max} = \frac{k}{m}A$.

$f = 4.00\text{ Hz} \Rightarrow \omega = \sqrt{\frac{k}{m}} = 2\pi f = 2\pi(4.00\text{ Hz}) = 25.13\text{ rad/s}$

13-15 (cont)

$$a_{max} = \omega^2 A = (25.13 \text{ rad/s})^2 (0.180 \text{ m}) = \underline{114 \text{ m/s}^2}$$

$$\tfrac{1}{2}mv^2 + \tfrac{1}{2}kx^2 = \tfrac{1}{2}kA^2$$

$v = v_{max}$ when $x=0 \Rightarrow \tfrac{1}{2}mv_{max}^2 = \tfrac{1}{2}kA^2$

$$v_{max} = \sqrt{\tfrac{k}{m}}\, A = \omega A = (25.13 \text{ rad/s})(0.180 \text{ m}) = \underline{4.52 \text{ m/s}}$$

b) $a = -\frac{k}{m}x = -\omega^2 x = -(25.13 \text{ rad/s})^2(0.090 \text{m}) = \underline{-57 \text{ m/s}^2}$

$$\tfrac{1}{2}mv^2 + \tfrac{1}{2}kx^2 = \tfrac{1}{2}kA^2 \Rightarrow v = \pm\sqrt{\tfrac{k}{m}}\sqrt{A^2 - x^2} = \pm\,\omega\sqrt{A^2-x^2}$$

$$v = \pm(25.13 \text{ rad/s})\sqrt{(0.180\text{m})^2 - (0.090\text{ m})^2} = \pm 3.92 \text{ m/s}$$

The speed is $\underline{3.9 \text{ m/s}}$.

c) $x = A\cos(\omega t + \phi)$

Let $\phi = -\frac{\pi}{2}$ so that $x=0$ at $t=0$.

Then $x = A\cos(\omega t - \frac{\pi}{2}) = A\sin(\omega t)$ [using $\cos(a - \frac{\pi}{2}) = \sin a$].

Find the time t that gives $x = 0.120$ m.

$$0.120 \text{ m} = (0.180\text{m})\sin(\omega t)$$

$$\sin \omega t = 0.6667$$

$$t = \frac{\arcsin(0.6667)}{\omega} = \frac{0.7297 \text{ rad}}{25.13 \text{ rad/s}} = \underline{0.0290 \text{s}}$$

Note: It takes one-fourth of a period for the object to go from $x=0$ to $x = A = 0.180$ m. So the time we have calculated should be less than $T/4$. $T = \frac{1}{f} = \frac{1}{4.00 \text{ Hz}} = 0.250 \text{ s}$; $\frac{T}{4} = 0.0625 \text{s}$, and the time we calculated is less than this.

13-17

Let d be the distance the spring stretches when the block hangs at rest.

d $a=0$ m

Free-body diagram for the block: Take +y to be downward.

kd (the spring force) x mg y

$$\Sigma F_y = ma_y$$

$$mg - kd = 0 \Rightarrow d = \frac{mg}{k}$$

$$T = 2\pi\sqrt{\frac{m}{k}} \Rightarrow \frac{m}{k} = \left(\frac{T}{2\pi}\right)^2$$

Thus $d = \left(\frac{T}{2\pi}\right)^2 g = \left(\frac{0.400 \text{ s}}{2\pi}\right)^2 (9.80 \text{ m/s}^2) = \underline{0.0397 \text{ m}}$.

13-19

Ticks four times each second $\Rightarrow$ 0.25 s per tick.
Each tick is half a period $\Rightarrow T = 0.50\,s$ and $f = \frac{1}{T} = \frac{1}{0.50\,s} = 2.00\,Hz$.

a) thin rim $\Rightarrow I = MR^2$ (from Table 9-2)
$I = (0.800 \times 10^{-3}\,kg)(0.65 \times 10^{-2}\,m)^2 = \underline{3.4 \times 10^{-8}\,kg \cdot m^2}$

b) $f = \frac{1}{2\pi}\sqrt{\frac{K}{I}}$
$\Rightarrow K = I(2\pi f)^2 = 3.4 \times 10^{-8}\,kg \cdot m^2\,[2\pi(2.00\,Hz)]^2 = \underline{5.4 \times 10^{-6}\,N \cdot m}$

13-21

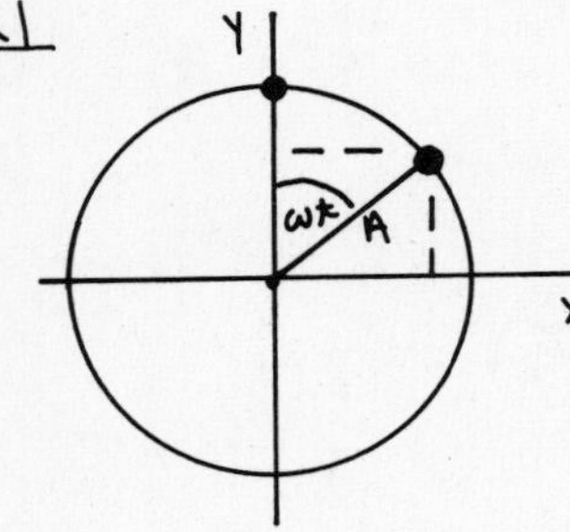

From the diagram, $x = A \sin \omega t$

$T = \frac{2\pi}{\omega} \Rightarrow \omega = \frac{2\pi}{T} = \frac{2\pi}{(\frac{\pi}{2}\,s)} = 4.00\,rad/s$

Then $t = \frac{\pi}{10}\,s \Rightarrow \omega t = (4.00\,rad/s)(\frac{\pi}{10}\,s) = 0.400\pi\,rad.$

$x = A \sin \omega t = (0.300\,m) \sin(0.400\pi\,rad) = \underline{0.285\,m}$

13-25

Let the period on earth be $T_E = 2\pi\sqrt{\frac{L}{g_E}}$, where $g_E = 9.80\,m/s^2$, the value on earth.

Let the period on the moon be $T_M = 2\pi\sqrt{\frac{L}{g_M}}$, where $g_M = 1.62\,m/s^2$, the value on the moon.

We can eliminate L, which we don't know, by taking a ratio:

$$\frac{T_M}{T_E} = 2\pi\sqrt{\frac{L}{g_M}}\,\frac{1}{2\pi}\sqrt{\frac{g_E}{L}} = \sqrt{\frac{g_E}{g_M}}$$

$$T_M = T_E\sqrt{\frac{g_E}{g_M}} = (1.20\,s)\sqrt{\frac{9.80\,m/s^2}{1.62\,m/s^2}} = \underline{2.95\,s}$$

13-27

The ornament is a physical pendulum; $T = 2\pi\sqrt{\frac{I}{mgd}}$ (Eq. 13-36).

$I = \frac{7MR^2}{5}$, the moment of inertia about an axis at the edge of the sphere
d is the distance from the axis to the center of gravity, which is at the center of the sphere, so $d = R$.

Thus $T = 2\pi\sqrt{\frac{7MR^2}{5MgR}} = 2\pi\sqrt{\frac{7}{5}}\sqrt{\frac{R}{g}} = 2\pi\sqrt{\frac{7}{5}}\sqrt{\frac{0.050\,m}{9.80\,m/s^2}} = \underline{0.531\,s}$.

13-29

a) Eq. (13-40) says $\omega' = \sqrt{\frac{k}{m} - \frac{b^2}{4m^2}}$

$$\omega' = \sqrt{\frac{300\ N/m}{0.400\ kg} - \frac{(6.00\ kg/s)^2}{4(0.400\ kg)^2}} = 26.34\ rad/s$$

$$f' = \frac{\omega'}{2\pi} = \frac{26.34\ rad/s}{2\pi} = \underline{4.19\ Hz}$$

b) critical damping $\Rightarrow b = 2\sqrt{km}$ (Eq. 13-41)

$$b = 2\sqrt{(300\ N/m)(0.400\ kg)} = \underline{21.9\ kg/s}$$

13-33

Eq. (13-47): $U(x) = \frac{1}{2}kx^2\left(1 - \frac{x}{L}\right)^2$

maximum $\Rightarrow \left.\frac{dU}{dx}\right|_{x=\frac{L}{2}} = 0$ and $\left.\frac{d^2U}{dx^2}\right|_{x=\frac{L}{2}} < 0$, so the graph of $U(x)$ will look like

U

x

L/2

Evaluate $\frac{dU}{dx}$ and $\frac{d^2U}{dx^2}$ at $x = \frac{L}{2}$ and see if they have the above values:

$$\frac{dU}{dx} = kx\left(1-\frac{x}{L}\right)^2 + \frac{1}{2}kx^2(2)\left(1-\frac{x}{L}\right)\left(-\frac{1}{L}\right) = kx\left(1-\frac{x}{L}\right)^2 - kx^2\left(1-\frac{x}{L}\right)\left(\frac{1}{L}\right)$$

$$\frac{dU}{dx} = kx\left(1-\frac{x}{L}\right)\left[1-\frac{x}{L}-\frac{x}{L}\right] = kx\left(1-\frac{x}{L}\right)\left(1-\frac{2x}{L}\right)$$

Then $\frac{d^2U}{dx^2} = k\left(1-\frac{x}{L}\right)\left(1-\frac{2x}{L}\right) - \frac{1}{L}kx\left(1-\frac{2x}{L}\right) - \frac{2}{L}kx\left(1-\frac{x}{L}\right)$

$$\frac{d^2U}{dx^2} = \left(1-\frac{2x}{L}\right)\left[k - \frac{kx}{L} - \frac{kx}{L}\right] - \frac{2kx}{L}\left(1-\frac{x}{L}\right) = k\left(1-\frac{2x}{L}\right)^2 - k\frac{2x}{L}\left(1-\frac{x}{L}\right)$$

$x = \frac{L}{2}$

$$\Rightarrow \left.\frac{dU}{dx}\right|_{x=\frac{L}{2}} = k\left(\frac{L}{2}\right)\left(1-\frac{L}{2L}\right)\left(1-\frac{2\left(\frac{L}{2}\right)}{L}\right) = \frac{kL}{2}\left(1-\frac{1}{2}\right)(1-1) = 0 \checkmark$$

$$\left.\frac{d^2U}{dx^2}\right|_{x=\frac{L}{2}} = k\left(1-\frac{2\left(\frac{L}{2}\right)}{L}\right)^2 - k\frac{2}{L}\left(\frac{L}{2}\right)\left(1-\frac{L}{2L}\right) = 0 - k\left(\frac{1}{2}\right) = -\frac{k}{2} < 0 \checkmark$$

Problems

13-37

a) $T = 2\pi\sqrt{\frac{m}{k}}$

We are given information about v at a particular x. The expression relating these two quantities comes from conservation of energy:

$$\frac{1}{2}mv^2 + \frac{1}{2}kx^2 = \frac{1}{2}kA^2$$

13-37 (cont)

We can solve this eq. for $\sqrt{\frac{m}{k}}$, and then use that result to calculate T.

$mv^2 = k(A^2 - x^2)$

$$\sqrt{\frac{m}{k}} = \frac{\sqrt{A^2-x^2}}{v} = \frac{\sqrt{(0.100\text{ m})^2 - (0.060\text{ m})^2}}{0.320\text{ m/s}} = 0.250\text{ s}$$

Then $T = 2\pi\sqrt{\frac{m}{k}} = 2\pi(0.250\text{ s}) = \underline{1.57\text{ s}}$

b) We are asked to relate x and v, so use the conservation of energy equation.

$$\tfrac{1}{2}mv^2 + \tfrac{1}{2}kx^2 = \tfrac{1}{2}kA^2$$

$$kx^2 = kA^2 - mv^2$$

$$x = \sqrt{A^2 - \frac{m}{k}v^2} = \sqrt{(0.100\text{ m})^2 - (0.250\text{ s})^2(0.120\text{ m/s})^2} = \underline{0.0954\text{ m}}$$

c) "small object whose mass is much less than the mass of the block"
$\Rightarrow$ doesn't alter the motion of the block.

For the block, $-kx = ma \Rightarrow a = -\frac{k}{m}x$
The maximum $|x|$ is A, so $a_{max} = \frac{k}{m}A$.

If the small object doesn't slip then the static friction force must be able to give it this much acceleration.
Free-body diagram for the small mass (mass m'):

a →; y; n; $f_s = \mu_s n$; x; m'g

$$\Sigma F_y = ma_y$$
$$n - m'g = 0$$
$$n = m'g$$

$$\Sigma F_x = ma_x$$
$$\mu_s n = m'a$$
$$\mu_s m'g = m'a$$
$$a = \mu_s g$$

But we require that $a = a_{max} = \frac{k}{m}A$, so $\frac{k}{m}A = \mu_s g$

$$\Rightarrow \mu_s = \frac{k}{m}\frac{A}{g} = \left(\frac{1}{0.250\text{ s}}\right)^2\left(\frac{0.100\text{ m}}{9.80\text{ m/s}^2}\right) = \underline{0.163}$$

13-39

a) Eq. (13-1): $F = -kx$

Eq. (11-12): $Y = \frac{l_0 F_\perp}{A\Delta l} \Rightarrow F_\perp = \frac{YA\Delta l}{l_0}$

$F_\perp$ is the force applied to the end of the wire; the force F with which the wire pulls back is $F = -F_\perp$. The displacement x of the end of the wire is $x = \Delta l$.

Thus $F = -\left(\frac{YA}{l_0}\right)x$ and $k = \frac{YA}{l_0}$.

b) From Table 11-1, Y(copper) $= 1.1 \times 10^{11}$ Pa

$$k = \frac{YA}{l_0} = \frac{(1.1\times10^{11}\text{ Pa})(\pi[0.75\times10^{-3}\text{ m}]^2)}{2.00\text{ m}} = \underline{9.7\times10^4\text{ N/m}}$$

13-41

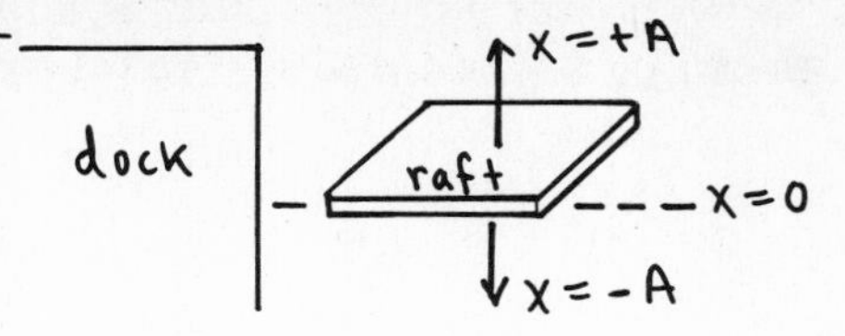

Let the raft be at $x = +A$ when $t = 0$. Then $\phi = 0$ and $x(t) = A\cos\omega t$. Calculate the time it takes the raft to move from $x = +A = +0.600\text{ m}$ to $x = A - 0.300\text{ m} = 0.300\text{ m}$.

Write the eq. for $x(t)$ in terms of T rather than ω:

$$\omega = \frac{2\pi}{T} \Rightarrow x(t) = A\cos\left(\frac{2\pi t}{T}\right)$$

$x = A$ at $t = 0$

$x = 0.300\text{ m} \Rightarrow 0.300\text{ m} = (0.600\text{ m})\cos\left(\frac{2\pi t}{T}\right)$

$$\cos\left(\frac{2\pi t}{T}\right) = 0.500 \Rightarrow \frac{2\pi t}{T} = \arccos(0.500) = 1.047\text{ rad}$$

$$t = \frac{T}{2\pi}(1.047\text{ rad}) = \frac{3.50\text{ s}}{2\pi}(1.047\text{ rad}) = 0.583\text{ s}$$

This is the time for the raft to move down from $x = 0.600\text{ m}$ to $x = 0.300\text{ m}$. But people can also get off while the raft is moving up from $x = 0.300\text{ m}$ to $x = 0.600\text{ m}$, so during each period of the motion the time the people have to get off is $2t = 2(0.583\text{ s}) = \underline{1.17\text{ s}}$

13-43

Measure x from the equilibrium position of the object, where the gravity and spring forces balance. Let $+x$ be downward.

a) Conservation of energy $\Rightarrow \frac{1}{2}mv^2 + \frac{1}{2}kx^2 = \frac{1}{2}kA^2$

$x = 0 \Rightarrow \frac{1}{2}mv^2 = \frac{1}{2}kA^2 \Rightarrow v = A\sqrt{\frac{k}{m}}$, just as for horizontal SHM.

We can use the period to calculate $\sqrt{\frac{k}{m}}$: $T = 2\pi\sqrt{\frac{m}{k}} \Rightarrow \sqrt{\frac{k}{m}} = \frac{2\pi}{T}$.

Thus $v = \frac{2\pi A}{T} = \frac{2\pi\,(0.100\text{ m})}{1.80\text{ s}} = \underline{0.349\text{ m/s}}$.

b) $ma = -kx \Rightarrow a = -\frac{k}{m}x$

$+x$ is downward, so here $x = -0.050\text{ m}$

$$\Rightarrow a = -\left(\frac{2\pi}{T}\right)^2(-0.050\text{ m}) = +\left(\frac{2\pi}{1.80\text{ s}}\right)^2(0.050\text{ m}) = \underline{0.609\text{ m/s}^2}\ (\text{positive} \Rightarrow \text{downward})$$

c) This is twice the time it takes to go from $x = 0$ to $x = +0.050\text{ m}$.

$x(t) = A\cos(\omega t + \phi)$

Let $\phi = -\frac{\pi}{2}$, so $x = 0$ at $t = 0$.

Then $x = A\cos\left(\omega t - \frac{\pi}{2}\right) = A\sin\omega t = A\sin\left(\frac{2\pi t}{T}\right)$

Find the time t that gives $x = +0.050\text{ m}$:

$$0.050\text{ m} = (0.100\text{ m})\sin\left(\frac{2\pi t}{T}\right) \Rightarrow \frac{2\pi t}{T} = \arcsin(0.50)$$

$$\frac{2\pi t}{T} = \frac{\pi}{6} \Rightarrow t = \frac{T}{12} = \frac{1.80\text{ s}}{12} = 0.150\text{ s}$$

The time asked for in the problem is twice this, $\underline{0.300\text{ s}}$.

13-43 (cont)

d) The problem is asking for the distance d that the spring stretches when the object hangs at rest from it.

Free-body diagram for the object:

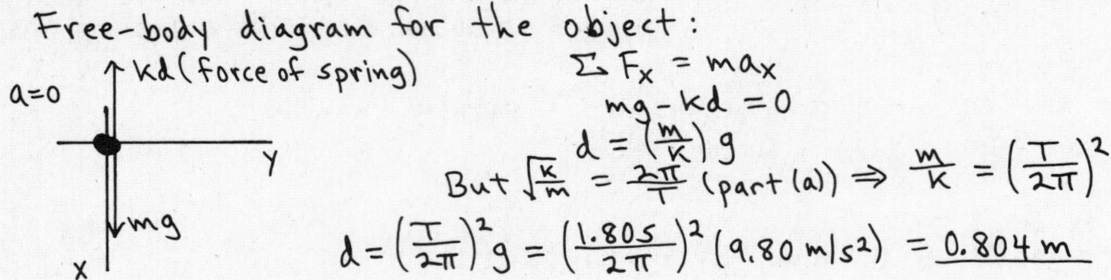

$$\Sigma F_x = ma_x$$

$$mg - kd = 0$$

$$d = \left(\frac{m}{k}\right) g$$

But $\sqrt{\frac{k}{m}} = \frac{2\pi}{T}$ (part (a)) $\Rightarrow \frac{m}{k} = \left(\frac{T}{2\pi}\right)^2$

$$d = \left(\frac{T}{2\pi}\right)^2 g = \left(\frac{1.80 s}{2\pi}\right)^2 (9.80 \text{ m/s}^2) = \underline{0.804 \text{ m}}$$

13-45

a) First find the speed of the steak just before it strikes the pan. Use a coordinate system with +y downward.

$v_{0y} = 0$ (released from rest)

$y - y_0 = 0.40$ m

$a_y = +9.80 \text{ m/s}^2$

$v_y = ?$

$$v_y^2 = v_{0y}^2 + 2a_y(y - y_0) \quad (v_{0y}^2 = 0)$$

$$v_y = +\sqrt{2a_y(y-y_0)} = +\sqrt{2(9.80 \text{ m/s}^2)(0.40 \text{ m})}$$

$$v_y = +2.80 \text{ m/s}$$

Apply conservation of momentum to the collision between the steak and the pan. After the collision the steak and the pan are moving together with the common velocity v_2. Let A be the steak and B be the pan.

P_y conserved $\Rightarrow m_A v_{A1y} + m_B v_{B1y} = m_A v_{A2y} + m_B v_{B2y}$

$$m_A v_{A1} = (m_A + m_B) v_2$$

$$v_2 = \left(\frac{m_A}{m_A + m_B}\right) v_{A1} = \left(\frac{1.8 \text{ kg}}{1.8 \text{ kg} + 0.20 \text{ kg}}\right)(2.80 \text{ m/s}) = \underline{2.52 \text{ m/s}}$$

b) Conservation of energy applied to the SHM gives

$$\tfrac{1}{2}mv_0^2 + \tfrac{1}{2}kx_0^2 = \tfrac{1}{2}kA^2,$$

where v_0 and x_0 are the initial speed and displacement of the object and where the displacement is measured from the equilibrium position of the object.

The weight of the steak will stretch the spring an additional distance d given by $kd = mg \Rightarrow d = \frac{mg}{k} = \frac{(1.8 \text{ kg})(9.80 \text{ m/s}^2)}{400 \text{ N/m}} = 0.0441$ m. So just after the steak hits the pan, before the pan has had time to move, the steak+pan is 0.0441 m above the equilibrium position of the combined object $\Rightarrow x_0 = 0.0441$ m.

13-45 (cont)

From part (a) $v_0 = 2.52\,m/s$, the speed of the combined object just after the collision.

Then $\frac{1}{2}mv_0^2 + \frac{1}{2}kx_0^2 = \frac{1}{2}kA^2$

$$\Rightarrow A = \sqrt{\frac{mv_0^2 + kx_0^2}{k}} = \sqrt{\frac{2.0\,kg\,(2.52\,m/s)^2 + (400\,N/m)(0.0441\,m)^2}{400\,N/m}} = \underline{0.184\,m}$$

c) $$T = 2\pi\sqrt{\frac{m}{k}} = 2\pi\sqrt{\frac{2.0\,kg}{400\,N/m}} = \underline{0.444\,s}$$

13-49

a) Measure x from the equilibrium position of the ball bearing and let +x be upward.

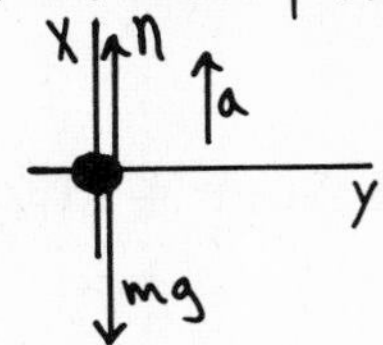

Let n be the normal force exerted on the ball bearing by the lens.

Free-body diagram for the ball bearing:

$$\Sigma F_x = ma_x$$
$$n - mg = ma$$
$$n = m(g+a)$$

The lens is moving with simple harmonic motion

$$\Rightarrow x(t) = A\cos(\omega t + \phi) = A\cos(2\pi f t + \phi)$$

and $\quad a = -\omega^2 x = -(2\pi f)^2 A\cos(2\pi f t + \phi)$

But this must also be a for the ball bearing. Use in the above eq. for n

$$\Rightarrow \boxed{n = m(g - (2\pi f)^2 A\cos(2\pi f t + \phi))}$$

b) The ball bounces when it loses contact with the lens during the motion.

Loses contact $\Rightarrow n = 0$, so $g - (2\pi f_b)^2 A\cos(2\pi f_b t + \phi) = 0$

The smallest frequency f_b where this happens is when $\cos(2\pi f_b t + \phi) = 1$, so

$$g - (2\pi f_b)^2 A = 0 \Rightarrow \boxed{g = (2\pi f_b)^2 A}$$

13-51

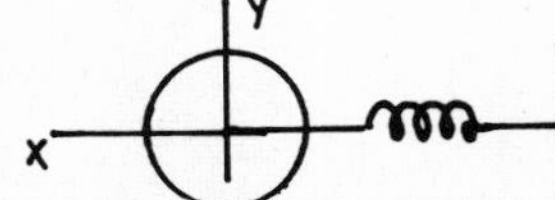

Let the origin of coordinates be at the center of the cylinder when it is at its equilibrium position.

Free-body diagram for the cylinder when it is displaced a distance x to the left:

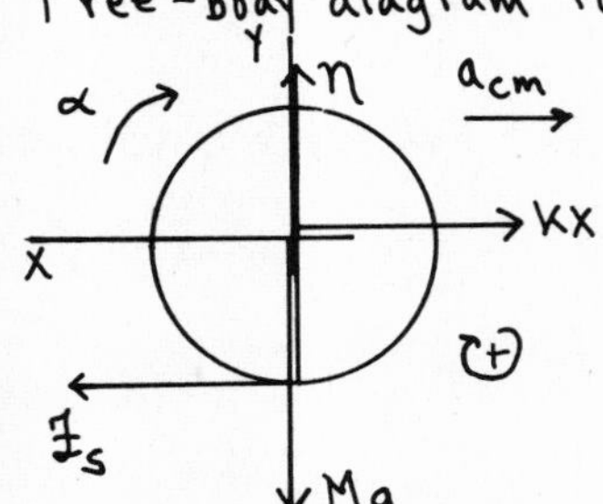

$$\Sigma \tau = I_{cm}\alpha$$
$$f_s R = (\tfrac{1}{2}MR^2)\alpha$$
$$f_s = \tfrac{1}{2}MR\alpha$$

But $R\alpha = a_{cm} \Rightarrow f_s = \frac{1}{2}Ma_{cm}$

$$\Sigma F_x = ma_x$$
$$f_s - kx = -Ma_{cm}$$
$$\tfrac{1}{2}Ma_{cm} - kx = -Ma_{cm}$$
$$kx = \tfrac{3}{2}Ma_{cm}$$
$$\left(\tfrac{2k}{3M}\right)x = a_{cm}$$

13-51 (cont)

Eq. (13-15): $a = -\omega^2 x$ (The minus sign says that x and a have opposite directions, as our diagram shows.)

Our result for a_{cm} is of this form, with $\omega^2 = \frac{2k}{3M} \Rightarrow \omega = \sqrt{\frac{2k}{3M}}$

$$T = \frac{2\pi}{\omega} = 2\pi\sqrt{\frac{3M}{2k}}$$

13-55

a) $T = 2\pi\sqrt{\frac{L}{g}} \Rightarrow L = g\left(\frac{T}{2\pi}\right)^2 = (9.80\text{ m/s}^2)\left(\frac{3.00\text{ s}}{2\pi}\right)^2 = \underline{2.23\text{ m}}$

b) Use a uniform slender rod of mass M and length $L = 0.50$ m. Pivot the rod about an axis that is a distance d above the center of the rod. The rod will oscillate as a physical pendulum with period $T = 2\pi\sqrt{\frac{I}{Mgd}}$. Chose d so that $T = 3.00$ s.

$$I = I_{cm} + Md^2 = \tfrac{1}{12}ML^2 + Md^2 = M\left(\tfrac{1}{12}L^2 + d^2\right)$$

$$T = 2\pi\sqrt{\frac{I}{Mgd}} = 2\pi\sqrt{\frac{M\left(\frac{1}{12}L^2 + d^2\right)}{Mgd}} = 2\pi\sqrt{\frac{\frac{1}{12}L^2 + d^2}{gd}}$$

Note that $T \to \infty$ as $d \to 0$ (pivot at center of rod) and that if pivot is at top of rod then $d = \frac{L}{2}$ and $T = 2\pi\sqrt{\frac{\left(\frac{1}{12} + \frac{1}{4}\right)L^2}{Lg/2}} = 2\pi\sqrt{\frac{L}{g}\frac{4}{6}} = 2\pi\sqrt{\frac{2L}{3g}}$

$T = 2\pi\sqrt{\frac{2(0.50\text{ m})}{3(9.80\text{ m/s}^2)}} = 1.16$ s, which is less than the desired 3.00 s. Thus it is reasonable to expect that there is a value of d between 0 and $\frac{L}{2}$ for which $T = 3.00$ s.

$$T = 2\pi\sqrt{\frac{\frac{1}{12}L^2 + d^2}{gd}}\text{; solve for } d:$$

$$gd\left(\frac{T}{2\pi}\right)^2 = \tfrac{1}{12}L^2 + d^2$$

$$d^2 - \left(\frac{T}{2\pi}\right)^2 gd + \frac{L^2}{12} = 0$$

$$d^2 - \left(\frac{3.00\text{ s}}{2\pi}\right)^2(9.80\text{ m/s}^2)\,d + \frac{(0.50\text{ m})^2}{12} = 0$$

$$d^2 - 2.2341d + 0.020833 = 0$$

The quadratic formula gives

$$d = \tfrac{1}{2}\left[2.2341 \pm \sqrt{(2.2341)^2 - 4(0.020833)}\,\right]\text{ m}$$

$$d = (1.1171 \pm 1.1077)\text{ m} \Rightarrow d = 2.225\text{ m or } d = 0.0094\text{ m}$$

The maximum value d can have is $\frac{L}{2} = 0.25$ m, so the answer we want is

$$d = 0.0094\text{ m} = 0.94\text{ cm}$$

Therefore, take a slender rod of length 0.50 m and pivot it about an axis that is 0.94 cm above its center.

CHAPTER 14

Exercises 3, 5, 9, 13, 15, 17, 19, 21, 23, 25, 31, 33

Problems 37, 39, 41, 43, 49, 51, 53, 57, 59, 61, 63, 67, 69

Exercises

14-3

$\rho = \frac{m}{V} \Rightarrow m = \rho V$

From Table 14-1, $\rho = 2.7 \times 10^3 \text{ kg/m}^3$.

For a cylinder of length L and radius R, $V = (\pi R^2)L = \pi(0.30 \times 10^{-2}\text{ m})^2(0.62\text{ m})$

$V = 1.753 \times 10^{-5}\text{ m}^3$

Then $m = \rho V = (2.7 \times 10^3\text{ kg/m}^3)(1.753 \times 10^{-5}\text{ m}^3) = \underline{0.0473\text{ kg}}$

14-5

a) gauge pressure $= p - p_0 = \rho g h$

From Table 14-1 the density of seawater is $1.03 \times 10^3\text{ kg/m}^3$, so

$p - p_0 = \rho g h = (1.03 \times 10^3\text{ kg/m}^3)(9.80\text{ m/s}^2)(800\text{ m}) = \underline{8.08 \times 10^6\text{ Pa}}$

b) The force on each side of the window is $F = pA$. Inside the pressure is p_0 and outside in the water the pressure is $p = p_0 + \rho g h$.

inside bell | outside bell

$F_1 = p_0 A$ → | ← $F_2 = (p_0 + \rho g h)A$

The net force is

$F_2 - F_1 = (p_0 + \rho g h)A - p_0 A = (\rho g h)A$

$= (8.08 \times 10^6\text{ Pa})\,\pi\,(7.5 \times 10^{-2}\text{ m})^2 = \underline{1.43 \times 10^5\text{ N}}$

14-9

$p_a = 970\text{ millibar}\left(\frac{1.013 \times 10^5\text{ Pa}}{1013\text{ millibar}}\right) = 9.70 \times 10^4\text{ Pa}$

a) Apply $p = p_0 + \rho g h$ to the right-hand tube. The top of this tube is open to the air so $p_0 = p_a$. The density of the liquid (mercury) is $13.6 \times 10^3\text{ kg/m}^3$.
Thus $p = 9.70 \times 10^4\text{ Pa} + (13.6 \times 10^3\text{ kg/m}^3)(9.80\text{ m/s}^2)(0.0700\text{ m}) = \underline{1.06 \times 10^5\text{ Pa}}$.

b) $p = p_0 + \rho g h = 9.70 \times 10^4\text{ Pa} + (13.6 \times 10^3\text{ kg/m}^3)(9.80\text{ m/s}^2)(0.0400\text{ m}) = \underline{1.02 \times 10^5\text{ Pa}}$.

c) Since $y_2 - y_1 = 4.00\text{ cm}$ the pressure at the mercury surface in the left-hand tube equals that calculated in part (b). Thus the absolute pressure of gas in the tank is $1.02 \times 10^5\text{ Pa}$.

d) $p - p_a = \rho g h = (13.6 \times 10^3\text{ kg/m}^3)(9.80\text{ m/s}^2)(0.0400\text{ m}) = \underline{5.33 \times 10^3\text{ Pa}}$

14-13

The floating object is the slab of ice plus the woman; the buoyant force must support both. The volume of water displaced equals the volume V_{ice} of the ice.

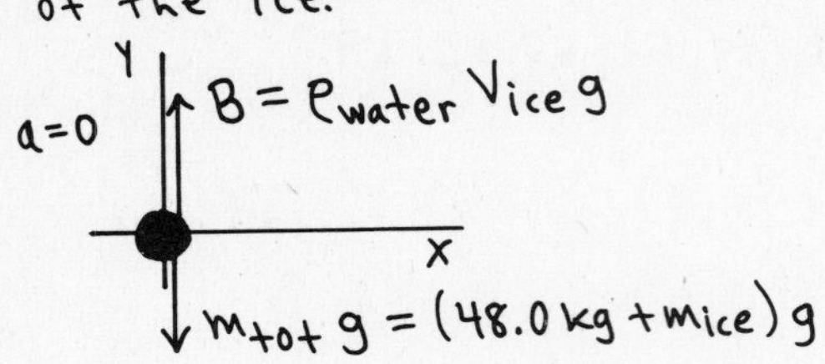

$$\Sigma F_y = ma_y$$

$$B - m_{tot}\, g = 0$$

$$\rho_{water} V_{ice} g = (48.0\text{ kg} + m_{ice}) g$$

But $\rho = \frac{m}{V} \Rightarrow m_{ice} = \rho_{ice} V_{ice}$

$$\Rightarrow \rho_{water} V_{ice} = 48.0\text{ kg} + \rho_{ice} V_{ice}$$

$$V_{ice} = \frac{48.0\text{ kg}}{\rho_{water} - \rho_{ice}} = \frac{48.0\text{ kg}}{1000\text{ kg/m}^3 - 920\text{ kg/m}^3} = \underline{0.600\text{ m}^3}$$

14-15

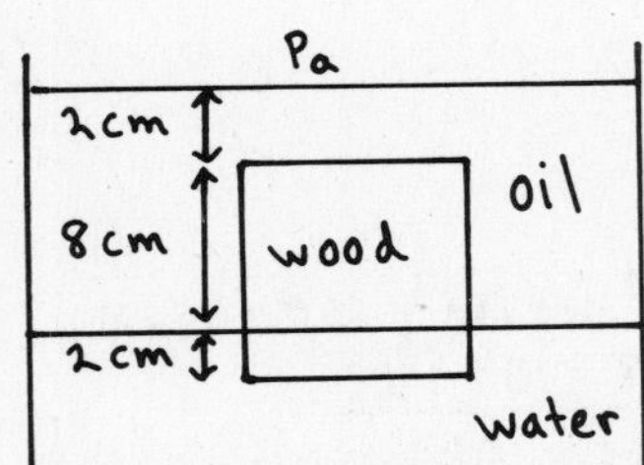

a) gauge pressure $= p - p_a = \rho g h$

The upper face is 2.00 cm below the top of the oil

$\Rightarrow p - p_a = (650\text{ kg/m}^3)(9.80\text{ m/s}^2)(0.0200\text{ m}) = \underline{127\text{ Pa}}$

b) The pressure at the interface is

$$p_{interface} = p_a + \rho_{oil}\, g\, (0.100\text{ m})$$

The lower face of the block is 2.00 cm below the interface, so the pressure there is $p = p_{interface} + \rho_{water}\, g\, (0.0200\text{ m})$.

Combining these two equations gives

$$p - p_a = \rho_{oil}\, g\,(0.100\text{ m}) + \rho_{water}\, g\,(0.0200\text{ m}) = [(650\text{ kg/m}^3)(0.100\text{ m}) + (1000\text{ kg/m}^3)(0.0200\text{ m})](9.80\text{ m/s}^2)$$

$$p - p_a = \underline{833\text{ Pa}}$$

c) Consider the forces on the block. The area of each face of the block is $A = (0.100\text{ m})^2 = 0.0100\text{ m}^2$. Let the absolute pressure at the top face be p_t and the pressure at the bottom face be p_b. Then the free-body diagram for the block is

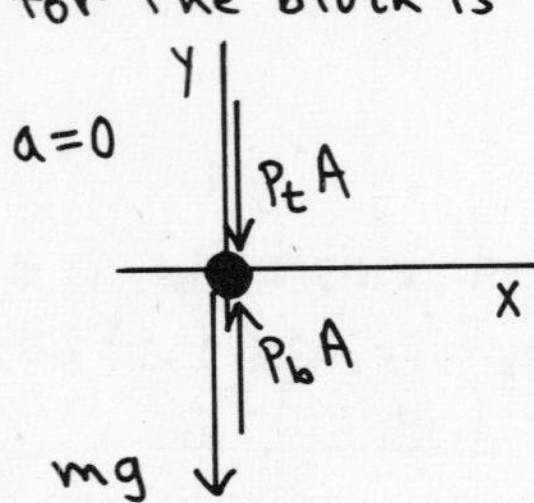

$$\Sigma F_y = ma_y$$

$$p_b A - p_t A - mg = 0$$

$$(p_b - p_t) A = mg$$

Note that $(p_b - p_t) = (p_b - p_a) - (p_t - p_a) = 833\text{ Pa} - 127\text{ Pa} = 706\text{ Pa}$; the difference in absolute pressures equals the difference in gauge pressures.

$$m = \frac{(p_b - p_t)A}{g} = \frac{(706\text{ Pa})(0.0100\text{ m}^2)}{9.80\text{ m/s}^2} = \underline{0.720\text{ kg}}$$

14-17

Use Eq. (14-12): $p - p_a = \frac{4\gamma}{R} = \frac{4(25.0\times10^{-3}\text{ N/m})}{1.50\times10^{-2}\text{ m}} = \underline{6.67\text{ Pa}}$

14-19

a) volume flow rate $= vA = v\pi r^2$

$\Rightarrow v = \dfrac{\text{volume flow rate}}{\pi r^2} = \dfrac{0.600\ \text{m}^3/\text{s}}{\pi(0.200\text{m})^2} = \underline{4.77\ \text{m/s}}$

b) The continuity equation, Eq. (14-14), says

$v_1 A_1 = v_2 A_2$

$v_1 \pi r_1^2 = v_2 \pi r_2^2 \Rightarrow r_2 = r_1\sqrt{\dfrac{v_1}{v_2}} = 0.200\text{m}\sqrt{\dfrac{4.77\ \text{m/s}}{3.80\ \text{m/s}}} = \underline{0.224\ \text{m}}$

14-21

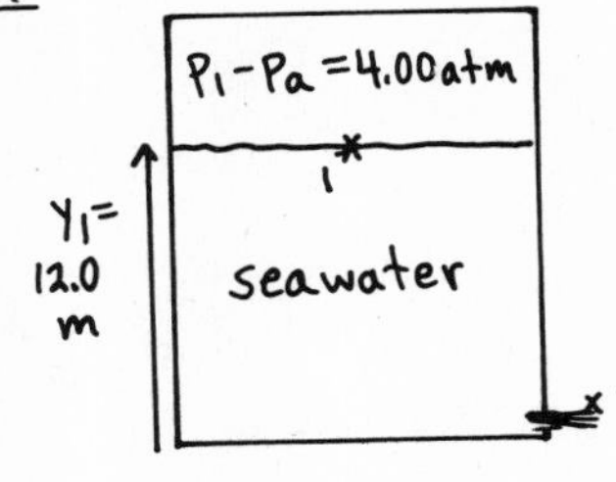

Apply Bernoulli's equation with points 1 and 2 chosen as shown in the sketch. Let $y=0$ at the bottom of the tank so $y_1 = 12.0\text{m}$ and $y_2 = 0$.

$P_2 = P_a$

$P_1 - P_a = 4.00\ \text{atm}\left(\dfrac{1.013\times10^5\ \text{Pa}}{1\ \text{atm}}\right) = 4.052\times10^5\ \text{Pa}$

$P_1 + \rho g y_1 + \frac{1}{2}\rho v_1^2 = P_2 + \rho g y_2 + \frac{1}{2}\rho v_2^2$

$A_1 v_1 = A_2 v_2$, so $v_1 = \left(\frac{A_2}{A_1}\right) v_2$. But the cross-section area of the tank (A_1) is much larger than the cross-section area of the hole (A_2) so $v_1 \ll v_2$ and the $\frac{1}{2}\rho v_1^2$ term can be neglected.

$\Rightarrow \frac{1}{2}\rho v_2^2 = (P_1 - P_2) + \rho g y_1$

Use $P_2 = P_a$ and solve for v_2:

$v_2 = \sqrt{\dfrac{2(P_1-P_a)}{\rho} + 2gy_1} = \sqrt{\dfrac{2(4.052\times10^5\ \text{Pa})}{1030\text{kg/m}^3} + 2(9.80\ \text{m/s}^2)(12.0\ \text{m})}$

$v_2 = \underline{32.0\ \text{m/s}}$

14-23

$P_1 + \rho g y_1 + \frac{1}{2}\rho v_1^2 = P_2 + \rho g y_2 + \frac{1}{2}\rho v_2^2$

horizontal $\Rightarrow y_1 = y_2 \Rightarrow P_1 + \frac{1}{2}\rho v_1^2 = P_2 + \frac{1}{2}\rho v_2^2$

If we solve for v_2 then we can use the discharge rate to calculate A_2.

Also, $v_1 A_1 = 4.00\times10^{-3}\text{m}^3/\text{s} \Rightarrow v_1 = \dfrac{4.00\times10^{-3}\ \text{m}^3/\text{s}}{1.00\times10^{-3}\ \text{m}^3} = 4.00\ \text{m/s}$

$\frac{1}{2}\rho v_2^2 = \frac{1}{2}\rho v_1^2 + (P_1 - P_2)$

$v_2 = \sqrt{v_1^2 + \dfrac{2(P_1-P_2)}{\rho}} = \sqrt{(4.00\ \text{m/s})^2 + \dfrac{2(1.60\times10^5\ \text{Pa} - 1.20\times10^5\ \text{Pa})}{1000\ \text{kg/m}^3}} = 9.80\ \text{m/s}$

Then $v_2 A_2 = 4.00\times10^{-3}\text{m}^3/\text{s} \Rightarrow A_2 = \dfrac{4.00\times10^{-3}\ \text{m}^3/\text{s}}{9.80\ \text{m/s}} = \underline{4.08\times10^{-4}\ \text{m}^2}$

14-25

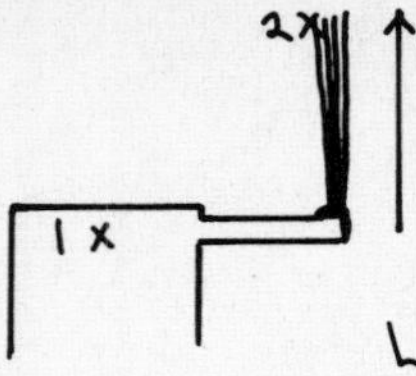

Apply Bernoulli's equation to points 1 and 2 as shown in the sketch. Point 1 is in the mains and point 2 is at the maximum height reached by the stream, so $v_2 = 0$.

$$p_1 + \rho g y_1 + \tfrac{1}{2}\rho v_1^2 = p_2 + \rho g y_2 + \tfrac{1}{2}\rho v_2^2 \rightarrow 0$$

Let $y_1 = 0$, so $y_2 = 25.0$ m. The mains have large diameter so $v_1 \approx 0$.

$$\Rightarrow p_1 = p_2 + \rho g y_2$$

But $p_2 = p_a$, so $p_1 - p_a = \rho g y_2 = (1000\ \text{kg/m}^3)(9.80\ \text{m/s}^2)(25.0\ \text{m}) = \underline{2.45 \times 10^5\ \text{Pa}}$

14-31

$\eta = 1.005$ centipoise $= 1.005 \times 10^{-3}$ N·s/m^2

a) The volume flow rate $\frac{dV}{dt}$ is given by Eq.(14-29):

$$\frac{dV}{dt} = \frac{\pi}{8}\left(\frac{R^4}{\eta}\right)\left(\frac{p_1 - p_2}{L}\right)$$

The absolute pressure at the pump is p_1 and $p_2 = p_a$ is the pressure at the open end of the pipe, so $p_1 - p_2 = p_1 - p_a$, the gauge pressure at the pump.

$$\frac{dV}{dt} = \frac{\pi}{8}\,\frac{(4.00 \times 10^{-2}\ \text{m})^4}{1.005 \times 10^{-3}\ \text{N·s/m}^2}\left(\frac{1400\ \text{Pa}}{15.0\ \text{m}}\right) = \underline{0.0934\ \text{m}^3/\text{s}}$$

b) For the same volume flow rate $R^4\,\Delta p$ must stay constant, where Δp is the gauge pressure maintained by the pump. Let $R_a = 8.00$ cm and $R_b = 4.00$ cm, so $\Delta p_a = 1400$ Pa and we are asked to calculate Δp_b.

$$R_a^4\,\Delta p_a = R_b^4\,\Delta p_b \Rightarrow \Delta p_b = \Delta p_a \left(\frac{R_a}{R_b}\right)^4 = 1400\ \text{Pa}\left(\frac{8.00\ \text{cm}}{4.00\ \text{cm}}\right)^4 = \underline{2.24 \times 10^4\ \text{Pa}}$$

c) Same R, Δp, and $L \Rightarrow \left(\frac{dV}{dt}\right)\eta = \frac{\pi}{8}R^4\frac{(p_1 - p_2)}{L} = \text{constant}$

$$\Rightarrow \left(\frac{dV}{dt}\right)_a \eta_a = \left(\frac{dV}{dt}\right)_b \eta_b$$ where a refers to 20°C and b to 60°C

$$\left(\frac{dV}{dt}\right)_b = \left(\frac{dV}{dt}\right)_a \frac{\eta_a}{\eta_b} = 0.0934\ \text{m}^3/\text{s}\left(\frac{1.005\ \text{centipoise}}{0.469\ \text{centipoise}}\right) = \underline{0.200\ \text{m}^3/\text{s}}$$

14-33

The viscous drag force is given by Eq.(14-30): $F = 6\pi\eta r v$.

To compare this to the weight of the sphere, express the weight in terms of the density ρ and radius r of the sphere:

$$m = \rho V = \rho\left(\tfrac{4}{3}\pi r^3\right) \Rightarrow w = \rho g\left(\tfrac{4}{3}\pi r^3\right)$$

$$F = \tfrac{1}{4}w \Rightarrow 6\pi\eta r v = \tfrac{1}{4}\rho g\left(\tfrac{4}{3}\pi r^3\right)$$

$$v = \frac{\rho r^2 g}{18\eta}$$

From Table 14-1, $\rho_{gold} = 19.3 \times 10^3$ kg/m^3.

Thus $$v = \frac{(19.3 \times 10^3\ \text{kg/m}^3)(6.00 \times 10^{-3}\ \text{m})^2(9.80\ \text{m/s}^2)}{18\,(0.986\ \text{N·s/m}^2)} = \underline{0.384\ \text{m/s}}$$

Problems

14-37

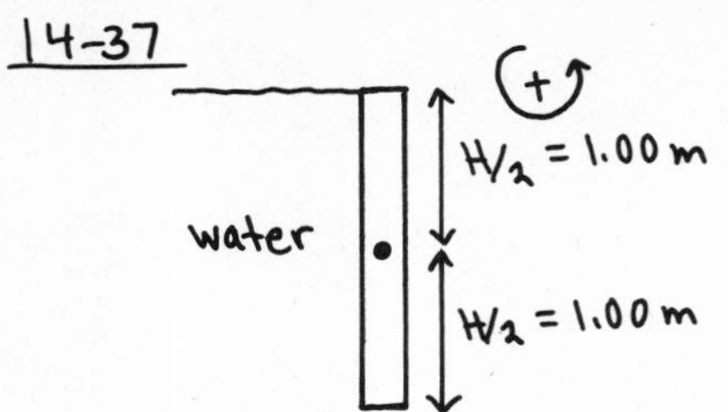

Let τ_u be the torque due to the net force of the water on the upper half of the gate and τ_l be the torque due to the force on the lower half. With the indicated sign convention τ_l is positive and τ_u is negative, so the net torque about the hinge is $\tau = \tau_l - \tau_u$. Let H be the height of the gate.

Upper-half of gate:

Calculate the torque due to the force on a narrow strip of height dy located a distance y below the top of the gate. Then integrate to get the total torque.

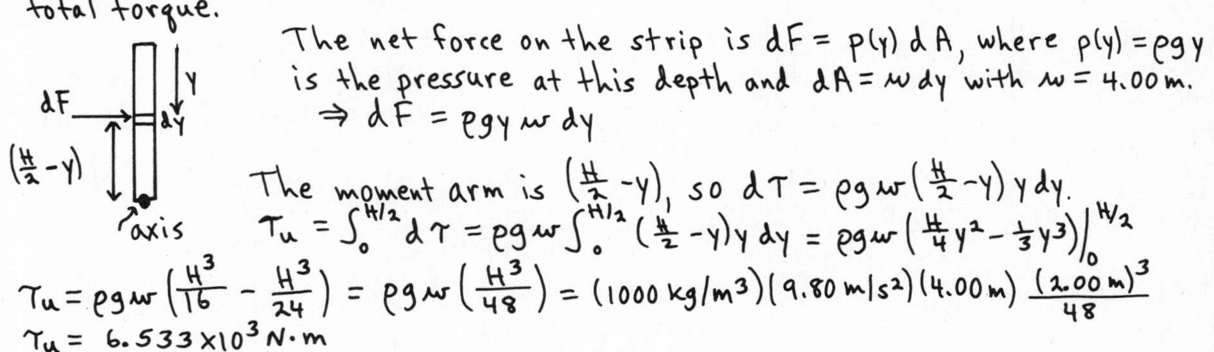

The net force on the strip is $dF = p(y)\,dA$, where $p(y) = \rho g y$ is the pressure at this depth and $dA = w\,dy$ with $w = 4.00$ m.

$\Rightarrow dF = \rho g y w\, dy$

The moment arm is $(\frac{H}{2} - y)$, so $d\tau = \rho g w (\frac{H}{2} - y) y\, dy$.

$$\tau_u = \int_0^{H/2} d\tau = \rho g w \int_0^{H/2} (\tfrac{H}{2} - y) y\, dy = \rho g w \left(\tfrac{H}{4}y^2 - \tfrac{1}{3}y^3\right)\Big|_0^{H/2}$$

$$\tau_u = \rho g w \left(\frac{H^3}{16} - \frac{H^3}{24}\right) = \rho g w \left(\frac{H^3}{48}\right) = (1000\ \text{kg/m}^3)(9.80\ \text{m/s}^2)(4.00\ \text{m})\,\frac{(2.00\ \text{m})^3}{48}$$

$\tau_u = 6.533 \times 10^3$ N·m

Lower-half of gate:

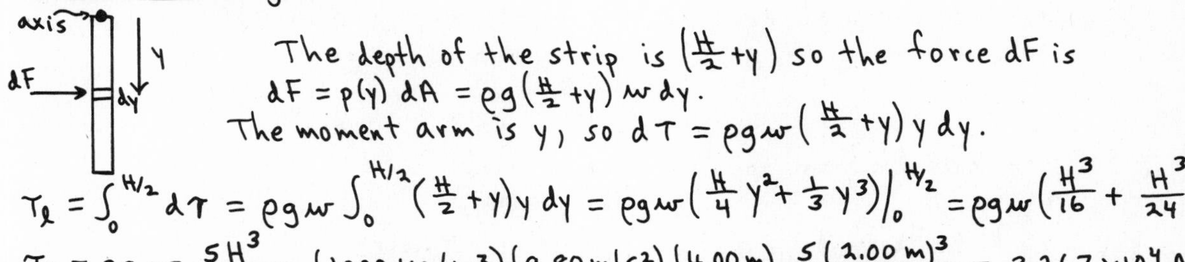

The depth of the strip is $(\frac{H}{2} + y)$ so the force dF is $dF = p(y)\,dA = \rho g (\frac{H}{2} + y) w\, dy$.

The moment arm is y, so $d\tau = \rho g w (\frac{H}{2} + y) y\, dy$.

$$\tau_l = \int_0^{H/2} d\tau = \rho g w \int_0^{H/2} (\tfrac{H}{2} + y) y\, dy = \rho g w \left(\tfrac{H}{4}y^2 + \tfrac{1}{3}y^3\right)\Big|_0^{H/2} = \rho g w \left(\frac{H^3}{16} + \frac{H^3}{24}\right)$$

$$\tau_l = \rho g w \frac{5H^3}{48} = (1000\ \text{kg/m}^3)(9.80\ \text{m/s}^2)(4.00\ \text{m})\,\frac{5(2.00\ \text{m})^3}{48} = 3.267 \times 10^4\ \text{N·m}$$

Then $\tau = \tau_l - \tau_u = 3.267 \times 10^4\ \text{N·m} - 6.533 \times 10^3\ \text{N·m} = \underline{2.61 \times 10^4\ \text{N·m}}$.

14-39

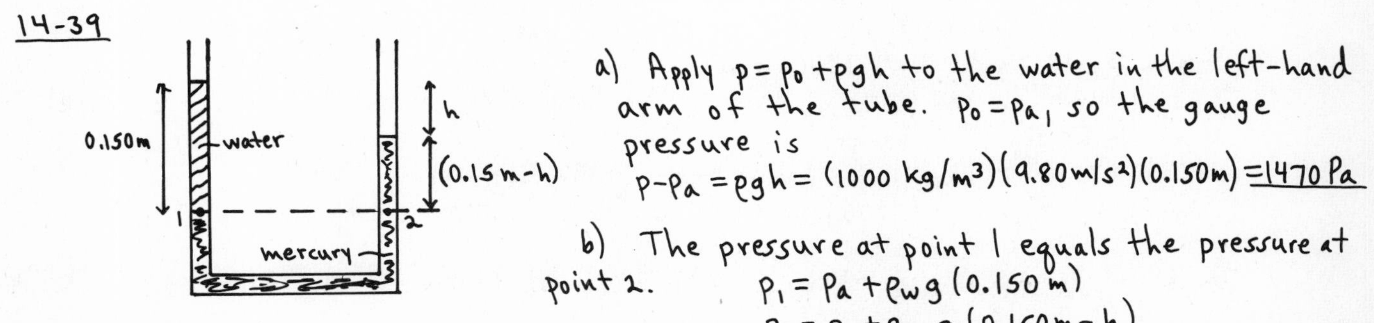

a) Apply $p = p_0 + \rho g h$ to the water in the left-hand arm of the tube. $p_0 = p_a$, so the gauge pressure is

$p - p_a = \rho g h = (1000\ \text{kg/m}^3)(9.80\ \text{m/s}^2)(0.150\ \text{m}) = \underline{1470\ \text{Pa}}$

b) The pressure at point 1 equals the pressure at point 2.

$p_1 = p_a + \rho_w g (0.150\ \text{m})$

$p_2 = p_a + \rho_{Hg} g (0.150\ \text{m} - h)$

14-39 (cont)

$$p_1 = p_2 \Rightarrow \rho_w g (0.150\text{ m}) = \rho_{Hg} g (0.150\text{ m} - h)$$

$$0.150\text{m} - h = \frac{\rho_w (0.150\text{ m})}{\rho_{Hg}} = \frac{(1000\text{ kg/m}^3)(0.150\text{ m})}{13.6\times10^3\text{ kg/m}^3} = 0.011\text{ m}$$

$$h = 0.150\text{ m} - 0.011\text{ m} = 0.139\text{ m} = \underline{13.9\text{ cm}}$$

14-41

a) Free-body diagram for the dirigible:
(The mass m_{tot} is 12,000 kg plus the mass m_{gas} of the gas that fills the dirigible. B is the buoyant force exerted by the air.)

y; $B = \rho_{air} V g$; $a = 0$; x; $m_{tot} g = (12{,}000\text{ kg} + m_{gas}) g$

$$\Sigma F_y = ma_y$$
$$B - m_{tot} g = 0$$
$$\rho_{air} V g = (12{,}000\text{kg} + m_{gas}) g$$

Write m_{gas} in terms of V:

$$m_{gas} = \rho_{gas} V$$
$$\Rightarrow \rho_{air} V = 12{,}000\text{ kg} + \rho_{gas} V$$

$$V = \frac{12{,}000\text{ kg}}{\rho_{air} - \rho_{gas}} = \frac{12{,}000\text{ kg}}{1.20\text{ kg/m}^3 - 0.0899\text{ kg/m}^3} = \underline{1.08\times10^4\text{ m}^3}$$

b) Let m_{lift} be the mass that could be lifted.
From part (a), $m_{lift} = (\rho_{air} - \rho_{gas}) V = (1.20\text{kg/m}^3 - 0.166\text{ kg/m}^3)(1.08\times10^4\text{m}^3) = \underline{11{,}200\text{Kg}}$

Hydrogen is not used because it is highly explosive in air.

14-43

Free-body diagram for the barge + coal:

y; $B = \rho_w V_{barge} g$; $a=0$; x; $(m_{barge} + m_{coal}) g$

$$\Sigma F_y = ma_y$$
$$B - (m_{barge} + m_{coal}) g = 0$$
$$\rho_w V_{barge} g = (m_{barge} + m_{coal}) g$$
$$m_{coal} = \rho_w V_{barge} - m_{barge}$$

$$V_{barge} = (22\text{m})(12\text{m})(40\text{ m}) = 1.056\times10^4\text{m}^3$$

The mass of the barge is $m_{barge} = \rho_s V_s$, where s refers to steel.
From Table 14-1, $\rho_s = 7800\text{ kg/m}^3$.
The volume V_s is 0.075 m times the total area of the five pieces of steel that make up the barge:

$$V_s = (0.075\text{ m})[2(22\text{m})(12\text{m}) + 2(40\text{ m})(12\text{ m}) + (22\text{ m})(40\text{ m})] = 177.6\text{ m}^3.$$

Therefore, $m_{barge} = (7800\text{ kg/m}^3)(177.6\text{ m}^3) = 1.385\times10^6\text{ kg}$

Thus $m_{coal} = \rho_w V_{barge} - m_{barge} = (1000\text{ kg/m}^3)(1.056\times10^4\text{m}^3) - 1.385\times10^6\text{kg} = \underline{9.18\times10^6\text{ kg}}$

14-43 (cont)

The volume of this mass of coal is $V_{coal} = \dfrac{m_{coal}}{\rho_{coal}} = \dfrac{9.18 \times 10^6\ \text{kg}}{1500\ \text{kg/m}^3} = 6120\ \text{m}^3$; this is less than V_{barge} so will fit into the barge.

14-49

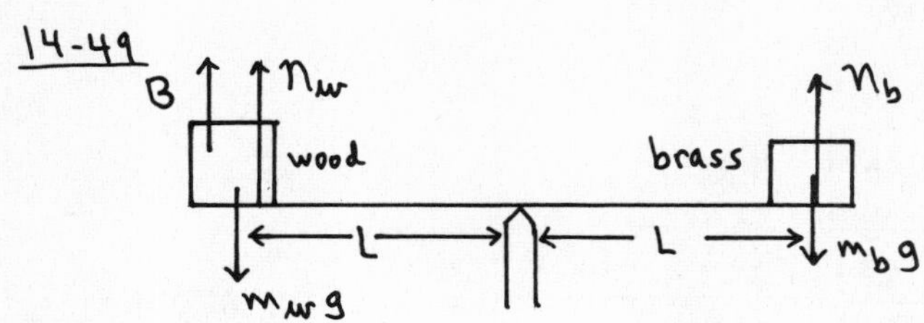

The buoyant force on the brass is neglected, but we include the buoyant force B on the block of wood. n_w and n_b are the normal forces exerted by the balance arm on which the objects sit.

Free-body diagram for the balance arm:

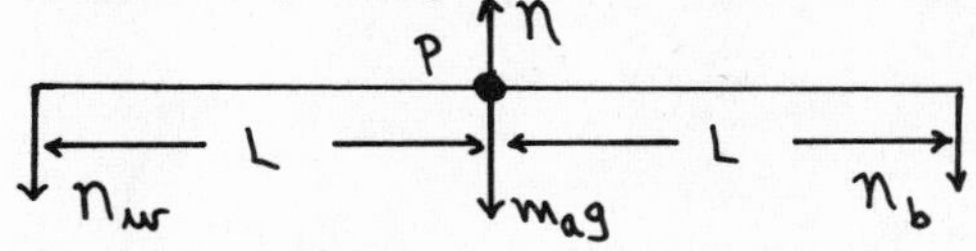

$$\Sigma \tau_P = 0$$
$$n_w L - n_b L = 0$$
$$n_w = n_b$$

Free-body diagram for the brass mass:

a=0

$$\Sigma F_y = ma_y$$
$$n_b - m_b g = 0$$
$$n_b = m_b g$$

Free-body diagram for the block of wood:

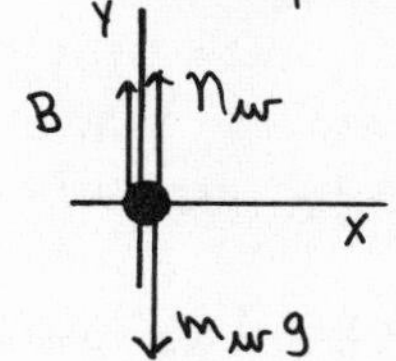

$$\Sigma F_y = ma_y$$
$$n_w + B - m_w g = 0$$
$$n_w = m_w g - B$$

But $n_b = n_w \Rightarrow m_b g = m_w g - B$

And $B = \rho_{air} V_w g = \rho_{air}\left(\dfrac{m_w}{\rho_w}\right) g \Rightarrow m_b g = m_w g - \rho_{air}\left(\dfrac{m_w}{\rho_w}\right) g$

$$m_w = \frac{m_b}{1 - \rho_{air}/\rho_w} = \frac{0.0750\ \text{kg}}{1 - 1.2\ \text{kg/m}^3 / 150\ \text{kg/m}^3} = \underline{0.0756\ \text{kg}}$$

14-51

Free-body diagram for the piece of alloy:

B, T = 32.0 N, $m_{tot} g = 45.0$ N

$$\Sigma F_y = ma_y$$
$$B + T - m_{tot} g = 0$$
$$B = m_{tot} g - T = 45.0\ \text{N} - 32.0\ \text{N} = 13.0\ \text{N}$$

Also, $m_{tot} g = 45.0\ \text{N} \Rightarrow m_{tot} = \dfrac{45.0\ \text{N}}{9.80\ \text{m/s}^2} = 4.59\ \text{kg}$

We can use the known value of the buoyant force to calculate the volume of the object:

$$B = \rho_w V_{obj}\, g = 13.0\ \text{N}$$
$$V_{obj} = \frac{13.0\ \text{N}}{\rho_w g} = \frac{13.0\ \text{N}}{(1000\ \text{kg/m}^3)(9.80\ \text{m/s}^2)} = 1.327 \times 10^{-3}\ \text{m}^3$$

We know two things:

(1) The mass m_g of the gold plus the mass m_a of the aluminum must add to m_{tot}:

$$m_g + m_a = m_{tot}$$

14-51 (cont)

We write this in terms of the volumes V_g and V_a of the gold and aluminum

$\Rightarrow$ $\boxed{\rho_g V_g + \rho_a V_a = m_{tot}}$

(2) The volumes V_a and V_g must add to give V_{obj}:

$V_a + V_g = V_{obj} \Rightarrow \boxed{V_a = V_{obj} - V_g}$

Use this in the first eq. to eliminate V_a

$\Rightarrow \rho_g V_g + \rho_a (V_{obj} - V_g) = m_{tot}$

$$V_g = \frac{m_{tot} - \rho_a V_{obj}}{\rho_g - \rho_a} = \frac{4.59\text{ kg} - (2.7\times10^3\text{ kg/m}^3)(1.327\times10^{-3}\text{ m}^3)}{(19.3\times10^3 - 2.7\times10^3)\text{ kg/m}^3} = 6.067\times10^{-5}\text{ m}^3$$

Then $w_g = m_g g = \rho_g V_g g = (19.3\times10^3\text{ kg/m}^3)(6.067\times10^{-5}\text{ m}^3)(9.80\text{ m/s}^2) = \underline{11.5\text{ N}}$

14-53

In both cases the total buoyant force must equal the weight of the barge plus the weight of the anchor. Thus the total amount of water displaced must be the same when the anchor is in the boat as when it is over the side. When the anchor is in the water the barge displaces less water, less by the amount the anchor displaces $\Rightarrow$ the barge <u>rises</u> in the water.

The volume of the anchor is $V_{anchor} = \frac{m}{\rho} = \frac{25.0\text{ kg}}{7860\text{ kg/m}^3} = 3.181\times10^{-3}\text{ m}^3$.
The barge rises in the water a vertical distance, h given by
$hA = 3.181\times10^{-3}\text{ m}^3$, where A is the area of the bottom of the barge.

$$h = \frac{3.181\times10^{-3}\text{ m}^3}{6.00\text{ m}^2} = \underline{5.30\times10^{-4}\text{ m}}$$

14-57

a) Free-body diagram for the brass block:

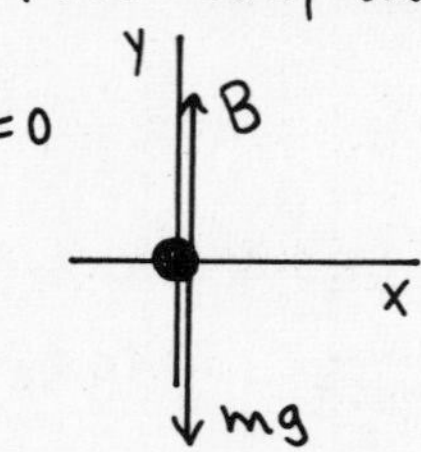

$\Sigma F_y = ma_y$

$B - mg = 0$

$\rho_{Hg} V_{sub} g = \rho_{brass} V_{tot} g$, where V_{sub} is the volume of the block that is below the surface of the mercury and V_{tot} is the total volume of the brass block.

The fraction submerged is $\frac{V_{sub}}{V_{tot}} = \frac{\rho_{brass}}{\rho_{Hg}} = \frac{8.6\times10^3\text{ kg/m}^3}{13.6\times10^3\text{ kg/m}^3} = 0.632.$

The fraction above the mercury surface is $1 - 0.632 = \underline{0.368}$.

b)

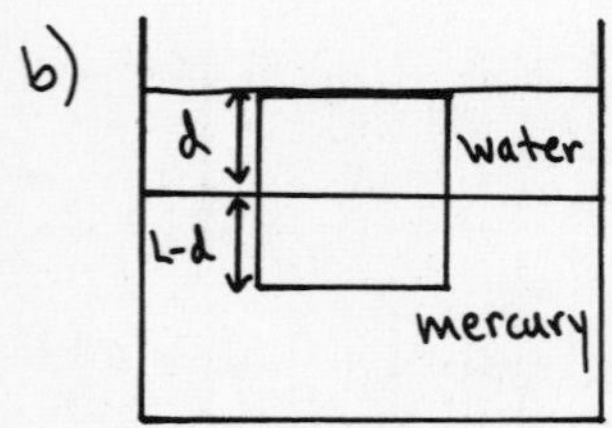

Let the depth of the water layer be d. Calculate the upward buoyant force B by calculating the gauge pressure $p - p_a$ at the lower face of the block:

$$p - p_a = \rho_w g d + \rho_{Hg}(L-d)g$$

$$B = (p - p_a)L^2 = \rho_w g d L^2 + \rho_{Hg}(L-d)g L^2$$

14-57 (cont)

We can write the weight of the block as $mg = \rho_{brass} V g = \rho_{brass} L^3 g$

Then $B - mg = 0 \Rightarrow \rho_w g d L^2 + \rho_{Hg}(L-d) g L^2 = \rho_{brass} L^3 g$

$$\rho_w d + \rho_{Hg}(L-d) = \rho_{brass} L$$

$$d = \left(\frac{\rho_{Hg} - \rho_{brass}}{\rho_{Hg} - \rho_w}\right) L = \left(\frac{13.6\times10^3\ kg/m^3 - 8.6\times10^3\ kg/m^3}{13.6\times10^3\ kg/m^3 - 1.0\times10^3\ kg/m^3}\right) L = \underline{0.397 L}$$

14-59

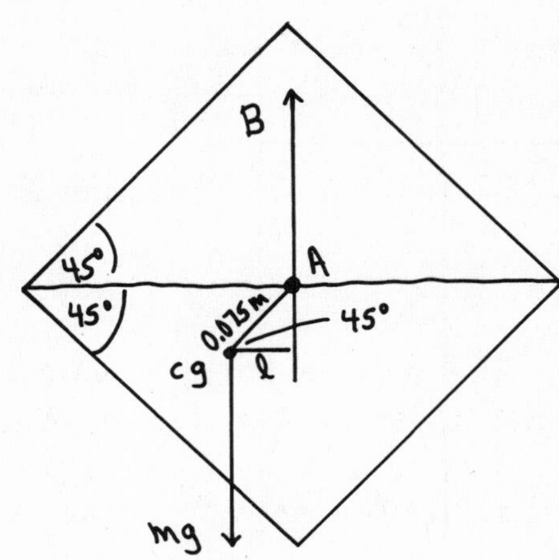

The resultant buoyant force acts at the geometrical center of the submerged portion of the object. The weight of the object acts at the center of gravity of the object. These two points are displaced from each other in Fig. 14-42(b), and this gives rise to a restoring torque about point A.

For an axis at point A

$\tau_B = 0$

$\tau_{mg} = mgl = mg(0.075\ m)\cos 45° = (0.053\ m)\,mg$

The block is floating $\Rightarrow B = mg$. Calculate B in order to find the weight mg of the block. Half of the volume of the block is submerged

$\Rightarrow V_{sub} = \frac{1}{2}(0.30\ m)^3 = 0.0135\ m^3$

$B = \rho_w V_{sub} g = (1000\ kg/m^3)(0.0135\ m^3)(9.80\ m/s^2) = 132.3\ N$

Then $\sum \tau_A = \tau_{mg} = (0.053\ m)(132.3\ N) = \underline{7.01\ N\cdot m}$

14-61

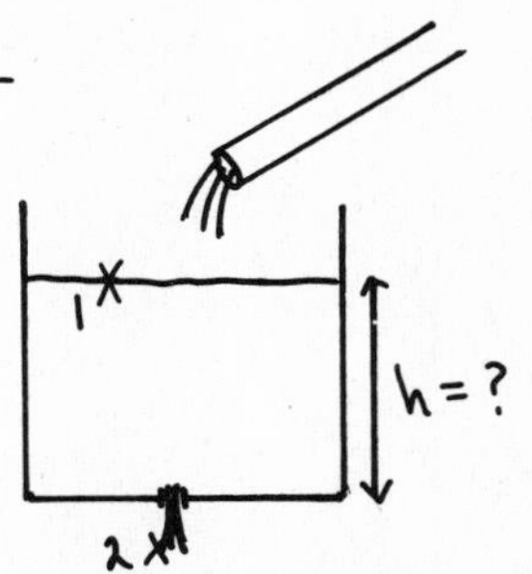

The water level in the vessel will rise until the volume flow rate into the vessel ($1.50\times10^{-4}\ m^3/s$) equals the volume flow rate out the hole in the bottom.

Let points 1 and 2 be chosen as in the sketch.

Bernoulli's equation $\Rightarrow p_1 + \rho g y_1 + \frac{1}{2}\rho v_1^2 = p_2 + \rho g y_2 + \frac{1}{2}\rho v_2^2$

Volume flow rate out hole equals volume flow rate from tube

$\Rightarrow v_2 A_2 = 1.50\times10^{-4}\ m^3/s \Rightarrow v_2 = \frac{1.50\times10^{-4}\ m^3/s}{1.00\times10^{-4}\ m^2} = 1.50\ m/s$

$A_1 \gg A_2$ and $v_1 A_1 = v_2 A_2 \Rightarrow \frac{1}{2}\rho v_1^2 \ll \frac{1}{2}\rho v_2^2$; neglect the $\frac{1}{2}\rho v_1^2$ term.

Measure y from the bottom of the bucket, so $y_2 = 0$ and $y_1 = h$.

$p_1 = p_2 = p_a$ (air pressure)

$\Rightarrow \cancel{p_a} + \rho g h = \cancel{p_a} + \frac{1}{2}\rho v_2^2$

$$h = \frac{v_2^2}{2g} = \frac{(1.50\ m/s)^2}{2(9.80\ m/s^2)} = \underline{0.115\ m}$$

14-63

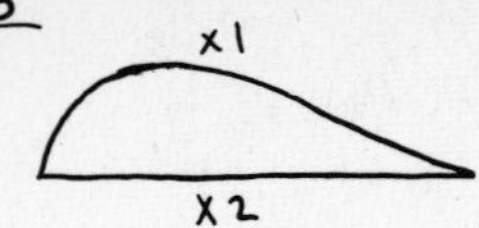

Apply Bernoulli's equation to points 1 and 2, where point 1 is just above the wing and point 2 is just below the wing:

$$p_1 + \rho g y_1 + \tfrac{1}{2}\rho v_1^2 = p_2 + \rho g y_2 + \tfrac{1}{2}\rho v_2^2$$

A "lift" of 1000 N/m² $\Rightarrow$ $p_2 - p_1 = 1000$ Pa

Solve for $v_1 \Rightarrow v_1 = \sqrt{\dfrac{2(p_2-p_1)}{\rho} + v_2^2 - 2g(y_1-y_2)}$

Note that $\dfrac{2(p_2-p_1)}{\rho} = \dfrac{2(1000\ \text{Pa})}{1.20\ \text{kg/m}^3} = 1667\ \text{m}^2/\text{s}^2$

We aren't given a value for $y_1 - y_2$, but it must be about 1 m or so.

For $y_1 - y_2 = 1$ m, $2g(y_1-y_2) = 19.6\ \text{m}^2/\text{s}^2$. So $2g(y_1-y_2) \ll \dfrac{2(p_2-p_1)}{\rho}$ and can be neglected.

Thus $v_1 = \sqrt{\dfrac{2(p_2-p_1)}{\rho} + v_2^2} = \sqrt{1667\ \text{m}^2/\text{s}^2 + (140\ \text{m/s})^2} = \underline{146\ \text{m/s}}$.

14-67

a)

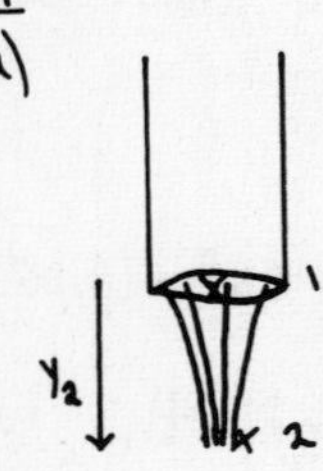

Let point 1 be at the end of the pipe and let point 2 be in the stream of liquid at a distance y_2 below the end of the tube.

Consider the free fall of the liquid. Take +y to be downward.

Free fall $\Rightarrow a = +g$

$v_2^2 = v_1^2 + 2a(y - y_0) \Rightarrow v_2^2 = v_1^2 + 2gy_2 \Rightarrow v_2 = \sqrt{v_1^2 + 2gy_2}$

Equation of continuity $\Rightarrow v_1 A_1 = v_2 A_2$

$A = \pi r^2 \Rightarrow v_1 \pi r_1^2 = v_2 \pi r_2^2 \Rightarrow v_2 = v_1\left(\dfrac{r_1^2}{r_2^2}\right)$

Use this in the above to eliminate $v_2 \Rightarrow v_1 \dfrac{r_1^2}{r_2^2} = \sqrt{v_1^2 + 2gy_2}$

$$r_2 = r_1 \frac{\sqrt{v_1}}{(v_1^2 + 2g y_2)^{1/4}}$$

Note that this equation says that r_2 decreases the farther point 2 is below the discharge end of the pipe.

b) $v_1 = 1.60$ m/s

We want the value of y_2 that gives $r_2 = \frac{1}{2} r_1 \Rightarrow r_1 = 2r_2$.

The result obtained in part (a) says $r_2^4(v_1^2 + 2gy_2) = r_1^4 v_1^2$

$$2gy_2 = \left(\frac{r_1}{r_2}\right)^4 v_1^2 - v_1^2$$

$$y_2 = \frac{[(r_1/r_2)^4 - 1]\, v_1^2}{2g} = \frac{[16-1]\,(1.60\ \text{m/s})^2}{2(9.80\ \text{m/s}^2)} = \underline{1.96\ \text{m}}$$

14-69

Free-body diagram for the bubble:

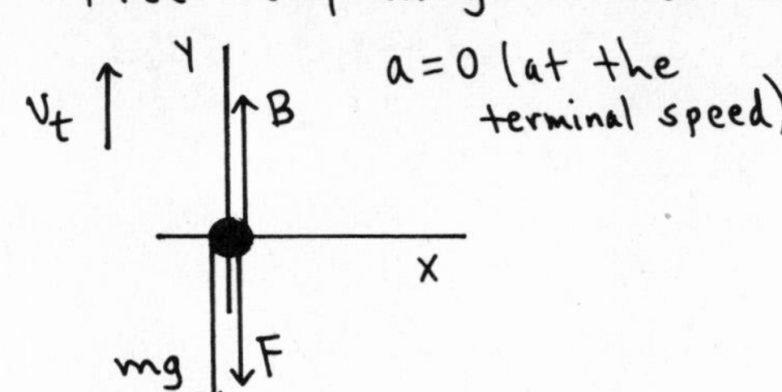

Note: The viscous drag force F is downward, since the bubble is traveling upward.

$$\Sigma F_y = ma_y$$
$$B - mg - F = 0$$
$$B = mg + F$$

$B = \rho' V g = \frac{4}{3}\pi r^3 \rho' g$, where ρ' is the density of the liquid

$mg = \rho V g = \frac{4}{3}\pi r^3 \rho g$, where ρ is the density of the air in the bubble

$F = 6\pi \eta r v_t$

Thus $B = mg + F \Rightarrow \frac{4}{3}\pi r^3 \rho' g = \frac{4}{3}\pi r^3 \rho g + 6\pi\eta r v_t$

$$6\eta v_t = \frac{4}{3} r^2 g (\rho' - \rho)$$

$$v_t = \frac{2r^2 g}{9\eta}(\rho' - \rho)$$

$$v_t = \frac{2(1.00 \times 10^{-3}\text{ m})^2 (9.80\text{ m/s}^2)}{9(0.150\text{ N}\cdot\text{s/m}^2)}(800\text{ kg/m}^3 - 1.20\text{ kg/m}^3) = 1.16 \times 10^{-2}\text{ m/s} = \underline{1.16\text{ cm/s}}$$

(Note: The precise value of $\rho = \rho_{air}$ that is used is unimportant since $\rho' \gg \rho$.)

b) $v_t = \frac{2r^2 g}{9\eta}(\rho' - \rho) = \frac{2(1.00 \times 10^{-3}\text{ m})^2 (9.80\text{ m/s}^2)}{9(1.005 \times 10^{-3}\text{ N}\cdot\text{s/m}^2)}(1000\text{ kg/m}^3 - 1.20\text{ kg/m}^3) = \underline{2.16\text{ m/s}}$

CHAPTER 15

Exercises 3, 11, 13, 17, 19, 25, 27, 29, 31, 33, 35, 39, 45, 49, 51, 53, 57

Problems 59, 61, 63, 67, 69, 71, 75, 77, 81, 83, 85, 87

Exercises

15-3

$$T_F = \tfrac{9}{5} T_c + 32°$$

$$T_F = T_c \Rightarrow T = \tfrac{9}{5}T + 32°$$

$$\tfrac{4}{5}T = -32° \Rightarrow T = -40°$$

$$-40°C = -40°F$$

15-11

Let $L_0 = 80.000\ cm$; $T_0 = 20.0°C$

$\Delta T = 40.0°C - 20.0°C = 20.0\ C°$ gives $\Delta L = 0.028\ cm$

Thus $\Delta L = \alpha L_0 \Delta T \Rightarrow \alpha = \dfrac{\Delta L}{L_0 \Delta T} = \dfrac{0.028\ cm}{(80.000\ cm)(20.0C°)} = \underline{1.75 \times 10^{-5}\ (C°)^{-1}}$

15-13

The diameter of the hole undergoes linear expansion just as does a length of brass. α_{brass} is given in Table 15-1.

$\Delta L = \alpha L_0 \Delta T = (2.0 \times 10^{-5}\ (C°)^{-1})(1.400cm)(200°C - 20°C) = 0.005\ cm$

$L = L_0 + \Delta L = 1.400\ cm + 0.005\ cm = \underline{1.405\ cm}$

15-17

Calculate ΔV for the ethanol. From Table 15-2, β for ethanol is $75 \times 10^{-5}\ K^{-1}$. $\Delta T = 10.0°C - 25.0°C = -15.0\ C° = -15.0\ K$

Then $\Delta V = \beta V_0 \Delta T = (75 \times 10^{-5} K^{-1})(1800\ L)(-15.0\ K) = -20\ L$.

The volume of the air space will be $\underline{20\ L} = 0.020\ m^3$.

15-19

a) $\Delta L = L_0 \alpha \Delta T$

$\Rightarrow \alpha = \dfrac{\Delta L}{L_0 \Delta T} = \dfrac{1.9 \times 10^{-2}\ m}{(3.00m)(520°C - 20°C)} = \underline{1.3 \times 10^{-5}\ (C°)^{-1}}$

b) Eq. (15-13): stress $\dfrac{F}{A} = -Y \alpha \Delta T$

$\Delta T = 20°C - 520°C = -500\ C°$ (ΔT always means final temperature minus initial temperature)

$\dfrac{F}{A} = -(2.0 \times 10^{11}\ Pa)(1.3 \times 10^{-5}\ (C°)^{-1})(-500\ C°) = +\underline{1.3 \times 10^9\ Pa}$

$\dfrac{F}{A}$ is positive $\Rightarrow$ the stress is a tensile (stretching) stress.

15-25

copper

$$Q = mc\Delta T$$

$c = 390\ \text{J/kg}\cdot\text{K}$ (from Table 15-3)

$$Q = (2.00\ \text{kg})(390\ \text{J/kg}\cdot\text{K})(80.0\,°\text{C} - 20.0\,°\text{C}) = 4.68\times10^4\ \text{J}$$

water

$$Q = mc\Delta T$$

$c = 4190\ \text{J/kg}\cdot\text{K}$ (from Table 15-3)

$$Q = (3.00\ \text{kg})(4190\ \text{J/kg}\cdot\text{K})(80.0\,°\text{C} - 20.0\,°\text{C}) = 7.54\times10^5\ \text{J}$$

Total $Q = 4.68\times10^4\ \text{J} + 7.54\times10^5\ \text{J} = \underline{8.01\times10^5\ \text{J}}$

15-27

$P = \frac{Q}{t}$, so the total heat transferred to the liquid is

$$Q = Pt = (65.0\ \text{W})(100\ \text{s}) = 6500\ \text{J}$$

$$Q = mc\Delta T \Rightarrow c = \frac{Q}{m\Delta T} = \frac{6500\ \text{J}}{0.530\ \text{kg}(20.77\,°\text{C} - 17.64\,°\text{C})} = \underline{3.92\times10^3\ \text{J/kg}\cdot\text{K}}$$

15-29

Heat must be added to do the following:

ice at $-10\,°\text{C} \rightarrow$ ice at $0\,°\text{C}$

$$Q_{ice} = mc_{ice}\Delta T = (6.00\times10^{-3}\ \text{kg})(2000\ \text{J/kg}\cdot\text{K})(0\,°\text{C} - (-10\,°\text{C})) = 120\ \text{J}$$

phase transition ice ($0\,°\text{C}$) $\rightarrow$ liquid water ($0\,°\text{C}$) (melting)

$$Q_{melt} = +mL_f = (6.00\times10^{-3}\ \text{kg})(334\times10^3\ \text{J/kg}) = 2.004\times10^3\ \text{J}$$

water at $0\,°\text{C}$ (from melted ice) $\rightarrow$ water at $100\,°\text{C}$

$$Q_{water} = mc_{water}\Delta T = (6.00\times10^{-3}\ \text{kg})(4190\ \text{J/kg}\cdot\text{K})(100\,°\text{C} - 0\,°\text{C}) = 2.514\times10^3\ \text{J}$$

phase transition water ($100\,°\text{C}$) $\rightarrow$ steam ($100\,°\text{C}$) (boiling)

$$Q_{boil} = +mL_v = (6.00\times10^{-3}\ \text{kg})(2256\times10^3\ \text{J/kg}) = 1.354\times10^4\ \text{J}$$

The total Q is $Q = 120\ \text{J} + 2.004\times10^3\ \text{J} + 2.514\times10^3\ \text{J} + 1.354\times10^4\ \text{J} = \underline{1.82\times10^4\ \text{J}}$

$$1.82\times10^4\ \text{J}\left(\frac{1\ \text{cal}}{4.186\ \text{J}}\right) = \underline{4.35\times10^3\ \text{cal}}$$

$$1.82\times10^4\ \text{J}\left(\frac{1\ \text{Btu}}{1055\ \text{J}}\right) = \underline{17.3\ \text{Btu}}$$

15-31

The heat that must be added to a lead bullet of mass m to melt it is

$$Q = mc\Delta T + mL_F$$

($mc\Delta T$ is the heat required to raise the temperature from $25\,°\text{C}$ to the melting point of $327.3\,°\text{C}$; mL_F is the heat required to make the solid $\rightarrow$ liquid phase change)

15-31 (cont)

The kinetic energy of the bullet if its speed is v is $K = \frac{1}{2}mv^2$.

Then $K = Q \Rightarrow \frac{1}{2}mv^2 = mc\Delta T + mL_F$

$$v = \sqrt{2(c\Delta T + L_F)} = \sqrt{2[(130\ \mathrm{J/kg\cdot K})(327.3°C - 25°C) + 24.5\times10^3\ \mathrm{J/kg}]}$$

$v = \underline{357\ \mathrm{m/s}}$

15-33

one-ton air conditioner $\Rightarrow$ 1 ton (2000 lb) of ice can be frozen from water at 0°C in 24 h.

Find the mass m that corresponds to 2000 lb (weight) of water:

$m = (2000\ \mathrm{lb})\left(\frac{1\ \mathrm{kg}}{2.205\ \mathrm{lb}}\right) = 907\ \mathrm{kg}$ (the kg $\leftrightarrow$ lb equivalence from Appendix E has been used.)

The heat that must be removed from the water to freeze it is

$Q = -mL_F = -(907\ \mathrm{kg})(334\times10^3\ \mathrm{J/kg}) = -3.03\times10^8\ \mathrm{J}$.

The power required if this is to be done in 1 h is

$$P = \frac{|Q|}{t} = \frac{3.03\times10^8\ \mathrm{J}}{(24\ \mathrm{h})\left(\frac{3600\ \mathrm{s}}{1\ \mathrm{h}}\right)} = \underline{3510\ \mathrm{W}}$$

or $P = 3510\ \mathrm{W}\left(\frac{1\ \mathrm{Btu/h}}{0.293\ \mathrm{W}}\right) = \underline{1.20\times10^4\ \mathrm{Btu/h}}$

15-35

$Q_{system} = 0$

Calculate Q for each component of the system:

aluminum can

$Q_{can} = mc\Delta T = (0.500\ \mathrm{kg})(910\ \mathrm{J/kg\cdot K})(T - 20.0°C) = (455\ \mathrm{J/K})T - 9100\ \mathrm{J}$

water

$Q_{water} = mc\Delta T = (0.130\ \mathrm{kg})(4190\ \mathrm{J/kg\cdot K})(T - 20.0°C) = (544.7\ \mathrm{J/K})T - 1.089\times10^4\ \mathrm{J}$

iron

$Q_{iron} = mc\Delta T = (0.200\ \mathrm{kg})(470\ \mathrm{J/kg\cdot K})(T - 75.0°C) = (94.0\ \mathrm{J/K})T - 7050\ \mathrm{J}$

$Q_{system} = 0 \Rightarrow Q_{can} + Q_{water} + Q_{iron} = 0$

$\Rightarrow (455\ \mathrm{J/K})T - 9100\ \mathrm{J} + (544.7\ \mathrm{J/K})T - 1.089\times10^4\ \mathrm{J} + (94.0\ \mathrm{J/K})T - 7050\ \mathrm{J} = 0$

$(1094\ \mathrm{J/K})T = 2.704\times10^4\ \mathrm{J}$

$$T = \frac{2.704\times10^4\ \mathrm{J}}{1094\ \mathrm{J/K}} = \underline{24.7°C}$$

15-39

$Q_{system} = 0$

Calculate Q for each component of the system.

(Beaker has small mass $\Rightarrow Q = mc\Delta T$ for beaker can be neglected.)

15-39 (cont)

0.500 kg of water (cools from 80.0°C to 40.0°C)

$Q_{water} = mc\Delta T = (0.500\ kg)(4190\ J/kg\cdot K)(40.0\,°C - 80.0°C) = -8.38\times10^4\ J$

ice (warms to 0°C; melts; water from melted ice warms to 40.0°C)

$Q_{ice} = mc_{ice}\Delta T + mL_f + mc_{water}\Delta T$

$Q_{ice} = m[(2000\ J/kg\cdot K)(0°C - (-20.0°C)) + 334\times10^3\ J/kg + (4190\ J/kg\cdot K)(40.0°C - 0°C)]$

$Q_{ice} = (5.416\times10^5\ J/kg)\,m$

$Q_{system} = 0 \Rightarrow Q_{water} + Q_{ice} = 0$

$-8.38\times10^4\ J + (5.416\times10^5\ J/kg)\,m = 0$

$m = \dfrac{8.38\times10^4\ J}{5.416\times10^5\ J/kg} = \underline{0.155\ kg}$

15-45

a)

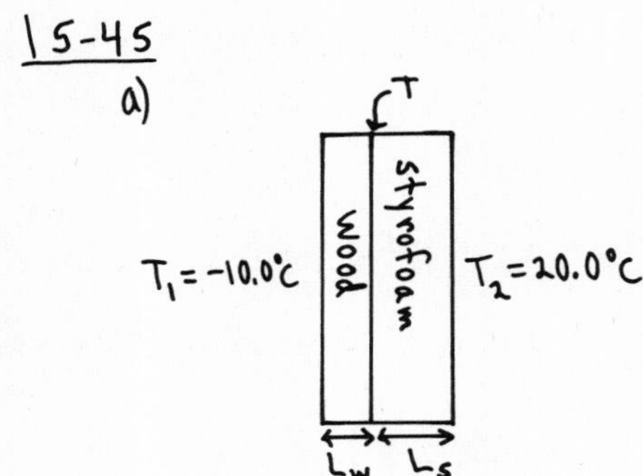

Call the temperature at the interface between the wood and styrofoam T. The heat current in each material is given by $H = kA\left(\frac{T_H - T_C}{L}\right)$.

Heat current through the wood:

$H_w = k_w \dfrac{A(T - T_1)}{L_w}$

Heat current through the styrofoam:

$H_s = k_s \dfrac{A(T_2 - T)}{L_s}$

In steady-state heat does not accumulate in either material. The same heat has to pass through both materials in succession, so $H_w = H_s$.

$\Rightarrow k_w \dfrac{A(T - T_1)}{L_w} = k_s \dfrac{A(T_2 - T)}{L_s}$

$k_w L_s (T - T_1) = k_s L_w (T_2 - T)$

$T = \dfrac{k_w L_s T_1 + k_s L_w T_2}{k_w L_s + k_s L_w} = \dfrac{(0.080\ W/m\cdot K)(0.030\ m)(-10.0°C) + (0.010\ W/m\cdot K)(0.020\ m)(20.0°C)}{(0.080\ W/m\cdot K)(0.030\ m) + (0.010\ W/m\cdot K)(0.020\ m)}$

$T = \dfrac{-0.024\ W\cdot°C/K + 0.004\ W\cdot°C/K}{0.0026\ W/K} = \underline{-7.7°C}$

b) Heat flow per square meter is $\frac{H}{A} = k\left(\frac{T_H - T_C}{L}\right)$. We can calculate this either for the wood or for the styrofoam; the results must be the same.

wood

$\dfrac{H_w}{A} = k_w \dfrac{T - T_1}{L_w} = (0.080\ W/m\cdot K)\dfrac{(-7.7°C - (-10.0°C))}{0.020\ m} = \underline{9.2\ W}$

or

styrofoam

$\dfrac{H_s}{A} = k_s \dfrac{T_2 - T}{L_s} = (0.010\ W/m\cdot K)\dfrac{20.0°C - (-7.7°C)}{0.030\ m} = 9.2\ W$

(Since $k_s < k_w$ it takes a larger temperature gradient $\frac{T_H - T_C}{L}$ across the stryofoam to set up the same heat current as in the wood.)

15-49

a) temperature gradient $\frac{T_H - T_C}{L} = \frac{100.0°C - 0.0°C}{0.200\ m} = 500\ C°/m = \underline{500\ K/m}$.

b) $H = kA\frac{T_H - T_C}{L}$

From Table 15-5, $k = 385\ W/m\cdot K$

$\Rightarrow H = (385\ W/m\cdot K)(1.20\times10^{-4}\ m^2)(500\ K/m) = \underline{23.1\ W}$

c) $H = 23.1\ W$ for all sections of the rod

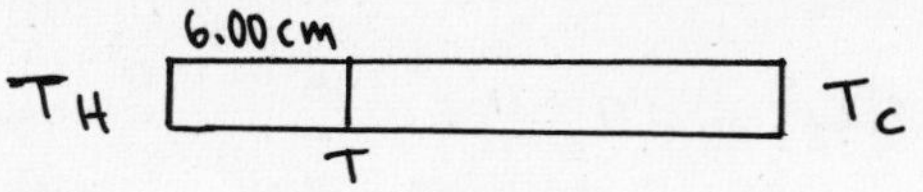

Apply $H = kA\frac{\Delta T}{L}$ to the 6 cm section:

$H = kA\frac{T_H - T}{L} \Rightarrow T_H - T = \frac{LH}{Ak}$

$T = T_H - \frac{LH}{Ak} = 100.0°C - \frac{(0.0600\ m)(23.1\ W)}{(1.20\times10^{-4}\ m^2)(385\ W/m\cdot K)} = \underline{70.0°C}$

15-51

$H_{net} = Ae\sigma(T^4 - T_s^4)$ (Eq. 15-27; T must be in kelvins)

Example 15-16 gives $A = 1.2\ m^2$, $e = 1.0$, and $T = 30°C = 303\ K$ (body surface temperature)

$T_s = 12.0°C = 285\ K$

$\Rightarrow H_{net} = (1.2\ m^2)(1.0)(5.67\times10^{-8}\ W/m^2\cdot K^4)((303\ K)^4 - (285\ K)^4)$

$H_{net} = 573.5\ W - 448.9\ W = \underline{125\ W}$

(Note that this is larger than H_{net} calculated in Example 15-16. The lower temperature of the surroundings increases the rate of heat loss by radiation.)

15-53

$H = Ae\sigma T^4 \Rightarrow A = \frac{H}{e\sigma T^4}$

60-W lamp and all electrical energy consumed is radiated $\Rightarrow H = 60\ W$

$A = \frac{60\ W}{(0.35)(5.67\times10^{-8}\ W/m^2\cdot K^4)(2450\ K)^4} = 8.39\times10^{-5}\ m^2\left(\frac{1\times10^4\ cm^2}{1\ m^2}\right) = \underline{0.839\ cm^2}$

15-57

Eq. (15-28): $H = \frac{T_{ic} - T_{amb}}{r_{th}} \Rightarrow T_{amb} = T_{ic} - Hr_{th}$

$T_{amb} = 120°C - (36\ W)(3.0\ K/W) = 120°C - 108\ C° = \underline{12°C}$

Problems

15-59

a) $V_0 = (0.100\text{ m})^3 = 1.00\times10^{-3}\text{ m}^3$

From Table 15-2, $\beta = 7.2\times10^{-5}\text{ K}^{-1}$

$\Delta V = V_0 \beta \Delta T = (1.00\times10^{-3}\text{ m}^3)(7.2\times10^{-5}\text{ K}^{-1})(60.0°\text{C} - 10.0°\text{C}) = 3.6\times10^{-6}\text{ m}^3$

(the volume increases)

b) $\rho = \frac{m}{V} \Rightarrow \rho_0 = \frac{m}{V_0}$ and $\rho = \frac{m}{V_0 + \Delta V} = \frac{m}{V_0}\left(\frac{1}{1+\Delta V/V_0}\right) = \rho_0\left(1 + \frac{\Delta V}{V_0}\right)^{-1}$

$\frac{\Delta V}{V_0} = \beta \Delta T = (7.2\times10^{-5}\text{ K}^{-1})(60.0°\text{C} - 10.0°\text{C}) = 3.6\times10^{-3}$

From Table 14-1, $\rho_0 = 2.7\times10^3\text{ kg/m}^3$

$$\Delta\rho = \rho - \rho_0 = \rho_0\left(1 + \frac{\Delta V}{V_0}\right)^{-1} - \rho_0$$

To avoid subtracting two numbers that are nearly equal, use the binomial theorem and the fact that $\frac{\Delta V}{V_0}$ is small to write

$\left(1+\frac{\Delta V}{V_0}\right)^{-1} = 1 - \frac{\Delta V}{V_0}$ + terms that can neglect

$\Rightarrow \Delta\rho = \rho_0\left(1 - \frac{\Delta V}{V_0}\right) - \rho_0 = -\frac{\Delta V}{V_0}\rho_0 = -(3.6\times10^{-3})(2.7\times10^3\text{ kg/m}^3) = -9.7\text{ kg/m}^3$

(the density decreases)

15-61

a) Heat the ring to make its diameter equal to 3.0050 in. The diameter of the ring undergoes linear expansion.

$\Delta L = L_0 \alpha \Delta T \Rightarrow \Delta T = \frac{\Delta L}{L_0 \alpha} = \frac{0.0050\text{ in.}}{(3.0000\text{ in.})(1.2\times10^{-5}(\text{C}°)^{-1})} = 139\text{ C}°$

$T = T_0 + \Delta T = 20.0°\text{C} + 139\text{ C}° = \underline{159°\text{C}}$

b) $L = L_0(1 + \alpha\Delta T)$

Want L_s (steel) $= L_b$ (brass) for the same ΔT for both materials

$\Rightarrow L_{0s}(1 + \alpha_s \Delta T) = L_{0b}(1 + \alpha_b \Delta T)$

$L_{0s} + L_{0s}\alpha_s \Delta T = L_{0b} + L_{0b}\alpha_b \Delta T$

$\Delta T = \frac{L_{0b} - L_{0s}}{L_{0s}\alpha_s - L_{0b}\alpha_b} = \frac{3.0050\text{ in.} - 3.0000\text{ in.}}{(3.0000\text{ in.})(1.2\times10^{-5}(\text{C}°)^{-1}) - (3.0050\text{ in.})(2.0\times10^{-5}(\text{C}°)^{-1})}$

$\Delta T = \frac{0.0050}{3.60\times10^{-5} - 6.01\times10^{-5}}\text{ C}° = -207.5\text{ C}°$

$T = T_0 + \Delta T = 20.0°\text{C} - 207.5\text{ C}° = \underline{-187°\text{C}}$

15-63

Call the metals A and B. Use the data given to calculate α for each metal. $\Delta L = L_0 \alpha \Delta T \Rightarrow \alpha = \frac{\Delta L}{L_0 \Delta T}$

metal A: $\alpha_A = \frac{\Delta L}{L_0 \Delta T} = \frac{0.0750\text{ cm}}{(30.0\text{ cm})(100\text{ C}°)} = 2.50\times10^{-5}(\text{C}°)^{-1}$

15-63 (cont)

metal B: $\alpha_B = \frac{\Delta L}{L_0 \Delta T} = \frac{0.0400\text{cm}}{(30.0\text{cm})(100C°)} = 1.33\times10^{-5}(C°)^{-1}$

Now consider the composite rod. Let L_A be the length of metal A in this rod.

$\leftarrow L_A \rightarrow \leftarrow 30.0\text{ cm} - L_A \rightarrow$

A	B

$\Delta T = 100C° \Rightarrow \Delta L = 0.065\text{ cm}$

$$\Delta L = \Delta L_A + \Delta L_B = (\alpha_A L_A + \alpha_B L_B)\Delta T$$

$$\frac{\Delta L}{\Delta T} = \alpha_A L_A + \alpha_B(0.300\text{m} - L_A)$$

$$L_A = \frac{\frac{\Delta L}{\Delta T} - (0.300\text{m})\alpha_B}{\alpha_A - \alpha_B} = \frac{\frac{0.065\times10^{-2}\text{m}}{100C°} - (0.300\text{m})(1.33\times10^{-5}(C°)^{-1})}{2.50\times10^{-5}(C°)^{-1} - 1.33\times10^{-5}(C°)^{-1}}$$

$$L_A = \frac{6.50\times10^{-6} - 3.99\times10^{-6}}{2.50\times10^{-5} - 1.33\times10^{-5}}\text{ m} = 0.215\text{ m} = \underline{21.5\text{ cm}}$$

$$L_B = 30.0\text{cm} - L_A = 30.0\text{ cm} - 21.5\text{ cm} = \underline{8.5\text{cm}}$$

15-67

Let β_l and β_m be the coefficients of volume expansion for the liquid and for the metal. Let ΔT be the (negative) change in temperature when the system is cooled to the new temperature.

Change in volume of cylinder when cool: $\Delta V_m = V_0 \beta_m \Delta T$ (negative)
Change in volume of liquid when cool: $\Delta V_l = V_0 \beta_l \Delta T$ (negative)
The difference $\Delta V_m - \Delta V_l$ must be equal to the positive volume change due to the decrease in pressure, which is $-\frac{\Delta p V_0}{B} = -k\,\Delta p\, V_0$.

$$\Rightarrow \Delta V_m - \Delta V_l = -k\,\Delta p\, V_0$$

$$\cancel{V_0}\,\Delta T(\beta_m - \beta_l) = -k\,\Delta p\,\cancel{V_0}$$

$$\Delta T = -\frac{k\,\Delta p}{\beta_m - \beta_l} = \frac{k\,\Delta p}{\beta_l - \beta_m} = \frac{(7.00\times10^{-10}\text{Pa}^{-1})(1.013\times10^5\text{Pa} - 1.013\times10^7\text{Pa})}{5.30\times10^{-4}\text{K}^{-1} - 4.00\times10^{-5}\text{K}^{-1}} = -14.3C°$$

$$T = T_0 + \Delta T = 60.0°\text{C} - 14.3C° = \underline{45.7°\text{C}}$$

15-69

a) The kinetic energy is $K = \frac{1}{2}mv^2$.
The heat energy required to raise its temperature by 600 C° (but not to melt it) is $Q = mc\Delta T$.

The ratio is $\frac{K}{Q} = \frac{\frac{1}{2}mv^2}{mc\Delta T} = \frac{v^2}{2c\Delta T} = \frac{(7000\text{ m/s})^2}{2(910\text{ J/kg}\cdot\text{K})(600C°)} = \underline{44.9}.$

b) The heat generated when friction work (due to the friction force exerted by the air) removes the kinetic energy of the satellite during reentry is very large, and could melt the satellite. Manned space vehicles must have heat shields made of very high melting temperature materials and reentry must be made slowly.

15-71

a) Eq. (15-15) $\Rightarrow dQ = mc\,dT$

But the problem gives the molar heat capacity $C = Mc$; $dQ = nC\,dT$ (Eq. 15-20)

$$Q = n\int_{T_1}^{T_2} C\,dT = n\int_{T_1}^{T_2} k\frac{T^3}{\Theta^3}\,dT = \frac{nk}{\Theta^3}\int_{T_1}^{T_2} T^3\,dT = \frac{nk}{\Theta^3}\left(\frac{1}{4}T^4\Big|_{T_1}^{T_2}\right)$$

$$Q = \frac{nk}{4\Theta^3}(T_2^4 - T_1^4) = \frac{(2.00\text{ mol})(1940\text{ J/mol}\cdot\text{K})}{4(281\text{ K})^3}\left((60.0\text{ K})^4 - (10.0\text{ K})^4\right) = \underline{566\text{ J}}$$

b) $C_{av} = \frac{1}{n}\frac{\Delta Q}{\Delta T} = \frac{1}{2.00\text{ mol}}\left(\frac{566\text{ J}}{60.0\text{ K} - 10.0\text{ K}}\right) = 5.66\text{ J/mol}\cdot\text{K}$

c) $C = k\left(\frac{T}{\Theta}\right)^3 = (1940\text{ J/mol}\cdot\text{K})\left(\frac{60.0\text{ K}}{281\text{ K}}\right)^3 = 18.9\text{ J/mol}\cdot\text{K}$

(C is increasing with T, so C at the upper end of the temperature interval is larger than its average value over the interval.)

15-75

Heat comes out of the 0.0°C water and into the ice cube to warm it to 0.0°C. The heat that comes out of the water causes a phase change liquid → solid.

Heat that goes into the ice cube:

$Q_{ice} = mc_{ice}\Delta T = (0.070\text{ kg})(2000\text{ J/kg}\cdot\text{K})(0.0°\text{C} - (-10.0°\text{C})) = 1400\text{ J}$

Heat that comes out of the water when mass m freezes:

$Q_{water} = -mL_f$

$Q_{system} = 0 \Rightarrow Q_{ice} + Q_{water} = 0$

$1400\text{ J} - mL_f = 0$

$m = \frac{1400\text{ J}}{L_f} = \frac{1400\text{ J}}{334\times10^3\text{ J/kg}} = 4.19\times10^{-3}\text{ kg} = \underline{4.19\text{ g}}$

15-77

Assume that all the ice melts and that all the steam condenses. If we calculate a final temperature T that is outside the range 0°C to 100°C then we know that this assumption is incorrect. Calculate Q for each piece of the system and then set the total $Q_{system} = 0$.

copper can (changes temperature from 0.0°C to T; no phase change)

$Q_{can} = mc\Delta T = (0.322\text{ kg})(390\text{ J/kg}\cdot\text{K})(T - 0.0°\text{C}) = (125.6\text{ J/K})T$

ice (melting phase change and then water produced warms to T)

$Q_{ice} = +mL_f + mc\Delta T = (0.046\text{ kg})(334\times10^3\text{ J/kg}) + (0.046\text{ kg})(4190\text{ J/kg}\cdot\text{K})(T - 0.0°\text{C})$

$Q_{ice} = 1.536\times10^4\text{ J} + (192.7\text{ J/K})T$

steam (condenses to liquid and then water produced cools to T)

$Q_{steam} = -mL_v + mc\Delta T = -(0.012\text{ kg})(2256\times10^3\text{ J/kg}) + (0.012\text{ kg})(4190\text{ J/kg}\cdot\text{K})(T - 100°\text{C})$

$Q_{steam} = -2.707\times10^4\text{ J} + 50.28T - 5028\text{ J} = -3.210\times10^4\text{ J} + 50.28T$

15-77 (cont)

$$Q_{system} = 0 \Rightarrow Q_{can} + Q_{ice} + Q_{steam} = 0$$

$$(125.6\ J/K)T + 1.536\times10^4\ J + (192.7\ J/K)T - 3.210\times10^4\ J + 50.28T = 0$$

$$(368.6\ J/K)T = 1.674\times10^4\ J$$

$$T = \frac{1.674\times10^4\ J}{368.6\ J/K} = \underline{45.4°C}$$

(This is between 0°C and 100°C so our assumptions about the phase changes being complete were correct.)

15-81

single pane

$$H_{single} = kA\left(\frac{T_H - T_C}{L}\right) = (0.80\ W/m\cdot K)(0.15\ m^2)\frac{\Delta T}{2.5\times10^{-3}\ m} = (48.0\ W/K)\Delta T$$

double pane

glass #2 glass #1

air

T_H T_2 T_1 T_C

Use H written in terms of the thermal resistance R:

$$H = \frac{A\Delta T}{R}$$

glass #1 $H = \frac{A}{R_{glass}}(T_1 - T_C) \Rightarrow T_1 - T_C = \frac{R_{glass}H}{A}$

air $H = \frac{A}{R_{air}}(T_2 - T_1) \Rightarrow T_2 - T_1 = \frac{R_{air}H}{A}$

glass #2 $H = \frac{A}{R_{glass}}(T_H - T_2) \Rightarrow T_H - T_2 = \frac{R_{glass}H}{A}$

(H is same through all three.)

Add these equations $\Rightarrow T_H - T_C = \frac{H}{A}(2R_{glass} + R_{air})$

$$H = \frac{A(T_H - T_C)}{2R_{glass} + R_{air}}$$

$$R_{glass} = \frac{L_{glass}}{k_{glass}} = \frac{2.5\times10^{-3}\ m}{0.80\ W/m\cdot K} = 3.125\times10^{-3}\ m^2\cdot K/W$$

$$R_{air} = \frac{L_{air}}{k_{air}} = \frac{5.0\times10^{-3}\ m}{0.024\ W/m\cdot K} = 0.2083\ m^2\cdot K/W$$

$$H_{double} = \frac{(0.15\ m^2)\Delta T}{2(3.125\times10^{-3}\ m^2\cdot K/W) + 0.2083\ m^2\cdot K/W} = (0.699\ W/K)\Delta T$$

The ratio is $\frac{H_{single}}{H_{double}} = \frac{(48.0\ W/K)\Delta T}{(0.699\ W/K)\Delta T} = \underline{69}.$

15-83

a) Consider a section of ice that has area A. At time t let the thickness be h. Consider a short time interval t to t+dt. Let the thickness that freezes in this time be dh. The mass of the section that freezes in dt is $dm = \rho dV = \rho A\, dh$. The heat that must be conducted away from this mass of water to freeze it is $dQ = dm\, L_f = \rho A L_f\, dh$.

$H = \frac{dQ}{dt} = kA\frac{\Delta T}{h}$, so the heat dQ conducted in time dt through the thickness h that is already there is $dQ = kA\left(\frac{T_H - T_C}{h}\right)dt$.

15-83 (cont)

Equate these two expressions for dQ

$$\Rightarrow \rho A L_f\, dh = kA\left(\frac{T_H - T_C}{h}\right) dt$$

$$h\, dh = \left(\frac{k(T_H - T_C)}{\rho L_f}\right) dt$$

Integrate from $t = 0$ to time t. At $t = 0$ the thickness h is zero.

$$\int_0^h h\, dh = \frac{k(T_H - T_C)}{\rho L_f} \int_0^t dt$$

$$\frac{1}{2}h^2 = \frac{k(T_H - T_C)}{\rho L_f}\, t \Rightarrow h = \sqrt{\frac{2k(T_H - T_C)}{\rho L_f}}\ \sqrt{t}$$

The thickness after time t is proportional to $\sqrt{t}$.

b) The expression in part (a) gives

$$t = \frac{h^2 \rho L_f}{2k(T_H - T_C)} = \frac{(0.30\text{ m})^2(920\text{ kg/m}^3)(334\times10^3\text{ J/kg})}{2(1.6\text{ W/m}\cdot\text{K})(0°\text{C} - (-10°\text{C}))} = 8.64\times10^5\text{ s} = \underline{240\text{ h}}$$

15-85

Work with a 1.00 m^2 area.

The heat current into a 1.00 m^2 area of ice due to the absorbed solar radiation is $H = (0.70)(600\text{ W/m}^2)(1.00\text{ m}^2) = 420\text{ W}$.

The heat required to melt a $h = 1.40$ cm thick layer of ice that is initially at 0°C is $Q = mL_f = \rho h A L_f = (920\text{ kg/m}^3)(0.0140\text{ m})(1.00\text{ m}^2)(334\times10^3\text{ J/kg}) = 4.302\times10^6\text{ J}$.

$H = \frac{Q}{t}$, so the time t it takes the heat current from solar radiation to input this amount of heat into the ice is $t = \frac{Q}{H} = \frac{4.302\times10^6\text{ J}}{420\text{ W}} = 1.02\times10^4\text{ s} = \underline{171\text{ min}}$.

15-87

Calculate the net rate of radiation of heat from the can:

$$H_{net} = Ae\sigma(T^4 - T_s^4).$$

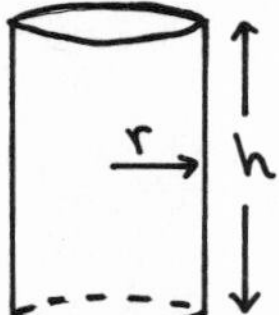

The surface area of the cylindrical can is

$$A = 2\pi rh + 2\pi r^2 = 2\pi r(h + r)$$

$$A = 2\pi(0.030\text{ m})(0.100\text{ m} + 0.030\text{ m}) = 0.0245\text{ m}^2$$

$$H_{net} = (0.0245\text{ m}^2)(0.200)(5.67\times10^{-8}\text{ W/m}^2\cdot\text{K}^4)((4.22\text{ K})^4 - (77.3\text{ K})^4)$$

$H_{net} = -9.92\times10^{-3}$ W (minus $\Rightarrow$ heat current into the can)

The heat that is put into the can by radiation in one hour is

$$Q = -(H_{net})t = (9.92\times10^{-3}\text{ W})(3600\text{ s}) = 35.7\text{ J}$$

This heat boils a mass m of helium according to the eq. $Q = mL_f$, so

$$m = \frac{Q}{L_f} = \frac{35.7\text{ J}}{2.09\times10^4\text{ J/kg}} = 1.71\times10^{-3}\text{ kg} = \underline{1.71\text{ g}}.$$

CHAPTER 16

Exercises 3, 7, 11, 17, 19, 23, 25, 27, 29

Problems 35, 37, 39, 41, 43, 47, 49, 51

Exercises

16-3

a) $n = \dfrac{m_{tot}}{M} = \dfrac{0.280\text{ kg}}{4.00\times10^{-3}\text{ kg/mol}} = \underline{70.0\text{ mol}}$

b) $pV = nRT \Rightarrow p = \dfrac{nRT}{V}$

T must be in kelvins; $T = (27+273)\text{ K} = 300\text{ K}$

$p = \dfrac{(70.0\text{ mol})(8.3145\text{ J/mol}\cdot\text{K})(300\text{ K})}{20.0\times10^{-3}\text{ m}^3} = \underline{8.73\times10^6\text{ Pa}}$

$p = (8.73\times10^6\text{ Pa})\left(\dfrac{1.00\text{ atm}}{1.013\times10^5\text{ Pa}}\right) = \underline{86.2\text{ atm}}$

16-7

$pV = nRT$

n, R constant $\Rightarrow \dfrac{pV}{T} = nR = \text{constant} \Rightarrow \dfrac{p_1V_1}{T_1} = \dfrac{p_2V_2}{T_2}$

$T_1 = (27+273)\text{ K} = 300\text{ K}$

$p_1 = 1.01\times10^5\text{ Pa}$

$p_2 = 2.25\times10^6\text{ Pa} + 1.01\times10^5\text{ Pa} = 2.35\times10^6\text{ Pa}$ (in the ideal-gas equation the pressures must be absolute not gauge pressures)

$T_2 = T_1\left(\dfrac{p_2}{p_1}\right)\left(\dfrac{V_2}{V_1}\right) = 300\text{ K}\left(\dfrac{2.35\times10^6\text{ Pa}}{1.01\times10^5\text{ Pa}}\right)\left(\dfrac{75.0\text{ cm}^3}{800\text{ cm}^3}\right) = 654\text{ K}$

$T_2 = (654-273)^\circ\text{C} = \underline{381^\circ\text{C}}$

(Note that the units cancel in the $\frac{V_2}{V_1}$ volume ratio, so it was not necessary to convert the volumes in cm^3 to m^3.)

16-11

$pV = nRT$; n, R, p are constant $\Rightarrow \dfrac{V}{T} = \dfrac{nR}{p} = \text{constant} \Rightarrow \dfrac{V_1}{T_1} = \dfrac{V_2}{T_2}$

$T_1 = (27+273)\text{ K} = 300\text{ K}$ (T must be in kelvins)

$V_2 = V_1\left(\dfrac{T_2}{T_1}\right) = (0.800\text{ L})\left(\dfrac{77.3\text{ K}}{300\text{ K}}\right) = \underline{0.206\text{ L}}$

16-17

a) Use the density and the mass of 1 mole to calculate the volume.

$\rho = \dfrac{m}{V} \Rightarrow V = \dfrac{m}{\rho}$, where $m = m_{tot}$, the mass of 1.00 mol of water.

$m_{tot} = nM = (1.00\text{ mol})(18.0\times10^{-3}\text{ kg/mol}) = 18.0\times10^{-3}\text{ kg}$

16-17 (cont)

$$V = \frac{m}{\rho} = \frac{18.0\times10^{-3}\,\text{kg}}{1000\ \text{kg/m}^3} = \underline{1.80\times10^{-5}\ \text{m}^3}$$

b) One mole contains $N_A = 6.022\times10^{23}$ molecules, so the volume occupied by one molecule is $\frac{1.80\times10^{-5}\ \text{m}^3/\text{mol}}{6.022\times10^{23}\ \text{molecules/mol}} = 2.989\times10^{-29}\ \text{m}^3/\text{molecule}$

$V = a^3$, where a is the length of each side of the cube occupied by a molecule. $a^3 = 2.989\times10^{-29}\ \text{m}^3 \Rightarrow a = \underline{3.10\times10^{-10}\ \text{m}}$

c) Atoms and molecules are on the order of 10^{-10} m in diameter, in agreement with the above estimates.

<u>16-19</u>

$v_{rms} = \sqrt{\frac{3RT}{M}}$ (Eq. 16-20) $\Rightarrow \frac{v_{rms}^2}{3R} = \frac{T}{M}$, where T must be in kelvins.

Same $v_{rms} \Rightarrow$ same $\frac{T}{M}$ for the two gases $\Rightarrow \frac{T_{N_2}}{M_{N_2}} = \frac{T_{H_2}}{M_{H_2}}$

$$T_{N_2} = T_{H_2}\left(\frac{M_{N_2}}{M_{H_2}}\right) = (27+273)\ \text{K}\left(\frac{28.01\ \text{g/mol}}{2.02\ \text{g/mol}}\right) = 4.16\times10^3\ \text{K}$$

$$T_{N_2} = (4160-273)\,^\circ\text{C} = \underline{3887\,^\circ\text{C}}$$

<u>16-23</u>

a) $\frac{1}{2}m(v^2)_{av} = \frac{3}{2}kT = \frac{3}{2}(1.381\times10^{-23}\ \text{J/molecule}\cdot\text{K})(300\ \text{K}) = \underline{6.21\times10^{-21}\ \text{J}}$

b) We need the mass m of one atom:

$$m = \frac{M}{N_A} = \frac{4.00\times10^{-3}\ \text{kg/mol}}{6.022\times10^{23}\ \text{molecules/mol}} = 6.64\times10^{-27}\ \text{kg/molecule}$$

Then $\frac{1}{2}m(v^2)_{av} = 6.21\times10^{-21}$ J (part (a)) gives

$$(v^2)_{av} = \frac{2(6.21\times10^{-21}\ \text{J})}{m} = \frac{2(6.21\times10^{-21}\ \text{J})}{6.64\times10^{-27}\ \text{kg}} = \underline{1.87\times10^{6}\ \text{m}^2/\text{s}^2}$$

c) $v_{rms} = \sqrt{(v^2)_{av}} = \sqrt{1.87\times10^6\ \text{m}^2/\text{s}^2} = \underline{1.37\times10^3\ \text{m/s}}$

d) $p = m v_{rms} = (6.64\times10^{-27}\ \text{kg})(1.37\times10^3\ \text{m/s}) = \underline{9.10\times10^{-24}\ \text{kg}\cdot\text{m/s}}$

e) Time between collisions with one wall is

$$t = \frac{0.20\ \text{m}}{v_{rms}} = \frac{0.20\ \text{m}}{1.37\times10^3\ \text{m/s}} = 1.46\times10^{-4}\ \text{s}$$

In a collision $\vec{v}$ changes direction, so $\Delta p = 2m v_{rms} = 2(9.10\times10^{-24}\ \text{kg}\cdot\text{m/s})$

$\Delta p = 1.82\times10^{-23}\ \text{kg}\cdot\text{m/s}$

$$F = \frac{dp}{dt} \Rightarrow F_{av} = \frac{\Delta p}{\Delta t} = \frac{1.82\times10^{-23}\ \text{kg}\cdot\text{m/s}}{1.46\times10^{-4}\ \text{s}} = \underline{1.25\times10^{-19}\ \text{N}}$$

f) pressure $= \frac{F}{A} = \frac{1.25\times10^{-19}\ \text{N}}{(0.10\ \text{m})^2} = \underline{1.25\times10^{-17}\ \text{Pa}}$ (due to one atom)

16-23 (cont)

g) pressure = 1 atm = 1.013×10^5 Pa

Number of atoms needed = $\dfrac{1.013 \times 10^5 \text{ Pa}}{1.25 \times 10^{-17} \text{ Pa/atom}} = \underline{8.10 \times 10^{21} \text{ atoms}}$

h) $pV = NkT$ (Eq. 16-19) $\Rightarrow N = \dfrac{pV}{kT} = \dfrac{(1.013 \times 10^5 \text{Pa})(0.10\text{m})^3}{(1.384 \times 10^{-23} \text{J/molecule}\cdot\text{K})(300\text{K})}$

$N = \underline{2.45 \times 10^{22} \text{ atoms}}$

i) From the factor of $\frac{1}{3}$ in $(v_x^2)_{av} = \frac{1}{3}(v^2)_{av}$.

16-25

a) $Q = nC_v \Delta T = n\left(\frac{5}{2}R\right)\Delta T = (4.00 \text{ mol})\left(\frac{5}{2}\right)(8.3145 \text{ J/mol}\cdot\text{K})(25.0\text{ K}) = \underline{2.08 \times 10^3 \text{ J}}$

b) $Q = nC_v \Delta T = n\left(\frac{3}{2}R\right)\Delta T = (4.00 \text{mol})\left(\frac{3}{2}\right)(8.3145 \text{ J/mol}\cdot\text{K})(25.0\text{K}) = \underline{1.25 \times 10^3 \text{ J}}$

16-27

$\frac{1}{2}R$ contribution to C_v for each degree of freedom $\Rightarrow C_v = 6\left(\frac{1}{2}R\right) = 3R$

$C_v = 3(8.3145 \text{ J/mol}\cdot\text{K}) = \underline{24.9 \text{ J/mol}\cdot\text{K}}$

For water vapor the specific heat capacity is $c = 2000$ J/kg·K. The molar heat capacity is $C = Mc = (18.0 \times 10^{-3} \text{kg/mol})(2000 \text{ J/kg}\cdot\text{K}) = 36.0 \text{J/mol}\cdot\text{K}$

The difference is 36.0 J/mol·K − 24.9 J/mol·K = 11.1 J/mol·K, which is about $2.7\left(\frac{1}{2}R\right)$; the vibrational degrees of freedom make a significant contribution.

16-29

Eq. (16-35): $f(v) = \dfrac{8\pi}{m}\left(\dfrac{m}{2\pi kT}\right)^{3/2} \epsilon e^{-\epsilon/kT}$

At the maximum of $f(\epsilon)$, $\dfrac{df}{d\epsilon} = 0$.

$\dfrac{df}{d\epsilon} = \dfrac{8\pi}{m}\left(\dfrac{m}{2\pi kT}\right)^{3/2} \dfrac{d}{d\epsilon}\left(\epsilon e^{-\epsilon/kT}\right) = 0$

So $\dfrac{df}{d\epsilon} = 0 \Rightarrow \dfrac{d}{d\epsilon}\left(\epsilon e^{-\epsilon/kT}\right) = 0 \Rightarrow e^{-\epsilon/kT} - \dfrac{\epsilon}{kT}e^{-\epsilon/kT} = 0$

$\left(1 - \dfrac{\epsilon}{kT}\right)e^{-\epsilon/kT} = 0 \Rightarrow 1 - \dfrac{\epsilon}{kT} = 0 \Rightarrow \epsilon = kT$, as was to be shown.

Problems

16-35

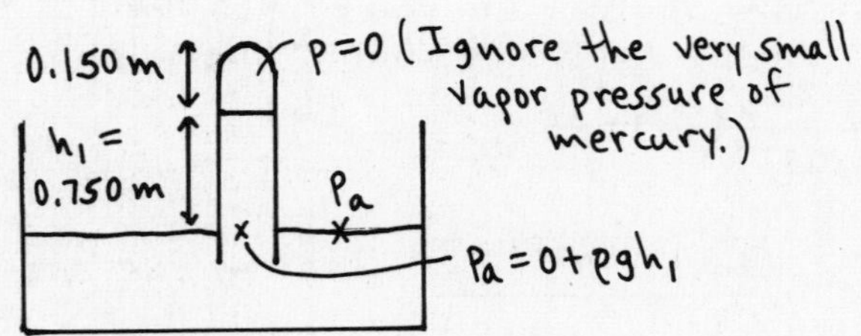

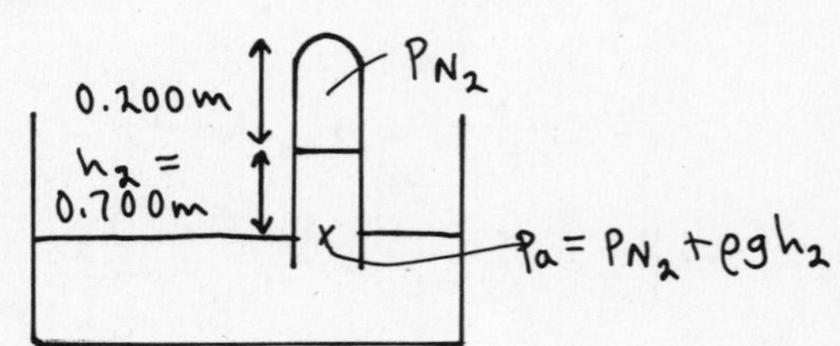

16-35 (cont)

The pressure at points level with the surface of the mercury outside the tube is equal to air pressure, p_a. In the first sketch this gives $p_a = \rho g h_1$ and in the second, $p_a = p_{N_2} + \rho g h_2$. Equating these two expressions for p_a gives $\rho g h_1 = p_{N_2} + \rho g h_2$.

$\Rightarrow p_{N_2} = \rho g (h_1 - h_2) = (13.6\times10^3\,kg/m^3)(9.80\,m/s^2)(0.750\,m - 0.700\,m) = 6.664\times10^3\,Pa$

We know p, V, and T; calculate the mass m_{tot} of nitrogen:

$pV = nRT$

$n = \frac{pV}{RT} = \frac{(6.664\times10^3\,Pa)(0.200\,m)(0.620\times10^{-4}\,m^2)}{(8.3145\,J/mol\cdot K)(300\,K)} = 3.313\times10^{-5}\,mol$

$m_{tot} = nM = (3.313\times10^{-5}\,mol)(28.0\times10^{-3}\,kg/mol) = \underline{9.28\times10^{-7}\,kg}$

16-37

a) $pV = nRT$

Consider the cooling of the flask, after the stopcock is closed.

n, R, V constant $\Rightarrow \frac{p}{T} = \frac{nR}{V} = \text{constant} \Rightarrow \frac{p_1}{T_1} = \frac{p_2}{T_2} \Rightarrow p_2 = \left(\frac{T_2}{T_1}\right) p_1$

At $T_1 = 400\,K$ the flask is open to the air $\Rightarrow p_1 = 1.013\times10^5\,Pa$

$p_2 = p_1\left(\frac{T_2}{T_1}\right) = (1.013\times10^5\,Pa)\left(\frac{300\,K}{400\,K}\right) = \underline{7.60\times10^4\,Pa}$

b) Apply $pV = nRT$ to the oxygen left in the flask at the end of the procedure.

$n = \frac{m_{tot}}{M} \Rightarrow pV = \frac{m_{tot}}{M} RT$

$m_{tot} = \frac{pVM}{RT} = \frac{(7.60\times10^4\,Pa)(1.20\times10^{-3}\,m^3)(32.0\times10^{-3}\,kg/mol)}{(8.3145\,J/mol\cdot K)(300\,K)} = 1.17\times10^{-3}\,kg$

$m_{tot} = \underline{1.17\,g}$

16-39

a) Consider the gas in one cylinder. Calculate the volume to which this quantity of gas expands when the pressure is decreased from 2.50×10^6 Pa to 1.01×10^5 Pa.

$pV = nRT$

n, R, T constant $\Rightarrow pV = nRT = \text{constant} \Rightarrow p_1 V_1 = p_2 V_2$

$V_2 = V_1\left(\frac{p_1}{p_2}\right) = (2.50\,m^3)\left(\frac{2.50\times10^6\,Pa}{1.01\times10^5\,Pa}\right) = 61.88\,m^3$

The number of cylinders required to fill a 500 m³ balloon is

$\frac{500\,m^3}{61.88\,m^3} = \underline{8.08\text{ cylinders}}$

b) The upward force on the balloon is given by Archimedes' principle (Chapter 13):

B = weight of air displaced by balloon $= \rho_{air} V g$

16-39 (cont)

Free-body diagram for the balloon:

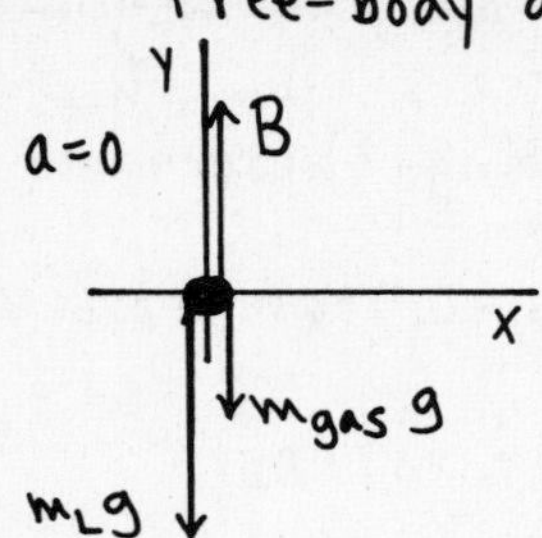

m_{gas} is the mass of the gas that is inside the balloon

m_L is the mass of the load that can be supported by the balloon

$$\Sigma F_y = ma_y$$

$$B - m_L g - m_{gas} g = 0$$

$$\rho_{air} V \cancel{g} - m_L \cancel{g} - m_{gas} \cancel{g} = 0$$

$$\boxed{m_L = \rho_{air} V - m_{gas}}$$

Calculate m_{gas}, the mass of hydrogen that occupies $500\ m^3$ at $0°C$ and $p = 1.01 \times 10^5\ Pa$.

$$pV = nRT = \frac{m_{gas}}{M} RT \Rightarrow m_{gas} = \frac{pVM}{RT} = \frac{(1.01\times10^5 Pa)(500\,m^3)(2.02\times10^{-3}\,kg/mol)}{(8.3145 J/mol\cdot K)(273\,K)} = 44.94\ kg$$

Then $m_L = (1.29\ kg/m^3)(500\ m^3) - 44.94\ kg = 600\ kg$, and the weight that can be supported is $w_L = m_L g = (600\ kg)(9.80\ m/s^2) = \underline{5880\ N}$

16-41

Calculate the number of water molecules N:

$$n = \frac{m_{tot}}{M} = \frac{60\ kg}{18.0\times10^{-3}\ kg/mol} = 3.333\times10^3\ mol$$

$$N = nN_A = (3.333\times10^3 mol)(6.022\times10^{23}\ molecules/mol) = 2.007\times10^{27}\ molecules$$

Each water molecule has three atoms, so the number of atoms is

$$3(2.007\times10^{27}) = \underline{6.02\times10^{27}\ atoms}$$

16-43

a) $v_{rms} = \sqrt{\frac{3RT}{M}} = \sqrt{\frac{3(8.3145\ J/mol\cdot K)(300K)}{28.0\times10^{-3}\ kg/mol}} = \underline{517\ m/s}$

b) $(v_x^2)_{av} = \frac{1}{3}(v^2)_{av} \Rightarrow \sqrt{(v_x^2)_{av}} = \frac{1}{\sqrt{3}}\sqrt{(v^2)_{av}} = \frac{1}{\sqrt{3}} v_{rms} = \frac{1}{\sqrt{3}}(517\ m/s) = \underline{298\ m/s}$

16-47

a) Apply conservation of energy: $K_1 + U_1 + W_{other} = K_2 + U_2$, where $U = -\frac{Gmm_E}{r}$. Let point #1 be at the surface of the earth where the projectile is launched, and let point #2 be far from the earth. Just barely escapes $\Rightarrow v_2 = 0$.

Only gravity does work $\Rightarrow W_{other} = 0$.

$U_1 = -\frac{Gmm_E}{R}$

$r_2 \to \infty \Rightarrow U_2 = 0$

$v_2 = 0 \Rightarrow K_2 = 0$

thus $K_1 - \frac{Gmm_E}{R} = 0 \Rightarrow K_1 = \frac{Gmm_E}{R}$

But $g = G\frac{m_E}{R^2} \Rightarrow \frac{Gm_E}{R} = Rg$ and $K_1 = mgR$, as was to be shown.

16-47 (cont)

b) $\frac{1}{2}m(v^2)_{av} = mgR$ (from part (a))

But also, $\frac{1}{2}(v^2)_{av} = \frac{3}{2}kT \Rightarrow mgR = \frac{3}{2}kT$

$$T = \frac{2mgR}{3k}$$

oxygen

From Example 16-5, $m_{O_2} = 53.1\times10^{-27}$ kg/molecule

$$T = \frac{2(53.1\times10^{-27}\text{ kg/molecule})(9.80\text{ m/s}^2)(6.38\times10^6\text{ m})}{3(1.381\times10^{-23}\text{ J/molecule}\cdot\text{K})} = \underline{1.60\times10^5\text{ K}}$$

hydrogen

From Example 16-5, $m_{H_2} = 2(1.674\times10^{-27}\text{ kg}) = 3.348\times10^{-27}$ kg/molecule

$$T = \frac{2(3.348\times10^{-27}\text{ kg/molecule})(9.80\text{ m/s}^2)(6.38\times10^6\text{ m})}{3(1.381\times10^{-23}\text{ J/molecule}\cdot\text{K})} = \underline{1.01\times10^4\text{ K}}$$

16-49

a) Eq. (16-34): $f(v) = 4\pi\left(\frac{m}{2\pi kT}\right)^{3/2} v^2 e^{-mv^2/2kT}$

$$\int_0^\infty f(v)\,dv = 4\pi\left(\frac{m}{2\pi kT}\right)^{3/2}\int_0^\infty v^2 e^{-mv^2/2kT}\,dv$$

Use the integral formula given in the problem, with $n=1$ and $a = \frac{m}{2kT}$

$$\Rightarrow \int_0^\infty f(v)\,dv = 4\pi\left(\frac{m}{2\pi kT}\right)^{3/2}\frac{1}{4}\,\frac{2kT}{m}\sqrt{\frac{2\pi kT}{m}} = 1 \checkmark$$

b) $\int_0^\infty v^2 f(v)\,dv = 4\pi\left(\frac{m}{2\pi kT}\right)^{3/2}\int_0^\infty v^4 e^{-mv^2/2kT}\,dv$

The integral formula with $n=2$ gives $\int_0^\infty v^4 e^{-av^2}dv = \frac{3}{8a^2}\sqrt{\frac{\pi}{a}}$.

Use with $a = \frac{m}{2kT}$,

$$\Rightarrow \int_0^\infty v^2 f(v)\,dv = 4\pi\left(\frac{m}{2\pi kT}\right)^{3/2}\frac{3}{8}\left(\frac{2kT}{m}\right)^2\sqrt{\frac{2\pi kT}{m}} = \frac{3}{2}\,\frac{2kT}{m} = \frac{3kT}{m}$$

Equation (16-17) says $\frac{1}{2}m(v^2)_{av} = \frac{3}{2}kT \Rightarrow (v^2)_{av} = \frac{3kT}{m}$, in agreement with our calculation.

16-51

$$\text{relative humidity} = \frac{\text{partial pressure of water vapor at temperature } T}{\text{vapor pressure of water at temperature } T}$$

The experiment shows that the dew point is 15°C, so the partial pressure of water vapor at 40°C is equal to the vapor pressure at 15°C, which is 1.69×10^3 Pa.

Thus rel. hum. $= \frac{1.69\times10^3\text{ Pa}}{7.34\times10^3\text{ Pa}} = 0.230 = \underline{23.0\%}$

CHAPTER 17

Exercises 3, 9, 13, 15, 19, 21

Problems 23, 25, 29, 31, 33

Exercises

17-3

$W = \int_{V_1}^{V_2} p\, dV$

Constant pressure $\Rightarrow W = p\int_{V_1}^{V_2} dV = p(V_2 - V_1)$

The problem gives T rather than p and V, so use the ideal gas law to rewrite the expression for W:

$pV = nRT \Rightarrow p_1V_1 = nRT_1,\ pV_2 = nRT_2$; subtract $\Rightarrow p(V_2 - V_1) = nR(T_2 - T_1)$

Thus $W = nR(T_2 - T_1)$ is an alternative expression for the work in a constant pressure process for an ideal gas.

Then $W = nR(T_2 - T_1) = (3.00\ \text{mol})(8.3145\ \text{J/mol}\cdot\text{K})(177°\text{C} - 27°\text{C}) = \underline{3.74 \times 10^3\ \text{J}}$

17-9

a) For the water $\Delta T > 0$, so by $Q = mc\Delta T$ heat has been added to the water. This heat energy comes from the burning fuel-oxygen mixture $\Rightarrow$ Q for the system (fuel and oxygen) is negative.

b) Constant volume $\Rightarrow W = 0$.

c) The 1st Law (Eq. 17-6) says $\Delta U = Q - W$.

$Q < 0,\ W = 0 \Rightarrow \Delta U < 0$.

The internal energy of the fuel-oxygen mixture decreased. In this process internal energy from the fuel-oxygen mixture was transferred to the water, raising its temperature.

17-13

a) $W = \int_{V_1}^{V_2} p\, dV = p(V_2 - V_1)$ for this constant pressure process.

$W = (2.00 \times 10^5\ \text{Pa})(0.80\ \text{m}^3 - 1.20\ \text{m}^3) = \underline{-8.0 \times 10^4\ \text{J}}$

(The volume decreases in the process, so W is negative.)

b) $\Delta U = Q - W$

$Q = -2.80 \times 10^5\ \text{J}$ (Q is negative since heat is removed from the gas.)

$\Delta U = Q - W = -2.80 \times 10^5\ \text{J} - (-8.0 \times 10^4\ \text{J}) = \underline{-2.00 \times 10^5\ \text{J}}$

c) $W = \int_{V_1}^{V_2} p\, dV = p(V_2 - V_1)$ (constant pressure) and $\Delta U = Q - W$ apply to <u>any</u> system, not just to an ideal gas. We did not use the ideal gas equation, either directly or indirectly, in any of the calculations, so the results are the same whether the gas is ideal or not.

17-15

a) $W = \int_{V_1}^{V_2} p\, dV$

$pV = nRT \Rightarrow p = \frac{nRT}{V}$

$W = \int_{V_1}^{V_2} \frac{nRT}{V} dV = nRT \int_{V_1}^{V_2} \frac{dV}{V} = nRT \ln\left(\frac{V_2}{V_1}\right)$

$W = (0.100\text{ mol})(8.3145\text{ J/mol}\cdot\text{K})(300\text{ K}) \ln\left(\frac{\frac{1}{8}V_1}{V_1}\right) = 249.4\text{ J} \ln\left(\frac{1}{8}\right) = \underline{-519\text{ J}}$

(W for the gas is negative, since the volume decreases.)

b) $\Delta U = nC_V \Delta T$ for any ideal gas process.

$\Delta T = 0$ (isothermal) $\Rightarrow \Delta U = 0$

c) $\Delta U = Q - W$

$\Delta U = 0 \Rightarrow Q = W = \underline{-519\text{ J}}$

(Q negative $\Rightarrow$ the gas liberates 519 J of heat to the surroundings.)

Note: $Q = nC_V\Delta T$ is only for a constant volume process so doesn't apply here.
$Q = nC_p \Delta T$ is only for a constant pressure process so doesn't apply here.

17-19

For an adiabatic process for an ideal gas

$T_1 V_1^{\gamma-1} = T_2 V_2^{\gamma-1}$, $p V_1^{\gamma} = p_2 V_2^{\gamma}$, and $pV = nRT$.

Air is mostly diatomic (O_2, N_2, H_2) $\Rightarrow \gamma = 1.4$ (Example 17-7)

$T_1 V_1^{\gamma-1} = T_2 V_2^{\gamma-1} \Rightarrow T_2 = T_1 \left(\frac{V_1}{V_2}\right)^{\gamma-1} = (293\text{ K})\left(\frac{V_1}{\frac{1}{3}V_1}\right)^{1.4-1} = (293\text{ K})(3)^{0.4}$

$T_2 = \underline{455\text{ K}} = \underline{182^\circ\text{C}}$

(Note: In the relation $T_1 V_1^{\gamma-1} = T_2 V_2^{\gamma-1}$ the temperature must be in kelvins.)

$p_1 V_1^{\gamma} = p_2 V_2^{\gamma} \Rightarrow p_2 = p_1 \left(\frac{V_1}{V_2}\right)^{\gamma} = (1.00\text{ atm})\left(\frac{V_1}{\frac{1}{3}V_1}\right)^{1.4} = (1.00\text{ atm})(3)^{1.4} = \underline{4.66\text{ atm}}$

Alternatively, can use $pV = nRT$ to calculate p_2:

n, R constant $\Rightarrow \frac{pV}{T} = nR = \text{constant} \Rightarrow \frac{p_1 V_1}{T_1} = \frac{p_2 V_2}{T_2}$

$p_2 = p_1 \left(\frac{V_1}{V_2}\right)\left(\frac{T_2}{T_1}\right) = (1.00\text{ atm})\left(\frac{V_1}{\frac{1}{3}V_1}\right)\left(\frac{455\text{ K}}{293\text{ K}}\right) = 4.66$ atm, which checks.

17-21

a) Adiabatic $\Rightarrow Q = 0$

Then $\Delta U = Q - W \Rightarrow W = -\Delta U = -nC_V \Delta T = nC_V(T_1 - T_2)$ (Eq. 17-27)

$C_V = 20.85\text{ J/mol}\cdot\text{K}$ (Table 17-1)

$W = (0.600\text{ mol})(20.85\text{ J/mol}\cdot\text{K})(300\text{ K} - 283\text{ K}) = \underline{+213\text{ J}}$

(W positive for $\Delta V > 0$ (expansion))

b) Adiabatic process $\Rightarrow \underline{Q = 0}$

Problems

17-23

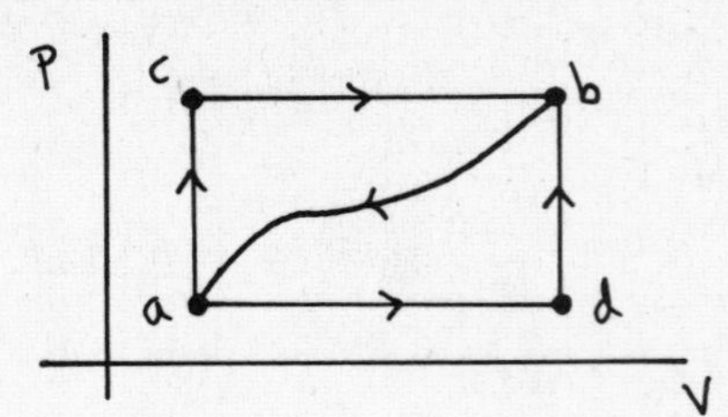

$Q_{acb} = +90.0$ J (positive since heat flows in)

$W_{acb} = +70.0$ J (positive since $\Delta V > 0$)

a) $\Delta U = Q - W$

ΔU is path independent ; Q and W depend on the path.

$\Delta U = U_b - U_a$

This can be calculated for any path from a to b, in particular for path acb:

$\Delta U_{a\to b} = Q_{acb} - W_{acb} = 90.0\text{ J} - 70.0\text{ J} = \underline{20.0\text{ J}}$

Now apply $\Delta U = Q - W$ to path adb; $\Delta U = 20.0$ J for this path also.

$W_{adb} = +10.0$ J (positive since $\Delta V > 0$)

$\Delta U_{a\to b} = Q_{adb} - W_{adb} \Rightarrow Q_{adb} = \Delta U_{a\to b} + W_{adb} = +20.0\text{ J} + 10.0\text{ J} = \underline{+30.0\text{ J}}$

b) Apply $\Delta U = Q - W$ to path ba:

$\Delta U_{b\to a} = Q_{ba} - W_{ba}$

$W_{ba} = -45.0$ J (negative since $\Delta V < 0$)

$\Delta U_{b\to a} = U_a - U_b = -(U_b - U_a) = -\Delta U_{a\to b} = -20.0$ J

$Q_{ba} = \Delta U_{b\to a} + W_{ba} = -20.0\text{ J} - 45.0\text{ J} = \underline{-65.0\text{ J}}$

($Q_{ba} < 0 \Rightarrow$ the system liberates heat)

c) $U_a = 0$, $U_d = 6.0$ J

$\Delta U_{a\to b} = U_b - U_a = +20.0\text{ J} \Rightarrow U_b = 20.0$ J

process $a\to d$

$\Delta U_{a\to d} = Q_{ad} - W_{ad}$

$\Delta U_{a\to d} = U_d - U_a = +6.0$ J

$W_{adb} = +10.0$ J and $W_{adb} = W_{ad} + W_{db}$. But the work W_{db} for the process $d\to b$ is zero since $\Delta V = 0$ for that process. Therefore $W_{ad} = W_{adb} = +10.0$ J

Then $Q_{ad} = \Delta U_{a\to d} + W_{ad} = +6.0\text{ J} + 10.0\text{ J} = \underline{+16.0\text{ J}}$ (positive $\Rightarrow$ heat absorbed)

process db

$\Delta U_{d\to b} = Q_{db} - W_{db}$

$W_{db} = 0$, as already noted.

$\Delta U_{d\to b} = U_b - U_d = 20.0\text{ J} - 6.0\text{ J} = 14.0$ J

Then $Q_{db} = W_{db} + \Delta U_{db} = \underline{+14.0\text{ J}}$ (positive $\Rightarrow$ heat absorbed)

17-25

$\Delta U = Q - W$

$Q = +(2.65 \times 10^5 \text{ J})$ (positive since heat energy goes into system)

$\Delta U = 0 \Rightarrow W = Q = 2.65 \times 10^5 \text{ J}$

constant pressure $\Rightarrow W = \int_{V_1}^{V_2} p\, dV = p(V_2 - V_1) = p\,\Delta V$

$$\Delta V = \frac{W}{p} = \frac{2.65 \times 10^5 \text{ J}}{6.90 \times 10^5 \text{ Pa}} = \underline{0.384 \text{ m}^3}$$

17-29

a)

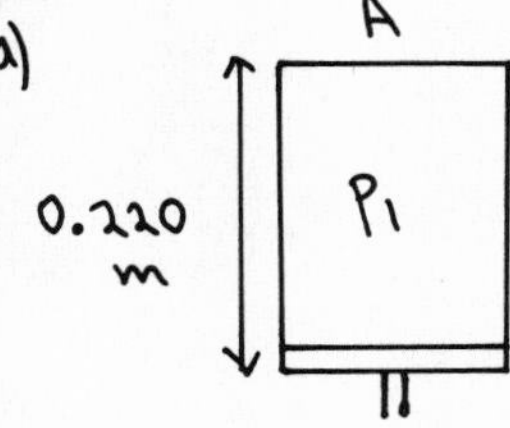

x

P_2

?

$P_1 = 1.01 \times 10^5 \text{ Pa}$

$P_2 = 1.01 \times 10^5 \text{ Pa} + 4.40 \times 10^5 \text{ Pa} = 5.41 \times 10^5 \text{ Pa}$

Use the equations for an adiabatic process to solve for V_2:

$P_1 V_1^\gamma = P_2 V_2^\gamma$; assume that $\gamma = 1.40$ for air (Example 17-7)

$V_2^\gamma = V_1^\gamma \left(\frac{P_1}{P_2}\right) \Rightarrow V_2 = V_1 \left(\frac{P_1}{P_2}\right)^{1/\gamma}$

But $V_1 = (0.22 \text{ m}) A$, where A is the cross-section area of the cylinder.

$V_2 = x A$

$x A = (0.22 \text{ m}) A \left(\frac{P_1}{P_2}\right)^{1/\gamma}$

$$x = (0.22 \text{ m})\left(\frac{P_1}{P_2}\right)^{1/\gamma} = (0.22 \text{ m})\left(\frac{1.01 \times 10^5 \text{ Pa}}{5.41 \times 10^5 \text{ Pa}}\right)^{1/1.40} = 0.0663 \text{ m}$$

The distance the piston has moved is $0.220 \text{ m} - x = 0.220 \text{ m} - 0.0663 \text{ m} = \underline{0.154 \text{ m}}$

b) $T_1 V_1^{\gamma-1} = T_2 V_2^{\gamma-1}$

$$T_2 = T_1 \left(\frac{V_1}{V_2}\right)^{\gamma-1} = (300 \text{ K})\left(\frac{(0.220 \text{ m}) A}{(0.0663 \text{ m}) A}\right)^{1.40-1} = (300 \text{ K})\left(\frac{0.220 \text{ m}}{0.0663 \text{ m}}\right)^{0.40} = 485 \text{ K} = \underline{212^\circ\text{C}}$$

c) Adiabatic process $\Rightarrow W = n C_V (T_1 - T_2)$ (Eq. 17-27)

$W = (30.0 \text{ mol})(20.8 \text{ J/mol}\cdot\text{K})(300 \text{ K} - 485 \text{ K}) = \underline{-1.15 \times 10^5 \text{ J}}$

(Negative since work is done on the gas and $\Delta V < 0$.)

17-31

a) isothermal ($\Delta T = 0$)

$\Delta U = Q - W$

$W = +600 \text{ J}$

For any process of an ideal gas, $\Delta U = n C_V \Delta T$. So for an ideal gas, if $\Delta T = 0$ then $\Delta U = 0$.

$\Rightarrow Q = W = \underline{+600 \text{ J}}$

b) adiabatic ($Q = 0$)

$\Delta U = Q - W$; $W = +600 \text{ J}$

$\underline{Q = 0} \Rightarrow \Delta U = -W = \underline{-600 \text{ J}}$

17-33

a) In the adiabatic expansion the pressure decreases.

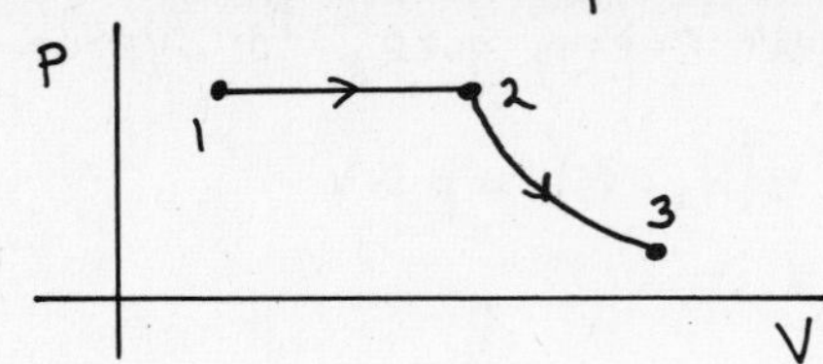

b) process 1→2 ($\Delta p = 0$)

$pV = nRT$ and p constant $\Rightarrow \frac{V_1}{T_1} = \frac{V_2}{T_2}$

$T_2 = T_1\left(\frac{V_2}{V_1}\right) = (300\,K)\left(\frac{2V_1}{V_1}\right) = 600\,K$

process 2→3 ($Q = 0$)

$T_2 = 600\,K$

Temperature returns to its initial value $\Rightarrow T_3 = T_1 = 300\,K$

$Q_{12} = nC_p\Delta T$, since $\Delta p = 0$ for this process.

$Q_{12} = (2.00\,mol)(20.78\,J/mol\cdot K)(600K - 300\,K) = 1.247\times10^4\,J$

$Q_{23} = 0$ since process 2→3 is adiabatic

$Q = Q_{12} + Q_{23} = \underline{1.25\times10^4\,J}$

c) $\Delta U_{12} = nC_V\Delta T = nC_V(T_2 - T_1)$

$\Delta U_{23} = nC_V\Delta T = nC_V(T_3 - T_2)$

$\Delta U_{tot} = \Delta U_{12} + \Delta U_{23} = nC_V(T_2 - T_1 + T_3 - T_2) = nC_V(T_3 - T_1) = 0$, since $T_1 = T_3$;

$\Delta U_{tot} = 0$ since the ideal gas ends up in the same temperature as at the start.

d) $\Delta U = Q - W$

$W = Q - \Delta U$

$\Delta U = 0 \Rightarrow W = Q = \underline{1.25\times10^4\,J}$

e) In process 1→2 the volume doubles, so $V_2 = 0.0800\,m^3$.

Process 2→3 is adiabatic and we know T_2 and T_3. So to find out what happens to the volume use $T_2V_2^{\gamma-1} = T_3V_3^{\gamma-1}$.

$$V_3^{\gamma-1} = V_2^{\gamma-1}\left(\frac{T_2}{T_3}\right) \Rightarrow V_3 = V_2\left(\frac{T_2}{T_3}\right)^{1/(\gamma-1)}$$

For helium, $\gamma = 1.67$ (from Table 17-1), so

$$V_3 = (0.0800\,m^3)\left(\frac{600\,K}{300\,K}\right)^{1/(1.67-1)} = \underline{0.225\,m^3}$$

(Note that $V_3 > V_2$ as it should be, since process 2→3 is an expansion.)

CHAPTER 18

Exercises 1, 7, 9, 13, 15, 19, 21, 25, 27

Problems 29, 31, 33, 35

Exercises

18-1

a) $e = \dfrac{\text{work output}}{\text{heat energy input}} = \dfrac{W}{Q_H} = \dfrac{2000\text{ J}}{8000\text{ J}} = 0.250 = \underline{25.0\%}$

b) $W = Q = |Q_H| - |Q_c|$

Heat discarded is $|Q_c| = |Q_H| - W = 8000\text{ J} - 2000\text{ J} = \underline{6000\text{ J}}$

c) Q_H is supplied by the burning fuel; $Q_H = mL_c$ where L_c is the heat of combustion

$m = \dfrac{Q_H}{L_c} = \dfrac{8000\text{ J}}{5.00\times10^4\text{ J/g}} = \underline{0.160\text{ g}}$

d) $W = 2000\text{ J}$ per cycle

In $t = 1.00\text{ s}$ the engine goes through 40 cycles.

$P = \dfrac{W}{t} = \dfrac{40(2000\text{ J})}{1.00\text{ s}} = \underline{8.00\times10^4\text{ W}}$

$P = (8.00\times10^4\text{ W})\left(\dfrac{1\text{ hp}}{746\text{ W}}\right) = \underline{107\text{ hp}}$

18-7

Process $a\to b$ is adiabatic $\Rightarrow T_a V_a^{\gamma-1} = T_b V_b^{\gamma-1}$

$V_b = V$, $V_a = rV$ (read from Fig. 18-3)

$T_a = 22.0^\circ\text{C} = 295\text{ K}$

$T_b = T_a\left(\dfrac{V_a}{V_b}\right)^{\gamma-1} = (295\text{ K})\left(\dfrac{rV}{V}\right)^{1.40-1} = (295\text{ K})(7.00)^{0.4} = 642\text{ K} = \underline{369^\circ\text{C}}$

18-9

a) Performance coefficient $K = \dfrac{Q_c}{|W|}$ (Eq. 18-9)

$\Rightarrow |W| = \dfrac{Q_c}{K} = \dfrac{3.00\times10^4\text{ J}}{2.00} = \underline{1.50\times10^4\text{ J}}$

b)

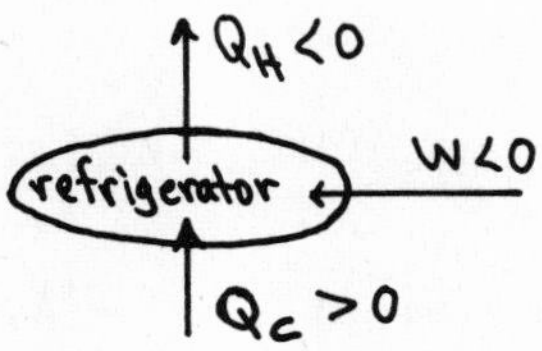

$W = Q_c + Q_H$

$Q_H = W - Q_c = -1.50\times10^4\text{ J} - 3.00\times10^4\text{ J} = \underline{-4.50\times10^4\text{ J}}$

(minus because heat goes out of the system)

18-13

a)

$T_H = 400\,K$

$Q_H > 0$

engine

$W > 0$

$Q_C < 0$

T_C

For a Carnot cycle,

$$\frac{|Q_C|}{|Q_H|} = \frac{T_C}{T_H} \quad \text{(Eq. 18-13)}$$

$$T_C = T_H \frac{|Q_C|}{|Q_H|} = 400\,K\left(\frac{335\,J}{480\,J}\right) = \underline{279\,K}$$

b) $e_{Carnot} = 1 - \frac{T_C}{T_H} = 1 - \frac{279\,K}{400\,K} = 0.302 = \underline{30.2\%}$

c) $W = Q_C + Q_H = -335\,J + 480\,J = \underline{145\,J}$

or $e = \frac{W}{Q_H} \Rightarrow W = eQ_H = (0.302)(480\,J) = 145\,J$, which checks

18-15

a)

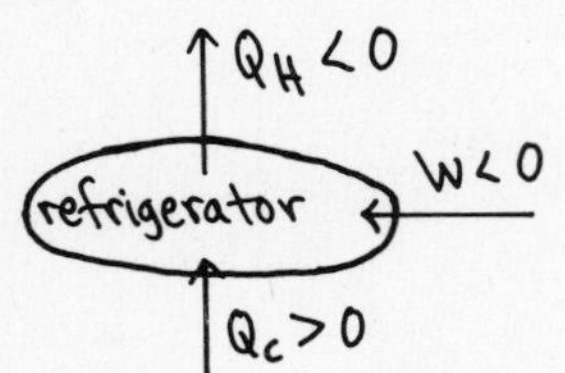

$T_H = 27.0°C = 300\,K$

$T_C = 0.0°C = 273\,K$

The amount of heat taken out of the water to make the liquid → solid phase change is $Q = -mL_f = -(40.0\,kg)(334 \times 10^3\,J/kg) = -1.336 \times 10^7\,J$. This amount of heat must go into the working substance of the refrigerator $\Rightarrow Q_C = +1.336 \times 10^7\,J$.

Carnot cycle $\Rightarrow \frac{|Q_C|}{|Q_H|} = \frac{T_C}{T_H}$

$$|Q_H| = |Q_C|\left(\frac{T_H}{T_C}\right) = 1.336 \times 10^7\,J\left(\frac{300\,K}{273\,K}\right) = \underline{1.47 \times 10^7\,J}$$

b) $W = Q_C + Q_H = +1.34 \times 10^7\,J - 1.47 \times 10^7\,J = \underline{-1.3 \times 10^6\,J}$

(W is negative because this much energy must be supplied to the refrigerator rather than obtained from it.)

18-19

First use $Q_{system} = 0$ to find the final temperature of the mixed water.

The 1.00 kg of water warms from 20.0°C to T:

$$Q = mc\,\Delta T = (1.00\,kg)(4190\,J/kg\cdot K)(T - 20.0°C)$$

The 2.00 kg of water cools from 60.0°C to T:

$$Q = mc\,\Delta T = (2.00\,kg)(4190\,J/kg\cdot K)(T - 60.0°C)$$

$$Q_{system} = 0 \Rightarrow (1.00\,kg)(\cancel{4190\,J/kg}\cdot K)(T - 20.0°C) + (2.00\,kg)(\cancel{4190\,J/kg}\cdot K)(T - 60.0°C) = 0$$

$$T - 20.0°C + 2(T - 60.0°C) = 0$$

$$3T = 140°C$$

$$T = 46.7°C$$

18-19 (cont)

Now we can calculate the entropy change.

The 1.00 kg of water:

$$\Delta S = \int_1^2 \frac{dQ}{T}$$

$dQ = mc\,dT \Rightarrow \Delta S = mc\int_1^2 \frac{dT}{T} = mc\ln\left(\frac{T_2}{T_1}\right)$ (Example 18-6)

$$\Delta S = (1.00\,\text{kg})(4190\,\text{J/kg}\cdot\text{K})\ln\left(\frac{(46.7+273)\,\text{K}}{(20.0+273)\,\text{K}}\right) = 365\,\text{J/K}$$

The 2.00 kg of water:

$$\Delta S = mc\ln\left(\frac{T_2}{T_1}\right) = (2.00\,\text{kg})(4190\,\text{J/kg}\cdot\text{K})\ln\left(\frac{(46.7+273)\,\text{K}}{(60.0+273)\,\text{K}}\right) = -342\,\text{J/K}$$

The total entropy change of the system is

$$\Delta S_{tot} = 365\,\text{J/K} - 342\,\text{J/K} = \underline{23\,\text{J/K}}$$

18-21

Reversible $\Rightarrow \Delta S = \int_1^2 \frac{dQ}{T}$

Isothermal $\Rightarrow$ T is constant $\Rightarrow \Delta S = \frac{Q}{T}$

Calculate Q:

$$\Delta U = Q - W$$

$\Delta U = nC_V\Delta T = 0$ since $\Delta T = 0$.

$W = \int_{V_1}^{V_2} p\,dV$; $pV = nRT \Rightarrow p = \frac{nRT}{V}$, where T is constant

$$W = nRT\int_{V_1}^{V_2}\frac{dV}{V} = nRT\ln\left(\frac{V_2}{V_1}\right) = (2.00\,\text{mol})(8.3145\,\text{J/mol}\cdot\text{K})(300\,\text{K})\ln\left(\frac{0.0500\,\text{m}^3}{0.0200\,\text{m}^3}\right)$$

$W = 4.571\times10^3\,\text{J}$

Then $Q = \Delta U + W = 0 + 4.571\times10^3\,\text{J} = 4.571\times10^3\,\text{J}$

$$\Delta S = \frac{Q}{T} = \frac{4.571\times10^3\,\text{J}}{300\,\text{K}} = \underline{+15.2\,\text{J/K}}$$

18-25

The power collected is $(0.50)(200\,\text{W/m}^2)(30.0\,\text{m}^2) = 3000\,\text{W}$.
The energy collected in 1.00 h is $(3000\,\text{W})(3600\,\text{s}) = 1.08\times10^7\,\text{J}$

This energy is inputed to the water, to heat it from 15.0°C to 60.0°C.
Calculate the mass of water that this much energy can heat:

$$Q = mc\Delta T \Rightarrow m = \frac{Q}{c\Delta T} = \frac{1.08\times10^7\,\text{J}}{(4190\,\text{J/kg}\cdot\text{K})(60.0^\circ\text{C} - 15.0^\circ\text{C})} = 57.28\,\text{kg}$$

Find the volume of this mass of water:

$$\rho = \frac{m}{V} \Rightarrow V = \frac{m}{\rho} = \frac{57.28\,\text{kg}}{1000\,\text{kg/m}^3} = 0.0573\,\text{m}^3 = \underline{57.3\,\text{L}}$$

18-27

a) Calculate the mass of water required and then from this the volume of water:

$Q = mc\Delta T \Rightarrow m = \frac{Q}{c\Delta T} = \frac{5.25\times10^9\ \text{J}}{(4190\ \text{J/kg}\cdot\text{K})(49.0^\circ\text{C} - 21.0^\circ\text{C})} = 4.475\times10^4\ \text{kg}$

$\rho = \frac{m}{V} \Rightarrow V = \frac{m}{\rho} = \frac{4.475\times10^4\ \text{kg}}{1.00\times10^3\ \text{kg/m}^3} = \underline{44.8\ \text{m}^3}$

b) Repeat the above calculation, but now with Glauber salt in place of water. The essential difference is that Glauber salt undergoes a phase transition in this temperature range (at 32.0 °C).

$Q = mc_{solid}(32.0^\circ\text{C} - 21.0^\circ\text{C}) + mL_f + mc_{liquid}(49.0^\circ\text{C} - 32.0^\circ\text{C})$

$Q = m[(1930\ \text{J/kg}\cdot\text{K})(11.0\ \text{K}) + 2.42\times10^5\ \text{J/kg} + (2850\ \text{J/kg}\cdot\text{K})(17.0\ \text{K})]$

$Q = m[3.117\times10^5\ \text{J/kg}]$

Then $m = \frac{Q}{3.117\times10^5\ \text{J/kg}} = \frac{5.25\times10^9\ \text{J}}{3.117\times10^5\ \text{J/kg}} = 1.684\times10^4\ \text{kg}$

$V = \frac{m}{\rho} = \frac{1.684\times10^4\ \text{kg}}{1600\ \text{kg/m}^3} = \underline{10.5\ \text{m}^3}$

(The space requirement is smaller by about a factor of 4 when Glauber salt is used instead of water.)

Problems

18-29

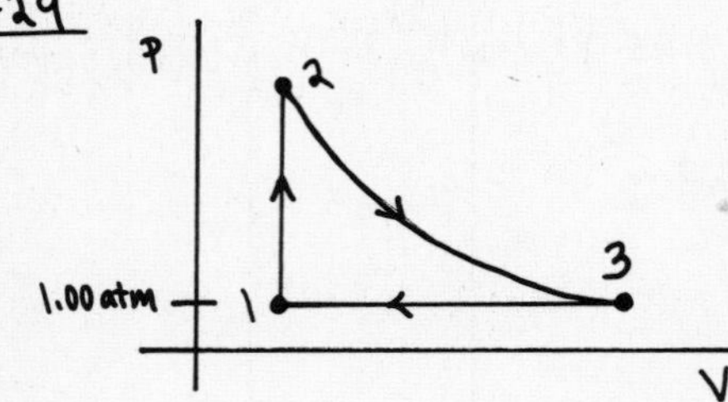

$T_1 = 300\ \text{K}$

$T_2 = 600\ \text{K}$

$T_3 = 455\ \text{K}$

$\gamma = 1.67$

$\Rightarrow C_V = 12.47\ \text{J/mol}\cdot\text{K}$

$C_p = 20.78\ \text{J/mol}\cdot\text{K}$

(from Table 17-1)

a) point 1

$p_1 = 1.00\ \text{atm} = \underline{1.013\times10^5\ \text{Pa}}$ (given)

$pV = nRT \Rightarrow V_1 = \frac{nRT_1}{p_1} = \frac{(0.300\ \text{mol})(8.3145\ \text{J/mol}\cdot\text{K})(300\ \text{K})}{1.013\times10^5\ \text{Pa}} = \underline{7.39\times10^{-3}\ \text{m}^3}$

point 2

process 1→2 is at constant volume $\Rightarrow V_2 = V_1 = \underline{7.39\times10^{-3}\ \text{m}^3}$

$pV = nRT$ and n, R, V constant $\Rightarrow \frac{p_1}{T_1} = \frac{p_2}{T_2}$

$p_2 = p_1\left(\frac{T_2}{T_1}\right) = 1.00\ \text{atm}\left(\frac{600\ \text{K}}{300\ \text{K}}\right) = 2.00\ \text{atm} = \underline{2.027\times10^5\ \text{Pa}}$

point 3

Consider the process 3→1, since it is simpler than 2→3.

Process 3→1 is at constant pressure $\Rightarrow p_3 = p_1 = 1.00\ \text{atm} = \underline{1.013\times10^5\ \text{Pa}}$

$pV = nRT$ and n, R, p constant $\Rightarrow \frac{V_1}{T_1} = \frac{V_3}{T_3}$

$V_3 = V_1\left(\frac{T_3}{T_1}\right) = (7.39\times10^{-3}\ \text{m}^3)\left(\frac{455\ \text{K}}{300\ \text{K}}\right) = \underline{1.12\times10^{-2}\ \text{m}^3}$

18-29 (cont)

b) process $1 \to 2$

Constant volume ($\Delta V = 0$)

$Q = nC_V\Delta T = (0.300\text{ mol})(12.47\text{ J/mol}\cdot\text{K})(600\text{ K} - 300\text{ K}) = 1122\text{ J}$

$\Delta V = 0 \Rightarrow W = 0$

Then $\Delta U = Q - W = 1122\text{ J}$.

process $2 \to 3$

adiabatic $\Rightarrow Q = 0$

$\Delta U = nC_V\Delta T$ (any process) $\Rightarrow \Delta U = (0.300\text{ mol})(12.47\text{ J/mol}\cdot\text{K})(455\text{K} - 600\text{K}) = -542\text{ J}$

Then $\Delta U = Q - W \Rightarrow W = Q - \Delta U = +542\text{ J}$

(It is correct for W to be positive, since ΔV is positive.)

process $3 \to 1$

constant pressure $\Rightarrow W = p\Delta V = (1.013 \times 10^5\text{ Pa})(7.39 \times 10^{-3}\text{ m}^3 - 1.12 \times 10^{-2}\text{ m}^3) = -386\text{ J}$

or $W = nR\Delta T = (0.300\text{ mol})(8.3145\text{ J/mol}\cdot\text{K})(300\text{ K} - 455\text{ K}) = -386\text{ J}$, which checks

(It is correct for W to be negative, since ΔV is negative for this process.)

$Q = nC_p\Delta T = (0.300\text{ mol})(20.78\text{ J/mol}\cdot\text{K})(300\text{ K} - 455\text{ K}) = -966\text{ K}$

$\Delta U = Q - W = -966\text{ J} - (-386\text{ J}) = -580\text{ J}$

or $\Delta U = nC_V\Delta T = (0.300\text{ mol})(12.47\text{ J/mol}\cdot\text{K})(300\text{K} - 455\text{ K}) = -580\text{ J}$, which checks

c) $W_{net} = W_{1\to2} + W_{2\to3} + W_{3\to1} = 0 + 542\text{ J} - 386\text{ J} = \underline{+156\text{ J}}$

d) $Q_{net} = Q_{1\to2} + Q_{2\to3} + Q_{3\to1} = 1122\text{ J} + 0 - 966\text{ J} = \underline{+156\text{ J}}$

Note: For a cycle $\Delta U = 0$, so by $\Delta U = Q - W$ it must be that $Q_{net} = W_{net}$ for a cycle. We can also check that $\Delta U_{net} = 0$:

$\Delta U_{net} = \Delta U_{1\to2} + \Delta U_{2\to3} + \Delta U_{3\to1} = 1122\text{ J} - 542\text{ J} - 580\text{ J} = 0$ ✓

e) $e = \dfrac{\text{work output}}{\text{heat energy input}} = \dfrac{W}{Q_H} = \dfrac{156\text{ J}}{1122\text{ J}} = 0.139 = \underline{13.9\%}$

18-31

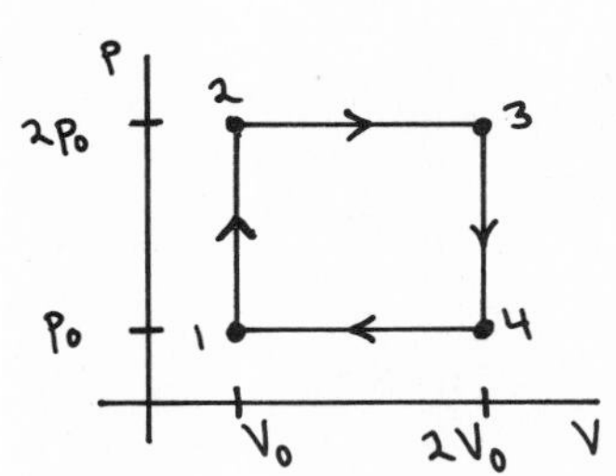

$C_V = 20.5\text{ J/mol}\cdot\text{K}$

For an ideal gas $C_p = C_V + R$.

Calculate Q and W for each process:

process $1 \to 2$

$\Delta V = 0 \Rightarrow W = 0$

$\Delta V = 0 \Rightarrow Q = nC_V\Delta T = nC_V(T_2 - T_1)$

But $pV = nRT$ and V constant $\Rightarrow p_1V = nRT_1$ and $p_2V = nRT_2$.

Thus $(p_2 - p_1)V = nR(T_2 - T_1) \Rightarrow V\Delta p = nR\Delta T$ (true when V is constant).

18-31 (cont)

Then $Q = nC_V \frac{V\Delta p}{nR} = \frac{C_V}{R} V\Delta p = \frac{C_V}{R} V_0(2p_0 - p_0) = \frac{C_V}{R} p_0 V_0$

($Q > 0 \Rightarrow$ heat is absorbed by the gas.)

process $2 \to 3$

$\Delta p = 0 \Rightarrow W = p\Delta V = p(V_3 - V_2) = 2p_0(2V_0 - V_0) = 2p_0 V_0$ (W is positive since V increases.)

$\Delta p = 0 \Rightarrow Q = nC_p \Delta T = nC_p(T_3 - T_2)$

But $pV = nRT$ and p constant $\Rightarrow pV_2 = nRT_2$ and $pV_3 = nRT_3$.

So $p(V_3 - V_2) = nR(T_3 - T_2) \Rightarrow p\Delta V = nR\Delta T$ (true when p is constant).

Thus $Q = nC_p \frac{p\Delta V}{nR} = \frac{C_p}{R} p\Delta V = \frac{C_p}{R}(2p_0)(2V_0 - V_0) = 2\frac{C_p}{R} p_0 V_0$.

($Q > 0 \Rightarrow$ heat is absorbed by the gas.)

process $3 \to 4$

$\Delta V = 0 \Rightarrow W = 0$.

$\Delta V = 0 \Rightarrow Q = nC_V \Delta T = nC_V \frac{V\Delta p}{nR} = \frac{C_V}{R}(2V_0)(p_0 - 2p_0) = -2\frac{C_V}{R} p_0 V_0$

($Q < 0 \Rightarrow$ heat rejected by the gas.)

process $4 \to 1$

$\Delta p = 0 \Rightarrow W = p\Delta V = p(V_1 - V_4) = p_0(V_0 - 2V_0) = -p_0 V_0$

(W is negative since V decreases.)

$\Delta p = 0 \Rightarrow Q = nC_p \Delta T = nC_p \frac{p\Delta V}{nR} = \frac{C_p}{R} p\Delta V = \frac{C_p}{R} p_0(V_0 - 2V_0) = -\frac{C_p}{R} p_0 V_0$

($Q < 0 \Rightarrow$ heat is rejected by the gas.)

total work performed by the gas during the cycle:

$W_{tot} = W_{1\to 2} + W_{2\to 3} + W_{3\to 4} = 0 + 2p_0V_0 + 0 - p_0V_0 = p_0 V_0$

(Note that W_{tot} equals the area enclosed by the cycle in the pV-diagram.)

total heat absorbed by the gas during the cycle (Q_H):

Heat is absorbed in processes $1 \to 2$ and $2 \to 3$.

$Q_H = Q_{1\to 2} + Q_{2\to 3} = \frac{C_V}{R} p_0 V_0 + 2\frac{C_p}{R} p_0 V_0 = \left(\frac{C_V + 2C_p}{R}\right) p_0 V_0$

But $C_p = C_V + R \Rightarrow Q_H = \frac{C_V + 2(C_V + R)}{R} p_0 V_0 = \left(\frac{3C_V + 2R}{R}\right) p_0 V_0$

total heat rejected by the gas during the cycle (Q_C):

Heat is rejected in processes $3 \to 4$ and $4 \to 1$.

$Q_C = Q_{3\to 4} + Q_{4\to 1} = -2\frac{C_V}{R} p_0 V_0 - \frac{C_p}{R} p_0 V_0 = -\left(\frac{2C_V + C_p}{R}\right) p_0 V_0$

$C_p = C_V + R \Rightarrow Q_C = -\left(\frac{2C_V + C_V + R}{R}\right) p_0 V_0 = -\left(\frac{3C_V + R}{R}\right) p_0 V_0$

Note as a check on the calculations that

$Q_C + Q_H = -\left(\frac{3C_V + R}{R}\right) p_0 V_0 + \left(\frac{3C_V + 2R}{R}\right) p_0 V_0 = p_0 V_0 = W$, as it should.

18-31 (cont)

efficiency

$$e = \frac{W}{Q_H} = \frac{p_0 V_0}{\left(\frac{3C_V + 2R}{R}\right)(p_0 V_0)} = \frac{R}{3C_V + 2R} = \frac{8.3145\ \text{J/mol}\cdot\text{K}}{2(20.5\ \text{J/mol}\cdot\text{K}) + 2(8.3145\ \text{J/mol}\cdot\text{K})} = 0.106 = \underline{10.6\%}$$

18-33

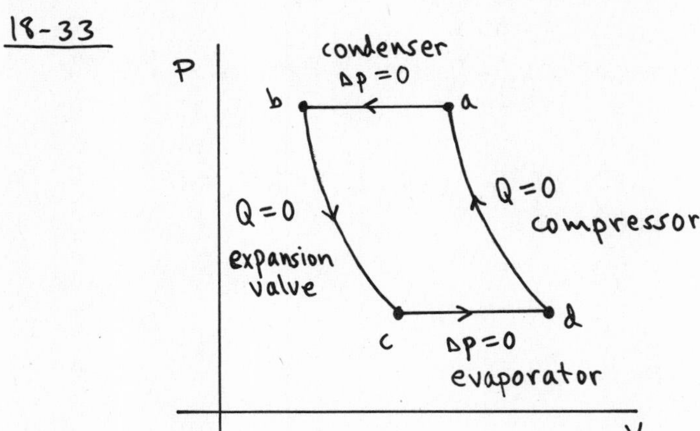

a) process $c \to d$

$\Delta U = U_d - U_c = 1657 \times 10^3\ \text{J} - 1005 \times 10^3\ \text{J} = 6.52 \times 10^5\ \text{J}$

$W = \int_{V_c}^{V_d} p\, dV = p\, \Delta V$ (since is constant pressure process)

$W = (363 \times 10^3\ \text{Pa})(0.4513\ \text{m}^3 - 0.2202\ \text{m}^3) = +8.389 \times 10^4\ \text{J}$ (positive since process is an expansion)

$\Delta U = Q - W \Rightarrow Q = \Delta U + W = 6.52 \times 10^5\ \text{J} + 8.389 \times 10^4\ \text{J} = \underline{7.36 \times 10^5\ \text{J}}$

(positive ⇒ heat goes into the coolant)

b) process $a \to b$

$\Delta U = U_b - U_a = 1171 \times 10^3\ \text{J} - 1969 \times 10^3\ \text{J} = -7.98 \times 10^5\ \text{J}$

$W = p\, \Delta V = (2305 \times 10^3\ \text{Pa})(0.00946\ \text{m}^3 - 0.0682\ \text{m}^3) = -1.354 \times 10^5\ \text{J}$

(negative since $\Delta V < 0$ for the process)

$Q = \Delta U + W = -7.98 \times 10^5\ \text{J} - 1.354 \times 10^5\ \text{J} = \underline{-9.33 \times 10^5\ \text{J}}$

(negative ⇒ heat comes out of coolant)

c) The coolant cannot be treated as an ideal gas, so we can't calculate W for the adiabatic processes. But $\Delta U = 0$ (cycle) ⇒ $W_{net} = Q_{net}$

$Q = 0$ for the two adiabatic processes, so

$Q_{net} = Q_{cd} + Q_{ab} = 7.36 \times 10^5\ \text{J} - 9.33 \times 10^5\ \text{J} = -1.97 \times 10^5\ \text{J}$

⇒ $W_{net} = -\underline{1.97 \times 10^5\ \text{J}}$ (negative since work is done on the coolant (working substance))

d) $K = \frac{Q_c}{|W|} = \frac{+7.36 \times 10^5\ \text{J}}{1.97 \times 10^5\ \text{J}} = \underline{3.74}$

18-35

First use the methods of Chapter 15 to calculate the final temperature T of the system:

0.500 kg of water (cools from 60.0°C to T)

$Q = mc\Delta T = (0.500\text{ kg})(4190\text{ J/kg}\cdot\text{K})(T - 60.0°\text{C}) = (2095\text{ J/K})T - 1.257\times10^5\text{ J}$

0.0600 kg of ice (warms to 0°C, melts, and water warms from 0°C to T)

$Q = mc_{ice}(0°\text{C} - (-15.0°\text{C})) + mL_f + mc_{water}(T - 0°\text{C})$

$Q = (0.0600\text{ kg})(2000\text{ J/kg}\cdot\text{K})(15.0\text{ K}) + (0.0600\text{ kg})(334\times10^3\text{ J/kg}) + (0.0600\text{ kg})(4190\text{ J/kg}\cdot\text{K})(T - 0°\text{C})$

$Q = 1800\text{ J} + 2.004\times10^4\text{ J} + (215.4\text{ J/K})T = 2.184\times10^4\text{ J} + (251.4\text{ J/K})T$

$Q_{system} = 0 \Rightarrow (2095\text{ J/K})T - 1.257\times10^5\text{ J} + 2.184\times10^4\text{ J} + (251.4\text{ J/K})T = 0$

$(2.346\times10^3\text{ J/K})T = 1.0386\times10^5\text{ J}$

$T = \dfrac{1.0386\times10^5\text{ J}}{2.346\times10^3\text{ J/K}} = 44.3°\text{C} = 317.3\text{ K}$

Now we can calculate the entropy changes:

ice

The process takes ice at −15.0°C and produces water at 44.3°C. Calculate ΔS for a reversible process between these two states, in which heat is added very slowly. ΔS is path independent, so ΔS for a reversible process is the same as ΔS for the actual (irreversible process) as long as the initial and final states are the same.

$\Delta S = \int_1^2 \frac{dQ}{T}$ (where T must be in kelvins)

For a temperature change $dQ = mc\,dT \Rightarrow \Delta S = \int_{T_1}^{T_2} \frac{mc\,dT}{T} = mc\ln\left(\frac{T_2}{T_1}\right)$.

For a phase change, since it occurs at constant T,

$\Delta S = \int \frac{dQ}{T} = \frac{Q}{T} = \frac{\pm mL}{T}$

Therefore $\Delta S_{ice} = mc_{ice}\ln\left(\frac{273\text{ K}}{258\text{ K}}\right) + \frac{mL_f}{273\text{ K}} + mc_{water}\ln\left(\frac{317.3\text{ K}}{273\text{ K}}\right)$

$\Delta S_{ice} = (0.0600\text{ kg})(2000\text{ J/kg}\cdot\text{K})\ln\left(\frac{273}{258}\right) + \frac{(0.0600\text{ kg})(334\times10^3\text{ J/kg})}{273\text{ K}} + (0.0600\text{ kg})(4190\text{ J/kg}\cdot\text{K})\ln\left(\frac{317.3}{273}\right)$

$\Delta S_{ice} = 6.78\text{ J/K} + 73.41\text{ J/K} + 37.80\text{ J/K} = 118.0\text{ J/K}$

water

$\Delta S_{water} = mc\ln\left(\frac{T_2}{T_1}\right) = (0.500\text{ kg})(4190\text{ J/kg}\cdot\text{K})\ln\left(\frac{317.3\text{ K}}{333\text{ K}}\right) = -101.2\text{ J/K}$

For the system, $\Delta S = \Delta S_{ice} + \Delta S_{water} = 118.0\text{ J/K} - 101.2\text{ J/K} = \underline{+16.8\text{ J/K}}$

Note: Our calculation gives $\Delta S > 0$, as it must for an irreversible process of an isolated system.

CHAPTER 19

Exercises 3, 5, 7, 11, 13, 15, 21, 23, 25

Problems 27, 31, 33

Exercises

19-3

a) $v = 344$ m/s
$v = \lambda f \Rightarrow \lambda = \frac{v}{f}$

$f = 20.0 \text{ Hz} \Rightarrow \lambda = \frac{344 \text{ m/s}}{20.0 \text{ Hz}} = \underline{17.2 \text{ m}}$

$f = 20{,}000 \text{ Hz} \Rightarrow \lambda = \frac{344 \text{ m/s}}{20{,}000 \text{ Hz}} = \underline{0.0172 \text{ m}}$

b) $v = 1480$ m/s

$f = 20.0 \text{ Hz} \Rightarrow \lambda = \frac{1480 \text{ m/s}}{20.0 \text{ Hz}} = \underline{74.0 \text{ m}}$

$f = 20{,}000 \text{ Hz} \Rightarrow \lambda = \frac{1480 \text{ m/s}}{20{,}000 \text{ Hz}} = \underline{0.074 \text{ m}}$

19-5

a)

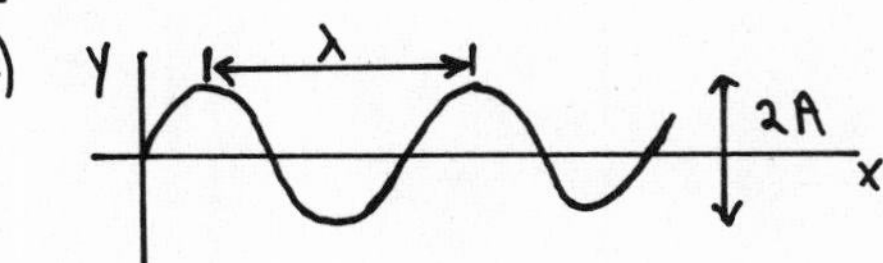

The distance between crests in the wavelength; $\lambda = 8.0$ m.
Time between largest upward displacement to largest downward displacement is one-half the period $\Rightarrow \frac{T}{2} = 3.0 \text{ s} \Rightarrow T = 6.0 \text{ s}$.
$f = \frac{1}{T} = \frac{1}{6.0 \text{ s}} = 0.1667$ Hz
Then $v = f\lambda = (0.1667 \text{ Hz})(8.0 \text{ m}) = \underline{1.33 \text{ m/s}}$

b) The distance between the highest and lowest points is twice the amplitude, so $2A = 0.800 \text{ m} \Rightarrow A = \underline{0.400 \text{ m}}$.

19-7

$v = 15.0$ m/s, $A = 0.0800$ m, $\lambda = 0.300$ m

a) $v = f\lambda \Rightarrow f = \frac{v}{\lambda} = \frac{15.0 \text{ m/s}}{0.300 \text{ m}} = 50.0$ Hz

$T = \frac{1}{f} = \frac{1}{50.0 \text{ Hz}} = \underline{0.0200 \text{ s}}$

$k = \frac{2\pi}{\lambda} = \frac{2\pi \text{ rad}}{0.300 \text{ m}} = 20.9$ rad/m

b) For a wave traveling in the +x-direction $y(x,t) = A \sin 2\pi\left(\frac{t}{T} - \frac{x}{\lambda}\right)$ (Eq. 19-4)
At $x = 0$, $y(0,t) = A \sin 2\pi\left(\frac{t}{T}\right)$, so $y = 0$ at $t = 0$ and y is positive for t slightly greater than zero. This equation describes the wave specified in the problem.

19-7 (cont)

Substitute in numerical values $\Rightarrow y(x,t) = (0.0800\text{ m})\sin 2\pi\left(\frac{t}{0.0200\text{ s}} - \frac{x}{0.300\text{ m}}\right)$.

Or, $y(x,t) = (0.0800\text{ m})\sin((314\text{ rad/s})t - (20.9\text{ m}^{-1})x)$.

c) The expression from part (b) $\Rightarrow y = (0.0800\text{ m})\sin 2\pi\left(\frac{0.200\text{ s}}{0.0200\text{ s}} - \frac{0.200\text{ m}}{0.300\text{ m}}\right)$

$y = (0.0800\text{ m})\sin(58.64\text{ rad}) = \underline{0.0693\text{ m}}$

19-11

a) The tension F in the string is the weight of the hanging mass;

$F = mg = (5.00\text{ kg})(9.80\text{ m/s}^2) = 49.0\text{ N}$

$v = \sqrt{\frac{F}{\mu}} = \sqrt{\frac{49.0\text{ N}}{0.0180\text{ kg/m}}} = \underline{52.2\text{ m/s}}$

b) $v = f\lambda \Rightarrow \lambda = \frac{v}{f} = \frac{52.2\text{ m/s}}{240\text{ Hz}} = \underline{0.218\text{ m}}$

19-13

Calculate the time it takes each sound wave to travel the $L = 120\text{ m}$ length of the pipe.

wave in air: $t = \frac{120\text{ m}}{344\text{ m/s}} = 0.3488\text{ s}$

wave in the metal: $v = \sqrt{\frac{Y}{\rho}} = \sqrt{\frac{2.00\times10^{11}\text{ Pa}}{7800\text{ kg/m}^3}} = 5.06\times10^3\text{ m/s}$

$\Rightarrow t = \frac{120\text{ m}}{5.06\times10^3\text{ m/s}} = 0.0237\text{ s}$

The time interval between the two sounds is $\Delta t = 0.3488\text{ s} - 0.0237\text{ s} = \underline{0.325\text{ s}}$

19-15

Calculate the wave speed: $v = f\lambda = (150\text{ Hz})(8.00\text{ m}) = 1200\text{ m/s}$

$v = \sqrt{\frac{B}{\rho}}$ (Eq. 19-21) $\Rightarrow B = \rho v^2 = (800\text{ kg/m}^3)(1200\text{ m/s})^2 = \underline{1.15\times10^9\text{ Pa}}$

19-21

Eq. (19-34): $P_{av} = \frac{1}{2}\sqrt{\mu F}\,\omega^2 A^2$

$v = \sqrt{\frac{F}{\mu}} \Rightarrow \sqrt{\mu} = \frac{\sqrt{F}}{v} \Rightarrow P_{av} = \frac{1}{2}\frac{\sqrt{F}}{v}\sqrt{F}\,\omega^2 A^2 = \frac{1}{2}F\frac{\omega^2 A^2}{v}$

$\omega = 2\pi f \Rightarrow \frac{\omega}{v} = 2\pi\frac{f}{v} = \frac{2\pi}{\lambda} = k$

$\Rightarrow P_{av} = \frac{1}{2}Fk\omega A^2$, as was to be shown.

19-23

a) group velocity $= \frac{d\omega}{dk} \Rightarrow$ group velocity is the slope of the tangent to the graph of ω versus k.

b) $\omega = 2\sqrt{\frac{k'}{m}} \sin\left(\frac{ka}{2}\right)$ (Eq. 19-38)

$\frac{d\omega}{dk} = 2\sqrt{\frac{k'}{m}}\left(\frac{d}{dk}\sin\left(\frac{ka}{2}\right)\right) = 2\sqrt{\frac{k'}{m}}\,\frac{a}{2}\cos\left(\frac{ka}{2}\right) = a\sqrt{\frac{k'}{m}}\cos\left(\frac{ka}{2}\right)$

c) $\frac{d\omega}{dk} = 0 \Rightarrow \cos\left(\frac{ka}{2}\right) = 0$

$\frac{ka}{2} = \frac{\pi}{2}, \frac{3\pi}{2}, \ldots \Rightarrow k = \frac{\pi}{a}, \frac{3\pi}{a}, \ldots$

For wave numbers $k=0$ to $k=\frac{\pi}{a}$ the group velocity is zero when $k=\frac{\pi}{a}$.

$\Rightarrow \frac{2\pi}{\lambda} = \frac{\pi}{a} \Rightarrow \underline{\lambda = 2a}$

d) $\frac{d\omega}{dk} = a\sqrt{\frac{k'}{m}}\cos\left(\frac{ka}{2}\right)$ (group velocity)

$\frac{\omega}{k} = \frac{2}{k}\sqrt{\frac{k'}{m}}\sin\left(\frac{ka}{2}\right)$ (phase velocity)

magnitude of group velocity equals magnitude of phase velocity $\Rightarrow \frac{d\omega}{dk} = \frac{\omega}{k}$

$a\sqrt{\frac{k'}{m}}\cos\left(\frac{ka}{2}\right) = \frac{2}{k}\sqrt{\frac{k'}{m}}\sin\left(\frac{ka}{2}\right)$

$\frac{ka}{2} = \tan\left(\frac{ka}{2}\right)$

$\tan x \simeq x$ when x is small, so this equation is approximately satisfied when $\frac{ka}{2} \ll 1 \Rightarrow \frac{2\pi}{\lambda}\frac{a}{2} \ll 1 \Rightarrow \underline{\lambda \gg a}$.

19-25

The energy of the phonon is $1.230\times10^{-19}\,\text{J} - 1.189\times10^{-19}\,\text{J} = 4.1\times10^{-21}\,\text{J}$

$E = hf \Rightarrow f = \frac{E}{h} = \frac{4.1\times10^{-21}\,\text{J}}{6.626\times10^{-34}\,\text{J}\cdot\text{s}} = \underline{6.2\times10^{12}\,\text{Hz}}$

$\omega = 2\pi f = 2\pi\,(6.2\times10^{12}\,\text{Hz}) = \underline{3.9\times10^{13}\,\text{rad/s}}$

Problems

19-27

$A = 0.0800\,\text{m},\ \lambda = 1.60\,\text{m},\ v = 100\,\text{m/s}$

a) $v = f\lambda \Rightarrow f = \frac{v}{\lambda} = \frac{100\,\text{m/s}}{1.60\,\text{m}} = \underline{62.5\,\text{Hz}}$

$\omega = 2\pi f = 2\pi\,(62.5\,\text{Hz}) = \underline{393\,\text{rad/s}}$

$k = \frac{2\pi}{\lambda} = \frac{2\pi\,\text{rad}}{1.60\,\text{m}} = \underline{3.93\,\text{rad/m}}$

b) Wave traveling to the right $\Rightarrow y(x,t) = A\sin(\omega t - kx)$.

But to have the $x=0$ end of the string move downward just after $t=0$, this equation must be changed to $y(x,t) = -A\sin(\omega t - kx)$.

Put in the numbers $\Rightarrow y(x,t) = -(0.0800\,\text{m})\sin((393\,\text{rad/s})t - (3.93\,\text{rad/m})x)$

19-27 (cont)

c) left hand end $\Rightarrow x=0$

Put this value into the equation of part (b) $\Rightarrow y(0,t) = -(0.0800\text{ m})\sin((393\text{ rad/s})t)$

d) Put $x = 1.20\text{ m}$ into the equation of part (b)

$\Rightarrow y(1.20\text{m}, t) = -(0.0800\text{m})\sin((393\text{ rad/s})t - (3.93\text{ rad/m})(1.20\text{ m}))$

$y(1.20\text{m}, t) = -(0.0800\text{ m})\sin((393\text{ rad/s})t - 4.72\text{ rad})$

$4.72\text{ rad} = \frac{3\pi}{2}\text{ rad}$; $\sin(\theta - \frac{3\pi}{2}) = \sin(\theta + \frac{\pi}{2} - 2\pi) = \sin(\theta + \frac{\pi}{2}) = \cos\theta$

Thus $y(1.20\text{m}, t) = -(0.0800\text{ m})\cos((393\text{ rad/s})t)$

e) $y = -A\sin(\omega t - kx)$ (part (b))

The transverse velocity is given by

$v_y = \frac{\partial y}{\partial t} = -A\frac{\partial}{\partial t}\sin(\omega t - kx) = -A\omega\cos(\omega t - kx)$

The maximum v_y is $A\omega = (0.0800\text{m})(393\text{ rad/s}) = \underline{31.4\text{ m/s}}$

f) $y(x,t) = -(0.0800\text{m})\sin((393\text{ rad/s})t - (3.93\text{ rad/m})x)$

$t = 0.0240\text{s}$ and $x = 1.20\text{m}$

$\Rightarrow y = -(0.0800\text{ m})\sin((393\text{ rad/s})(0.0240\text{ s}) - (3.93\text{ rad/m})(1.20\text{ m})) = \underline{+0.0800\text{ m}}$

$v_y = -A\omega\cos(\omega t - kx) = -(31.4\text{ m/s})\cos((393\text{ rad/s})t - (3.93\text{ rad/m})x)$

$t = 0.0240\text{s}$ and $x = 1.20\text{m}$

$\Rightarrow v_y = -(31.4\text{ m/s})\cos((393\text{ rad/s})(0.0240\text{ s}) - (3.93\text{ rad/m})(1.20\text{ m})) = 0.0$ (approximately)

(the displacement is a maximum and the transverse velocity is zero.)

19-31

Calculate the wave speed for transverse waves on the rubber tube.

The tension in the tube is equal to the weight of the suspended mass:

$F = mg = (15.0\text{ kg})(9.80\text{ m/s}^2) = 147\text{ N}$

The mass per unit length for the tube is $\mu = \frac{m}{L} = \frac{0.800\text{ kg}}{12.0\text{ m}} = 0.0667\text{ kg/m}$.

$v = \sqrt{\frac{F}{\mu}} = \sqrt{\frac{147\text{ N}}{0.0667\text{ kg/m}}} = 46.9\text{ m/s}$

The time is the distance traveled divided by the speed $\Rightarrow t = \frac{12.0\text{m}}{46.9\text{m/s}} = \underline{0.256\text{ s}}$

19-33

The transverse wave speed is given by Eq. (19-13): $v_t = \sqrt{\frac{F}{\mu}}$

The longitudinal wave speed is given by Eq. (19-21): $v_\ell = \sqrt{\frac{Y}{\rho}}$

$v_\ell = 50 v_t \Rightarrow \sqrt{\frac{Y}{\rho}} = 50\sqrt{\frac{F}{\mu}} \Rightarrow \boxed{\frac{Y}{\rho} = 2500\frac{F}{\mu}}$

$\mu = \frac{m}{L}$, $\rho = \frac{m}{V} = \frac{m}{LA}$ where L is the length of the wire and A is its cross-section area. Thus $\rho = \frac{\mu}{A} \Rightarrow \mu = A\rho$

Use this in the above eq. $\Rightarrow \frac{Y}{\rho} = 2500\frac{F}{A\rho} \Rightarrow \frac{F}{A} = \frac{Y}{2500}$.

CHAPTER 20

Exercises 3, 5, 7, 11, 13, 17

Problems 19, 21, 23, 25

Exercises

20-3

Eq. (20-1): $y = (2A \cos \omega t) \sin kx$

a) node $\Rightarrow y = 0$ for all $t \Rightarrow \sin kx = 0 \Rightarrow kx = n\pi$, $n = 0, 1, 2, \ldots$

$$x = \frac{n\pi}{k} = \frac{n\pi}{1.67\pi \text{ rad/m}} = (0.60 \text{ m})\, n$$

x can't be larger than 2.40 m, the length of the wire

n	x	n	x
0	0 (the left-hand end)	3	1.80 m
1	0.60 m	4	2.40 m (the right-hand end of the wire)
2	1.20 m		

b) antinode $\Rightarrow \sin kx = \pm 1$, so y will have maximum amplitude

$\sin kx = \pm 1 \Rightarrow kx = \frac{n\pi}{2}$, $n = 1, 3, 5, \ldots$

$$x = \frac{n\pi}{2k} = \frac{n\pi}{2(1.67\pi \text{ rad/m})} = (0.30 \text{ m})\, n$$

n	x	n	x
1	0.30 m	5	1.50 m
3	0.90 m	7	2.10 m

20-5

a)

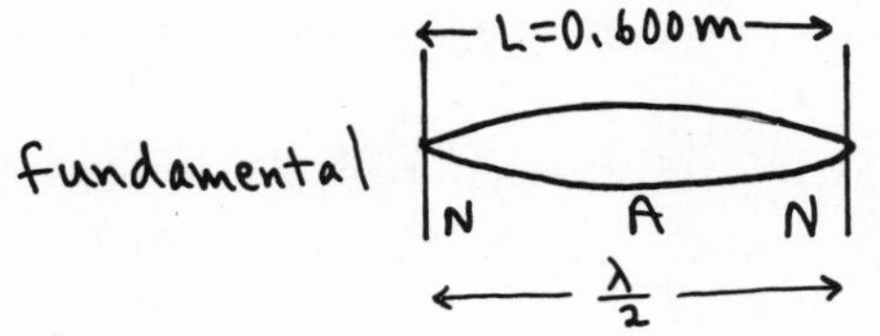

$f = 40.0$ Hz

From the sketch, $\frac{\lambda}{2} = L \Rightarrow \lambda = 2L = 1.20$ m

$v = f\lambda = (40.0 \text{ Hz})(1.20 \text{ m}) = \underline{48.0 \text{ m/s}}$

b) The tension is related to the wave speed by Eq. (19-13)

$v = \sqrt{\frac{F}{\mu}} \Rightarrow F = \mu v^2$

$\mu = \frac{m}{L} = \frac{0.0500 \text{ kg}}{0.600 \text{ m}} = 0.0833$ kg/m

$F = \mu v^2 = (0.0833 \text{ kg/m})(48.0 \text{ m/s})^2 = \underline{192 \text{ N}}$

20-7

a) Assume that in each case the string vibrates in its fundamental mode. $f = 440$ Hz when a length $L = 0.600$ m vibrates; use this information to calculate the speed v of waves on the string.

20-7 (cont)

fundamental $\Rightarrow \frac{\lambda}{2} = L \Rightarrow \lambda = 2L = 2(0.600\ \text{m}) = 1.20\ \text{m}$

Then $v = f\lambda = (440\ \text{Hz})(1.20\ \text{m}) = 528\ \text{m/s}$.

Now find the length L=x of the string that makes f = 528 Hz.

$$\lambda = \frac{v}{f} = \frac{528\ \text{m/s}}{528\ \text{Hz}} = 1.00\ \text{m}$$

$L = \frac{\lambda}{2} = 0.500\ \text{m}$, so $x = 0.500\ \text{m} = \underline{50.0\ \text{cm}}$

b) No retuning $\Rightarrow$ same wave speed as in part (a). Find the length of vibrating string needed to produce f = 294 Hz.

$$\lambda = \frac{v}{f} = \frac{528\ \text{m/s}}{294\ \text{Hz}} = 2.00\ \text{m}$$

$L = \frac{\lambda}{2} = 1.00\ \text{m}$; string is shorter than this. No, **not** possible.

20-11

a) The placement of the displacement nodes and antinodes along the pipe is as sketched below. The open end is a displacement antinode, and the closed end is a displacement node.

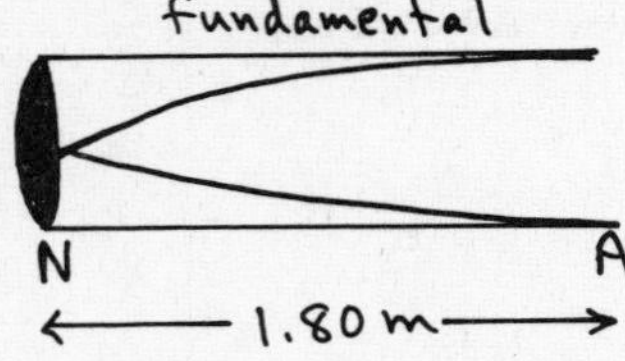

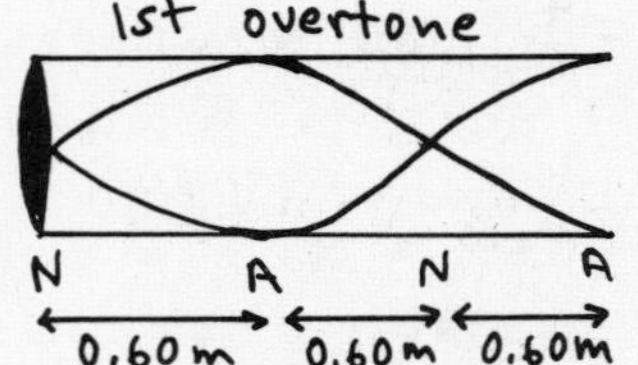

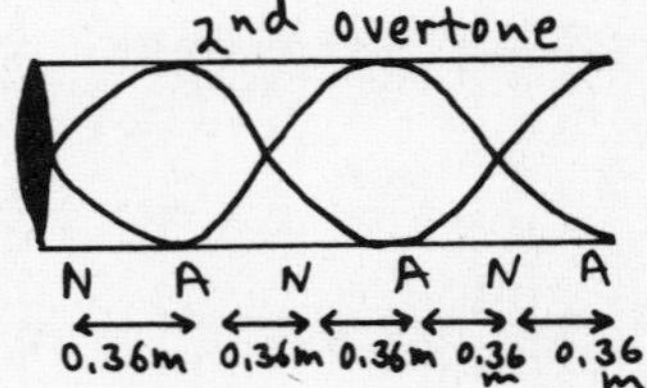

Location of the displacement antinodes (A), measured from the closed end.

fundamental	1.80 m
1st overtone	0.60 m, 1.80 m
2nd overtone	0.36 m, 1.08 m, 1.80 m

b) The pressure antinodes correspond to the displacement nodes shown in the sketch. Location of the pressure antinodes (displacement nodes (N)) measured from closed end:

fundamental	0
1st overtone	0, 1.20 m
2nd overtone	0, 0.72 m, 1.44 m

20-13

a) open at both ends $\Rightarrow$ displacement antinode at each end

fundamental

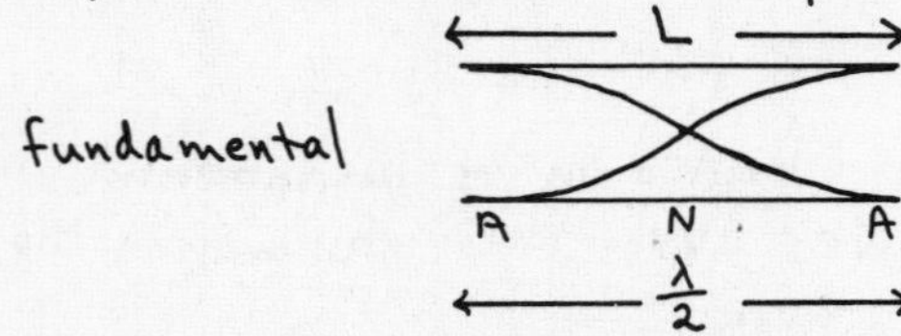

$\frac{\lambda}{2} = L \Rightarrow \lambda = 2L$

$$f = \frac{v}{\lambda} = \frac{v}{2L} = \frac{344\ \text{m/s}}{2(4.88\ \text{m})} = \underline{35.2\ \text{Hz}}$$

20-13 (cont)

b) closed at one end ⇒ displacement node at closed end

fundamental

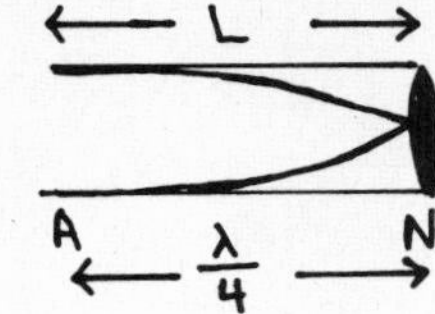

$\frac{\lambda}{4} = L \Rightarrow \lambda = 4L$

$f = \frac{v}{\lambda} = \frac{v}{4L} = \frac{344 \text{ m/s}}{4(4.88 \text{ m})} = \underline{17.6 \text{ Hz}}$

20-17

A ◁ B ◁ · Q

← 2.00 m → ← 1.00 m →

a) Path difference from points A and B to point Q is 3.00 m − 1.00 m = 2.00 m.
Reinforcement (constructive interference) ⇒ path difference = $n\lambda$, $n = 1, 2, 3, \ldots$

$2.00 \text{ m} = n\lambda \Rightarrow \lambda = \frac{2.00 \text{ m}}{n}$

$f = \frac{v}{\lambda} = \frac{nv}{2.00 \text{ m}} = \frac{n(350 \text{ m/s})}{2.00 \text{ m}} = n(175 \text{ Hz}), \quad n = 1, 2, 3, \ldots$

$\Rightarrow f = 175 \text{ Hz}, 350 \text{ Hz}, 525 \text{ Hz}, \ldots$

b) Destructive interference ⇒ path difference $= \left(\frac{n}{2}\right)\lambda$, $n = 1, 3, 5, \ldots$

$2.00 \text{ m} = \left(\frac{n}{2}\right)\lambda \Rightarrow \lambda = \frac{4.00 \text{ m}}{n}$

$f = \frac{v}{\lambda} = \frac{nv}{4.00 \text{ m}} = \frac{n(350 \text{ m/s})}{4.00 \text{ m}} = n(87.5 \text{ Hz}), \quad n = 1, 3, 5, \ldots$

$\Rightarrow f = 88 \text{ Hz}, 262 \text{ Hz}, 438 \text{ Hz}, \ldots$

Problems

20-19

a) Plank oscillates with maximum amplitude at its center ⇒ it is oscillating in its fundamental mode.

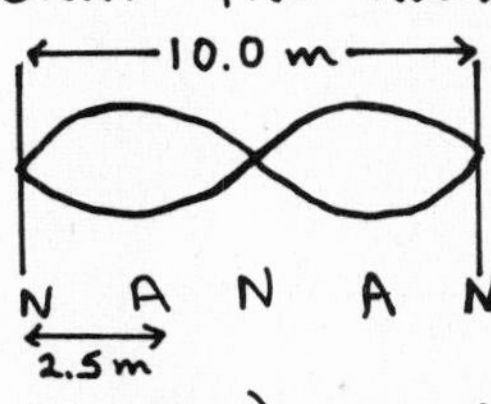

$\frac{\lambda}{2} = L \Rightarrow \lambda = 2L = 2(10.0 \text{ m}) = 20.0 \text{ m}$

Student jumps upward two times per second ⇒ $f = 2.00$ Hz

$v = f\lambda = (2.00 \text{ Hz})(20.0 \text{ m}) = \underline{40.0 \text{ m/s}}$

b) There now must be an antinode 2.5 m (a distance of $\frac{L}{4}$) from one end. The nodal structure for the standing wave must be

10.0 m

N A N A N

2.5 m

λ

v depends on the properties of the plank, so is the same as in part (a).

Now $\lambda = L = 10.0$ m.

$f = \frac{v}{\lambda} = \frac{40.0 \text{ m/s}}{10.0 \text{ m}} = 4.00 \text{ Hz}$

The student now has to jump four times each second.

20-21

a) For an open pipe the standing wave frequencies are

$f_n = \frac{nv}{2L}$, $n = 1, 2, 3, \ldots$ (Eq. 20-11)

If one harmonic is labeled with n then the next harmonic (next higher frequency) is labeled with $n+1$; $f_{n+1} = \frac{(n+1)v}{2L}$.

The difference between these two frequencies is

$$\Delta f = f_{n+1} - f_n = \frac{(n+1)v}{2L} - \frac{nv}{2L} = \frac{v}{2L}$$

The problem gives $f_n = 240$ Hz and $f_{n+1} = 280$ Hz $\Rightarrow \Delta f = 280\text{ Hz} - 240\text{ Hz} = 40$ Hz

$$\Delta f = \frac{v}{2L} \Rightarrow L = \frac{v}{2\Delta f} = \frac{344\text{ m/s}}{2(40\text{ Hz})} = 4.30\text{ m}$$

b) $f_n = \frac{nv}{2L} = 240\text{ Hz} \Rightarrow n = \frac{2L(240\text{ Hz})}{v} = \frac{2(4.30\text{ m})(240\text{ Hz})}{344\text{ m/s}} = 6$

These two harmonics are the 6th and 7th.

(Note that $f_{n+1} = f_7 = \frac{7v}{2L} = \frac{7(344\text{ m/s})}{2(4.30\text{ m})} = 280$ Hz, which checks.)

20-23

a) The tension F is related to the wave speed by $v = \sqrt{\frac{F}{\mu}}$ (Eq. 19-13), so use the information given to calculate v.

fundamental (L; N A N; $\frac{\lambda}{2}$)

$$\frac{\lambda}{2} = L \Rightarrow \lambda = 2L = 2(0.680\text{ m}) = 1.36\text{ m}$$

$$v = f\lambda = (220\text{ Hz})(1.36\text{ m}) = 299\text{ m/s}$$

$$\mu = \frac{m}{L} = \frac{1.42\times10^{-3}\text{ kg}}{0.680\text{ m}} = 2.088\times10^{-3}\text{ kg/m}$$

Then $v = \sqrt{\frac{F}{\mu}} \Rightarrow F = \mu v^2 = (2.088\times10^{-3}\text{ kg/m})(299\text{ m/s})^2 = \underline{187\text{ N}}$

b) $F = \mu v^2$ and $v = f\lambda \Rightarrow F = \mu f^2\lambda^2$

μ is a property of the string so is constant.

λ is determined by the length of the string so stays constant.

μ, λ constant $\Rightarrow \frac{F}{f^2} = \mu\lambda^2 = \text{constant} \Rightarrow \frac{F_1}{f_1^2} = \frac{F_2}{f_2^2}$

$$F_2 = F_1\left(\frac{f_2}{f_1}\right)^2 = (187\text{ N})\left(\frac{233\text{ Hz}}{220\text{ Hz}}\right)^2 = 210\text{ N}$$

The percent change in F is $\frac{F_2 - F_1}{F_1} = \frac{210\text{ N} - 187\text{ N}}{187\text{ N}} = 0.12 = 12\%$

20-25

a)

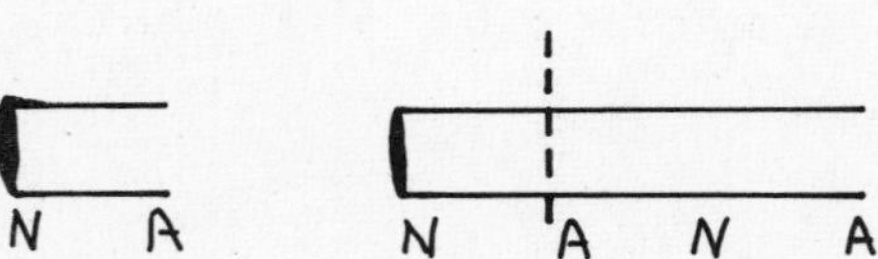

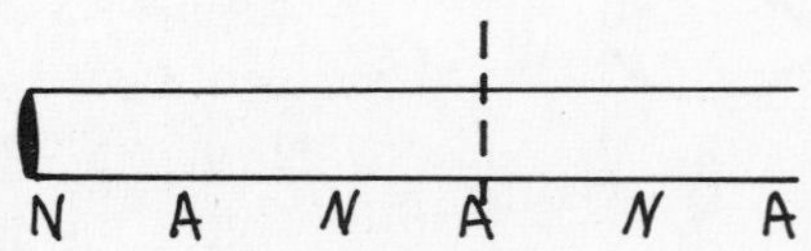

The successive lengths differ by an additional $\frac{\lambda}{2}$ length (the A to A distance) $\Rightarrow \frac{\lambda}{2} = 55.5\text{ cm} - 18.0\text{ cm} = 93.0\text{ cm} - 55.5\text{ cm} = 37.5\text{ cm}$

$$\lambda = 2(37.5\text{ cm}) = 75.0\text{ cm} = 0.750\text{ m}$$

20-25 (cont)

$$v = f\lambda = (500\ \text{Hz})(0.750\ \text{m}) = \underline{375\ \text{m/s}}$$

b) $v = \sqrt{\dfrac{\gamma RT}{M}}$ (Eq. 19-27)

$$\gamma = \frac{Mv^2}{RT} = \frac{(28.8 \times 10^{-3}\ \text{kg/mol})(375\ \text{m/s})^2}{(8.3145\ \text{J/mol}\cdot\text{K})(350\ \text{K})} = \underline{1.39}$$

CHAPTER 21

Exercises 1, 7, 9, 11, 13, 15, 17

Problems 19, 23, 27

Exercises

21-1

$p_{max} = BkA$ (Eq. 21-5)

As computed in Example 21-1, the adiabatic bulk modulus for air is $B = 1.42 \times 10^5$ Pa.

a) $f = 500$ Hz

Need to calculate k: $\lambda = \frac{v}{f}$ and $k = \frac{2\pi}{\lambda} \Rightarrow k = \frac{2\pi f}{v} = \frac{(2\pi \text{ rad})(500 \text{ Hz})}{344 \text{ m/s}} = 9.133$ rad/m.

Then $p_{max} = BkA = (1.42 \times 10^5 \text{ Pa})(9.133 \text{ rad/m})(0.0125 \times 10^{-3} \text{ m}) = \underline{16.2 \text{ Pa}}$.

This is below the pain threshold of 30 Pa.

b) $f = 20{,}000$ Hz

$k = \frac{2\pi f}{v} = \frac{(2\pi \text{ rad})(20{,}000 \text{ Hz})}{344 \text{ m/s}} = 365.3$ rad/m

Then $p_{max} = BkA = (1.42 \times 10^5 \text{ Pa})(365.3 \text{ rad/m})(0.0125 \times 10^{-3} \text{ m}) = \underline{648 \text{ Pa}}$.

This is well above the pain threshold.

21-7

$I = \frac{1}{2}\omega B k A^2$ (Eq. 21-8)

$\omega = 2\pi f = (2\pi \text{ rad})(400 \text{ Hz}) = 2.513 \times 10^3$ rad/s

$k = \frac{2\pi}{\lambda} = \frac{2\pi f}{v} = \frac{(2\pi \text{ rad})(400 \text{ Hz})}{344 \text{ m/s}} = 7.306$ rad/m

$B = 1.42 \times 10^5$ Pa (Example 21-1)

Then $I = \frac{1}{2}\omega B k A^2 = \frac{1}{2}(2.513 \times 10^3 \text{ rad/s})(1.42 \times 10^5 \text{ Pa})(7.306 \text{ rad/m})(6.00 \times 10^{-6} \text{ m})^2$

$I = \underline{0.0469 \text{ W/m}^2}$

Eq. (21-13): $\beta = (10 \text{ dB}) \log\left(\frac{I}{I_0}\right)$, with $I_0 = 1 \times 10^{-12}$ W/m².

$\Rightarrow \beta = (10 \text{ dB}) \log\left(\frac{0.0469 \text{ W/m}^2}{1 \times 10^{-12} \text{ W/m}^2}\right) = \underline{107 \text{ dB}}$

21-9

Let 1 refer to the mother and 2 to the father.

$\beta_1 = (10 \text{ dB}) \log\left(\frac{I_1}{I_0}\right)$, $\beta_2 = (10 \text{ dB}) \log\left(\frac{I_2}{I_0}\right)$

Then $\beta_2 - \beta_1 = (10 \text{ dB})[\log I_2 - \cancel{\log I_0} - \log I_1 + \cancel{\log I_0}] = (10 \text{ dB}) \log\left(\frac{I_2}{I_1}\right)$

(This same result was derived in Example 21-6.)

21-9 (cont)

Eq. (21-12): $\frac{I_1}{I_2} = \frac{r_2^2}{r_1^2} \Rightarrow \frac{I_2}{I_1} = \frac{r_1^2}{r_2^2}$

$\Delta\beta = (10\,dB)\log\left(\frac{I_2}{I_1}\right) = (10\,dB)\log\left(\frac{r_1}{r_2}\right)^2 = (20\,dB)\log\left(\frac{r_1}{r_2}\right)$

$\Delta\beta = (20\,dB)\log\left(\frac{3.00\,m}{0.40\,m}\right) = \underline{17.5\,dB}$

21-11

$f_{beat} = f_1 - f_2$ (Eq. 21-14)

3.2 beats per second $\Rightarrow f_1 - f_2 = 3.2\,Hz$

One frequency is 440 Hz $\Rightarrow$ the other is 440 Hz + 3.2 Hz = $\underline{443.2\,Hz}$ or 440 Hz − 3.2 Hz = $\underline{436.8\,Hz}$.

21-13

Note: When the source is at rest $\lambda = \frac{v}{f_s} = \frac{344\,m/s}{400\,Hz} = 0.860\,m$

a) Eq. (21-17): $\lambda = \frac{v - v_s}{f_s} = \frac{344\,m/s - 30.0\,m/s}{400\,Hz} = \underline{0.785\,m}$

b) Eq. (21-18): $\lambda = \frac{v + v_s}{f_s} = \frac{344\,m/s + 30.0\,m/s}{400\,Hz} = \underline{0.935\,m}$

c) $f_L = \frac{v}{\lambda}$ (since $v_L = 0$) $\Rightarrow f_L = \frac{344\,m/s}{0.785\,m} = \underline{438\,Hz}$

d) $f_L = \frac{v}{\lambda} = \frac{344\,m/s}{0.935\,m} = \underline{368\,Hz}$

21-15

a) S $\rightarrow$ $v_s = -30.0\,m/s$; L $\leftarrow$ $v_L = +15.0\,m/s$; + $\leftarrow$ L to S

$f_L = \left(\frac{v + v_L}{v + v_s}\right) f_s = \left(\frac{344\,m/s + 15.0\,m/s}{344\,m/s - 30.0\,m/s}\right)(400\,Hz) = \underline{457\,Hz}$

(Listener and source approaching $\Rightarrow f_L > f_s$.)

b) L $\leftarrow$ $v_L = -15.0\,m/s$; S $\rightarrow$ $v_s = +30.0\,m/s$; + $\rightarrow$ L to S

$f_L = \left(\frac{v + v_L}{v + v_s}\right) f_s = \left(\frac{344\,m/s - 15.0\,m/s}{344\,m/s + 30.0\,m/s}\right)(400\,Hz) = \underline{352\,Hz}$

(Listener and source moving away from each other $\Rightarrow f_L < f_s$.)

21-17

a) $\sin\alpha = \frac{v}{v_s}$ (Eq. 21-21)

Mach 2.00 $\Rightarrow \frac{v_s}{v} = 2.00$

$\sin\alpha = \frac{1}{2.00} = 0.500 \Rightarrow \alpha = \underline{30.0°}$

21-17 (cont)

b)

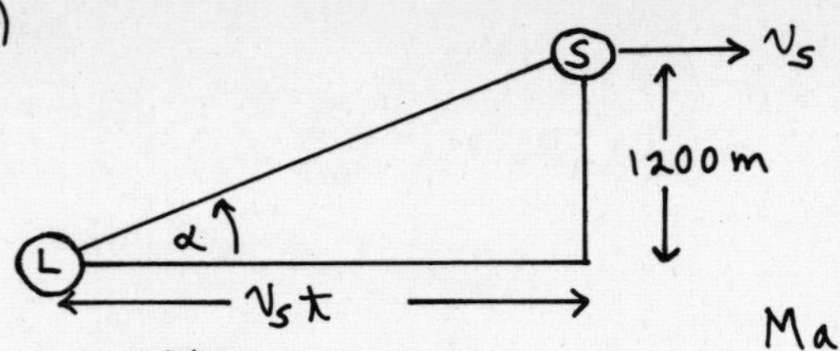

$\tan\alpha = \dfrac{1200\text{ m}}{v_s t}$

$t = \dfrac{1200\text{ m}}{v_s \tan\alpha}$

From part (a) $\alpha = 30.0°$

Mach 2.00 $\Rightarrow v_s = 2v = 2(340\text{ m/s}) = 680\text{ m/s}$

$t = \dfrac{1200\text{ m}}{(680\text{ m/s})(\tan 30.0°)} = \underline{3.06\text{ s}}$

Problems

21-19

a) Use the intensity level β to calculate I at this distance.

$\beta = (10\text{ dB})\log\left(\frac{I}{I_0}\right)$

$50.0\text{ dB} = (10\text{ dB})\log\left(\frac{I}{10^{-12}\text{ W/m}^2}\right)$

$\log\left(\frac{I}{10^{-12}\text{ W/m}^2}\right) = 5.00 \Rightarrow I = 1.00\times10^{-7}\text{ W/m}^2$

Then use Eq. (21-11) to calculate p_{max}:

$I = \frac{p_{max}^2}{2\rho v} \Rightarrow p_{max} = \sqrt{2\rho v I}$

From Example 21-2, $\rho = 1.20\text{ kg/m}^3$ for air at 20°C.

$p_{max} = \sqrt{2\rho v I} = \sqrt{2(1.20\text{ kg/m}^3)(344\text{ m/s})(1.00\times10^{-7}\text{ W/m}^2)} = \underline{9.09\times10^{-3}\text{ Pa}}$

b) Eq. (21-5): $p_{max} = BkA \Rightarrow A = \frac{p_{max}}{Bk}$

For air $B = 1.42\times10^5\text{ Pa}$ (Example 21-1).

$k = \frac{2\pi}{\lambda} = \frac{2\pi f}{v} = \frac{(2\pi\text{ rad})(440\text{ Hz})}{344\text{ m/s}} = 8.037\text{ rad/m}$

$A = \frac{p_{max}}{Bk} = \frac{9.09\times10^{-3}\text{ Pa}}{(1.42\times10^5\text{ Pa})(8.037\text{ rad/m})} = \underline{7.96\times10^{-9}\text{ m}}$

c) $\beta_2 - \beta_1 = (10\text{ dB})\log\left(\frac{I_2}{I_1}\right)$ (Example 21-6)

Eq. (21-12): $\frac{I_1}{I_2} = \frac{r_2^2}{r_1^2} \Rightarrow \frac{I_2}{I_1} = \frac{r_1^2}{r_2^2}$

$\beta_2 - \beta_1 = (10\text{ dB})\log\left(\frac{r_1}{r_2}\right)^2 = (20\text{ dB})\log\left(\frac{r_1}{r_2}\right)$

Let $\beta_2 = 50.0\text{ dB}$ and $r_2 = 6.00\text{ m}$. Then $\beta_1 = 30.0\text{ dB} \Rightarrow r_1 = ?$

$50.0\text{ dB} - 30.0\text{ dB} = (20\text{ dB})\log\left(\frac{r_1}{r_2}\right)$

$20.0\text{ dB} = (20\text{ dB})\log\left(\frac{r_1}{r_2}\right)$

$\log\left(\frac{r_1}{r_2}\right) = 1.00 \Rightarrow r_1 = 10r_2 = \underline{60.0\text{ m}}$

21-23

Apply Eq. (21-19), along with the sign convention that the direction from listener to source is positive, to the situations where the car is approaching the listener and where it is moving away.

car approaching

$v_{car} \rightarrow$ [S] $\xleftarrow[\text{L to S}]{+}$ $v_L = 0$ L f_{L1}

$v_S = -v_{car}$

$$f_L = \left(\frac{v + v_L}{v + v_S}\right) f_S$$

$$f_{L1} = \left(\frac{v}{v - v_{car}}\right) f_S$$

car moving away

$v_L = 0$ f_{L2} $\xrightarrow[\text{L to S}]{+}$ [S] $\rightarrow v_{car}$ f_S

$v_S = +v_{car}$

$$f_{L2} = \left(\frac{v + v_L}{v + v_S}\right) f_S = \left(\frac{v}{v + v_{car}}\right) f_S$$

The pitch of the car's horn drops a half-tone $\Rightarrow \dfrac{f_{L1}}{f_{L2}} = 1.059$

But from the above

$$\frac{f_{L1}}{f_{L2}} = \frac{\left(\frac{v}{v - v_{car}}\right) f_S}{\left(\frac{v}{v + v_{car}}\right) f_S} = \frac{v + v_{car}}{v - v_{car}} = 1.059$$

$$v + v_{car} = 1.059\, v - 1.059\, v_{car}$$

$$2.059\, v_{car} = 0.059\, v$$

$$v_{car} = \frac{0.059\, v}{2.059} = \frac{0.059(344 \text{ m/s})}{2.059} = \underline{9.86 \text{ m/s}}$$

21-27

a) $\lambda = \dfrac{v}{f} = \dfrac{1480 \text{ m/s}}{30.0 \times 10^3 \text{ Hz}} = \underline{0.0493 \text{ m}}$

b) Problem-Solving Strategy 3. on p. 595 describes how to do this problem. The frequency of the directly radiated waves is $f_S = 30{,}000$ Hz. The moving whale, first plays the role of a moving listener, receiving waves with frequency f_L'. The whale then acts as a moving source, emitting waves with the same frequency $f_S' = f_L'$ with which they were received. Let the speed of the whales be v_W.

whale receives waves

$\leftarrow v_W$ L f_L' $\xrightarrow[\text{L to S}]{+}$ $v_S = 0$ [S] f_S

$v_L = -v_W$

$$f_L' = f_S\left(\frac{v + v_L}{v + v_S}\right) = f_S\left(\frac{v - v_W}{v}\right)$$

21-27 (cont)

whale re-emits the waves

$\leftarrow v_w$ $\quad$ + $\quad$ $\overleftarrow{\text{L to S}}$ $\quad$ $v_L = 0$

S $\quad$ L

$f_S' = f_L'$ $\quad$ f_L

$v_S = +v_w$

$$f_L = f_S'\left(\frac{v+v_L}{v+v_S}\right) = f_S'\left(\frac{v}{v+v_w}\right)$$

But $f_S' = f_L'$ $\Rightarrow$ $f_L = f_S\left(\frac{v-v_w}{v}\right)\left(\frac{v}{v+v_w}\right) = f_S\left(\frac{v-v_w}{v+v_w}\right)$

$$f_L = (3.00\times10^4\ \text{Hz})\left(\frac{1480\ \text{m/s} - 6.95\ \text{m/s}}{1480\ \text{m/s} + 6.95\ \text{m/s}}\right) = 2.972\times10^4\ \text{Hz}$$

(Listener and source moving away from each other $\Rightarrow$ frequency is lowered)

The difference between the frequencies of the reflected and directly radiated waves is $3.00\times10^4\ \text{Hz} - 2.972\times10^4\ \text{Hz} =$ 280 Hz.

CHAPTER 22

Exercises 1, 5, 9, 11, 17, 19, 21, 25, 29

Problems 35, 37, 39, 43, 45, 47, 49

Exercises

22-1

$N = nN_A = (2.00 \text{ mol})(6.022\times10^{23} \text{ atoms/mol}) = 12.04\times10^{23} \text{ atoms}$

1 proton per hydrogen atom and each proton has charge $+e = 1.602\times10^{-19}$ C

$\Rightarrow Q = (1.602\times10^{-19} \text{ C/proton})(12.04\times10^{23} \text{ protons}) = \underline{1.93\times10^{5} \text{ C}}$

(This is a huge amount of charge.)

22-5

a) $q_1 = q_2 = q$

$F = \frac{1}{4\pi\epsilon_0}\frac{|q_1q_2|}{r^2} = \frac{q^2}{4\pi\epsilon_0 r^2} \Rightarrow q = r\sqrt{\frac{F}{(1/4\pi\epsilon_0)}} = (0.400 \text{ m})\sqrt{\frac{0.250 \text{ N}}{8.988\times10^9 \text{ N}\cdot\text{m}^2/\text{C}^2}}$

$q = \underline{2.11\times10^{-6} \text{ C}}$ (on each)

b) $q_2 = 2q_1$

$F = \frac{1}{4\pi\epsilon_0}\frac{|q_1q_2|}{r^2} = \frac{2q_1^2}{4\pi\epsilon_0 r^2} \Rightarrow q_1 = r\sqrt{\frac{F}{2(1/4\pi\epsilon_0)}} = \frac{1}{\sqrt{2}}(2.11\times10^{-6} \text{ C}) = \underline{1.49\times10^{-6} \text{ C}}$

and $q_2 = 2q_1 = \underline{2.98\times10^{-6} \text{ C}}$

22-9

y; $q_1 = +3.80$ nC; 0.600 m; $q_2 = -2.50$ nC; x; 0.400 m; $q_3 = +5.00$ nC

Like charges repel and unlike attract, so the free-body diagram for q_3 is

y; x; F_2; F_1

$F_1 = \frac{1}{4\pi\epsilon_0}\frac{|q_1q_3|}{r_{13}^2}$

$F_1 = (8.988\times10^9 \text{ N}\cdot\text{m}^2/\text{C}^2)\frac{(3.80\times10^{-9} \text{ C})(5.00\times10^{-9} \text{ C})}{(1.00 \text{ m})^2}$

$F_1 = 1.708\times10^{-7}$ N

$F_2 = \frac{1}{4\pi\epsilon_0}\frac{|q_2q_3|}{r_{23}^2}$

$F_2 = (8.988\times10^9 \text{ N}\cdot\text{m}^2/\text{C}^2)\frac{(2.50\times10^{-9} \text{ C})(5.00\times10^{-9} \text{ C})}{(0.400 \text{ m})^2} = 7.022\times10^{-7} \text{ N}$

The resultant force is $\vec{R} = \vec{F}_1 + \vec{F}_2$

$\Rightarrow R_x = 0$

$R_y = F_2 - F_1 = 7.022\times10^{-7} \text{ N} - 1.708\times10^{-7} \text{ N} = 5.31\times10^{-7} \text{ N}$

The resultant force has magnitude $\underline{5.31\times10^{-7} \text{ N}}$ and is in the $\underline{+y\text{-direction}}$.

22-11

Calculate the force exerted <u>by</u> the ring <u>on</u> the point charge. By Newton's 3rd law the force of the charge on the ring is equal and opposite.

Divide the ring up into small segments of charge dQ.

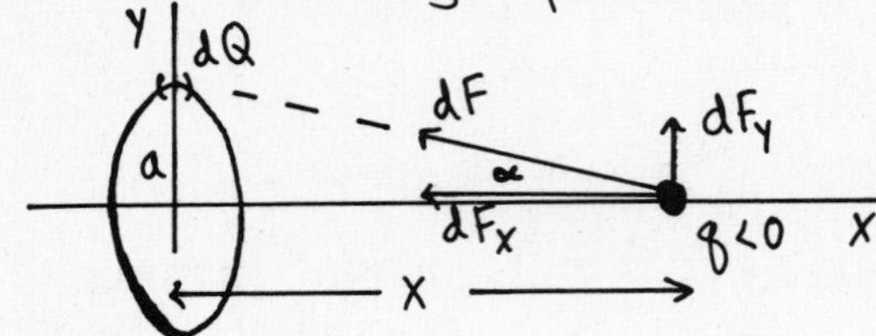

The force $d\vec{F}$ that the segment exerts on the point charge is attractive since one charge is positive and the other is negative.

The distance between the charges is $r = \sqrt{x^2 + a^2}$. The magnitude of dF is

$$dF = \frac{1}{4\pi\epsilon_0}\frac{dQ\,|q|}{r^2} = \frac{1}{4\pi\epsilon_0}\frac{|q|\,dQ}{x^2+a^2}$$

We can see from symmetry that dF_y changes direction for different dQ segments of the ring so that $F_y = \int dF_y = 0$. But the x-components from all dQ segments are in the same $-x$-direction and are given by

$$dF_x = -dF\cos\alpha = -\frac{1}{4\pi\epsilon_0}\frac{|q|\,dQ}{x^2+a^2}\left(\frac{x}{\sqrt{x^2+a^2}}\right) = -\frac{1}{4\pi\epsilon_0}\frac{dQ\,|q|\,x}{(x^2+a^2)^{3/2}}$$

$$F_x = \int dF_x = -\frac{1}{4\pi\epsilon_0}\frac{|q|\,x}{(x^2+a^2)^{3/2}}\int dQ = -\frac{1}{4\pi\epsilon_0}\frac{Q\,|q|\,x}{(x^2+a^2)^{3/2}}$$

(Note that a and x are the same for all the dQ segments of the ring.)

Putting in the numerical values

$$F_x = -(8.988\times10^{9}\ \mathrm{N\cdot m^2/C^2})\frac{(8.40\times10^{-6}\ \mathrm{C})(2.50\times10^{-6}\ \mathrm{C})(0.500\ \mathrm{m})}{((0.500\ \mathrm{m})^2+(0.250\ \mathrm{m})^2)^{3/2}} = -0.540\ \mathrm{N}$$

This is the force of the ring on the point charge. The force exerted by the charge q on the ring has magnitude 0.540 N and is in the <u>+x-direction</u>.

22-17

The electrical force has magnitude $F_E = |q|E = eE$.

The weight of an electron is $w = mg$

$F_E = w \Rightarrow eE = mg$

$$E = \frac{mg}{e} = \frac{(9.109\times10^{-31}\ \mathrm{kg})(9.80\ \mathrm{m/s^2})}{1.602\times10^{-19}\ \mathrm{C}} = \underline{5.57\times10^{-11}\ \mathrm{N/C}}$$

(This is a very small electric field.)

22-19

2.00 cm
0.50 cm
v_0
$\vec{E}$

electron $\Rightarrow q = -e$

$\vec{F} = q\vec{E}$ and q negative $\Rightarrow \vec{F}$ and $\vec{E}$ are in opposite directions $\Rightarrow \vec{F}$ is upward

y
F=eE
a
x

$\Sigma F_y = ma_y$

$eE = ma$

22-19 (cont)

Solve the kinematics to find the acceleration of the electron:

Just misses upper plate ⇒ $x - x_0 = 2.00$ cm when $y - y_0 = +0.500$ cm.

x-component

$v_{0x} = v_0 = 7.00 \times 10^6$ m/s

$a_x = 0$

$x - x_0 = 0.0200$ m

$t = ?$

$$x - x_0 = v_{0x} t + \tfrac{1}{2} a_x t^2 \quad (a_x = 0)$$

$$t = \frac{x - x_0}{v_{0x}} = \frac{0.0200\text{ m}}{7.00 \times 10^6\text{ m/s}} = 2.857 \times 10^{-9}\text{ s}$$

In this same time t the electron travels 0.0050 m vertically:

y-component

$t = 2.857 \times 10^{-9}$ s

$v_{0y} = 0$

$y - y_0 = +0.0050$ m

$$y - y_0 = v_{0y} t + \tfrac{1}{2} a_y t^2 \quad (v_{0y} = 0)$$

$$a_y = \frac{2(y - y_0)}{t^2} = \frac{2(0.0050\text{ m})}{(2.857 \times 10^{-9}\text{ s})^2} = 1.225 \times 10^{15}\text{ m/s}^2$$

(This analysis is very similar to that used in Chapter 3 for projectile motion, except that here the acceleration is upward rather than downward.)

This acceleration must be produced by the electric-field force:

$$eE = ma$$

$$E = \frac{ma}{e} = \frac{(9.109 \times 10^{-31}\text{ kg})(1.225 \times 10^{15}\text{ m/s}^2)}{1.602 \times 10^{-19}\text{ C}} = \underline{6.97 \times 10^3\text{ N/C}}$$

Note that the acceleration produced by the electric field is <u>much</u> larger than g, the acceleration produced by gravity, so it is perfectly ok to neglect the gravity force on the electron in this problem.

22-21

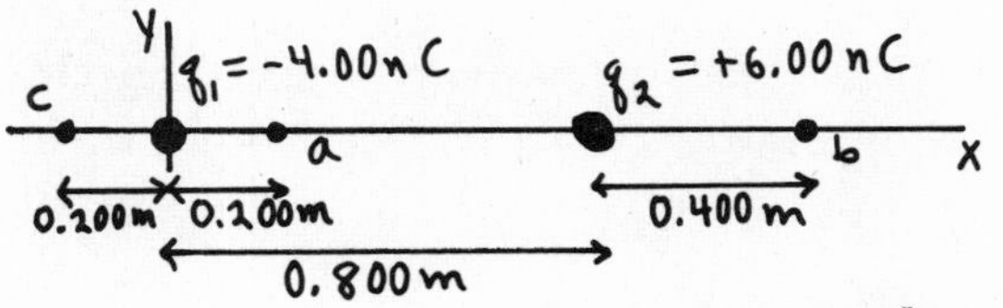

The electric field of a point charge is directed away from the point charge if the charge is positive and toward the point charge if its charge is negative. The magnitude of the electric field is $E = \frac{1}{4\pi\epsilon_0}\frac{|q|}{r^2}$, where r is the distance of the point where the field is calculated from the point charge.

a) At point a the fields $\vec{E}_1$ of q_1 and $\vec{E}_2$ of q_2 are:

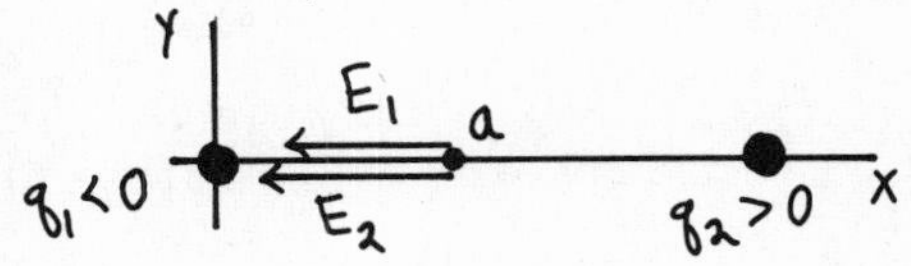

$$E_1 = \frac{1}{4\pi\epsilon_0}\frac{|q_1|}{r_1^2} = (8.988 \times 10^9\text{ N}\cdot\text{m}^2/\text{C}^2)\frac{4.00 \times 10^{-9}\text{ C}}{(0.200\text{ m})^2}$$

$E_1 = 898.8$ N/C

$$E_2 = \frac{1}{4\pi\epsilon_0}\frac{|q_2|}{r_2^2} = (8.988 \times 10^9\text{ N}\cdot\text{m}^2/\text{C}^2)\frac{6.00 \times 10^{-9}\text{ C}}{(0.600\text{ m})^2}$$

$E_2 = 149.8$ N/C

$E_{1x} = -898.8$ N/C, $E_{1y} = 0$

$E_{2x} = -149.8$ N/C, $E_{2y} = 0$

$E_x = E_{1x} + E_{2x} = -898.8\text{ N/C} - 149.8\text{ N/C} = -1050$ N/C

$E_y = E_{1y} + E_{2y} = 0$

The resultant field at point a has magnitude 1050 N/C and is in the $-x$-direction.

22-21 (cont)

b) At point b the fields $\vec{E}_1$ and $\vec{E}_2$ are:

Y, E_1, E_2, $q_1 < 0$, $q_2 > 0$, b, X

$$E_1 = \frac{1}{4\pi\epsilon_0}\frac{|q_1|}{r_1^2} = (8.988\times10^9\ \text{N}\cdot\text{m}^2/\text{C}^2)\frac{4.00\times10^{-9}\ \text{C}}{(1.20\ \text{m})^2} = 25.0\ \text{N/C}$$

$$E_2 = \frac{1}{4\pi\epsilon_0}\frac{|q_2|}{r_2^2} = (8.988\times10^9\ \text{N}\cdot\text{m}^2/\text{C}^2)\frac{6.00\times10^{-9}\ \text{C}}{(0.40\ \text{m})^2} = 337.0\ \text{N/C}$$

$E_{1x} = -25.0\ \text{N/C}$, $E_{1y} = 0$

$E_{2x} = +337.0\ \text{N/C}$, $E_{2y} = 0$

$E_x = E_{1x} + E_{2x} = -25.0\ \text{N/C} + 337.0\ \text{N/C} = +312\ \text{N/C}$

$E_y = E_{1y} + E_{2y} = 0$

The resultant field at point b has magnitude 312 N/C and is in the +x-direction.

c) At point c the fields $\vec{E}_1$ and $\vec{E}_2$ are:

E_2, E_1, Y, $q_2 > 0$, c, $q_1 < 0$, X

$$E_1 = \frac{1}{4\pi\epsilon_0}\frac{|q_1|}{r_1^2} = (8.988\times10^9\ \text{N}\cdot\text{m}^2/\text{C}^2)\frac{4.00\times10^{-9}\ \text{C}}{(0.200\ \text{m})^2} = 898.8\ \text{N/C}$$

$$E_2 = \frac{1}{4\pi\epsilon_0}\frac{|q_2|}{r_2^2} = (8.988\times10^9\ \text{N}\cdot\text{m}^2/\text{C}^2)\frac{6.00\times10^{-9}\ \text{C}}{(1.00\ \text{m})^2} = 53.9\ \text{N/C}$$

$E_{1x} = +898.8\ \text{N/C}$, $E_{1y} = 0$

$E_{2x} = -53.9\ \text{N/C}$, $E_{2y} = 0$

$E_x = E_{1x} + E_{2x} = 898.8\ \text{N/C} - 53.9\ \text{N/C} = +845\ \text{N/C}$

$E_y = E_{1y} + E_{2y} = 0$

The resultant field at point c has magnitude 845 N/C and is in the +x-direction.

22-25

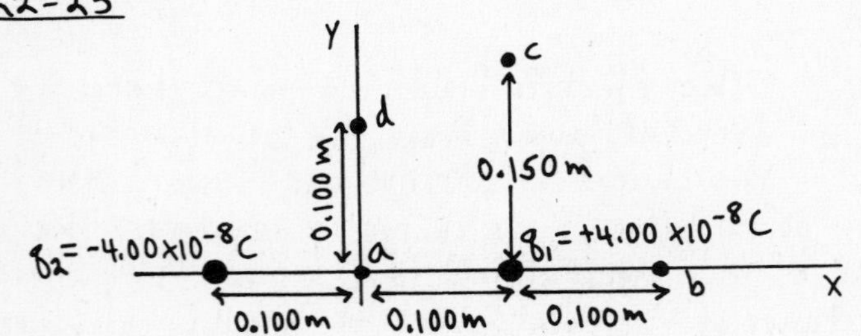

The resultant electric field is the vector sum of the field $\vec{E}_1$ of q_1 and $\vec{E}_2$ of q_2.

a) At point a (the origin) the electric fields $\vec{E}_1$ and $\vec{E}_2$ are:

Y, E_1, a, $q_2 < 0$, E_2, $q_1 > 0$, X

$$E_1 = \frac{1}{4\pi\epsilon_0}\frac{|q_1|}{r_1^2} = (8.988\times10^9\ \text{N}\cdot\text{m}^2/\text{C}^2)\frac{4.00\times10^{-8}\ \text{C}}{(0.100\ \text{m})^2}$$

$E_1 = 3.595\times10^4\ \text{N/C}$

$$E_2 = \frac{1}{4\pi\epsilon_0}\frac{|q_2|}{r_2^2} = (8.988\times10^9\ \text{N}\cdot\text{m}^2/\text{C}^2)\frac{4.00\times10^{-8}\ \text{C}}{(0.100\ \text{m})^2}$$

$E_2 = 3.595\times10^4\ \text{N/C}$

$E_{1x} = -3.595\times10^4\ \text{N/C}$, $E_{1y} = 0$

$E_{2x} = -3.595\times10^4\ \text{N/C}$, $E_{2y} = 0$

$E_x = E_{1x} + E_{2x} = 2(-3.595\times10^4\ \text{N/C}) = -7.19\times10^4\ \text{N/C}$

$E_y = E_{1y} + E_{2y} = 0$

The resultant field at point a has magnitude 7.19×10^4 N/C and is in the −x-direction.

22-25 (cont)

At point b the electric fields $\vec{E}_1$ and $\vec{E}_2$ are:

$$E_1 = \frac{1}{4\pi\epsilon_0}\frac{|q_1|}{r_1^2} = (8.988\times10^9\,\text{N}\cdot\text{m}^2/\text{C}^2)\frac{4.00\times10^{-8}\,\text{C}}{(0.100\,\text{m})^2}$$

$$E_1 = 3.595\times10^4\ \text{N/C}$$

$$E_2 = \frac{1}{4\pi\epsilon_0}\frac{|q_2|}{r_2^2} = (8.988\times10^9\,\text{N}\cdot\text{m}^2/\text{C}^2)\frac{4.00\times10^{-8}\,\text{C}}{(0.300\,\text{m})^2}$$

$$E_2 = 3.995\times10^3\ \text{N/C}$$

$E_{1x} = +3.595\times10^4\ \text{N/C}$, $E_{1y} = 0$

$E_{2x} = -3.995\times10^3\ \text{N/C}$, $E_{2y} = 0$

$E_x = E_{1x} + E_{2x} = +3.595\times10^4\ \text{N/C} - 3.995\times10^3\ \text{N/C} = 3.20\times10^4\ \text{N/C}$

$E_y = E_{1y} + E_{2y} = 0$

The resultant field at point b has magnitude 3.20×10^4 N/C and is in the +x-direction.

c)

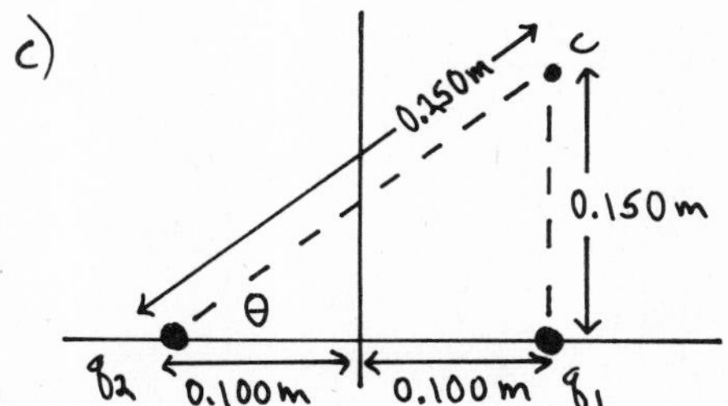

$$E_1 = \frac{1}{4\pi\epsilon_0}\frac{|q_1|}{r_1^2} = (8.988\times10^9\,\text{N}\cdot\text{m}^2/\text{C}^2)\frac{4.00\times10^{-8}\,\text{C}}{(0.150\,\text{m})^2}$$

$$E_1 = 1.598\times10^4\,\text{N/C}$$

$$E_2 = \frac{1}{4\pi\epsilon_0}\frac{|q_2|}{r_2^2} = (8.988\times10^9\,\text{N}\cdot\text{m}^2/\text{C}^2)\frac{4.00\times10^{-8}\,\text{C}}{(0.250\,\text{m})^2}$$

$$E_2 = 5.752\times10^3\ \text{N/C}$$

$\sin\theta = \dfrac{0.150\,\text{m}}{0.250\,\text{m}} = 0.600$; $\cos\theta = \dfrac{0.200\,\text{m}}{0.250\,\text{m}} = 0.800$

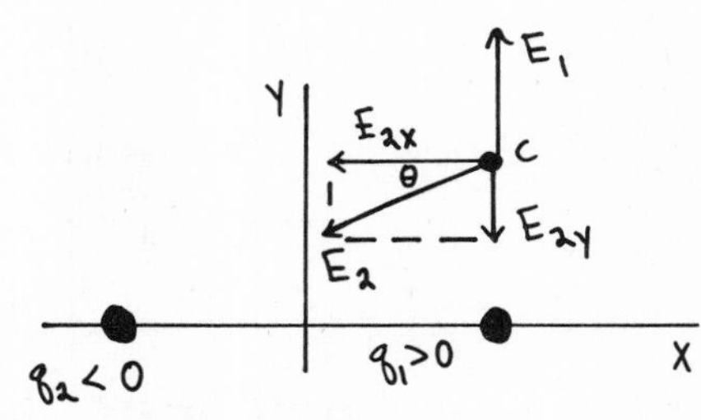

$E_{1x} = 0$, $E_{1y} = +1.598\times10^4\ \text{N/C}$

$E_{2x} = -E_2\cos\theta = -(5.752\times10^3\,\text{N/C})(0.800) = -4.602\times10^3\,\text{N/C}$

$E_{2y} = -E_2\sin\theta = -(5.752\times10^3\,\text{N/C})(0.600) = -3.451\times10^3\ \text{N/C}$

$E_x = E_{1x} + E_{2x} = 0 - 4.602\times10^3\,\text{N/C} = -4.602\times10^3\,\text{N/C}$

$E_y = E_{1y} + E_{2y} = +1.598\times10^4\,\text{N/C} - 3.451\times10^3\,\text{N/C}$

$E_y = 1.253\times10^4\ \text{N/C}$

$$E = \sqrt{E_x^2 + E_y^2} = 1.33\times10^4\,\text{N/C}$$

$$\tan\alpha = \frac{E_y}{E_x} = \frac{1.253\times10^4\,\text{N/C}}{-4.602\times10^3\,\text{N/C}} = -2.723 \Rightarrow \alpha = 110^\circ$$

The resultant field at point c has magnitude 1.33×10^4 N/C and is directed at an angle of 110° counterclockwise from the +x-axis.

d)

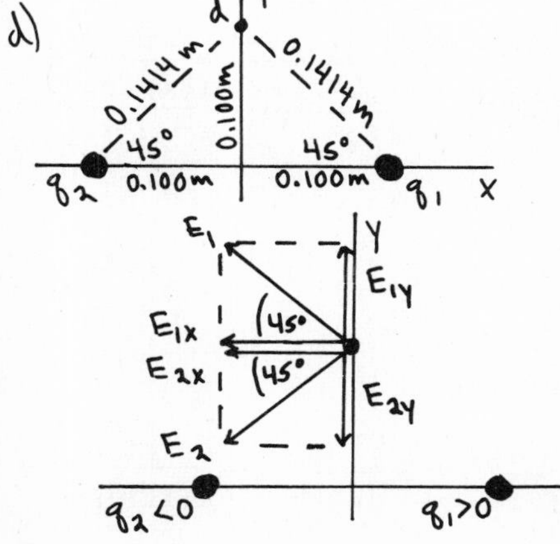

$$E_1 = \frac{1}{4\pi\epsilon_0}\frac{|q_1|}{r_1^2} = 8.988\times10^9\,\text{N}\cdot\text{m}^2/\text{C}^2\ \frac{4.00\times10^{-8}\,\text{C}}{(0.1414\,\text{m})^2}$$

$$E_1 = 1.798\times10^4\ \text{N/C}$$

$$E_2 = \frac{1}{4\pi\epsilon_0}\frac{|q_2|}{r_2^2} = 1.798\times10^4\ \text{N/C}$$

$E_{1x} = -E_1\cos45^\circ = -1.271\times10^4\ \text{N/C}$

$E_{1y} = +E_1\sin45^\circ = +1.271\times10^4\ \text{N/C}$

$E_{2x} = -E_2\cos45^\circ = -1.271\times10^4\ \text{N/C}$

$E_{2y} = -E_2\sin45^\circ = -1.271\times10^4\,\text{N/C}$

22-25 (cont)

$$E_x = E_{1x} + E_{2x} = -1.271\times10^4\ \text{N/C} - 1.271\times10^4\ \text{N/C} = -2.542\times10^4\ \text{N/C}$$

$$E_y = E_{1y} + E_{2y} = 1.271\times10^4\ \text{N/C} - 1.271\times10^4\ \text{N/C} = 0$$

The resultant field has magnitude 2.54×10^4 N/C and is in the $-x$-direction.

22-29

a) $U(\phi) = -\vec{p}\cdot\vec{E} = -pE\cos\phi$, where ϕ is the angle between $\vec{p}$ and $\vec{E}$.

parallel $\Rightarrow$ $\phi = 0$; $U(0°) = -pE$

perpendicular $\Rightarrow$ $\phi = 90°$; $U(90°) = 0$

$$\Delta U = U(90°) - U(0°) = +pE = (5.0\times10^{-30}\ \text{C}\cdot\text{m})(3.0\times10^5\ \text{N/C}) = \underline{1.5\times10^{-24}\ \text{J}}$$

b) $\frac{3}{2}kT = 1.5\times10^{-24}$ J

$$T = \frac{2(1.5\times10^{-24}\ \text{J})}{3k} = \frac{2(1.5\times10^{-24}\ \text{J})}{3(1.381\times10^{-23}\ \text{J/K})} = \underline{0.0724\ \text{K}}$$

Problems

22-35

a)

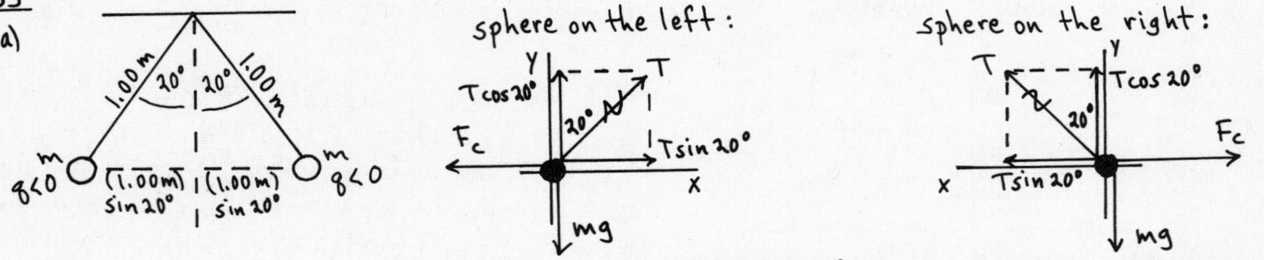

(F_c is the repulsive Coulomb force exerted by one sphere on the other.)

b) From either force diagram in part (a):

$\sum F_y = ma_y$

$T\cos 20° - mg = 0$

$T = \frac{mg}{\cos 20°}$

$\sum F_x = ma_x$

$T\sin 20° - F_c = 0$

$F_c = T\sin 20°$

Use the first equation to eliminate T in the second

$\Rightarrow F_c = \frac{mg}{\cos 20°}(\sin 20°) = mg\tan 20°$

$$F_c = \frac{1}{4\pi\epsilon_0}\frac{|q_1q_2|}{r^2} = \frac{1}{4\pi\epsilon_0}\frac{q^2}{r^2} = \frac{1}{4\pi\epsilon_0}\frac{q^2}{[2(1.00\ \text{m})\sin 20°]^2}$$

$$F_c = mg\tan 20° \Rightarrow mg\tan 20° = \frac{1}{4\pi\epsilon_0}\frac{q^2}{[2(1.00\ \text{m})\sin 20°]^2}$$

$$q = (2.00\ \text{m})\sin 20°\sqrt{\frac{mg\tan 20°}{(1/4\pi\epsilon_0)}} = (2.00\ \text{m})\sin 20°\sqrt{\frac{(16.0\times10^{-3}\ \text{kg})(9.80\ \text{m/s}^2)\tan 20°}{8.988\times10^9\ \text{N}\cdot\text{m}^2/\text{C}^2}}$$

$$q = \underline{1.72\times10^{-6}\ \text{C}}$$

c) The separation between the two spheres is now $2l\sin\theta$. $q = 1.72\ \mu$C as found in part (b).

$F_c = \frac{1}{4\pi\epsilon_0}\frac{q^2}{(2l\sin\theta)^2}$ and $F_c = mg\tan\theta$

$$\Rightarrow \frac{1}{4\pi\epsilon_0}\frac{q^2}{(2l\sin\theta)^2} = mg\tan\theta$$

$$(\sin\theta)^2\tan\theta = \frac{1}{4\pi\epsilon_0}\frac{q^2}{4l^2mg} = (8.988\times10^9\ \text{N}\cdot\text{m}^2/\text{C}^2)\frac{(1.72\times10^{-6}\ \text{C})^2}{4(0.500\ \text{m})^2(16.0\times10^{-3}\ \text{kg})(9.80\ \text{m/s}^2)} = 0.1696$$

22-35 (cont)

$$(\sin\theta)^2 \tan\theta = 0.1696$$

Solve this equation by trial and error. This will go quicker if we can make a good estimate of the value of θ that solves this equation. θ small gives $\tan\theta \approx \sin\theta$. With this approximation the equation becomes

$\sin^3\theta = 0.1696 \Rightarrow \sin\theta = 0.5535$ and $\theta = 33.6°$

Now refine this guess:

θ	$\sin^2\theta \tan\theta$	θ	$\sin^2\theta \tan\theta$
33.6°	0.2035	32.0°	0.1755
33.0°	0.1926	31.5°	0.1673
30.0°	0.1443	31.6°	0.1689 $\Rightarrow \theta = 31.6°$
31.0°	0.1594	31.7°	0.1705

22-37

a)

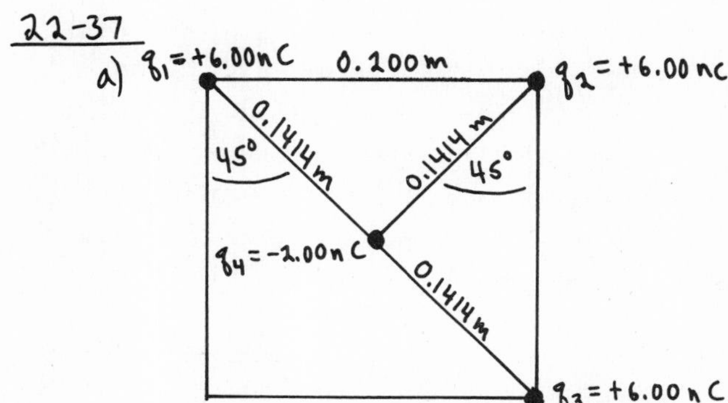

Forces on q_4:

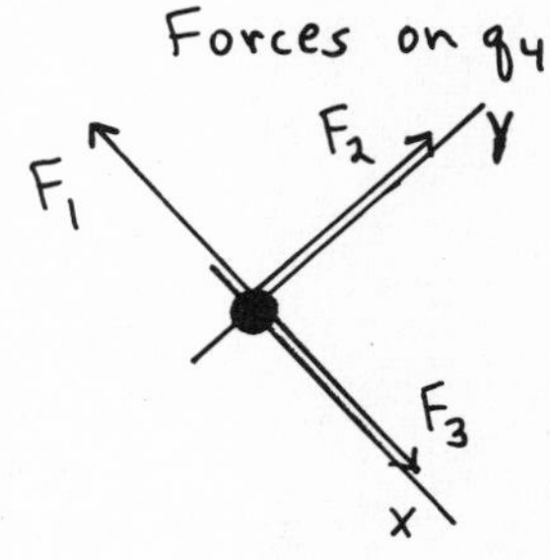

Use coordinates that are along the diagonals of the square. Forces on q_4 are all attractive.

$$F_1 = F_2 = F_3 = \frac{1}{4\pi\epsilon_0}\frac{|q_1 q_4|}{r^2} = (8.988\times10^9\ \text{N}\cdot\text{m}^2/\text{C}^2)\frac{(6.00\times10^{-9}\ \text{C})(2.00\times10^{-9}\ \text{C})}{(0.1414\ \text{m})^2} = 5.39\times10^{-6}\ \text{N}$$

$\vec{F}_1$ and $\vec{F}_3$ are equal in magnitude and opposite in direction, so $\vec{F}_1 + \vec{F}_3 = 0$. The resultant force $\vec{R}$ on q_4 equals $\vec{F}_2$; $R = 5.39\times10^{-6}$ N and is directed away from the vacant corner (the corner of the square where there is no charge).

b)

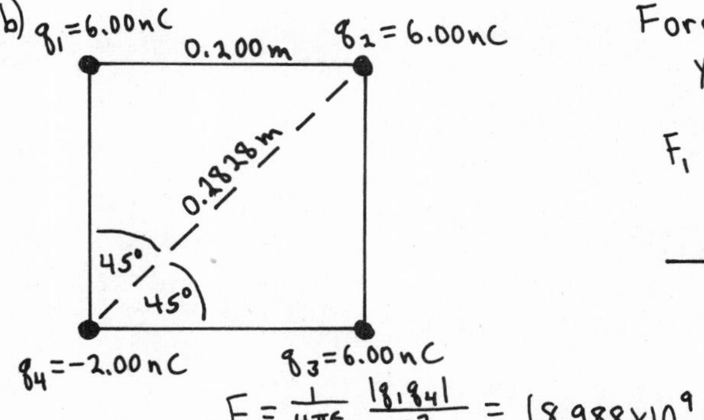

Forces on q_4:

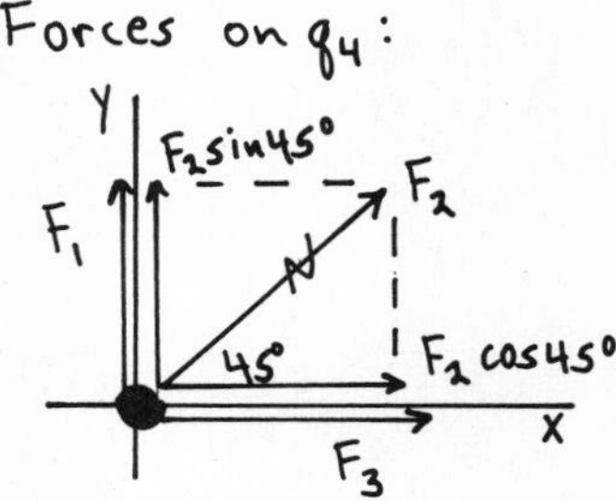

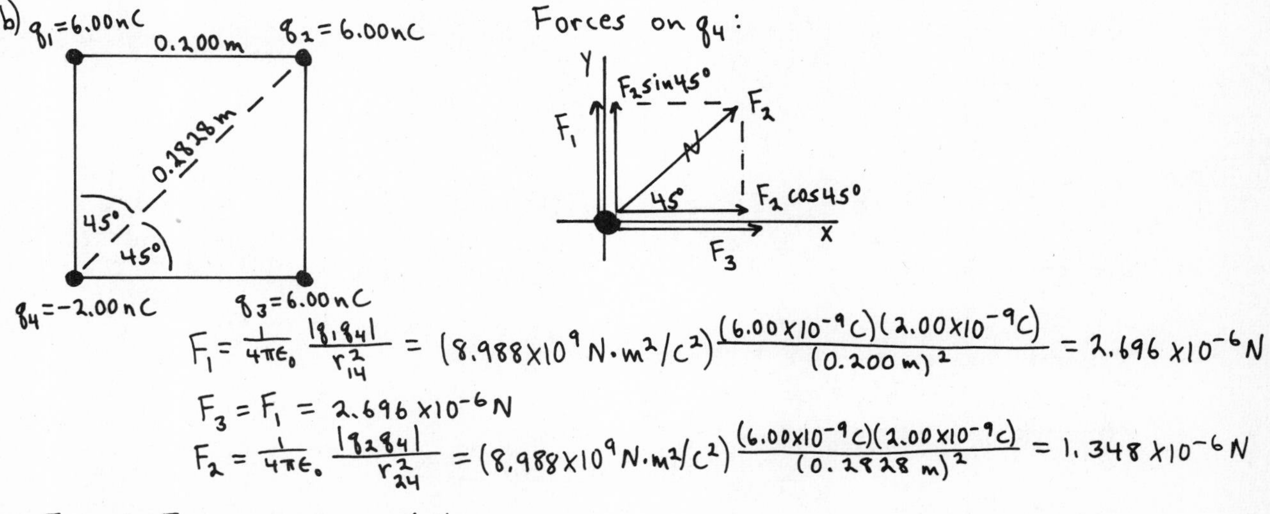

$$F_1 = \frac{1}{4\pi\epsilon_0}\frac{|q_1 q_4|}{r_{14}^2} = (8.988\times10^9\ \text{N}\cdot\text{m}^2/\text{C}^2)\frac{(6.00\times10^{-9}\ \text{C})(2.00\times10^{-9}\ \text{C})}{(0.200\ \text{m})^2} = 2.696\times10^{-6}\ \text{N}$$

$$F_3 = F_1 = 2.696\times10^{-6}\ \text{N}$$

$$F_2 = \frac{1}{4\pi\epsilon_0}\frac{|q_2 q_4|}{r_{24}^2} = (8.988\times10^9\ \text{N}\cdot\text{m}^2/\text{C}^2)\frac{(6.00\times10^{-9}\ \text{C})(2.00\times10^{-9}\ \text{C})}{(0.2828\ \text{m})^2} = 1.348\times10^{-6}\ \text{N}$$

$F_{1x} = 0$, $F_{1y} = +2.696\times10^{-6}$ N

$F_{2x} = F_2 \cos 45° = (1.348\times10^{-6}\ \text{N})\cos 45° = 0.9532\times10^{-6}$ N; $F_{2y} = F_2 \sin 45° = 0.9532\times10^{-6}$ N

22-37 (cont)

$$F_{3x} = 2.696\times10^{-6}\,N \;;\quad F_{3y} = 0$$

$$R_x = F_{1x} + F_{2x} + F_{3x} = 0 + 0.9532\times10^{-6}\,N + 2.696\times10^{-6}\,N = 3.649\times10^{-6}\,N$$

$$R_y = F_{1y} + F_{2y} + F_{3y} = 2.696\times10^{-6}\,N + 0.9532\times10^{-6}\,N + 0 = 3.649\times10^{-6}\,N$$

$$R = \sqrt{R_x^2 + R_y^2} = \sqrt{2(3.649\times10^{-6}\,N)^2} = 5.16\times10^{-6}\,N$$

$$\tan\theta = \frac{R_y}{R_x} = 1.00 \Rightarrow \theta = 45°$$

The resultant force has magnitude 5.16×10^{-6} N and is directed toward the opposite corner of the square.

22-39

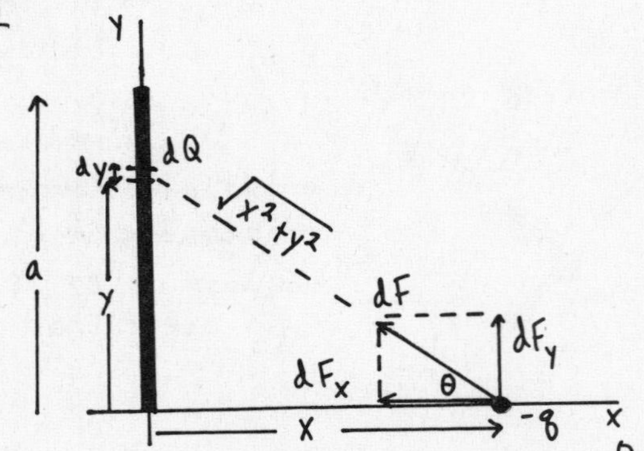

Slice the charge distribution up into small pieces of length dy. The charge dQ in each slice is $dQ = Q\frac{dy}{a}$. The force this slice exerts on the point charge is $d\vec{F}$. Calculate the components of $d\vec{F}$ and then integrate over the charge distribution to find the components of the total force.

$$dF = \frac{1}{4\pi\epsilon_0}\frac{|q\,dQ|}{r^2} = \frac{1}{4\pi\epsilon_0}\frac{q\frac{Q}{a}dy}{x^2+y^2} = \frac{1}{4\pi\epsilon_0}\frac{qQ}{a}\frac{dy}{x^2+y^2}$$

$$dF_x = -dF\cos\theta = -dF\frac{x}{\sqrt{x^2+y^2}} = -\frac{1}{4\pi\epsilon_0}\frac{qQx}{a}\frac{dy}{(x^2+y^2)^{3/2}}$$

$$dF_y = +dF\sin\theta = dF\frac{y}{\sqrt{x^2+y^2}} = \frac{1}{4\pi\epsilon_0}\frac{qQ}{a}\frac{y\,dy}{(x^2+y^2)^{3/2}}$$

$$F_x = \int dF_x = -\frac{1}{4\pi\epsilon_0}\frac{qQx}{a}\int_0^a\frac{dy}{(x^2+y^2)^{3/2}} = -\frac{qQx}{4\pi\epsilon_0 a}\left[\frac{1}{x^2}\frac{y}{\sqrt{x^2+y^2}}\Big|_0^a\right] = -\frac{qQ}{4\pi\epsilon_0 x}\frac{1}{\sqrt{x^2+a^2}}$$

$$F_y = \int dF_y = \frac{1}{4\pi\epsilon_0}\frac{qQ}{a}\int_0^a\frac{y\,dy}{(x^2+y^2)^{3/2}} = \frac{qQ}{4\pi\epsilon_0 a}\left[-\frac{1}{\sqrt{x^2+y^2}}\Big|_0^a\right] = \frac{qQ}{4\pi\epsilon_0 a}\left[\frac{1}{x} - \frac{1}{\sqrt{x^2+a^2}}\right]$$

22-43

$$\vec{F} = q\vec{E}\;;\; q = -e \;\Rightarrow\; \vec{F} = -e\vec{E}$$

The electric field is upward so the electric force on the negatively charged electron is downward, and has magnitude $F = eE$.

Free-body diagram for the electron:

$$\Sigma F_y = ma_y$$

$$-eE = ma$$

$$a = -\frac{eE}{m} = -\frac{(1.602\times10^{-19}\,C)(500\,N/C)}{9.109\times10^{-31}\,kg} = -8.794\times10^{13}\,m/s^2$$

(This is much larger in magnitude than g, so gravity can be neglected.)

22-43 (cont)

a) $v_y = 0$ (at maximum height)

$v_{0y} = v_0 \sin 30° = (4.00\times10^6 \text{ m/s}) \sin 30°$

$v_{0y} = 2.00\times10^6 \text{ m/s}$

$a_y = -8.794\times10^{13} \text{ m/s}^2$

$y - y_0 = ?$

$v_y^2 = v_{0y}^2 + 2a_y(y-y_0)$

$y - y_0 = \dfrac{v_y^2 - v_{0y}^2}{2a_y} = \dfrac{0 - (2.00\times10^6 \text{ m/s})^2}{2(-8.794\times10^{13} \text{ m/s}^2)}$

$y - y_0 = 0.0227 \text{ m} = \underline{2.27 \text{ cm}}$

b) Want $x - x_0$ when $y - y_0 = 0$:

$y - y_0 = 0$

$v_{0y} = 2.00\times10^6 \text{ m/s}$

$a_y = -8.794\times10^{13} \text{ m/s}^2$

$t = ?$

$\cancel{y - y_0}^{\,0} = v_{0y}t + \frac{1}{2}a_y t^2$

$-\frac{1}{2}a_y t^2 = v_{0y} t$

$t = \dfrac{-2v_{0y}}{a_y} = \dfrac{-2(2.00\times10^6 \text{ m/s})}{-8.794\times10^{13} \text{ m/s}^2} = 4.549\times10^{-8} \text{ s}$

Then

$x - x_0 = ?$

$v_{0x} = v_0 \cos 30° = (4.00\times10^6 \text{ m/s}) \cos 30° = 3.464\times10^6 \text{ m/s}$

$a_x = 0$

$t = 4.549\times10^{-8} \text{ s}$

$x - x_0 = v_{0x}t + \cancel{\frac{1}{2}a_x t^2}$

$x - x_0 = (3.464\times10^6 \text{ m/s})(4.549\times10^{-8} \text{ s})$

$x - x_0 = 0.158 \text{ m} = \underline{15.8 \text{ cm}}$

c) a_y is constant and downward; $a_x = 0$. The motion is just like projectile motion, just with a different numerical value for a_y. Thus the path of the electron is a parabola:

22-45

The electric field of q_1 at the origin has magnitude $E_1 = \frac{1}{4\pi\epsilon_0}\frac{|q_1|}{r_1^2} = (8.988\times10^9 \text{ N}\cdot\text{m}^2/\text{C}^2)\frac{5.00\times10^{-9}\text{ C}}{(1.20\text{ m})^2}$ $= 31.2$ N/C and is in the +x-direction.

The resultant field at the origin is given by

$\vec{E} = \vec{E}_1 + \vec{E}_2 \Rightarrow \vec{E}_2 = \vec{E} - \vec{E}_1$ is the field at the origin due to charge q_2.

$\Rightarrow E_{2x} = E_x - E_{1x}$

a) $E_x = +60.0$ N/C, $E_{1x} = +31.2$ N/C $\Rightarrow E_{2x} = 60.0 \text{ N/C} - 31.2 \text{ N/C} = +28.8$ N/C

$\vec{E}_2$ is in the +x-direction $\Rightarrow q_2$ is positive

$E_2 = \dfrac{1}{4\pi\epsilon_0}\dfrac{|q_2|}{r_2^2} \Rightarrow |q_2| = \dfrac{r_2^2 E_2}{(1/4\pi\epsilon_0)} = \dfrac{(0.60\text{ m})^2(28.8\text{ N/C})}{8.988\times10^9 \text{ N}\cdot\text{m}^2/\text{C}^2} = 1.15\times10^{-9}\text{ C}$

$q_2 = \underline{+1.15\times10^{-9}\text{ C}}$

b) $E_x = -60.0$ N/C, $E_{1x} = +31.2$ N/C

$E_{2x} = E_x - E_{1x} = -60.0 \text{ N/C} - 31.2 \text{ N/C} = -91.2$ N/C

$\vec{E}_2$ is in the −x-direction $\Rightarrow q_2$ is negative

$E_2 = \dfrac{1}{4\pi\epsilon_0}\dfrac{|q_2|}{r_2^2} \Rightarrow |q_2| = \dfrac{r_2^2 E_2}{(1/4\pi\epsilon_0)} = \dfrac{(0.60\text{ m})^2(91.2\text{ N/C})}{(8.988\times10^9 \text{ N}\cdot\text{m}^2/\text{C}^2)} = 3.65\times10^{-9}\text{ C}$

$q_2 = \underline{-3.65\times10^{-9}\text{ C}}$

22-47

y; ←3.00 m→ ←3.00 m→ ←2.00 m→; $q_1 = +16.0\text{ nC}$; q_2; $q_3 = +12.0\text{ nC}$; P; x

At point P the resultant field has magnitude 18.0 N/C and is in the +x-direction, so $E_x = +18.0\text{ N/C}$.

$$\vec{E} = \vec{E}_1 + \vec{E}_2 + \vec{E}_3 \Rightarrow \vec{E}_2 = \vec{E} - \vec{E}_1 - \vec{E}_3$$

$E_{2x} = E_x - E_{1x} - E_{3x}$; calculate E_{2x} and from it deduce the sign and magnitude of q_2.

The electric field $\vec{E}_1$ of q_1 at point P has magnitude

$E_1 = \dfrac{1}{4\pi\epsilon_0}\dfrac{|q_1|}{r_1^2} = (8.988\times10^9\text{ N}\cdot\text{m}^2/\text{C}^2)\dfrac{16.0\times10^{-9}\text{ C}}{(8.00\text{ m})^2} = 2.247\text{ N/C}$ and is in the +x-direction $\Rightarrow E_{1x} = +2.247\text{ N/C}$.

The electric field $\vec{E}_3$ of q_3 at point P has magnitude

$E_3 = \dfrac{1}{4\pi\epsilon_0}\dfrac{|q_3|}{r_3^2} = (8.988\times10^9\text{ N}\cdot\text{m}^2/\text{C}^2)\dfrac{12.0\times10^{-9}\text{ C}}{(2.00\text{ m})^2} = 26.96\text{ N/C}$ and is in the +x-direction $\Rightarrow E_{3x} = +26.96\text{ N/C}$.

Thus $E_{2x} = E_x - E_{1x} - E_{3x} = +18.0\text{ N/C} - 2.25\text{ N/C} - 26.96\text{ N/C} = -11.21\text{ N/C}$

$\Rightarrow \vec{E}_2$ is in the −x-direction at P $\Rightarrow q_2$ is negative.

$$E_2 = \frac{1}{4\pi\epsilon_0}\frac{|q_2|}{r_2^2} \Rightarrow |q_2| = \frac{r_2^2 E_2}{(1/4\pi\epsilon_0)} = \frac{(5.00\text{ m})^2(11.21\text{ N/C})}{(8.988\times10^9\text{ N}\cdot\text{m}^2/\text{C}^2)} = 3.12\times10^{-8}\text{ C}$$

$\Rightarrow q_2 = \underline{-31.2\text{ nC}}$

22-49

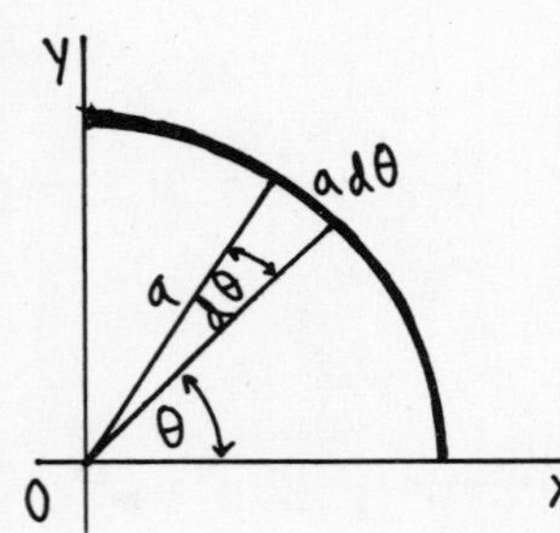

Divide the charge distribution into small segments, use the point charge formula for the electric field due to each small segment, and integrate over the charge distribution to find the x and y components of the total field.

A small segment that subtends angle $d\theta$ has length $a\,d\theta$ and contains charge $dQ = \left(\dfrac{a\,d\theta}{\frac{1}{2}\pi a}\right)Q = \dfrac{2Q}{\pi}d\theta$ ($\frac{1}{2}\pi a$ is the total length of the charge distribution). The charge is negative, so the field at the origin is directed toward the small segment. The small segment is located at angle θ as shown in the sketch.

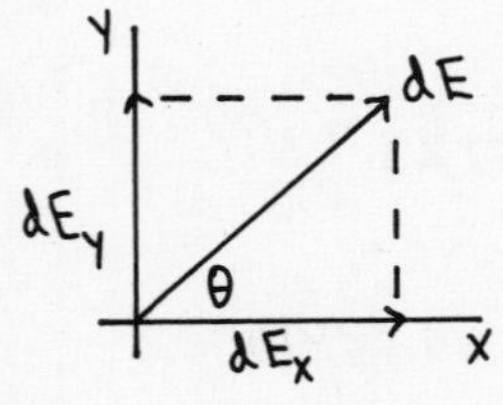

$$dE = \frac{1}{4\pi\epsilon_0}\frac{|dQ|}{r^2} = \frac{1}{4\pi\epsilon_0}\frac{|dQ|}{a^2} = \frac{Q}{2\pi^2\epsilon_0 a^2}d\theta$$

$$dE_x = dE\cos\theta = \frac{Q}{2\pi^2\epsilon_0 a^2}\cos\theta\,d\theta$$

$$dE_y = dE\sin\theta = \frac{Q}{2\pi^2\epsilon_0 a^2}\sin\theta\,d\theta$$

$$E_x = \int dE_x = \frac{Q}{2\pi^2\epsilon_0 a^2}\int_0^{\pi/2}\cos\theta\,d\theta = \frac{Q}{2\pi^2\epsilon_0 a^2}\left(\sin\theta\Big|_0^{\pi/2}\right) = \frac{Q}{2\pi^2\epsilon_0 a^2}$$

$$E_y = \int dE_y = \frac{Q}{2\pi^2\epsilon_0 a^2}\int_0^{\pi/2}\sin\theta\,d\theta = \frac{Q}{2\pi^2\epsilon_0 a^2}\left(-\cos\theta\Big|_0^{\pi/2}\right) = \frac{Q}{2\pi^2\epsilon_0 a^2}$$

CHAPTER 23

Exercises 1, 5, 7, 9
Problems 13, 17, 21, 23, 25

Exercises

23-1

a) λ ℓ r

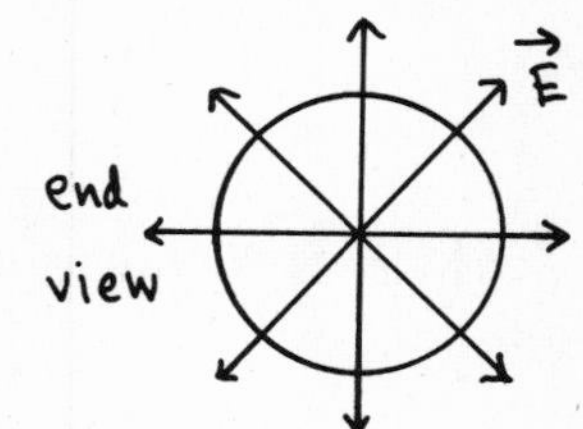

The area of the curved part of the cylinder is $A = 2\pi r\ell$.

The electric field is parallel to the end caps of the cylinder, so $\vec{E}\cdot\vec{A} = 0$ for the ends and the flux through the cylinder end caps is zero.

The electric field is normal to the curved surface of the cylinder and has the same magnitude $E = \frac{\lambda}{2\pi\epsilon_0 r}$ at all points on this surface. Thus $\phi = 0°$ and

$\Phi_E = EA\cos\phi = EA = \left(\frac{\lambda}{2\pi\epsilon_0 r}\right)(2\pi r\ell) = \frac{\lambda\ell}{\epsilon_0} = \frac{(5.00\times10^{-6}\ \text{C/m})(0.400\ \text{m})}{8.854\times10^{-12}\ \text{C}^2/\text{N}\cdot\text{m}^2}$

$\Phi_E = \underline{2.26\times10^{5}\ \text{N}\cdot\text{m}^2/\text{C}}$

b) In the calculation in part (a) the radius r of the cylinder divided out, so the flux is the same, $\Phi_E = \underline{2.26\times10^{5}\ \text{N}\cdot\text{m}^2/\text{C}}$.

c) $\Phi_E = \frac{\lambda\ell}{\epsilon_0} = \frac{(5.00\times10^{-6}\ \text{C/m})(0.800\ \text{m})}{8.854\times10^{-12}\ \text{C}^2/\text{N}\cdot\text{m}^2} = 4.52\times10^{5}\ \text{N}\cdot\text{m}^2/\text{C}$ (twice the flux calculated in parts (b) and (c))

23-5

It is rather difficult to calculate the flux directly by $\Phi_E = \oint \vec{E}\cdot d\vec{A}$ since the magnitude of $\vec{E}$ and its angle with $d\vec{A}$ varies over the surface of the cube. A much easier approach is to use Gauss's law to calculate the total flux through the cube:

$$\Phi_E = \frac{Q_{encl}}{\epsilon_0} = \frac{3.00\times10^{-9}\ \text{C}}{8.854\times10^{-12}\ \text{C}^2/\text{N}\cdot\text{m}^2} = 338.8\ \text{N}\cdot\text{m}^2/\text{C}$$

By symmetry the flux is the same through each of the six faces, so the flux through one face is $\frac{1}{6}(338.8\ \text{N}\cdot\text{m}^2/\text{C}) = \underline{56.5\ \text{N}\cdot\text{m}^2/\text{C}}$.

(Note that the size of the cube did not enter into the calculation.)

23-7

Example 23-5 derived that the electric field just outside the surface of a spherical conductor that has net charge q is $E = \frac{1}{4\pi\epsilon_0}\frac{q}{R^2}$.

$q = \frac{R^2 E}{(1/4\pi\epsilon_0)} = \frac{(0.090\ \text{m})^2(1300\ \text{N/C})}{8.988\times10^{9}\ \text{N}\cdot\text{m}^2/\text{C}^2} = 1.172\times10^{-9}\ \text{C}$

Each electron has a charge of magnitude $e = 1.602\times10^{-19}\ \text{C}$, so the number of excess electrons needed is $\frac{1.172\times10^{-9}\ \text{C}}{1.602\times10^{-19}\ \text{C}} = \underline{7.32\times10^{9}}$.

23-9

Let the cylindrical conductor have radius R and surface charge per unit area σ. By symmetry the electric field is perpendicular to the cylinder axis and is directed outward if the charge on the conductor is positive.

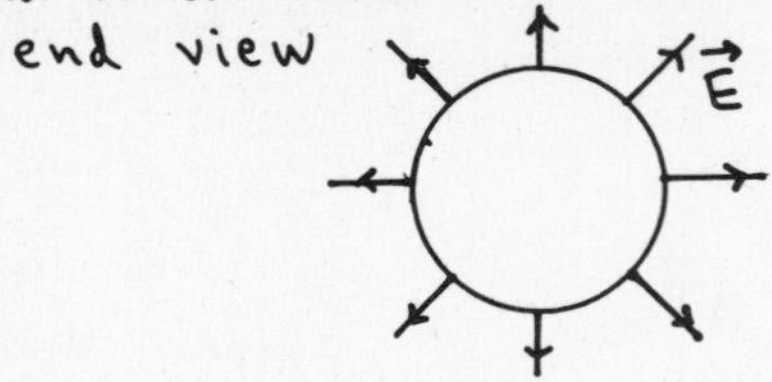

Apply Gauss's law to a cylindrical gaussian surface of length l and radius $r > R$.

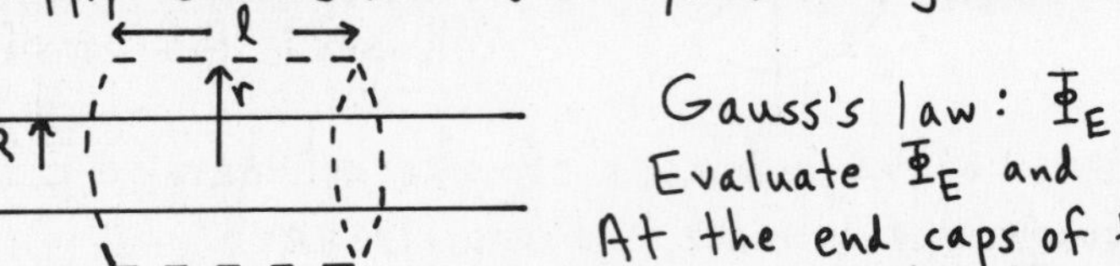

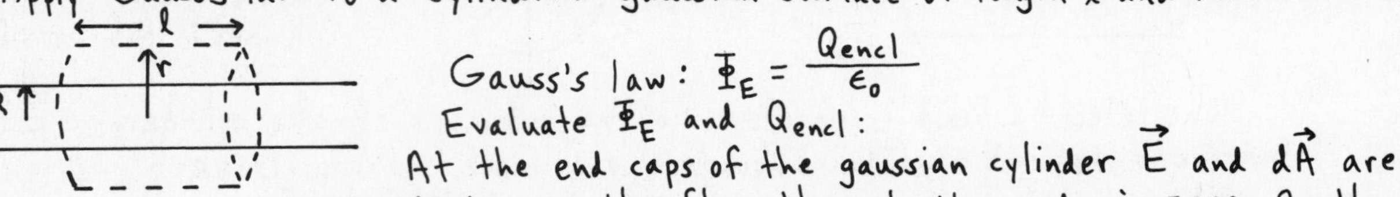

Gauss's law: $\Phi_E = \dfrac{Q_{encl}}{\epsilon_0}$

Evaluate Φ_E and Q_{encl}:

At the end caps of the gaussian cylinder $\vec{E}$ and $d\vec{A}$ are perpendicular, so the flux through the ends is zero. On the curved surface $\vec{E}$ and $d\vec{A}$ are parallel. By symmetry E depends only on the distance r from the axis of the charge distribution so E is constant at all points on the curved part of the gaussian cylinder. Thus

$$\Phi_E = EA = E(2\pi r l).$$

Q_{encl} equals the charge on the portion of the conductor that is inside the gaussian cylinder. The gaussian surface encloses a length l of the conductor, and this length has surface area $2\pi R l$. Thus $Q_{encl} = \sigma(2\pi R l)$.

Put this into Gauss's law

$$\Rightarrow \quad E(2\pi r l) = \frac{\sigma(2\pi R l)}{\epsilon_0}$$

$$\boxed{E = \frac{\sigma R}{r \epsilon_0}}$$

The electric field due to a line of charge is $E = \frac{1}{2\pi\epsilon_0}\frac{\lambda}{r}$ (Eq. 23-10), where λ is the charge per unit length. To compare these two expressions we must relate σ and λ. For the cylindrical conductor the charge on a length l can be written either as λl or as $\sigma(2\pi R l)$, so

$$\lambda l = \sigma(2\pi R l) \Rightarrow \sigma = \frac{\lambda}{2\pi R}$$

This can be used to rewrite our result as

$E = \dfrac{\sigma R}{r\epsilon_0} = \left(\dfrac{\lambda}{2\pi R}\right)\dfrac{R}{r\epsilon_0} = \dfrac{\lambda}{2\pi\epsilon_0 r}$, which is identical to Eq. (23-10) for the field due to a line of charge.

Problems

23-13

Find the net flux through the parallelepiped surface, and then use that in Gauss's law to find the net charge within. Flux out of the surface is positive and flux into the surface is negative.

23-13 (cont)

$\vec{E}_1$ gives flux out of the surface:

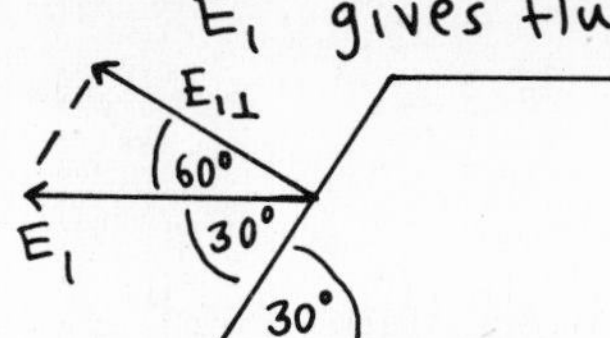

$$\Phi_{E_1} = +E_{1\perp} A$$

$$A = (0.0600\,\text{m})(0.0500\,\text{m}) = 3.00\times10^{-3}\,\text{m}^2$$

$$E_{1\perp} = E_1 \cos 60^\circ = (3.50\times10^4\,\text{N/C})\cos 60^\circ = 1.75\times10^4\,\text{N/C}$$

$$\Phi_{E_1} = +E_{1\perp} A = +(1.75\times10^4\,\text{N/C})(3.00\times10^{-3}\,\text{m}^2) = +52.5\,\text{N}\cdot\text{m}^2/\text{C}$$

$\vec{E}_2$ gives (negative) flux into the surface:

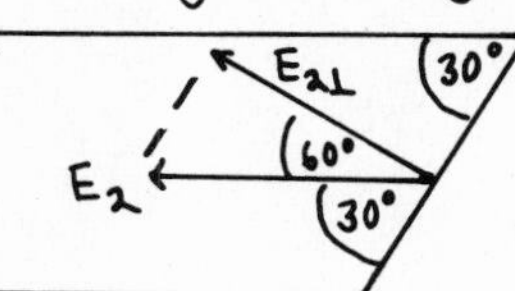

$$\Phi_{E_2} = -E_{2\perp} A$$

$$E_{2\perp} = E_2 \cos 60^\circ = (5.00\times10^4\,\text{N/C})\cos 60^\circ = 2.50\times10^4\,\text{N/C}$$

$$\Phi_{E_2} = -E_{2\perp} A = -(2.50\times10^4\,\text{N/C})(3.00\times10^{-3}\,\text{m}^2) = -75.0\,\text{N}\cdot\text{m}^2/\text{C}$$

The net flux is $\Phi_E = \Phi_{E_1} + \Phi_{E_2} = +52.5\,\text{N}\cdot\text{m}^2/\text{C} - 75.0\,\text{N}\cdot\text{m}^2/\text{C} = -22.5\,\text{N}\cdot\text{m}^2/\text{C}$.
The net flux is negative (inward), so the net charge enclosed is negative.

Apply Gauss's law: $\Phi_E = \dfrac{Q_{encl}}{\epsilon_0}$

$$Q_{encl} = \Phi_E \epsilon_0 = (-22.5\,\text{N}\cdot\text{m}^2/\text{C})(8.854\times10^{-12}\,\text{C}^2/\text{N}\cdot\text{m}^2) = \underline{-1.99\times10^{-10}\,\text{C}}$$

23-17

a) (i) $r < a$

Apply Gauss's law to a spherical gaussian surface with radius $r < a$.

$$\Phi_E = EA = E(4\pi r^2)$$

$Q_{encl} = 0$; no charge is enclosed

$$\Phi_E = \frac{Q_{encl}}{\epsilon_0} \Rightarrow E(4\pi r^2) = 0$$

$$E = 0$$

(ii) $a < r < b$

Points in this region are in the conductor of the small shell $\Rightarrow E = 0$.

(iii) $b < r < c$

Apply Gauss's law to a spherical gaussian surface with radius $b < r < c$.

$$\Phi_E = EA = E(4\pi r^2)$$

The gaussian surface encloses all of the small shell and none of the larger shell. $\Rightarrow Q_{encl} = +2q$

$$\Phi_E = \frac{Q_{encl}}{\epsilon_0} \Rightarrow E(4\pi r^2) = \frac{+2q}{\epsilon_0}$$

$$E = \frac{2q}{4\pi\epsilon_0 r^2}$$

(iv) $c < r < d$

Points in this region are in the conductor of the larger shell $\Rightarrow E = 0$.

23-17 (cont)

(v) $r > d$

Apply Gauss's law to a spherical gaussian surface with radius $r > d$.

$\Phi_E = EA = E(4\pi r^2)$

The gaussian surface encloses all of the small shell and all of the larger shell $\Rightarrow Q_{encl} = 2q + 4q = 6q$

$$\Phi_E = \frac{Q_{encl}}{\epsilon_0}$$

$$E(4\pi r^2) = \frac{6q}{\epsilon_0}$$

$$E = \frac{6q}{4\pi\epsilon_0 r^2}$$

b) (i) charge on inner surface of the small shell

Apply Gauss's law to a spherical gaussian surface with radius $a < r < b$. This surface lies within the conductor of the small shell, where $E = 0$, so $\Phi_E = 0$. Thus by Gauss's law $Q_{encl} = 0$, so there is zero charge on the inner surface of the small shell.

(ii) charge on outer surface of the small shell

The total charge on the small shell is $+2q$. We found in part (i) that there is zero charge on the inner surface of the shell, so all $+2q$ must reside on the outer surface.

(iii) charge on inner surface of large shell

Apply Gauss's law to a spherical gaussian surface with radius $c < r < d$. This surface lies within the conductor of the large shell, where $E = 0$, so $\Phi_E = 0$. Thus by Gauss's law $Q_{encl} = 0$. The surface encloses the $+2q$ on the small shell so there must be charge $-2q$ on the inner surface of the larger shell, to make the total enclosed charge zero.

(iv) charge on outer surface of large shell

The total charge on the larger shell is $+4q$. We showed in part (iii) that the charge on the inner surface is $-2q$, so there must be $+6q$ on the outer surface.

23-21

a)

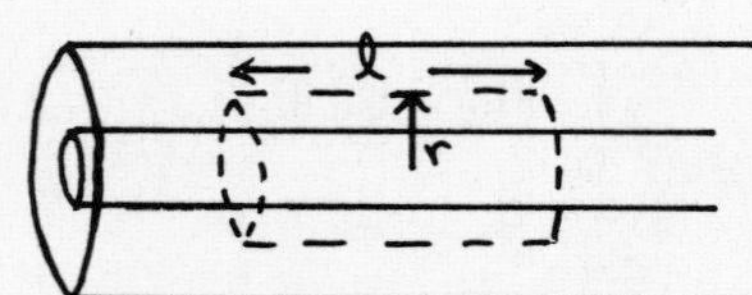

Apply Gauss's law to a gaussian cylinder of length l and radius r, where $a < r < b$:

$$\Phi_E = \frac{Q_{encl}}{\epsilon_0}$$

$\Phi_E = E(2\pi r l)$

$Q_{encl} = \lambda l$ (the charge on the length l of the inner conductor that is inside the gaussian surface)

$$\Rightarrow E(2\pi r l) = \frac{\lambda l}{\epsilon_0}$$

$$E = \frac{\lambda}{2\pi\epsilon_0 r}$$

23-21 (cont)

b)

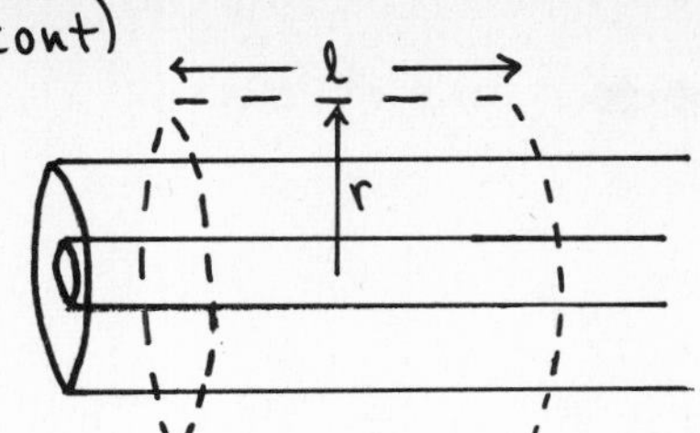

Apply Gauss's law to a gaussian cylinder of length l and radius r, where $r > c$:

$$\Phi_E = \frac{Q_{encl}}{\epsilon_0}$$

$$\Phi_E = E(2\pi r l)$$

$Q_{encl} = \lambda l$ (the charge on the length l of the inner conductor that is inside the gaussian surface; the outer conductor carries no net charge)

$$\Rightarrow E(2\pi r l) = \frac{\lambda l}{\epsilon_0}$$

$$E = \frac{\lambda}{2\pi\epsilon_0 r}$$

c) $E = 0$ within a conductor.

Thus $E = 0$ for $r < a$; $E = \frac{\lambda}{2\pi\epsilon_0 r}$ for $a < r < b$; $E = 0$ for $b < r < c$; $E = \frac{\lambda}{2\pi\epsilon_0 r}$ for $r > c$.

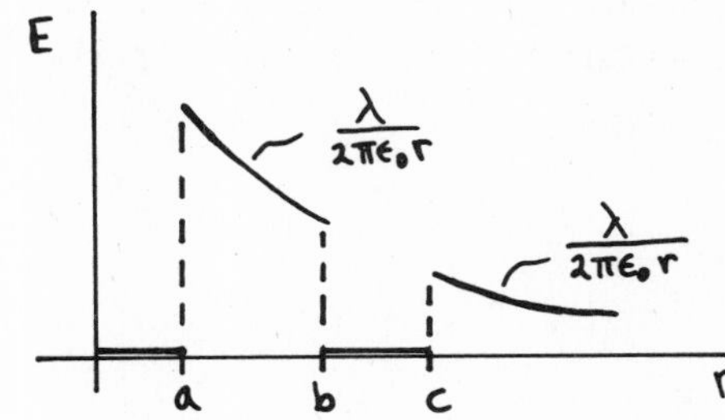

d) <u>inner surface</u>

Apply Gauss's law to gaussian cylinder with radius r, where $b < r < c$. This surface lies within the conductor of the outer cylinder, where $E = 0$, so $\Phi_E = 0$. Thus by Gauss's law $Q_{encl} = 0$. The surface encloses charge λl on the inner conductor, so it must enclose charge $-\lambda l$ on the inner surface of the outer conductor. The charge per unit length on the inner surface of the outer cylinder is $-\lambda$.

<u>outer surface</u>

The outer cylinder carries no net charge. So if there is charge per unit length $-\lambda$ on its inner surface there must be charge per unit length $+\lambda$ on the outer surface.

23-23

a)

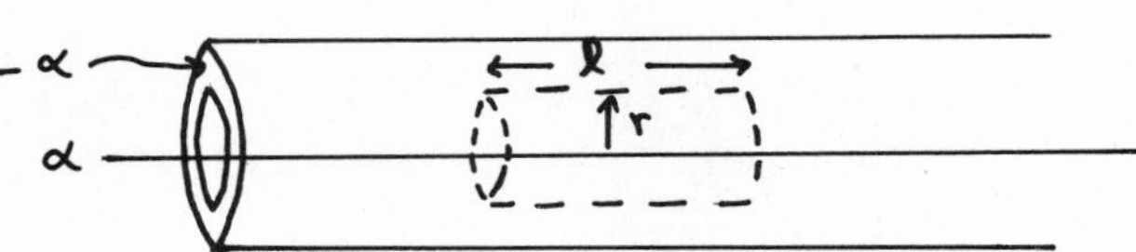

(i) <u>$r < a$</u> Apply Gauss's law to a cylindrical gaussian surface of length l and radius r, where $r < a$:

$$\Phi_E = \frac{Q_{encl}}{\epsilon_0}$$

$$\Phi_E = E(2\pi r l)$$

23-23 (cont)

$Q_{encl} = \alpha l$ (the charge on the length l of the line of charge)

$$\Rightarrow E(2\pi r l) = \frac{\alpha l}{\epsilon_0}$$

$$E = \frac{\alpha}{2\pi\epsilon_0 r}$$

(ii) $a < r < b$

Points in this region are within the conducting tube, so $E = 0$.

(iii) $r > b$

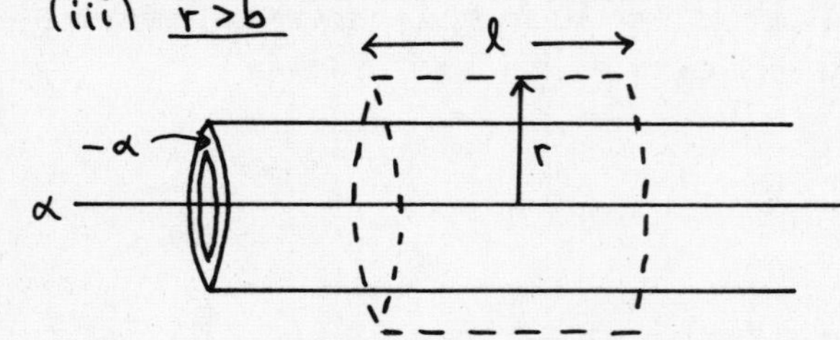

Apply Gauss's law to a cylindrical gaussian surface of length l and radius r, where $r > b$:

$$\Phi_E = \frac{Q_{encl}}{\epsilon_0}$$

$\Phi_E = E(2\pi r l)$

$Q_{encl} = \alpha l$ (charge on length l of the line of charge) $- \alpha l$ (charge on length l of tube)

$\Rightarrow Q_{encl} = 0$

Thus $E(2\pi r l) = 0 \Rightarrow E = 0$.

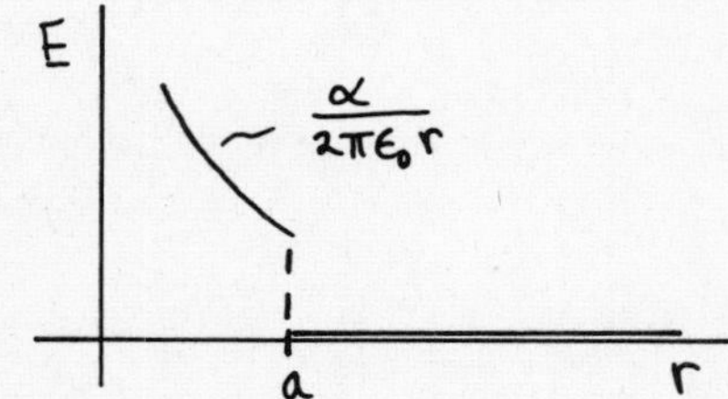

b) (i) inner surface

Apply Gauss's law to a cylindrical gaussian surface of length l and radius r, where $a < r < b$. This surface lies within the conductor of the tube, where $E = 0$, so $\Phi_E = 0$. Then by Gauss's law $Q_{encl} = 0$. The surface encloses charge αl on the line of charge so must enclose charge $-\alpha l$ on the inner surface of the tube. The charge per unit length on the inner surface of the tube is $-\alpha$.

(ii) outer surface

The net charge per unit length on the tube is $-\alpha$. We have shown above that this must all reside on the inner surface, so there is no net charge on the outer surface of the tube.

23-25

$\rho(r) = \rho_0(1 - r/R)$ for $r \le R$ where $\rho_0 = \frac{3Q}{\pi R^3}$

$\rho = 0$ for $r \ge R$

a) The charge density varies with r inside the spherical volume.

Divide the volume up into thin concentric shells, of radius r and thickness dr.

The volume of such a shell is $dV = 4\pi r^2\,dr$

The charge in the shell is

$$dq = \rho(r)\,dV = 4\pi r^2 \rho_0 (1 - \tfrac{r}{R})\,dr$$

23-25 (cont)

The total charge Q in the charge distribution is obtained by integrating dq over all such shells into which the sphere can be subdivided:

$$Q = \int dq = \int_0^R 4\pi r^2 \rho_0 \left(1 - \frac{r}{R}\right) dr = 4\pi\rho_0 \int_0^R \left(r^2 - \frac{1}{R} r^3\right) dr = 4\pi\rho_0 \left(\frac{r^3}{3} - \frac{r^4}{4R}\right)\Big|_0^R$$

$$Q = 4\pi\rho_0 \left(\frac{R^3}{3} - \frac{R^4}{4R}\right) = 4\pi\rho_0 \left(\frac{R^3}{12}\right) = 4\pi\left(\frac{3Q}{\pi R^3}\right)\left(\frac{R^3}{12}\right) = Q,$$ as was to be shown.

b) Apply Gauss's law to a spherical surface of radius r, where $r > R$:

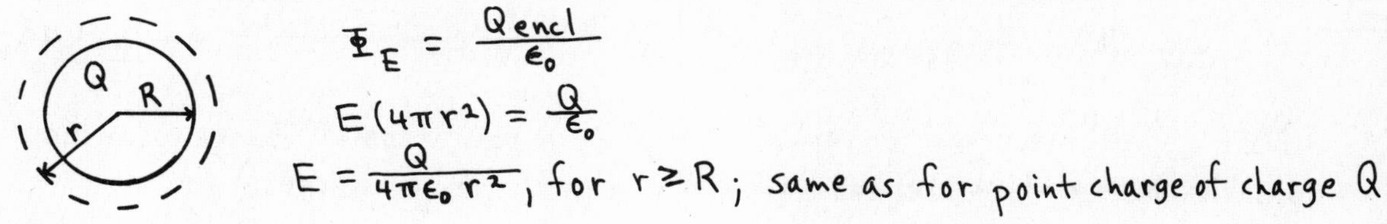

$$\Phi_E = \frac{Q_{encl}}{\epsilon_0}$$

$$E(4\pi r^2) = \frac{Q}{\epsilon_0}$$

$$E = \frac{Q}{4\pi\epsilon_0 r^2},$$ for $r \geq R$; same as for point charge of charge Q.

c) Apply Gauss's law to a spherical surface of radius r, where $r < R$:

$$\Phi_E = \frac{Q_{encl}}{\epsilon_0}$$

$$\Phi_E = E(4\pi r^2)$$

To calculate the enclosed charge Q_{encl} use the same technique as in part (a), except integrate dq out to r rather than R. (We want only that charge inside radius r.)

$$Q_{encl} = \int_0^r 4\pi r'^2 \rho_0 \left(1 - \frac{r'}{R}\right) dr' = 4\pi\rho_0 \int_0^r \left(r'^2 - \frac{r'^3}{R}\right) dr' = 4\pi\rho_0 \left(\frac{r'^3}{3} - \frac{r'^4}{4R}\right)\Big|_0^r$$

$$Q_{encl} = 4\pi\rho_0 \left(\frac{r^3}{3} - \frac{r^4}{4R}\right) = 4\pi\rho_0 r^3 \left(\frac{1}{3} - \frac{1}{4}\frac{r}{R}\right)$$

$$\rho_0 = \frac{3Q}{\pi R^3} \Rightarrow Q_{encl} = 12Q\frac{r^3}{R^3}\left(\frac{1}{3} - \frac{1}{4}\frac{r}{R}\right) = Q\left(\frac{r^3}{R^3}\right)\left(4 - 3\frac{r}{R}\right)$$

Thus $$E(4\pi r^2) = \frac{Q}{\epsilon_0}\left(\frac{r^3}{R^3}\right)\left(4 - 3\frac{r}{R}\right)$$

$$E = \frac{Qr}{4\pi\epsilon_0 R^3}\left(4 - 3\frac{r}{R}\right), \quad r \leq R$$

d) With $r = R$ in the $r \geq R$ expression (part (b)):

$$E(R) = \frac{Q}{4\pi\epsilon_0 R^2}$$

With $r = R$ in the $r \leq R$ expression (part (c)):

$$E(R) = \frac{Q}{4\pi\epsilon_0 R^2}(4 - 3) = \frac{Q}{4\pi\epsilon_0 R^2}$$

So these two results agree at the surface of the sphere.

CHAPTER 24

Exercises 1, 3, 5, 9, 11, 13, 15, 19, 21, 27, 31

Problems 37, 39, 41, 45, 47, 49, 53, 55

Exercises

24-1

a) $U = \frac{1}{4\pi\epsilon_0}\frac{qq'}{r} = (8.988\times10^9\ \text{N}\cdot\text{m}^2/\text{C}^2)\frac{(+8.00\times10^{-6}\text{C})(-0.500\times10^{-6}\text{C})}{0.800\text{m}} = \underline{-0.0449\ \text{J}}$

(Note: The electrical potential energy of a pair of point charges can be either positive or negative, depending on the signs of the charges.)

b) Y, Q, $v_b = ?$, b, $v_a = 0$, a, q, X, $r_b = 0.200$ m, $r_a = 0.800$ m

Use conservation of energy:
$K_a + U_a + W_{other} = K_b + U_b$
Only the electric force does work $\Rightarrow W_{other} = 0$ and $U = \frac{1}{4\pi\epsilon_0}\frac{qQ}{r}$.

$K_a = 0$ (released from rest)
$U_a = -0.0449$ J (from part (a))
$K_b = \frac{1}{2}mv_b^2$
$U_b = \frac{1}{4\pi\epsilon_0}\frac{qQ}{r_b} = (8.988\times10^9\ \text{N}\cdot\text{m}^2/\text{C}^2)\frac{(+8.00\times10^{-6}\text{C})(-0.500\times10^{-6}\text{C})}{0.200\ \text{m}} = -0.1798\ \text{J}$

Then $K_a + U_a + W_{other} = K_b + U_b$ gives
$U_a = K_b + U_b$
$K_b = U_a - U_b$
$\frac{1}{2}mv_b^2 = U_a - U_b \Rightarrow v_b = \sqrt{\frac{2(U_a - U_b)}{m}} = \sqrt{\frac{2(-0.0449\ \text{J} - (-0.1798\ \text{J}))}{3.00\times10^{-4}\ \text{kg}}} = \underline{30.0\ \text{m/s}}$

24-3

Y, q_1, q_2, a, b, X, $r_a = 0.160$ m, $r_b = 0.280$ m

$W_{a\to b} = U_a - U_b$

$U_a = \frac{1}{4\pi\epsilon_0}\frac{q_1q_2}{r_a} = (8.988\times10^9\ \text{N}\cdot\text{m}^2/\text{C}^2)\frac{(-6.00\times10^{-6}\text{C})(+4.00\times10^{-6}\text{C})}{0.160\ \text{m}}$
$U_a = -1.348$ J

$U_b = \frac{1}{4\pi\epsilon_0}\frac{q_1q_2}{r_b} = (8.988\times10^9\ \text{N}\cdot\text{m}^2/\text{C}^2)\frac{(-6.00\times10^{-6}\text{C})(+4.00\times10^{-6}\text{C})}{0.280\ \text{m}}$
$U_b = -0.770$ J

$W_{a\to b} = U_a - U_b = -1.348\ \text{J} - (-0.770\ \text{J}) = \underline{-0.578\ \text{J}}$

The attractive force on q_2 is in the -x-direction, so it does negative work on q_2 when q_2 moves to larger x.

24-5

a) $v_a = 22.0$ m/s, $v_b = ?$

a ○ q_2 b ○ q_2 ○ q_1

$r_a = 0.800$ m

$r_b = 0.500$ m

Use conservation of energy:
$K_a + U_a + W_{other} = K_b + U_b$
Let point a be where q_2 is 0.800 m from q_1 and point b be where q_2 is 0.500 m from q_1.

Only the electric force does work $\Rightarrow W_{other} = 0$ and $U = \frac{1}{4\pi\epsilon_0}\frac{q_1 q_2}{r}$.

$K_a = \frac{1}{2}mv_a^2 = \frac{1}{2}(2.00\times10^{-3}\text{ kg})(22.0\text{ m/s})^2 = 0.4840\text{ J}$

$U_a = \frac{1}{4\pi\epsilon_0}\frac{q_1 q_2}{r_a} = (8.988\times10^{9}\text{ N}\cdot\text{m}^2/\text{C}^2)\frac{(+6.00\times10^{-6}\text{C})(+2.50\times10^{-6}\text{C})}{0.800\text{ m}} = +0.1685\text{ J}$

$K_b = \frac{1}{2}mv_b^2$

$U_b = \frac{1}{4\pi\epsilon_0}\frac{q_1 q_2}{r_b} = (8.988\times10^{9}\text{ N}\cdot\text{m}^2/\text{C}^2)\frac{(+6.00\times10^{-6}\text{C})(+2.50\times10^{-6}\text{C})}{0.500\text{ m}} = +0.2696\text{ J}$

The conservation of energy equation then gives

$K_b = K_a + (U_a - U_b)$

$\frac{1}{2}mv_b^2 = +0.4840\text{ J} + (0.1685\text{ J} - 0.2696\text{ J}) = 0.3829\text{ J}$

$v_b = \sqrt{\frac{2(0.3829\text{ J})}{2.00\times10^{-3}\text{ kg}}} = \underline{19.6\text{ m/s}}$

(The potential energy increases when the two positively charged spheres get closer together, so the kinetic energy and speed decrease.)

b) Let point c be where q_2 has its speed momentarily reduced to zero. Apply conservation of energy to points a and c: $K_a + U_a + W_{other} = K_c + U_c$.

$v_a = 22.0$ m/s, $v_c = 0$

a ○ q_2 c ○ q_2 ○ q_1

$r_a = 0.800$ m

$r_c = ?$

$K_a = +0.4840$ J (from part (a))
$U_a = +0.1685$ J (from part (a))
$K_c = 0$ (at distance of closest approach the speed is zero)
$U_c = \frac{1}{4\pi\epsilon_0}\frac{q_1 q_2}{r_c}$

$\Rightarrow K_a + U_a = U_c$

$\frac{1}{4\pi\epsilon_0}\frac{q_1 q_2}{r_c} = +0.4840\text{ J} + 0.1685\text{ J} = 0.6525\text{ J}$

$r_c = \frac{1}{4\pi\epsilon_0}\frac{q_1 q_2}{+0.6525\text{ J}} = (8.988\times10^{9}\text{ N}\cdot\text{m}^2/\text{C}^2)\frac{(+6.00\times10^{-6}\text{C})(2.50\times10^{-6}\text{C})}{+0.6525\text{ J}} = \underline{0.207}$

24-9

a) The direction of $\vec{E}$ is always from high potential to low potential $\Rightarrow$ point a is at higher potential.

b) $V_a - V_b = -\int_b^a \vec{E}\cdot d\vec{l} = -\int_b^a E\,dx = -E(x_a - x_b)$

$E = -\frac{V_a - V_b}{x_a - x_b} = -\frac{+500\text{ V}}{0.80\text{m} - 1.20\text{m}} = \underline{1250\text{ V/m}}$

c) $W_{b\to a} = q(V_b - V_a) = (-0.200\times10^{-6}\text{C})(-500\text{ V}) = +\underline{1.00\times10^{-4}\text{ J}}$
(The electric force does positive work on a negative charge when the negative charge moves from low potential (point b) to high potential (point a).)

24-11

$$W_{a\to b} = q'\int_a^b \vec{E}\cdot d\vec{l}$$

Use coordinates where +y is upward and +x is to the right. Then $\vec{E} = E\vec{j}$ with $E = 5.00\times10^4$ N/C.

a)

$$d\vec{l} = dx\,\vec{i}$$

$$\vec{E}\cdot d\vec{l} = (E\vec{j})\cdot(dx\,\vec{i}) = 0 \Rightarrow W_{a\to b} = q'\int_a^b \vec{E}\cdot d\vec{l} = 0$$

The electric force on the positive charge is upward (in the direction of the electric field) and does no work for a horizontal displacement of the charge.

b)

$$d\vec{l} = dy\,\vec{j}$$

$$\vec{E}\cdot d\vec{l} = (E\vec{j})\cdot(dy\,\vec{j}) = E\,dy$$

$\Rightarrow W_{a\to b} = q'\int_a^b \vec{E}\cdot d\vec{l} = q'E\int_a^b dy = q'E(y_b - y_a)$; $y_b - y_a = -0.800$ m, minus since the displacement is downward and we have taken +y to be upward.

$$W_{a\to b} = q'E(y_b - y_a) = (+3.00\times10^{-8}\,\text{C})(5.00\times10^4\,\text{N/C})(-0.800\,\text{m}) = \underline{-1.20\times10^{-3}\,\text{J}}$$

The electric force on the positive charge is upward so it does negative work for a downward displacement of the charge.

c) $d\vec{l} = dx\,\vec{i} + dy\,\vec{j}$ (The displacement has both horizontal and vertical components.)

$\vec{E}\cdot d\vec{l} = (E\vec{j})\cdot(dx\,\vec{i} + dy\,\vec{j}) = E\,dy$ (only the vertical component of the displacement contributes to the work)

$$W_{a\to b} = q'\int_a^b \vec{E}\cdot d\vec{l} = q'E\int_a^b dy = q'E(y_b - y_a)$$

$y_a = 0$

$y_b = r\cos\theta = (2.60\,\text{m})\sin 45° = 1.838$ m

(The vertical component of the 2.60 m displacement is 1.838 m upward.)

$$W_{a\to b} = q'E(y_b - y_a) = (+3.00\times10^{-8}\,\text{C})(5.00\times10^4\,\text{N/C})(+1.838\,\text{m}) = \underline{2.76\times10^{-3}\,\text{J}}$$

24-13

$$V = \frac{1}{4\pi\epsilon_0}\sum_i \frac{q_i}{r_i}$$

$q_1 = +4.00$ nC, $q_2 = -3.00$ nC; distances 0.050 m, 0.050 m, 0.080 m, 0.060 m.

a) $V_A = \frac{1}{4\pi\epsilon_0}\left(\frac{q_1}{r_{A1}} + \frac{q_2}{r_{A2}}\right) = (8.988\times10^9\,\text{N}\cdot\text{m}^2/\text{C}^2)\left(\frac{+4.00\times10^{-9}\,\text{C}}{0.050\,\text{m}} + \frac{-3.00\times10^{-9}\,\text{C}}{0.050\,\text{m}}\right)$

$V_A = +180$ V

b) $V_B = \frac{1}{4\pi\epsilon_0}\left(\frac{q_1}{r_{B1}} + \frac{q_2}{r_{B2}}\right) = (8.988\times10^9\,\text{N}\cdot\text{m}^2/\text{C}^2)\left(\frac{+4.00\times10^{-9}\,\text{C}}{0.080\,\text{m}} + \frac{-3.00\times10^{-9}\,\text{C}}{0.060\,\text{m}}\right)$

$V_B = 0$

24-13 (cont)

c) $W_{B \to A} = q'(V_B - V_A) = (2.50 \times 10^{-9}\,C)(0 - 180\,V) = \underline{-4.50 \times 10^{-7}\,J}$

The electric force does negative work on the positive charge when it moves from low potential (point B) to high potential (point A).

24-15

a) 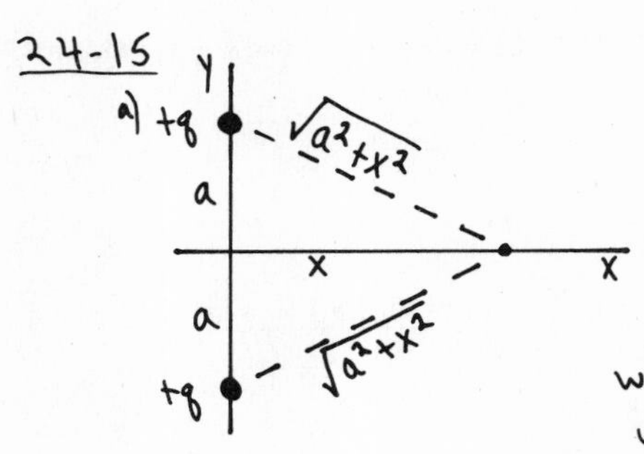

b) $V_0 = \frac{1}{4\pi\epsilon_0}\sum_i \frac{q_i}{r_i} = \frac{1}{4\pi\epsilon_0}\left(\frac{q}{a}+\frac{q}{a}\right) = \frac{q}{2\pi\epsilon_0 a}$

c) $V = \frac{1}{4\pi\epsilon_0}\sum_i \frac{q_i}{r_i} = \frac{1}{4\pi\epsilon_0}\left(\frac{q}{\sqrt{a^2+x^2}}+\frac{q}{\sqrt{a^2+x^2}}\right) = \frac{q}{2\pi\epsilon_0\sqrt{a^2+x^2}}$

d) V is symmetric about $x=0$. It has the value $V_0 = \frac{q}{2\pi\epsilon_0 a}$ when $x=0$ and $V = \frac{q}{2\pi\epsilon_0}\frac{1}{\sqrt{a^2+(4a)^2}} = \frac{q}{2\pi\epsilon_0 a}\frac{1}{\sqrt{17}} = 0.243\,V_0$ when $x = \pm 4a$.

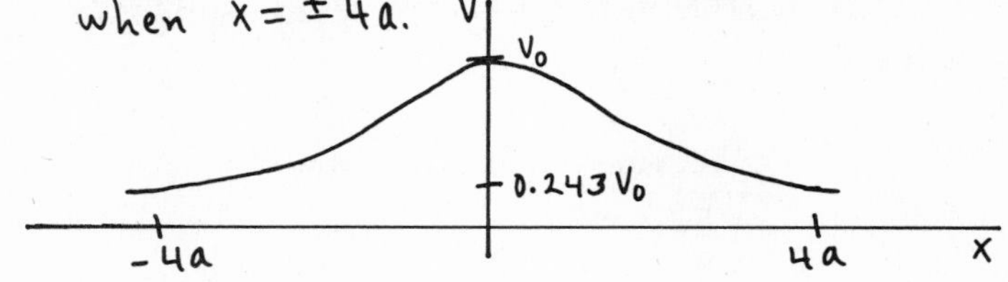

e) $V = \frac{1}{2}V_0 \Rightarrow \frac{q}{2\pi\epsilon_0\sqrt{a^2+x^2}} = \frac{1}{2}\frac{q}{2\pi\epsilon_0 a}$

$\frac{1}{\sqrt{a^2+x^2}} = \frac{1}{2a} \Rightarrow a^2 + x^2 = 4a^2$

$x^2 = 3a^2 \Rightarrow x = \underline{\pm\sqrt{3}\,a}$

24-19

a) Eq. (24-19): $E = \frac{V_{ab}}{d} = \frac{500\,V}{0.0800\,m} = \underline{6250\,V/m}$

b) $F = |q|E = (2.00 \times 10^{-9}\,C)(6250\,V/m) = \underline{1.25 \times 10^{-5}\,N}$

c) [Diagram: plate a with + charges on top, plate b with − charges on bottom, $\vec{E}$ arrows pointing down from a to b]

The plate with positive charge (plate a) is at higher potential. The electric field is directed from high potential toward low potential (or $\vec{E}$ is from + charge to − charge) $\Rightarrow$ $\vec{E}$ points from a to b. Hence the force that $\vec{E}$ exerts on the positive charge is from a to b, so it does positive work.

$W = \int_a^b \vec{F}\cdot d\vec{l} = Fd$, where d is the separation between the plates.

$W = Fd = (1.25 \times 10^{-5}\,N)(0.0800\,m) = \underline{+1.00 \times 10^{-6}\,J}$

d) $V_a - V_b = +500\,V$ (plate a is at higher potential)

$\Delta U = U_a - U_b = q(V_a - V_b) = (2.00 \times 10^{-9}\,C)(500\,V) = +1.00 \times 10^{-6}\,J$; $W_{a\to b} = U_a - U_b$

24-21

$E = \frac{V_{ab}}{d} \Rightarrow d = \frac{V_{ab}}{E} = \frac{2.50 \times 10^3\,V}{3.00 \times 10^6\,N/C} = 8.33 \times 10^{-4}\,m = \underline{0.833\,mm}$

24-27

a) No electric field ⇒ no electric force. Thus the only forces on the falling drop are gravity (downward) and the viscous air resistance force F (upward, opposite to the direction of the velocity).

Free-body diagram for the drop:

Y; $F = 6\pi\eta r v_t$ (Eq. 14-31); v_t; X; mg

The terminal speed v_t of the drop is the speed for which $a=0$.

$$\Sigma F_y = ma_y$$
$$F - mg = 0$$
$$6\pi\eta r v_t - mg = 0$$

$$v_t = \frac{mg}{6\pi\eta r} = \frac{(3.00\times10^{-14}\ \text{kg})(9.80\ \text{m/s}^2)}{6\pi(180\times10^{-7}\ \text{N}\cdot\text{s/m}^2)(2.00\times10^{-6}\ \text{m})} = 4.33\times10^{-4}\ \text{m/s} = \underline{0.433\ \text{mm/s}}$$

b) $\vec{F}_E = q\vec{E}$; $\vec{E}$ downward and q negative gives $\vec{F}_E$ upward. The magnitude of F_E is $F_E = |q|E = 8eE$.

Y; F_E; F; v_t; X; mg

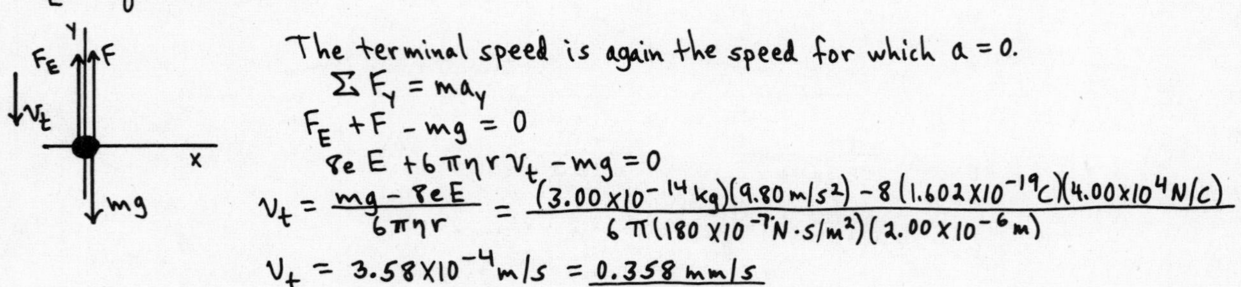

The terminal speed is again the speed for which $a=0$.

$$\Sigma F_y = ma_y$$
$$F_E + F - mg = 0$$
$$8eE + 6\pi\eta r v_t - mg = 0$$

$$v_t = \frac{mg - 8eE}{6\pi\eta r} = \frac{(3.00\times10^{-14}\ \text{kg})(9.80\ \text{m/s}^2) - 8(1.602\times10^{-19}\ \text{C})(4.00\times10^{4}\ \text{N/C})}{6\pi(180\times10^{-7}\ \text{N}\cdot\text{s/m}^2)(2.00\times10^{-6}\ \text{m})}$$

$$v_t = 3.58\times10^{-4}\ \text{m/s} = \underline{0.358\ \text{mm/s}}$$

24-31

a) Calculate the acceleration of the electron produced by the electric force: $\vec{F}_E = q\vec{E}$. Since $q=-e$ is negative $\vec{F}_E$ and $\vec{E}$ are in opposite directions; $\vec{E}$ is upward so $\vec{F}_E$ is downward. The magnitude of F_E is $F_E = |q|E = eE$.

X; a; $F_E = eE$; Y

$$\Sigma F_y = ma_y$$
$$F_E = ma$$
$$eE = ma$$

$$a = \frac{eE}{m} = \frac{(1.602\times10^{-19}\ \text{C})(6.00\times10^{3}\ \text{N/C})}{9.109\times10^{-31}\ \text{kg}}$$
$$a = 1.055\times10^{15}\ \text{m/s}^2$$

(This is much larger than $g = 9.80\ \text{m/s}^2$, so the gravity force on the electron can be neglected.)

The acceleration is constant and downward, so the motion is like that of a projectile. Use the horizontal motion to find the time and then use the time to find the vertical displacement.

<u>X-component</u>

$v_{0x} = v_0 = 8.00\times10^{6}\ \text{m/s}$

$a_x = 0$

$x - x_0 = 0.040\ \text{m}$

$t = ?$

$$x - x_0 = v_{0x}t + \tfrac{1}{2}a_x t^2 \to 0$$

$$t = \frac{x-x_0}{v_{0x}} = \frac{0.040\ \text{m}}{8.00\times10^{6}\ \text{m/s}} = 5.0\times10^{-9}\ \text{s}$$

24-31 (cont)

y-component

$v_{oy} = 0$

$a_y = 1.055 \times 10^{15}\ \text{m/s}^2$

$t = 5.0 \times 10^{-9}\ \text{s}$

$y - y_0 = ?$

$y - y_0 = \cancel{v_{oy}t} + \frac{1}{2} a_y t^2$

$y - y_0 = \frac{1}{2}(1.055 \times 10^{15}\ \text{m/s}^2)(5.0 \times 10^{-9}\ \text{s})^2 = 0.0132\ \text{m} = \underline{1.32\ \text{cm}}$

b)

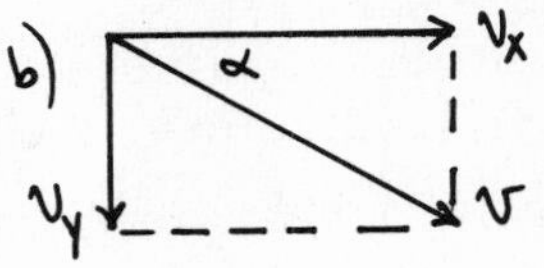

$v_x = v_{ox} = 8.00 \times 10^6\ \text{m/s}$ (since $a_x = 0$)

$v_y = v_{oy} + a_y t = 0 + (1.055 \times 10^{15}\ \text{m/s}^2)(5.0 \times 10^{-9}\ \text{s}) = 5.275 \times 10^6\ \text{m/s}$

$\tan\alpha = \dfrac{v_y}{v_x} = \dfrac{5.275 \times 10^6\ \text{m/s}}{8.00 \times 10^6\ \text{m/s}} = 0.6594 \Rightarrow \alpha = \underline{33.4^\circ}$

c) Consider the motion of the electron after it leaves the region between the plates. Outside the plates there is no electric field, so $a = 0$. (Gravity still can be neglected.) Use the horizontal motion to find the time it takes the electron to travel 0.120 m horizontally to the screen. From this time find the distance downward that the electron travels.

x-component

$a_x = 0$

$v_{ox} = 8.00 \times 10^6\ \text{m/s}$

$x - x_0 = 0.120\ \text{m}$

$t = ?$

$x - x_0 = v_{ox} t + \cancel{\frac{1}{2} a_x t^2}$

$t = \dfrac{x - x_0}{v_{ox}} = \dfrac{0.120\ \text{m}}{8.00 \times 10^6\ \text{m/s}} = 1.50 \times 10^{-8}\ \text{s}$

y-component

$t = 1.50 \times 10^{-8}\ \text{s}$

$a_y = 0$

$v_{oy} = 5.275 \times 10^6\ \text{m/s}$ (from part (b))

$y - y_0 = ?$

$y - y_0 = v_{oy} t + \cancel{\frac{1}{2} a_y t^2}$

$y - y_0 = (5.275 \times 10^6\ \text{m/s})(1.50 \times 10^{-8}\ \text{s}) = 0.0791\ \text{m}$

$y - y_0 = 7.91\ \text{cm}$

The electron travels downward a distance 1.32 cm while it is between the plates and a distance 7.91 cm while traveling from the edge of the plates to the screen. The total downward deflection is $1.32\ \text{cm} + 7.91\ \text{cm} = \underline{9.23\ \text{cm}}$.

Problems

24-37

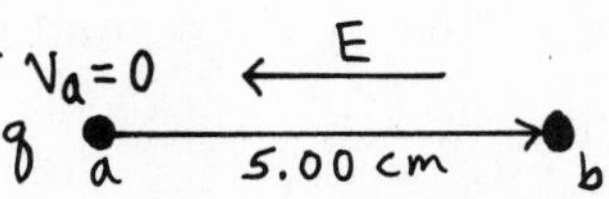

a) $W_{tot} = \Delta K = K_b - K_a = K_b$

$W_{tot} = K_b = 7.00 \times 10^{-5}\ \text{J}$

The electric force F_E and the additional force F both do work, so that

$W_{tot} = W_{F_E} + W_F$

$W_{F_E} = W_{tot} - W_F = 7.00 \times 10^{-5}\ \text{J} - 9.00 \times 10^{-5}\ \text{J} = \underline{-2.00 \times 10^{-5}\ \text{J}}$

The electric force is to the left (in the direction of the

24-37 (cont)

electric field since the particle has positive charge). The displacement is to the right, so the electric force does negative work. The additional force F is in the direction of the displacement, so it does positive work.

b) For the work done by the electric force,

$$W_{a\to b} = q(V_a - V_b) \Rightarrow V_a - V_b = \frac{W_{a\to b}}{q} = \frac{-2.00\times10^{-5}\ \text{J}}{3.00\times10^{-9}\ \text{C}} = \underline{-6.67\times10^{3}\ \text{V}}$$

The starting point (point a) is at 6.67×10^{3} V lower potential than the ending point (point b). We know that $V_b > V_a$ because the electric field always points from high potential toward low potential.

c) Since the electric field is uniform and directed opposite to the displacement $W_{a\to b} = -F_E d = -qEd$, where $d = 5.00$ cm is the displacement of the particle.

$$E = -\frac{W_{a\to b}}{qd} = -\frac{V_a - V_b}{d} = -\frac{-6.67\times10^{3}\ \text{V}}{0.0500\ \text{m}} = \underline{1.33\times10^{5}\ \text{V/m}}$$

24-39

a)

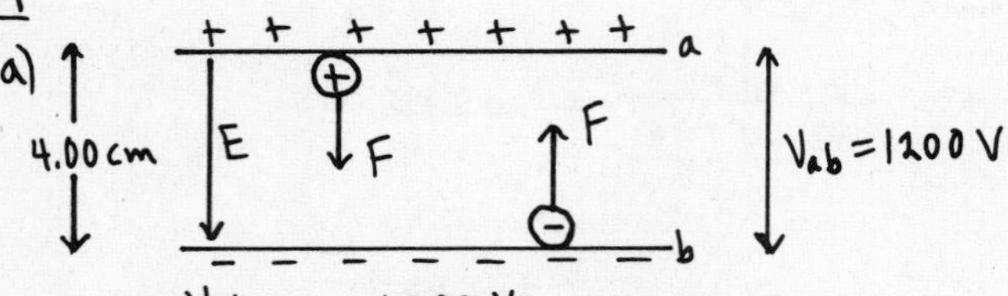

Calculate the electric field E, from that the force, and then the acceleration of each particle. Then we can use the constant acceleration kinematic equations.

$$E = \frac{V_{ab}}{d} = \frac{1200\ \text{V}}{0.0400\ \text{m}} = 3.00\times10^{4}\ \text{V/m}$$

$$F = |q|E = eE = (1.602\times10^{-19}\ \text{C})(3.00\times10^{4}\ \text{V/m}) = 4.806\times10^{-15}\ \text{N}$$

For the electron $a_e = \frac{F}{m_e} = \frac{4.806\times10^{-15}\ \text{N}}{9.109\times10^{-31}\ \text{kg}} = 5.276\times10^{15}\ \text{m/s}^2$

For the proton $a_p = \frac{F}{m_p} = \frac{4.806\times10^{-15}\ \text{N}}{1.673\times10^{-27}\ \text{kg}} = 2.873\times10^{12}\ \text{m/s}^2$

Consider the motion of each object:

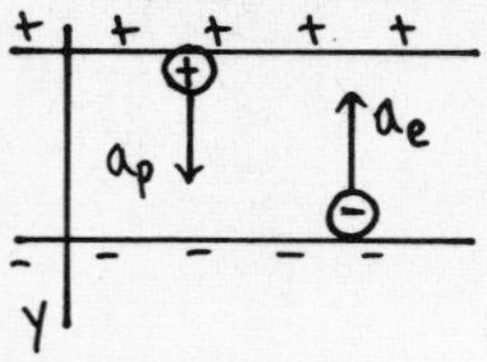

Use a coordinate system with the origin at the positive plate and the +y-direction toward the negative plate. At $t=0$ the proton is at $y_0 = 0$ and the electron is at $y_0 = +0.0400$ m.

proton

$v_{0y} = 0$

$y - y_0 = d$

$a_y = a_p$

$y - y_0 = \cancel{v_{0y}t} + \frac{1}{2}a_y t^2$

$\boxed{d = \frac{1}{2}a_p t^2}$

(d is the distance from the positive plate where they pass each other)

electron

$v_{0y} = 0$

$y = d$ (puts electron and proton at same place)

$y - y_0 = d - 0.0400$ m

$a_y = -a_e$

$y - y_0 = \cancel{v_{0y}t} + \frac{1}{2}a_y t^2$

$\boxed{d - 0.0400\ \text{m} = -\frac{1}{2}a_e t^2}$

24-39 (cont)

Combine these two equations to eliminate t:

$\frac{1}{2}t^2 = \frac{d}{a_p}$ and $\frac{1}{2}t^2 = \frac{0.0400\text{ m} - d}{a_e} \Rightarrow \frac{d}{a_p} = \frac{0.0400\text{ m} - d}{a_e}$

$$d = \left(\frac{a_p}{a_p + a_e}\right)(0.0400\text{ m}) = \left(\frac{2.873\times10^{12}\text{ m/s}^2}{2.873\times10^{12}\text{ m/s}^2 + 5.276\times10^{15}\text{ m/s}^2}\right)(0.0400\text{ m}) = \underline{2.18\times10^{-5}\text{ m}}$$

b) $K_1 + qV_1 = K_2 + qV_2$

released from rest $\Rightarrow K_1 = 0$, so $K_2 = q(V_1 - V_2)$

proton: $q = +e$, $V_1 = V_a$, $V_2 = V_b$ (starts at a, goes to b)

$K_2(\text{proton}) = +e(V_a - V_b)$

electron: $q = -e$, $V_1 = V_b$, $V_2 = V_a$ (starts at b, goes to a)

$K_2(\text{electron}) = -e(V_b - V_a) = e(V_a - V_b)$

Thus the kinetic energies are the same.

$K = \frac{1}{2}mv^2 \Rightarrow m_e v_e^2 = m_p v_p^2$

$$\frac{v_e}{v_p} = \sqrt{\frac{m_p}{m_e}} = \sqrt{\frac{1.673\times10^{-27}\text{ kg}}{9.109\times10^{-31}\text{ kg}}} = 42.9,$$ with the electron speed being larger

c) As derived in part (b) the kinetic energies are the same.

24-41

a) $V = Cx^{4/3}$

$$C = \frac{V}{x^{4/3}} = \frac{160\text{ V}}{(12.0\times10^{-3}\text{ m})^{4/3}} = \underline{5.82\times10^4\text{ V/m}^{4/3}}$$

b) $E_x = -\frac{\partial V}{\partial x} = -\frac{4}{3}Cx^{1/3}$

(The minus sign means that E_x is in the $-x$-direction, which means that $\vec{E}$ points from the positive anode toward the negative cathode.)

c) $\vec{F} = q\vec{E} \Rightarrow F_x = -eE_x = +\frac{4}{3}eCx^{1/3}$

Halfway between the electrodes $\Rightarrow x = 6.00\times10^{-3}$ m

$$F_x = \frac{4}{3}(1.602\times10^{-19}\text{ C})(5.82\times10^4\text{ V/m}^{4/3})(6.00\times10^{-3}\text{ m})^{1/3} = \underline{2.26\times10^{-15}\text{ N}}$$

F_x is positive $\Rightarrow$ the force is directed toward the positive anode.

24-45

a) Problem 24-43 derived that $E = \frac{V_{ab}}{\ln(b/a)}\frac{1}{r}$, where a is the radius of the inner cylinder (wire) and b is the radius of the outer hollow cylinder. The potential difference between the two cylinders is V_{ab}. Midway between the wire and the cylinder wall $\Rightarrow r = \frac{a+b}{2} = \frac{80.0\times10^{-6}\text{ m} + 0.120\text{ m}}{2} = 0.06004$ m.

$$\Rightarrow E = \frac{60.0\times10^3\text{ V}}{\ln\left(\frac{0.120\text{ m}}{80.0\times10^{-6}\text{ m}}\right)(0.06004\text{ m})} = \underline{1.37\times10^5\text{ V/m}}$$

24-45 (cont)

b) $F_E = 10mg$

$|q|E = 10mg$

$\Rightarrow |q| = \dfrac{10mg}{E} = \dfrac{10(30.0\times10^{-9}\text{ kg})(9.80\text{ m/s}^2)}{1.37\times10^5\text{ V/m}} = \underline{2.15\times10^{-11}\text{ C}}$

24-47

a) From Problem 23-24, $E(r) = \dfrac{\lambda r}{2\pi\epsilon_0 R^2}$ for $r \le R$ (inside the cylindrical charge distribution)

and $E(r) = \dfrac{\lambda}{2\pi\epsilon_0 r}$ for $r \ge R$. Let $V=0$ at $r=R$ (at the surface of the cylinder). Use $V_a - V_b = \int_a^b \vec{E}\cdot d\vec{l}$.

$\underline{r \ge R}$

Take point a to be at R and point b to be at r, where $r>R$. Let $d\vec{l} = d\vec{r}$. $\vec{E}$ and $d\vec{r}$ both radially outward $\Rightarrow \vec{E}\cdot d\vec{r} = E\,dr$. Thus $V_R - V_r = \int_R^r E\,dr$. $V_R = 0$ gives $V_r = -\int_R^r E\,dr$. In this interval $(r>R)$ $E(r) = \dfrac{\lambda}{2\pi\epsilon_0 r}$, so

$V_r = -\int_R^r \dfrac{\lambda}{2\pi\epsilon_0 r}\,dr = -\dfrac{\lambda}{2\pi\epsilon_0}\int_R^r \dfrac{dr}{r} = -\dfrac{\lambda}{2\pi\epsilon_0}\ln\left(\dfrac{r}{R}\right)$.

Note: This expression gives $V_r = 0$ when $r=R$ and the potential decreases (becomes a larger negative number) with increasing distance from the cylinder.

$\underline{r \le R}$

Take point a at r, where $r<R$, and point b at R. $\vec{E}\cdot d\vec{r} = E\,dr$ as before.

$\Rightarrow V_r - V_R = \int_r^R E\,dr$. $V_R = 0$ then gives $V_r = \int_r^R E\,dr$. In this interval $(r<R)$ $E(r) = \dfrac{\lambda r}{2\pi\epsilon_0 R^2}$ and $V_r = \int_r^R \dfrac{\lambda r}{2\pi\epsilon_0 R^2}\,dr = \dfrac{\lambda}{2\pi\epsilon_0 R^2}\int_r^R r\,dr = \dfrac{\lambda}{2\pi\epsilon_0 R^2}\left(\dfrac{R^2}{2} - \dfrac{r^2}{2}\right)$

$V_r = \dfrac{\lambda}{4\pi\epsilon_0}\left(1 - \left(\dfrac{r}{R}\right)^2\right)$

Note: This expression also gives $V_r = 0$ when $r=R$. The potential is $+\dfrac{\lambda}{4\pi\epsilon_0}$ at $r=0$ and decreases with increasing r.

b)

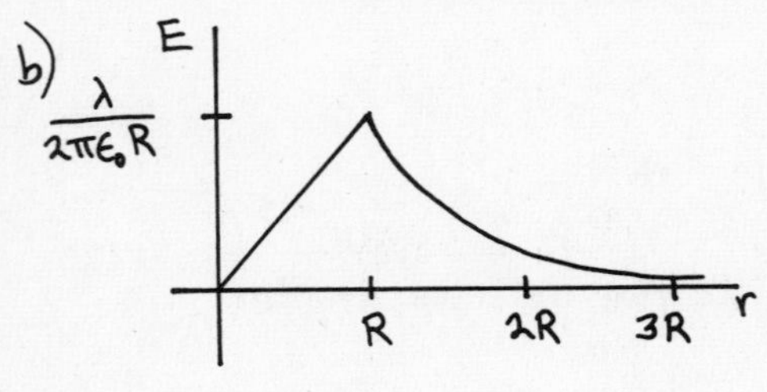

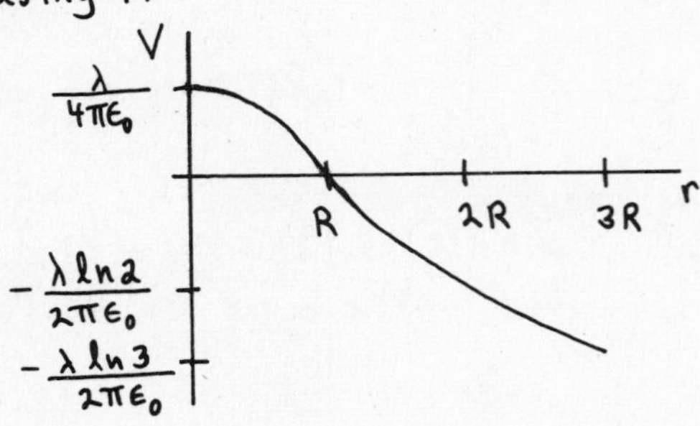

24-49

a) y′, x′, $r = x + a - x'$, dx′, dQ, P, x, a

Use coordinates with the origin at the left-hand end of the rod and one axis along the rod. Call the axes x' and y' so as not to confuse them with the x distance given in the problem.

24-49 (cont)

Slice the charged rod up into thin slices of width dx'. Each slice has charge $dQ = Q\frac{dx'}{a}$ and is a distance $r = x+a-x'$ from point P. The potential at P due to the small slice dQ is $dV = \frac{1}{4\pi\epsilon_0}\frac{dQ}{r} = \frac{1}{4\pi\epsilon_0}\frac{Q}{a}\frac{dx'}{x+a-x'}$.

Compute the total V at P due to the whole rod by integrating dV over the length of the rod ($x'=0$ to $x'=a$):

$$V = \int dV = \frac{Q}{4\pi\epsilon_0 a}\int_0^a \frac{dx'}{(x+a-x')} = \frac{Q}{4\pi\epsilon_0 a}\left(-\ln(x+a-x')\Big|_0^a\right) = \frac{Q}{4\pi\epsilon_0 a}\ln\left(\frac{x+a}{x}\right).$$

Note: As $x\to\infty$, $V \to \frac{Q}{4\pi\epsilon_0 a}\ln\left(\frac{x}{x}\right) = 0$.

b) 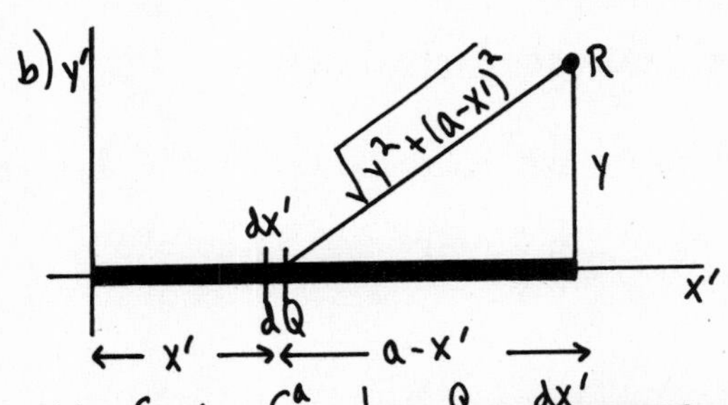

$dQ = \frac{Q}{a}dx'$ as in part (a)

Each slice dQ is a distance $r = \sqrt{y^2+(a-x')^2}$ from point R.

The potential dV at R due to the small slice dQ is $dV = \frac{1}{4\pi\epsilon_0}\frac{dQ}{r} = \frac{1}{4\pi\epsilon_0}\frac{Q}{a}\frac{dx'}{\sqrt{y^2+(a-x')^2}}$

$$V = \int dV = \int_0^a \frac{1}{4\pi\epsilon_0}\frac{Q}{a}\frac{dx'}{\sqrt{y^2+(a-x')^2}} = \frac{Q}{4\pi\epsilon_0 a}\int_0^a \frac{dx'}{\sqrt{y^2+(a-x')^2}}$$

In the integral make the change of variable $u = a - x'$; $du = -dx'$

$$V = -\frac{Q}{4\pi\epsilon_0 a}\int_a^0 \frac{du}{\sqrt{y^2+u^2}} = -\frac{Q}{4\pi\epsilon_0 a}\left[\ln(u+\sqrt{y^2+u^2})\Big|_a^0\right] = -\frac{Q}{4\pi\epsilon_0 a}\left[\ln y - \ln(a+\sqrt{y^2+a^2})\right]$$

$V = \frac{Q}{4\pi\epsilon_0 a}\left[\ln\frac{a+\sqrt{a^2+y^2}}{y}\right]$ (The expression for the integral was found in Appendix B.)

Note: As $y\to\infty$, $V \to \frac{Q}{4\pi\epsilon_0 a}\ln\left(\frac{y}{y}\right) = 0$.

c) part (a): $V = \frac{Q}{4\pi\epsilon_0 a}\ln\left(\frac{x+a}{x}\right) = \frac{Q}{4\pi\epsilon_0 a}\ln\left(1+\frac{a}{x}\right)$

From Appendix B, $\ln(1+u) = u - \frac{u^2}{2} + \dots \Rightarrow \ln\left(1+\frac{a}{x}\right) = \frac{a}{x} - \frac{a^2}{2x^2} \to \frac{a}{x}$ when x is large.

Thus $V \to \frac{Q}{4\pi\epsilon_0 a}\left(\frac{a}{x}\right) = \frac{Q}{4\pi\epsilon_0 x}$.

For large x, V becomes the potential of a point charge.

part (b): $V = \frac{Q}{4\pi\epsilon_0 a}\ln\left[\frac{a+\sqrt{a^2+y^2}}{y}\right] = \frac{Q}{4\pi\epsilon_0 a}\ln\left(\frac{a}{y}+\sqrt{1+\frac{a^2}{y^2}}\right)$

From the binomial theorem (Appendix B) $\sqrt{1+\frac{a^2}{y^2}} = \left(1+\frac{a^2}{y^2}\right)^{1/2} = 1 + \frac{a^2}{2y^2} + \dots$

$\frac{a}{y} + \sqrt{1+\frac{a^2}{y^2}} \to 1 + \frac{a}{y} + \frac{a^2}{2y^2} \to 1 + \frac{a}{y}$ (neglect $\frac{a^2}{2y^2}$)

Thus $V \to \frac{Q}{4\pi\epsilon_0 a}\ln\left(1+\frac{a}{y}\right) \to \frac{Q}{4\pi\epsilon_0 a}\left(\frac{a}{y}\right) = \frac{Q}{4\pi\epsilon_0 y}$.

For large y, V becomes the potential of a point charge.

24-53

a) The potential at the surface of a charged conducting sphere is given by Eq. (24-17): $V = \frac{1}{4\pi\epsilon_0}\frac{q}{R}$. For spheres A and B this gives $V_A = \frac{Q_A}{4\pi\epsilon_0 R_A}$ and $V_B = \frac{Q_B}{4\pi\epsilon_0 R_B}$. $V_A = V_B \Rightarrow \frac{Q_A}{4\pi\epsilon_0 R_A} = \frac{Q_B}{4\pi\epsilon_0 R_B}$.

24-53 (cont) $\dfrac{Q_B}{Q_A} = \dfrac{R_B}{R_A}$

But $R_A = 3R_B \Rightarrow \dfrac{Q_B}{Q_A} = \dfrac{1}{3}$

b) The electric field at the surface of a charged conducting sphere is given by Eq. (24-18): $E = \dfrac{1}{4\pi\epsilon_0}\dfrac{|q|}{R^2}$. For spheres A and B this gives $E_A = \dfrac{|Q_A|}{4\pi\epsilon_0 R_A^2}$ and $E_B = \dfrac{|Q_B|}{4\pi\epsilon_0 R_B^2}$. $\dfrac{E_B}{E_A} = \dfrac{|Q_B|}{4\pi\epsilon_0 R_B^2}\,\dfrac{4\pi\epsilon_0 R_A^2}{|Q_A|} = \left|\dfrac{Q_B}{Q_A}\right|\left(\dfrac{R_A}{R_B}\right)^2 = \frac{1}{3}(3)^2 = 3$

24-55

Problem 23-25 dealt with the spherical charge distribution
$\rho = \rho_0(1 - r/R)$ for $r \le R$
and $\rho = 0$ for $r \ge R$; $\rho_0 = \dfrac{3Q}{\pi R^3}$ where Q is the total charge in the distribution.
It was shown that
$E(r) = \dfrac{Qr}{4\pi\epsilon_0 R^3}\left(4 - 3\dfrac{r}{R}\right)$ for $r \le R$ and $E(r) = \dfrac{Q}{4\pi\epsilon_0 r^2}$ for $r \ge R$.

a) $V_a - V_b = \int_a^b \vec{E}\cdot d\vec{l}$

Take b to be at r and a to be at ∞. Then $V_a = V_\infty = 0$. Take $d\vec{l} = d\vec{r}$, so $\vec{E}\cdot d\vec{r} = E\,dr$. In this region $(r > R)$, $E = \dfrac{Q}{4\pi\epsilon_0 r^2}$.

$$\cancel{V_\infty}^{\,0} - V_r = \int_\infty^r E\,dr \Rightarrow V_r = \int_r^\infty E\,dr = \frac{Q}{4\pi\epsilon_0}\int_r^\infty \frac{dr}{r^2} = \frac{Q}{4\pi\epsilon_0}\left(-\frac{1}{r}\Big|_r^\infty\right) = \frac{Q}{4\pi\epsilon_0 r}.$$

This is the same expression as for a point charge Q.
Note: when $r = R$, $V_r = \dfrac{Q}{4\pi\epsilon_0 R}$.

b) $V_a - V_b = \int_a^b \vec{E}\cdot d\vec{l}$

Take b to be at R and a to be at r, where $r < R$.
From part (a), $V_b = V_R = \dfrac{Q}{4\pi\epsilon_0 R}$

$\vec{E}\cdot d\vec{r} = E\,dr$ as before, and in this region $(r<R)$ $E = \dfrac{Qr}{4\pi\epsilon_0 R^3}\left(4 - 3\dfrac{r}{R}\right)$.

$$V_r - V_R = \int_r^R E\,dr$$

$$V_r - \frac{Q}{4\pi\epsilon_0 R} = \frac{Q}{4\pi\epsilon_0 R^3}\int_r^R r\left(4 - \frac{3r}{R}\right)dr = \frac{Q}{4\pi\epsilon_0 R^3}\left[2r^2 - \frac{r^3}{R}\right]\Big|_r^R$$

$$V_r - \frac{Q}{4\pi\epsilon_0 R} = \frac{Q}{4\pi\epsilon_0 R^3}\left(2R^2 - R^2 - 2r^2 + \frac{r^3}{R}\right) = \frac{Q}{4\pi\epsilon_0 R}\left(1 - 2\left(\frac{r}{R}\right)^2 + \left(\frac{r}{R}\right)^3\right)$$

$$V_r = \frac{Q}{4\pi\epsilon_0 R}\left(2 - 2\left(\frac{r}{R}\right)^2 + \left(\frac{r}{R}\right)^3\right) = \frac{Q}{2\pi\epsilon_0 R}\left(1 - \left(\frac{r}{R}\right)^2 + \frac{1}{2}\left(\frac{r}{R}\right)^3\right)$$

Note: For $r = R$, $V_r = \dfrac{Q}{2\pi\epsilon_0 R}\left(1 - 1 + \frac{1}{2}\right) = \dfrac{Q}{4\pi\epsilon_0 R}$, the same result as obtained in part (a) when $r = R$.

CHAPTER 25

Exercises 1, 5, 7, 13, 15, 17, 19, 21

Problems 25, 27, 29, 31, 35

Exercises

25-1

a) $C = \dfrac{Q}{V_{ab}} \Rightarrow V_{ab} = \dfrac{Q}{C} = \dfrac{0.200 \times 10^{-6}\text{ C}}{500 \times 10^{-12}\text{ F}} = \underline{400\text{ V}}$

b) $C = \dfrac{\epsilon_0 A}{d} \Rightarrow A = \dfrac{Cd}{\epsilon_0} = \dfrac{(500 \times 10^{-12}\text{ F})(0.400 \times 10^{-3}\text{ m})}{8.854 \times 10^{-12}\text{ C}^2/\text{N}\cdot\text{m}^2} = \underline{0.0226\text{ m}^2}$

c) $V_{ab} = Ed \Rightarrow E = \dfrac{V_{ab}}{d} = \dfrac{400\text{ V}}{0.400 \times 10^{-3}\text{ m}} = \underline{1.00 \times 10^{6}\text{ V/m}}$

d) $E = \dfrac{\sigma}{\epsilon_0} \Rightarrow \sigma = E\epsilon_0 = (1.00 \times 10^6\text{ V/m})(8.854 \times 10^{-12}\text{ C}^2/\text{N}\cdot\text{m}^2) = \underline{8.85 \times 10^{-6}\text{ C/m}^2}$

or $\sigma = \dfrac{Q}{A} = \dfrac{0.200 \times 10^{-6}\text{ C}}{0.0226\text{ m}^2} = 8.85 \times 10^{-6}\text{ C/m}^2$, which checks

25-5

Do parts (a) and (b) together.

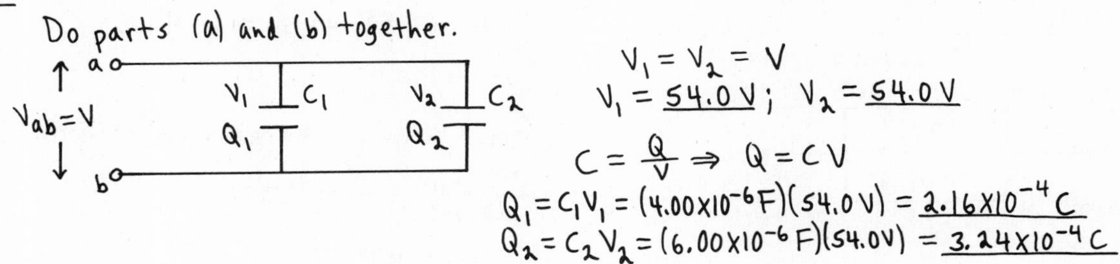

$V_1 = V_2 = V$

$V_1 = \underline{54.0\text{ V}}$; $V_2 = \underline{54.0\text{ V}}$

$C = \dfrac{Q}{V} \Rightarrow Q = CV$

$Q_1 = C_1V_1 = (4.00 \times 10^{-6}\text{ F})(54.0\text{ V}) = \underline{2.16 \times 10^{-4}\text{ C}}$

$Q_2 = C_2V_2 = (6.00 \times 10^{-6}\text{ F})(54.0\text{ V}) = \underline{3.24 \times 10^{-4}\text{ C}}$

25-7

Do parts (a) and (b) together.

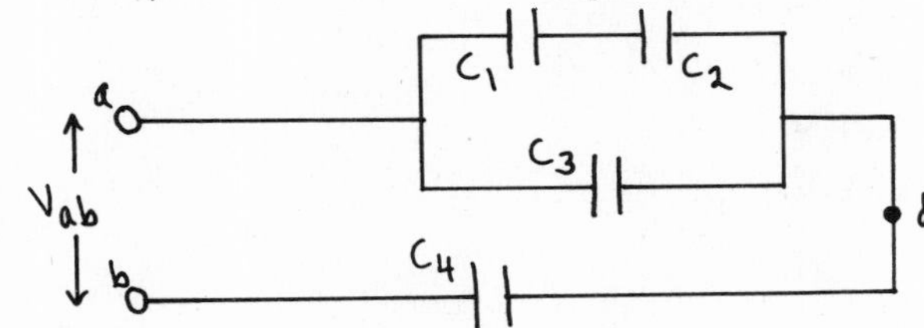

$C_1 = C_2 = C_3 = C_4 = 2.00\ \mu\text{F}$

$V_{ab} = 36.0\text{ V}$

Simplify the circuit by replacing the capacitor combinations by their equivalents:

C_1 and C_2 in series $=$ C_{12}

$\dfrac{1}{C_{12}} = \dfrac{1}{C_1} + \dfrac{1}{C_2} = \dfrac{C_1 + C_2}{C_1C_2}$

$C_{12} = \dfrac{C_1C_2}{C_1 + C_2} = \dfrac{(2.00 \times 10^{-6}\text{ F})(2.00 \times 10^{-6}\text{ F})}{2.00 \times 10^{-6}\text{ F} + 2.00 \times 10^{-6}\text{ F}} = 1.00 \times 10^{-6}\text{ F}$

C_{12} and C_3 in parallel $=$ C_{123}

$C_{123} = C_{12} + C_3 = 1.00 \times 10^{-6}\text{ F} + 2.00 \times 10^{-6}\text{ F} = 3.00 \times 10^{-6}\text{ F}$

25-7 (cont)

$$\frac{1}{C_{1234}} = \frac{1}{C_{123}} + \frac{1}{C_4} = \frac{C_{123}+C_4}{C_{123}C_4}$$

$$C_{1234} = \frac{C_{123}C_4}{C_{123}+C_4} = \frac{(3.00\times10^{-6}\,F)(2.00\times10^{-6}\,F)}{3.00\times10^{-6}\,F+2.00\times10^{-6}\,F} = 1.20\times10^{-6}\,F$$

The circuit is then equivalent to

$$V_{1234} = V = 36.0\,V$$

$$Q_{1234} = C_{1234}V = (1.20\times10^{-6}\,F)(36.0\,V) = 43.2\,\mu C$$

Now build back up the original circuit, step by step:

$Q_{123} = Q_4 = Q_{1234} = 43.2\,\mu C$ (charge same for capacitors in series)

Then $V_{123} = \frac{Q_{123}}{C_{123}} = \frac{43.2\,\mu C}{3.00\,\mu F} = 14.4\,V$,

$$V_4 = \frac{Q_4}{C_4} = \frac{43.2\,\mu C}{2.00\,\mu F} = 21.6\,V.$$

Note that $V_4 + V_{123} = 21.6\,V + 14.4\,V = 36.0\,V$, as it should.

$$V_3 = V_{12} = 36.0\,V - V_4 = 14.4\,V$$

$$Q_3 = C_3V_3 = (2.00\times10^{-6}\,F)(14.4\,V) = 28.8\,\mu C$$

$$Q_{12} = C_{12}V_{12} = (1.00\times10^{-6}\,F)(14.4\,V) = 14.4\,\mu C$$

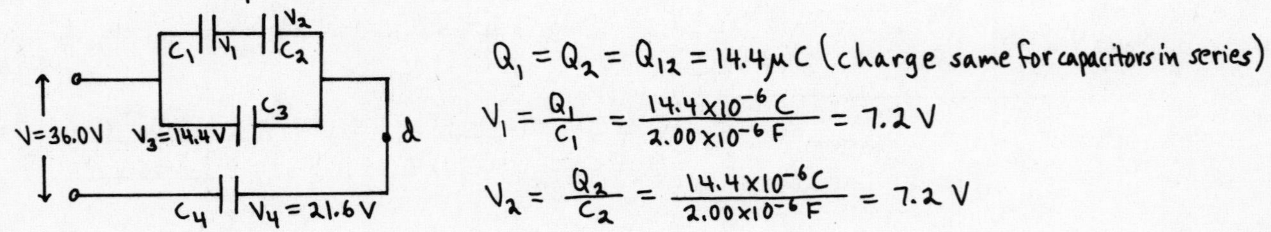

$Q_1 = Q_2 = Q_{12} = 14.4\,\mu C$ (charge same for capacitors in series)

$$V_1 = \frac{Q_1}{C_1} = \frac{14.4\times10^{-6}\,C}{2.00\times10^{-6}\,F} = 7.2\,V$$

$$V_2 = \frac{Q_2}{C_2} = \frac{14.4\times10^{-6}\,C}{2.00\times10^{-6}\,F} = 7.2\,V$$

Note that $V_1 + V_2 = 14.4\,V$, which equals V_3 as it should.

Summary: $Q_1 = 14.4\,\mu C$, $V_1 = 7.2\,V$
$Q_2 = 14.4\,\mu C$, $V_2 = 7.2\,V$
$Q_3 = 28.8\,\mu C$, $V_3 = 14.4\,V$
$Q_4 = 43.2\,\mu C$, $V_4 = 21.6\,V$

d) $V_{ad} = V_3 = \underline{14.4\,V}$

25-13

The energy density is given by Eq. (25-11): $u = \frac{1}{2}\epsilon_0 E^2$

Calculate E: $E = \frac{V}{d} = \frac{400\,V}{4.00\times10^{-3}\,m} = 1.00\times10^5\,V/m$.

Then $u = \frac{1}{2}\epsilon_0 E^2 = \frac{1}{2}(8.854\times10^{-12}\,C^2/N\cdot m^2)(1.00\times10^5\,V/m)^2 = \underline{0.0443\,J/m^3}$

25-15

a) $U = \frac{Q^2}{2C}$; $Q = q$ and $C = \frac{\epsilon_0 A}{x} \Rightarrow U = \frac{xq^2}{2\epsilon_0 A}$

b) $x \rightarrow x + dx \Rightarrow U = \frac{(x+dx)q^2}{2\epsilon_0 A}$

$dW = \frac{(x+dx)q^2}{2\epsilon_0 A} - \frac{xq^2}{2\epsilon_0 A} = \frac{dx\, q^2}{2\epsilon_0 A}$

But also $dW = F\,dx \Rightarrow F = \frac{q^2}{2\epsilon_0 A}$

c) [diagram: plates with + charges on top, − charges on bottom, field E arrows pointing down]

$E = \frac{\sigma}{\epsilon_0} = \frac{q}{\epsilon_0 A} \Rightarrow F = \frac{1}{2}qE$, not qE

The reason is that E is the field due to both plates. If we consider the positive plate only and calculate its electric field using Gauss's law:

[diagram: single positive plate with Gaussian surface of area A; field E arrows pointing up above and down below]

$\oint \vec{E}\cdot d\vec{A} = \frac{Q_{encl}}{\epsilon_0}$

$2EA = \frac{\sigma A}{\epsilon_0} \Rightarrow E = \frac{\sigma}{2\epsilon_0} = \frac{q}{2\epsilon_0 A}$

The force this field exerts on the other plate, that has charge $-q$, is $F = \frac{q^2}{2\epsilon_0 A}$.

25-17

$C = K\epsilon_0 \frac{A}{d}$. Minimum $A \Rightarrow$ smallest possible d.

d is limited by the requirement that E be less than 2.00×10^7 V/m when V is as large as 6000 V.

$V = Ed \Rightarrow d = \frac{V}{E} = \frac{6000\text{ V}}{2.00\times 10^7\text{ V/m}} = 3.00\times 10^{-4}\text{ m}$

Then $A = \frac{Cd}{K\epsilon_0} = \frac{(1.50\times 10^{-9}\text{ F})(3.00\times 10^{-4}\text{ m})}{(3.40)(8.854\times 10^{-12}\text{ C}^2/\text{N}\cdot\text{m}^2)} = \underline{0.0149\text{ m}^2}$

25-19

$E = \frac{E_0}{K} \Rightarrow K = \frac{E_0}{E} = \frac{3.60\times 10^5\text{ V/m}}{1.20\times 10^5\text{ V/m}} = 3.00$

a) $\sigma_i = \sigma\left(1 - \frac{1}{K}\right)$

$\sigma = \epsilon_0 E_0 = (8.854\times 10^{-12}\text{ C}^2/\text{N}\cdot\text{m}^2)(3.60\times 10^5\text{ N/C}) = 3.187\times 10^{-6}\text{ C/m}^2$

$\sigma_i = (3.187\times 10^{-6}\text{ C/m}^2)\left(1 - \frac{1}{3.00}\right) = \underline{2.12\times 10^{-6}\text{ C/m}^2}$

b) As calculated above, $K = \underline{3.00}$.

25-21

a) [diagram: capacitor plates +Q (top) and −Q (bottom) with dielectric K between; field E arrows pointing down; dotted Gaussian surface of area A′ enclosing top plate]

Apply Eq. (25-22) to the dotted surface:

$\oint K\vec{E}\cdot d\vec{A} = \frac{Q_{encl}}{\epsilon_0}$

$\oint K\vec{E}\cdot d\vec{A} = KEA'$ since $E = 0$ outside the plates

$Q_{encl} = \sigma A' = \frac{Q}{A}A'$

25-21 (cont)

$$\text{Thus } KEA' = \frac{Q A'}{\epsilon_0} \Rightarrow E = \frac{Q}{\epsilon_0 A K}$$

b) $V = Ed = \frac{Qd}{\epsilon_0 AK}$

c) $C = \frac{Q}{V} = \frac{Q}{\frac{Qd}{\epsilon_0 AK}} = K\frac{\epsilon_0 A}{d} = KC_0 \Rightarrow K = \frac{C}{C_0}$, which is Eq. (25-12)

Problems

25-25

a) $C = \frac{\epsilon_0 A}{d} = \frac{(8.854\times10^{-12}\ C^2/N\cdot m^2)(0.200\ m)^2}{1.20\times10^{-2}\ m} = 2.95\times10^{-11}\ F = \underline{29.5\ pF}$

b) Remains connected to the battery $\Rightarrow$ V stays 50.0 V
$Q = CV = (2.95\times10^{-11}\ F)(50.0\ V) = \underline{1.48\times10^{-9}\ C}$

c) $E = \frac{V}{d} = \frac{50.0\ V}{1.20\times10^{-2}\ m} = \underline{4.17\times10^{3}\ V/m}$

d) $U = \frac{1}{2}QV = \frac{1}{2}(1.48\times10^{-9}\ C)(50.0\ V) = \underline{3.70\times10^{-8}\ J}$

25-27

a)

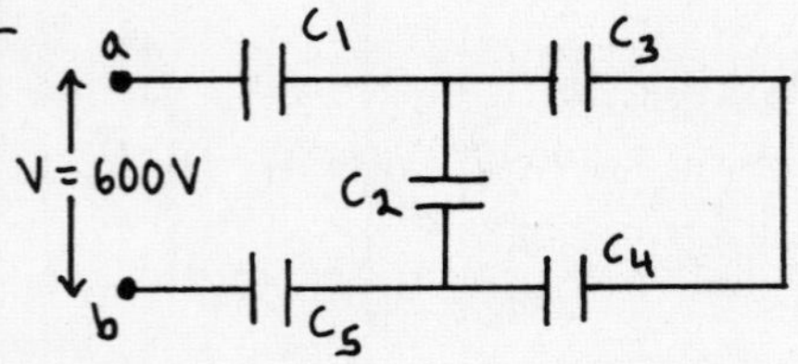

$C_1 = C_5 = 3.00\ \mu F$

$C_2 = C_3 = C_4 = 2.00\ \mu F$

Simplify the circuit by replacing the capacitor combinations by their equivalents:

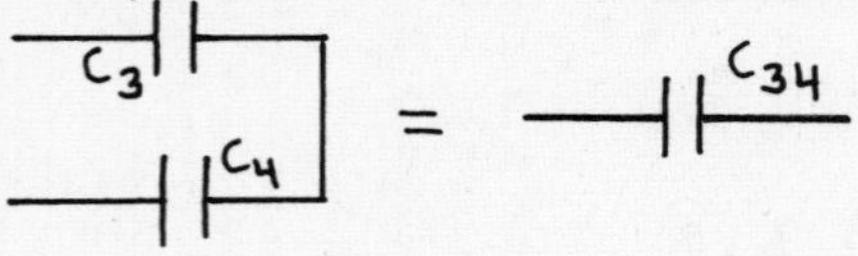

$$\frac{1}{C_{34}} = \frac{1}{C_3} + \frac{1}{C_4} = \frac{C_3 + C_4}{C_3 C_4}$$

$$C_{34} = \frac{C_3 C_4}{C_3 + C_4} = \frac{(2.00\times10^{-6}\ F)(2.00\times10^{-6}\ F)}{2.00\times10^{-6}\ F + 2.00\times10^{-6}\ F} = 1.00\times10^{-6}\ F$$

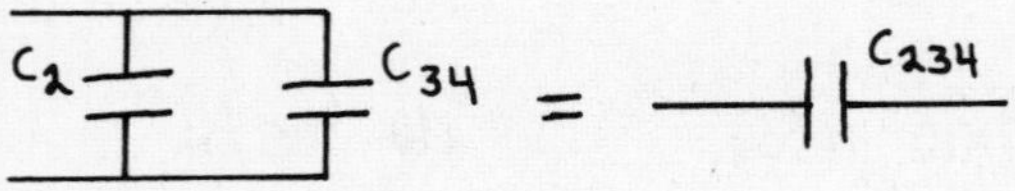

$C_{234} = C_2 + C_{34} = 2.00\ \mu F + 1.00\ \mu F = 3.00\ \mu F$

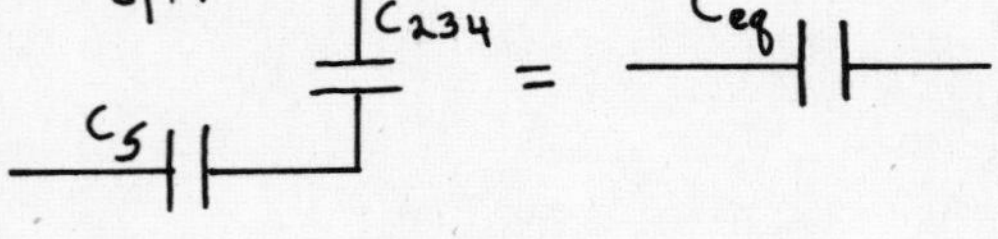

$$\frac{1}{C_{eq}} = \frac{1}{C_1} + \frac{1}{C_5} + \frac{1}{C_{234}} = \frac{3}{3.00\ \mu F}$$

$C_{eq} = \underline{1.00\ \mu F}$

b) Now build the original circuit back up, piece by piece:

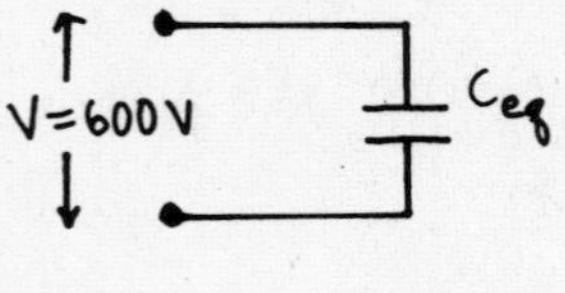

$Q_{eq} = C_{eq}V = (1.00\ \mu F)(600\ V) = 600\ \mu C$

$Q_1 = Q_5 = Q_{234} = 600\ \mu C$ (capacitors in series have same charge)

$V_1 = \frac{Q_1}{C_1} = \frac{600\ \mu C}{3.00\ \mu F} = 200\ V$

25-27 (cont)

$$V_5 = \frac{Q_5}{C_5} = \frac{600\,\mu C}{3.00\,\mu F} = 200\,V$$

$$V_{234} = \frac{Q_{234}}{C_{234}} = \frac{600\,\mu C}{3.00\,\mu F} = 200\,V$$

$V_2 = V_{34} = V_{234} = 200\,V$ (capacitors in parallel have the same potential)

$$Q_2 = C_2 V_2 = (2.00\,\mu F)(200\,V) = 400\,\mu C$$

$$Q_{34} = C_{34} V_{34} = (1.00\,\mu F)(200\,V) = 200\,\mu C$$

$Q_3 = Q_4 = Q_{34} = 200\,\mu C$ (capacitors in series have same charge)

$$V_3 = \frac{Q_3}{C_3} = \frac{200\,\mu C}{2.00\,\mu F} = 100\,V$$

$$V_4 = \frac{Q_4}{C_4} = \frac{200\,\mu C}{2.00\,\mu F} = 100\,V$$

Summary:
$Q_1 = 600\,\mu C$, $V_1 = 200\,V$
$Q_2 = 400\,\mu C$, $V_2 = 200\,V$
$Q_3 = 200\,\mu C$, $V_3 = 100\,V$
$Q_4 = 200\,\mu C$, $V_4 = 100\,V$
$Q_5 = 600\,\mu C$, $V_5 = 200\,V$

Note: $V_3 + V_4 = V_2$
$V_1 + V_2 + V_5 = 600\,V$
$Q_1 = Q_2 + Q_3$
$Q_5 = Q_2 + Q_4$

25-29

$C_1 = 9.00\,\mu F$
$C_2 = 6.00\,\mu F$

a) Simplify the network by replacing the capacitor combinations by their equivalents:

$$\frac{1}{C_{eq}} = \frac{3}{C_1} \Rightarrow C_{eq} = \frac{C_1}{3} = \frac{9.00\,\mu F}{3} = 3.00\,\mu F$$

$$C_{eq} = 3.00\,\mu F + C_2 = 3.00\,\mu F + 6.00\,\mu F = 9.00\,\mu F$$

$$\frac{1}{C_{eq}} = \frac{2}{C_1} + \frac{1}{9.00\,\mu F} = \frac{3}{9.00\,\mu F}$$

$$C_{eq} = 3.00\,\mu F$$

25-29 (cont)

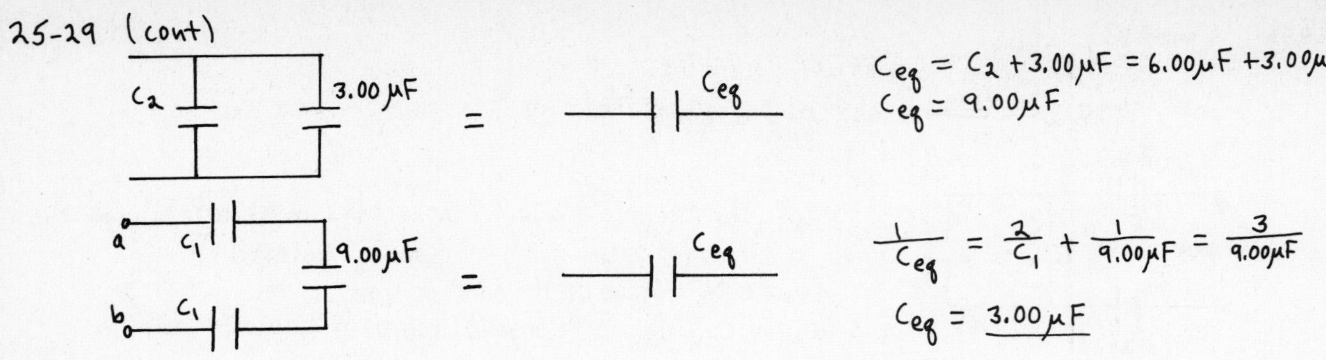

$C_{eq} = C_2 + 3.00\mu F = 6.00\mu F + 3.00\mu F$
$C_{eq} = 9.00\mu F$

$\frac{1}{C_{eq}} = \frac{2}{C_1} + \frac{1}{9.00\mu F} = \frac{3}{9.00\mu F}$

$C_{eq} = \underline{3.00\mu F}$

b)

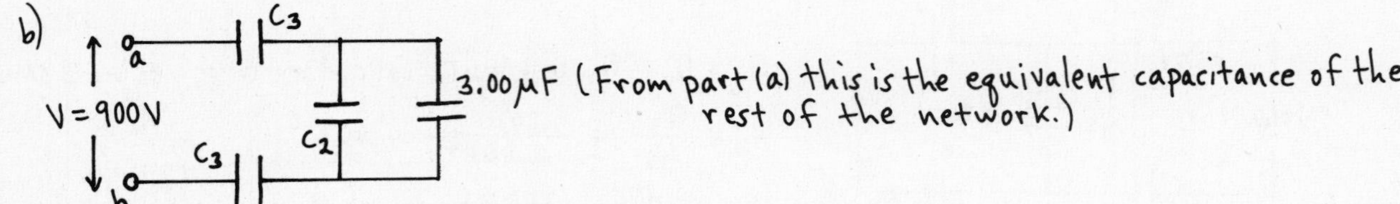

3.00μF (From part (a) this is the equivalent capacitance of the rest of the network.)

The equivalent network is

$C_3 = 9.00\mu F$
V = 900 V
9.00μF
$C_3 = 9.00\mu F$

Series ⇒ all three capacitors have same Q. But here all three have same C, so by $V = \frac{Q}{C}$ all three must have same V. The three V's must add to 900V, so each capacitor has V = 300 V. The 9.00μF capacitor is the equivalent of C_2 and the 3.00μF capacitor in parallel, so $V_2 = 300$ V (capacitors in parallel have same potential difference).

$Q_3 = C_3V_3 = (9.00\mu F)(300 V) = \underline{2.70 \times 10^{-3} C}$
$Q_2 = C_2V_2 = (6.00\mu F)(300 V) = \underline{1.80 \times 10^{-3} C}$

c) From the potentials deduced in part (b) we have

a
C_1
$C_1 = 9.00\mu F$
c
300V
V = 900 V
300 V
C_2
9.00 μF (From part (a) this is the equivalent capacitance of the rest of the network.)
d
b
300V
C_1
$C_1 = 9.00\mu F$

The three right-most capacitors are in series and therefore have the same charge. But their capacitances also are equal, so by $V = \frac{Q}{C}$ they each have the same potential difference. Their potentials must sum to 300 V, so the potential across each is 100 V; $V_{cd} = \underline{100 V.}$

25-31

a)

$C_1 = 8.00\mu F$
$C_2 = 8.00\mu F$
V = 36.0V
$C_3 = 4.00\mu F$
C_{eq}

$\frac{1}{C_{eq}} = \frac{1}{C_1} + \frac{1}{C_2} + \frac{1}{C_3} = \frac{4}{8.00\mu F}$

$C_{eq} = \frac{8.00\mu F}{4} = 2.00\mu F$

25-31 (cont)

$Q = C_{eq}V = (2.00\mu F)(36.0\,V) = 72.0\mu C$

The three capacitors are in series so they each have the same charge;

$Q_1 = Q_2 = Q_3 = \underline{72.0\mu C}$

b) $U = \frac{1}{2}Q_1V_1 + \frac{1}{2}Q_2V_2 + \frac{1}{2}Q_3V_3$

But $Q_1 = Q_2 = Q_3 = Q \Rightarrow U = \frac{1}{2}Q(V_1+V_2+V_3)$

But also $V_1+V_2+V_3 = V = 36.0\,V$

$\Rightarrow U = \frac{1}{2}QV = \frac{1}{2}(72.0\times10^{-6}\,C)(36.0\,V) = \underline{1.30\times10^{-3}\,J}$

c) C_1 + Q_{01}, C_2 + Q_{02}, + C_3 Q_{03} → C_1 + V_1 Q_1, C_2 + V_2 Q_2, C_3 + V_3 Q_3

The total positive charge that is available to be distributed on the upper plates of the three capacitors is $Q_0 = Q_{01} + Q_{02} + Q_{03} = 3(72.0\mu C) = 216\mu C$.

Thus $\boxed{Q_1 + Q_2 + Q_3 = 216\mu C.}$

After the circuit is completed the charge distributes to make $V_1 = V_2 = V_3$.

$V = \frac{Q}{C}$ and $V_1 = V_2 \Rightarrow \frac{Q_1}{C_1} = \frac{Q_2}{C_2}$ and then $C_1 = C_2$ says that $Q_1 = Q_2$.

$V_1 = V_3 \Rightarrow \frac{Q_1}{C_1} = \frac{Q_3}{C_3} \Rightarrow Q_1 = Q_3\left(\frac{C_1}{C_3}\right) = Q_3\left(\frac{8.00\mu F}{4.00\mu F}\right) = 2Q_3$

Use $Q_2 = Q_1$ and $Q_1 = 2Q_3$ in the above equation $\Rightarrow 2Q_3 + 2Q_3 + Q_3 = 216\mu C$

$5Q_3 = 216\mu C \Rightarrow Q_3 = 43.2\mu C$

$Q_1 = Q_2 = 86.4\mu C$

Then $V_1 = \frac{Q_1}{C_1} = \frac{86.4\mu C}{8.00\mu F} = 10.8\,V$

$V_2 = \frac{Q_2}{C_2} = \frac{86.4\mu C}{8.00\mu F} = 10.8\,V$

$V_3 = \frac{Q_3}{C_3} = \frac{43.2\mu C}{4.00\mu F} = 10.8\,V$

The voltage across each capacitor in the parallel combination is $\underline{10.8\,V}$.

d) $U = \frac{1}{2}Q_1V_1 + \frac{1}{2}Q_2V_2 + \frac{1}{2}Q_3V_3$

But $V_1 = V_2 = V_3 \Rightarrow U = \frac{1}{2}V_1(Q_1+Q_2+Q_3) = \frac{1}{2}(10.8\,V)(216\mu C) = \underline{1.17\times10^{-3}\,J}$.

This is less than the original energy of $1.30\times10^{-3}\,J$.

<u>25-35</u>

a +Q; K_1, $\frac{d}{2}$; b; K_2, $\frac{d}{2}$; c −Q

$C = \frac{Q}{V}$, so we need to calculate the effect of the dielectrics on the potential difference between the plates. Let the potential of the positive plate be V_a, the potential of the negative plate be V_c, and the potential midway between the plates where the dielectrics meet be V_b.

Then $C = \frac{Q}{V_a - V_c} = \frac{Q}{V_{ac}}$.

25-35 (cont)

$$V_{ac} = V_{ab} + V_{bc}$$

The electric field in the absence of any dielectric is $E_0 = \frac{Q}{\epsilon_0 A}$. In the first dielectric the electric field is reduced to $E_1 = \frac{E_0}{K_1} = \frac{Q}{K_1 \epsilon_0 A}$.

Thus $V_{ab} = E_1\left(\frac{d}{2}\right) = \frac{Qd}{K_1 2\epsilon_0 A}$.

In the second dielectric the electric field is reduced to $E_2 = \frac{E_0}{K_2} = \frac{Q}{K_2 \epsilon_0 A}$.

Thus $V_{bc} = E_2\left(\frac{d}{2}\right) = \frac{Qd}{K_2 2\epsilon_0 A}$.

Then $V_{ac} = V_{ab} + V_{bc} = \frac{Qd}{K_1 2\epsilon_0 A} + \frac{Qd}{K_2 2\epsilon_0 A} = \frac{Qd}{2\epsilon_0 A}\left(\frac{1}{K_1} + \frac{1}{K_2}\right)$

$$V_{ac} = \frac{Qd}{2\epsilon_0 A}\left(\frac{K_1 + K_2}{K_1 K_2}\right).$$

This gives $C = \frac{Q}{V_{ac}} = Q\,\frac{2\epsilon_0 A}{Qd}\left(\frac{K_1 K_2}{K_1 + K_2}\right) = \frac{2\epsilon_0 A}{d}\left(\frac{K_1 K_2}{K_1 + K_2}\right)$.

CHAPTER 26

Exercises 3, 7, 9, 13, 15, 17, 21, 25, 27, 29, 31

Problems 33, 35, 37, 39, 43

Exercises

26-3

a) $dQ = I\,dt$

$Q = \int_0^{10.0s} (4.00\text{A} + (0.600\ \text{A/s}^2)t^2)\,dt = [(4.00\text{A})t + (0.200\ \text{A/s}^2)t^3]\Big|_0^{10.0s}$

$Q = (4.00\text{A})(10.0s) + (0.200\ \text{A/s}^2)(10.0\,s)^3 = 40.0\text{C} + 200\text{C} = \underline{240\text{C}}$

b) $I = \frac{Q}{t} = \frac{240\text{C}}{10.0s} = \underline{24.0\text{ A}}$

26-7

a) Eq. (26-7): $\rho = \frac{E}{J} \Rightarrow J = \frac{E}{\rho}$

From Table 26-1 the resistivity for aluminum is $2.63 \times 10^{-8}\,\Omega\cdot\text{m}$.

$J = \frac{E}{\rho} = \frac{0.520\text{ V/m}}{2.63\times10^{-8}\Omega\cdot\text{m}} = 1.977\times10^{7}\text{ A/m}^2$

$I = JA = J\pi r^2 = (1.977\times10^7\text{ A/m}^2)\,\pi\,(0.400\times10^{-3}\text{m})^2 = \underline{9.94\text{A}}$

b) $V = EL = (0.520\text{ V/m})(12.0\text{ m}) = \underline{6.24\text{ V}}$

c) We can use Ohm's law (Eq. 26-13):

$V = IR \Rightarrow R = \frac{V}{I} = \frac{6.24\text{V}}{9.94\text{A}} = \underline{0.628\Omega}$

Or, we can calculate R from the resistivity and the dimensions of the wire (Eq. 26-12)

$R = \frac{\rho L}{A} = \frac{\rho L}{\pi r^2} = \frac{(2.63\times10^{-8}\Omega\cdot\text{m})(12.0\text{m})}{\pi\,(0.400\times10^{-3}\text{ m})^2} = 0.628\,\Omega$, which checks.

26-9

a) $E = \frac{V}{L} = \frac{7.20\text{ V}}{8.00\text{ m}} = \underline{0.900\text{ V/m}}$

b) $E = \rho J \Rightarrow \rho = \frac{E}{J} = \frac{0.900\text{ V/m}}{3.40\times10^7\text{ A/m}^2} = \underline{2.65\times10^{-8}\,\Omega\cdot\text{m}}$

26-13

$R_T = R_0\,[1 + \alpha(T - T_0)]$

From Table 26-2, for carbon $\alpha = -0.0005\ (\text{C}°)^{-1}$.

Let $T_0 = 4.0\ °\text{C}$, so $R_0 = 217.3\,\Omega$ and $R_T = 213.6\,\Omega$

$T - T_0 = \frac{R_T - R_0}{R_0\,\alpha} = \frac{213.6\,\Omega - 217.3\,\Omega}{(217.3\Omega)(-0.0005\,(\text{C}°)^{-1})} = +34\text{C}°$; $T = T_0 + 34\text{C}° = 4°\text{C} + 34\text{C}° = \underline{38°\text{C}}$

26-15

The voltmeter reads the potential difference V_{ab} between the terminals of the battery.

open circuit $\Rightarrow I = 0$

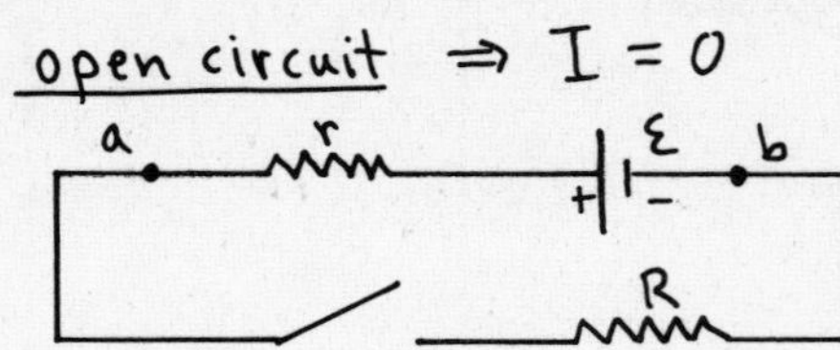

$$V_{ab} = \mathcal{E} = \underline{1.48\,\text{V}}$$

switch closed

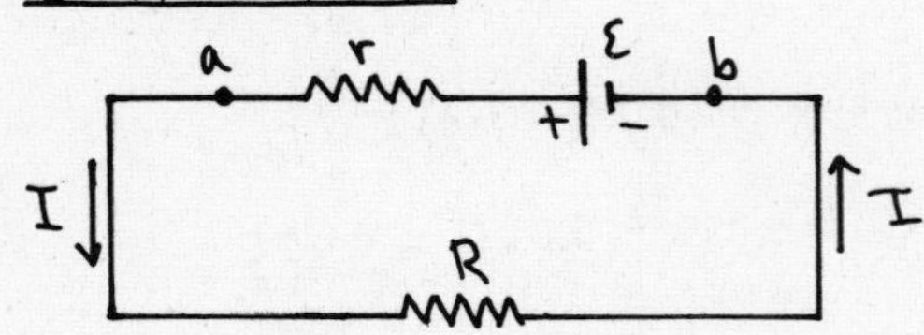

$$V_{ab} = \mathcal{E} - Ir = 1.37\,\text{V}$$

$$r = \frac{\mathcal{E} - 1.37\,\text{V}}{I} = \frac{1.48\,\text{V} - 1.37\,\text{V}}{1.30\,\text{A}} = \underline{0.085\,\Omega}$$

26-17

a)

$$\mathcal{E} - Ir = 0$$

$$r = \frac{\mathcal{E}}{I} = \frac{1.50\,\text{V}}{15.0\,\text{A}} = \underline{0.100\,\Omega}$$

b) $r = \frac{\mathcal{E}}{I} = \frac{1.50\,\text{V}}{5.00\,\text{A}} = \underline{0.300\,\Omega}$

c) $r = \frac{\mathcal{E}}{I} = \frac{12.0\,\text{V}}{1000\,\text{A}} = \underline{0.0120\,\Omega}$

26-21

By definition $p = \frac{P}{LA}$

a) E is related to V and J is related to I, so use

$$P = VI \Rightarrow p = \frac{VI}{LA}$$

$$\frac{V}{L} = E \text{ and } \frac{I}{A} = J \Rightarrow p = EJ$$

b) J is related to I and ρ is related to R, so use

$$P = I^2 R \Rightarrow p = \frac{I^2 R}{LA}$$

$$I = JA \text{ and } R = \frac{\rho L}{A} \Rightarrow p = \frac{J^2 A^2 \rho L}{LA^2} = \rho J^2$$

c) E is related to V and ρ is related to R, so use

$$P = \frac{V^2}{R} \Rightarrow p = \frac{V^2}{RLA}$$

$$V = EL \text{ and } R = \frac{\rho L}{A} \Rightarrow p = \frac{E^2 L^2}{LA}\left(\frac{A}{\rho L}\right) = \frac{E^2}{\rho}$$

26-25

a) $P = VI = (12\text{ V})(50\text{ A}) = 600\text{ W}$

The battery can provide this for 1.0 h, so the energy the battery has stored is

$U = Pt = (600\text{ W})(3600\text{ s}) = \underline{2.16\times10^6\text{ J}}$

b) For gasoline the heat of combustion is $L_c = 46\times10^6\text{ J/kg}$.

The mass of gasoline that supplies 2.16×10^6 J is

$$m = \frac{2.16\times10^6\text{ J}}{46\times10^6\text{ J/kg}} = 0.0470\text{ kg}$$

The volume of this mass of gasoline is

$$V = \frac{m}{\rho} = \frac{0.0470\text{ kg}}{900\text{ kg/m}^3} = 5.22\times10^{-5}\text{ m}^3\left(\frac{1000\text{ L}}{1\text{ m}^3}\right) = \underline{0.0522\text{ L}}$$

c) $U = Pt$

$$t = \frac{U}{P} = \frac{2.16\times10^6\text{ J}}{400\text{ W}} = 5400\text{ s} = 90\text{ min} = \underline{1.5\text{ h}}$$

26-27

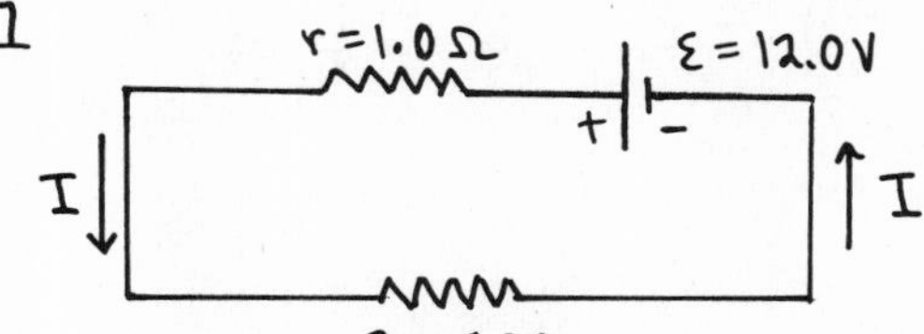

Compute I:

$\mathcal{E} - Ir - IR = 0$

$$I = \frac{\mathcal{E}}{r+R} = \frac{12.0\text{ V}}{1.0\,\Omega + 5.0\,\Omega} = 2.00\text{ A}$$

a) The rate of conversion of chemical energy to electrical energy in the emf of the battery is $P = \mathcal{E}I = (12.0\text{ V})(2.00\text{ A}) = \underline{24.0\text{ W}}$.

b) The rate of dissipation of electrical energy in the internal resistance of the battery is $P = I^2r = (2.00\text{ A})^2(1.0\,\Omega) = \underline{4.0\text{ W}}$

c) The rate of dissipation of electrical energy in the external resistor R is $P = I^2R = (2.00\text{ A})^2(5.0\,\Omega) = \underline{20.0\text{ W}}$

Note: The rate of production of electrical energy in the circuit is 24.0 W. The total rate of consumption of electrical energy in the circuit is 4.0 W + 20.0 W = 24.0 W. Equal rates of production and consumption of electrical energy are required by energy conservation.

26-29

Table 26-1 gives $\rho = 1.47\times10^{-8}\,\Omega\cdot\text{m}$ for silver at room temperature.

$$\text{Eq.(26-28)}: \rho = \frac{m}{ne^2\tau} \Rightarrow \tau = \frac{m}{ne^2\rho} = \frac{9.109\times10^{-31}\text{ kg}}{(5.80\times10^{23}\text{ m}^{-3})(1.602\times10^{-19}\text{ C})^2(1.47\times10^{-8}\,\Omega\cdot\text{m})}$$

$\tau = \underline{4.16\times10^{-9}\text{ s}}$

26-31

a) $I = \frac{V}{R}$

The total resistance R along the current path is the $R_{ps} = 2000\,\Omega$ of the power supply plus the $R_b = 10\times10^3\,\Omega$ body resistance; $R = R_{ps} + R_b = 2000\,\Omega + 10\times10^3\,\Omega$

$$I = \frac{20\times10^3\ \text{V}}{2000\,\Omega + 10\times10^3\,\Omega} = \underline{1.67\ \text{A}}$$

b) $P_b = I^2 R_b = (1.67\ \text{A})^2(10\times10^3\,\Omega) = \underline{2.79\times10^4\ \text{W}}$

c) $I = 1.00\ \text{mA} \Rightarrow R = \frac{V}{I} = \frac{20\times10^3\ \text{V}}{1.00\times10^{-3}\ \text{A}} = 2.0\times10^7\,\Omega$

$R = R_b + R_{ps} \Rightarrow R_{ps} = R - R_b = 2.0\times10^7\,\Omega - 10\times10^3\,\Omega = \underline{2.0\times10^7\,\Omega}$

Problems

26-33

a) $R = \frac{\rho L}{A} \Rightarrow \rho = \frac{RA}{L} = \frac{(0.0640\,\Omega)(2.00\times10^{-3}\ \text{m})^2}{12.0\ \text{m}} = \underline{2.13\times10^{-8}\,\Omega\cdot\text{m}}$

b) Use $V = IR \Rightarrow I = \frac{V}{R}$

$V = EL = (1.20\ \text{V/m})(12.0\ \text{m}) = 14.4\ \text{V}$

$I = \frac{V}{R} = \frac{14.4\ \text{V}}{0.0640\,\Omega} = \underline{225\ \text{A}}$

Or

$E = \rho J \Rightarrow J = \frac{E}{\rho} = \frac{1.20\ \text{V/m}}{2.13\times10^{-8}\,\Omega\cdot\text{m}} = 5.63\times10^7\ \text{A/m}^2$

$I = JA = (5.63\times10^7\ \text{A/m}^2)(2.00\times10^{-3}\ \text{m})^2 = 225\ \text{A}$, which checks

c) $J = n|q|v_d = nev_d$

$v_d = \frac{J}{ne} = \frac{5.63\times10^7\ \text{A/m}^2}{(8.5\times10^{28}\ \text{m}^{-3})(1.602\times10^{-19}\ \text{C})} = 4.13\times10^{-3}\ \text{m/s} = \underline{4.13\ \text{mm/s}}$

26-35

Use $E = \rho J$ to calculate the current density between the plates. Let A be the area of each plate; then $I = JA$.

$J = \frac{E}{\rho}$

$E = \frac{\sigma}{K\epsilon_0} = \frac{Q}{KA\epsilon_0}$

$J = \frac{E}{\rho} = \frac{Q}{KA\epsilon_0\rho}$; $I = JA = \frac{Q}{K\epsilon_0\rho}$, as was to be shown.

26-37

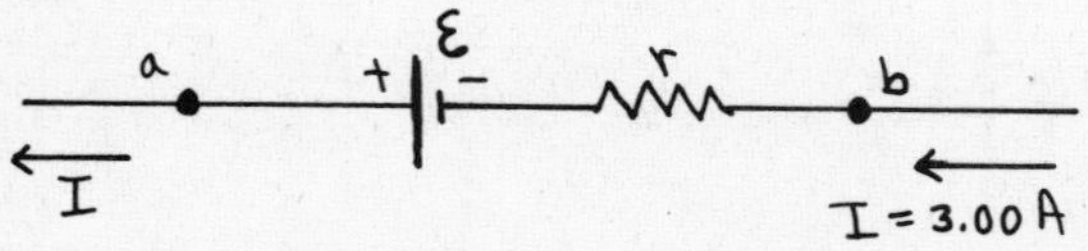

$V_{ab} = 9.2\ \text{V}$

$V_{ab} = \mathcal{E} - Ir$

$$\boxed{\mathcal{E} - (3.00\ \text{A})r = 9.2\ \text{V}}$$

26-37 (cont)

a, $\mathcal{E}$, r, b

$I = 2.00\text{ A}$ $\quad$ I

$V_{ab} = 11.2\text{ V}$

$V_{ab} = \mathcal{E} + Ir$

$$\boxed{\mathcal{E} + (2.00\text{ A})\,r = 11.2\text{ V}}$$

a) Solve the first equation for $\mathcal{E}$ and use that result in the second equation:

$$\mathcal{E} = 9.2\text{ V} + (3.00\text{ A})\,r$$

$$9.2\text{ V} + (3.00\text{ A})\,r + (2.00\text{ A})\,r = 11.2\text{ V}$$

$$(5.00\text{ A})\,r = 2.0\text{ V} \;;\; r = \frac{2.0\text{ V}}{5.00\text{ A}} = \underline{0.40\,\Omega}$$

b) Then $\mathcal{E} = 9.2\text{ V} + (3.00\text{ A})\,r = 9.2\text{ V} + (3.00\text{ A})(0.40\,\Omega) = \underline{10.4\text{ V}}$

26-39

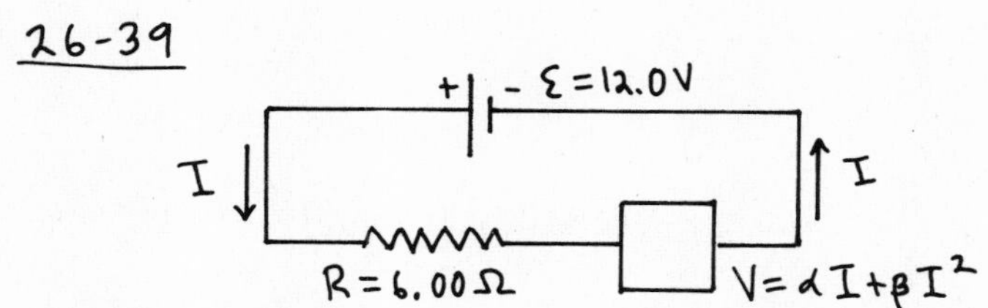

$$\mathcal{E} - IR - V = 0$$

$$\mathcal{E} - IR - \alpha I - \beta I^2 = 0$$

$$\beta I^2 + (R + \alpha)\,I - \mathcal{E} = 0$$

The quadratic formula gives

$$I = \frac{1}{2\beta}\left[-(R+\alpha) \pm \sqrt{(R+\alpha)^2 + 4\beta\mathcal{E}}\right]$$

I must be positive, so take the + sign

$$\Rightarrow I = \frac{1}{2\beta}\left[-(R+\alpha) + \sqrt{(R+\alpha)^2 + 4\beta\mathcal{E}}\right] = \frac{1}{2(1.20\,\Omega/\text{A})}\left[-10.0\,\Omega + \sqrt{(10.0\,\Omega)^2 + 4(1.20\,\Omega/\text{A})(12.0\text{ V})}\right]$$

$$I = -4.17\text{ A} + 5.23\text{ A} = \underline{1.06\text{ A}}$$

26-43

a)

I, $\mathcal{E}_1 = 12.0\text{ V}$, $r_1 = 1.0\,\Omega$, I, $R = 8.0\,\Omega$, $\mathcal{E}_2 = 8.0\text{ V}$, $r_2 = 1.0\,\Omega$

$$\mathcal{E}_1 - \mathcal{E}_2 - I(r_1 + r_2 + R) = 0$$

$$I = \frac{\mathcal{E}_1 - \mathcal{E}_2}{r_1 + r_2 + R} = \frac{12.0\text{ V} - 8.0\text{ V}}{1.0\,\Omega + 1.0\,\Omega + 8.0\,\Omega} = \underline{0.40\text{ A}}$$

b) $P = I^2R + I^2r_1 + I^2r_2 = I^2(R + r_1 + r_2) = (0.40\text{ A})^2(8.0\,\Omega + 1.0\,\Omega + 1.0\,\Omega) = \underline{1.6\text{ W}}$

c) Chemical → electrical energy in a battery when the current goes through the battery from the − to + terminal, so the electrical energy of the charges increases as the current passes through. This happens in the 12.0 V battery, and the rate of production of electrical energy is $P = \mathcal{E}_1 I = (12.0\text{ V})(0.40\text{ A}) = \underline{4.8\text{ W}}$.

d) Electrical → chemical energy in a battery when the current goes through the battery from the + to the − terminal, so the electrical energy of the charges decreases as the current passes through. This happens in the 8.0 V battery and the rate of consumption of electrical energy is $P = \mathcal{E}_2 I = (8.0\text{ V})(0.40\text{ A}) = \underline{3.2\text{ W}}$.

e) Total rate of production of electrical energy = 4.8 W.
Total rate of consumption of electrical energy = 1.6 W + 3.2 W = 4.8 W, which checks.

CHAPTER 27

Exercises 3, 5, 7, 9, 11, 13, 15, 21, 25, 27

Problems 29, 33, 35, 37, 39, 41, 43, 49, 51

Exercises

27-3

a)

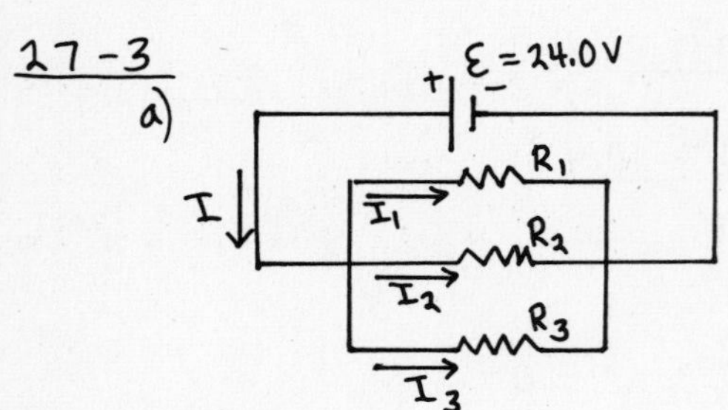

parallel $\Rightarrow \frac{1}{R_{eq}} = \frac{1}{R_1} + \frac{1}{R_2} + \frac{1}{R_3}$

$$\frac{1}{R_{eq}} = \frac{1}{2.00\Omega} + \frac{1}{3.00\Omega} + \frac{1}{4.00\Omega} = \frac{6+4+3}{12.0\Omega}$$

$$R_{eq} = \frac{12.0\Omega}{13} = \underline{0.923\Omega}$$

b) For resistors in parallel the voltage is the same across each and equal to the applied voltage $\Rightarrow V_1 = V_2 = V_3 = \mathcal{E} = 24.0\text{ V}$.

$$V = IR \Rightarrow I_1 = \frac{V_1}{R_1} = \frac{24.0\text{V}}{2.00\Omega} = 12.00\text{ A}$$

$$I_2 = \frac{V_2}{R_2} = \frac{24.0\text{V}}{3.00\Omega} = 8.00\text{ A}$$

$$I_3 = \frac{V_3}{R_3} = \frac{24.0\text{V}}{4.00\Omega} = 6.00\text{ A}$$

c) The currents through the resistors add to give the current through the battery:
$I = I_1 + I_2 + I_3 = 12.0\text{A} + 8.0\text{ A} + 6.0\text{ A} = 26.0\text{ A}$

Alternatively, we can use the equivalent resistance R_{eq}:

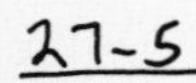

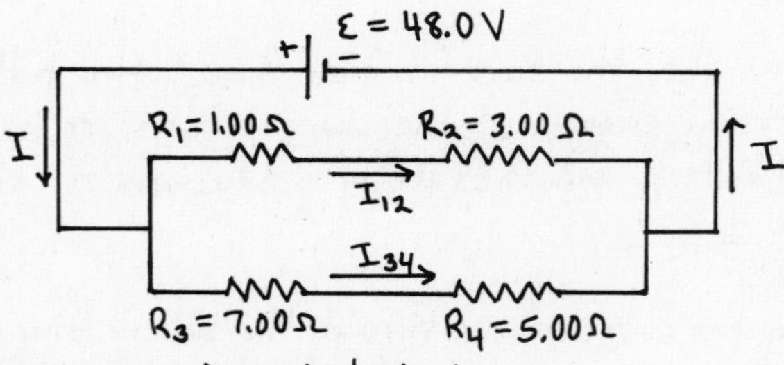

$$\mathcal{E} - I R_{eq} = 0$$

$$I = \frac{\mathcal{E}}{R_{eq}} = \frac{24.0\text{ V}}{0.923\Omega} = 26.0\text{ A, which checks}$$

27-5

R_1 and R_2 in series have an equivalent resistance of
$R_{12} = R_1 + R_2 = 4.00\Omega$
R_3 and R_4 in series have an equivalent resistance of
$R_{34} = R_3 + R_4 = 12.0\Omega$

The circuit is equivalent to

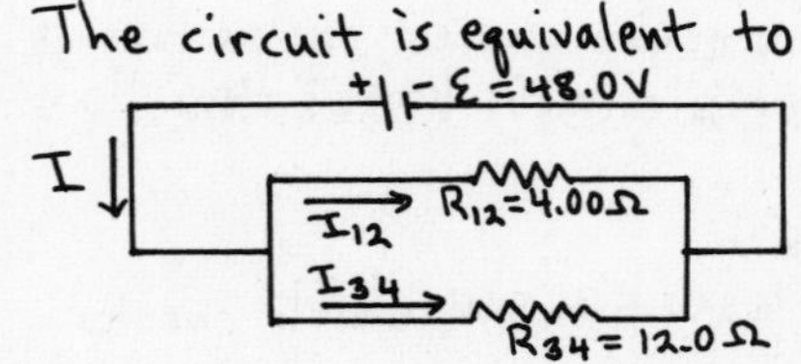

R_{12} and R_{34} in parallel are equivalent to R_{eq} given by

$$\frac{1}{R_{eq}} = \frac{1}{R_{12}} + \frac{1}{R_{34}} = \frac{R_{12} + R_{34}}{R_{12} R_{34}}$$

$$R_{eq} = \frac{R_{12} R_{34}}{R_{12} + R_{34}} = \frac{(4.00\Omega)(12.0\Omega)}{4.00\Omega + 12.0\Omega} = \underline{3.00\Omega}$$

27-5 (cont)

The voltage across each branch of the parallel combination is $\mathcal{E}$, so

$$\mathcal{E} - I_{12}R_{12} = 0$$
$$I_{12} = \frac{\mathcal{E}}{R_{12}} = \frac{48.0\ \text{V}}{4.00\ \Omega} = 12.0\ \text{A}$$

$$\mathcal{E} - I_{34}R_{34} = 0$$
$$I_{34} = \frac{\mathcal{E}}{R_{34}} = \frac{48.0\ \text{V}}{12.0\ \Omega} = 4.0\ \text{A}$$

The current is 12.0 A through the 1.00 Ω and 3.00 Ω resistors, and it is 4.0 A through the 7.00 Ω and 5.00 Ω resistors.

Note: The current through the battery is $I = I_{12} + I_{34} = 12.0\ \text{A} + 4.0\ \text{A} = 16.0\ \text{A}$, and this is $\frac{\mathcal{E}}{R_{eq}} = \frac{48.0\ \text{V}}{3.00\ \Omega} = 16.0\ \text{A}$.

27-7

a) $P = \frac{V^2}{R}$ (Eq. 26-20) $\Rightarrow V = \sqrt{PR} = \sqrt{(4.00\ \text{W})(10{,}000\ \Omega)} = \underline{200\ \text{V}}$

b) $P = \frac{V^2}{R} = \frac{(220\ \text{V})^2}{20{,}000\ \Omega} = \underline{2.42\ \text{W}}$

27-9

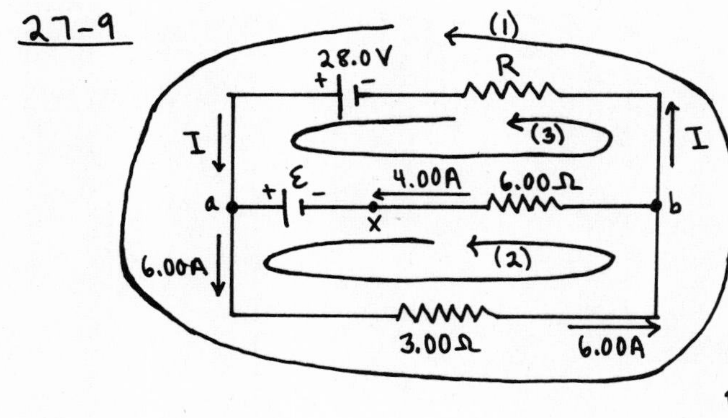

a) Apply Kirchhoff's point rule to point a:
$\Sigma I = 0 \Rightarrow I + 4.00\ \text{A} - 6.00\ \text{A} = 0$
$I = 2.00\ \text{A}$ (in the direction shown in the diagram)

b) Apply Kirchhoff's loop rule to loop (1):
$-(6.00\ \text{A})(3.00\ \Omega) - (2.00\ \text{A})R + 28.0\ \text{V} = 0$
$-18.0\ \text{V} - (2.00\ \text{A})R + 28.0\ \text{V} = 0$
$R = \frac{28.0\ \text{V} - 18.0\ \text{V}}{2.00\ \text{A}} = \underline{5.00\ \Omega}$

c) Apply Kirchhoff's loop rule to loop (2):
$-(6.00\ \text{A})(3.00\ \Omega) - (4.00\ \text{A})(6.00\ \Omega) + \mathcal{E} = 0$
$\mathcal{E} = 18.0\ \text{V} + 24.0\ \text{V} = \underline{42.0\ \text{V}}$

Note: Can check that the loop rule is satisfied for loop (3), as a check of our work:
$28.0\ \text{V} - \mathcal{E} + (4.00\ \text{A})(6.00\ \Omega) - (2.00\ \text{A})R = 0$
$28.0\ \text{V} - 42.0\ \text{V} + 24.0\ \text{V} - (2.00\ \text{A})(5.00\ \Omega) = 0$
$52.0\ \text{V} = 42.0\ \text{V} + 10.0\ \text{V}$
$52.0\ \text{V} = 52.0\ \text{V}$ ✓

d) If the circuit is broken at point x there can be no current in the 6.00 Ω resistor. There is now only a single current path:

27-9 (cont)

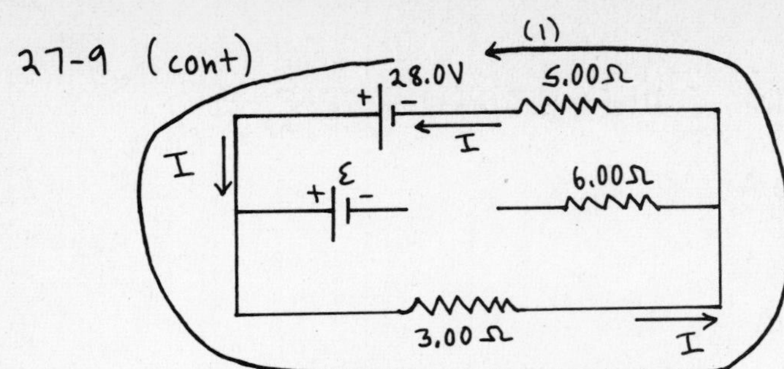

Apply the loop rule:

$$+28.0\,V - (3.00\,\Omega)I - (5.00\,\Omega)I = 0$$

$$I = \frac{28.0\,V}{8.00\,\Omega} = \underline{3.50\,A}$$

27-11

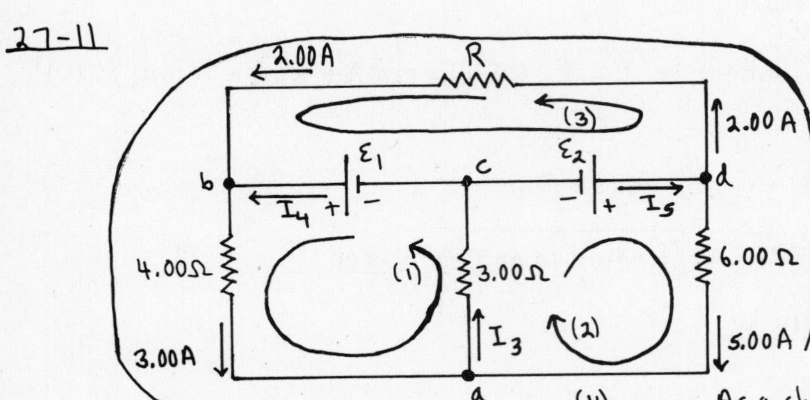

a) Apply the junction rule to point a:

$$3.00\,A + 5.00\,A - I_3 = 0$$

$$I_3 = \underline{8.00\,A}$$

Apply the junction rule to point b:

$$2.00\,A + I_4 - 3.00\,A = 0$$

$$I_4 = 3.00\,A - 2.00\,A = 1.00\,A$$

Apply the junction rule to point c:

$$I_3 - I_4 - I_5 = 0$$

$$I_5 = I_3 - I_4 = 8.00\,A - 1.00\,A = 7.00\,A$$

As a check, apply the junction rule to point d:

$$I_5 - 2.00\,A - 5.00\,A = 0$$

$$I_5 = 7.00\,A \checkmark$$

b) Apply the loop rule to loop (1):

$$\varepsilon_1 - (3.00\,A)(4.00\,\Omega) - I_3(3.00\,\Omega) = 0$$

$$\varepsilon_1 = 12.0\,V + (8.00\,A)(3.00\,\Omega) = \underline{36.0\,V}$$

Apply the loop rule to loop (2):

$$\varepsilon_2 - (5.00\,A)(6.00\,\Omega) - I_3(3.00\,\Omega) = 0$$

$$\varepsilon_2 = 30.0\,V + (8.00\,A)(3.00\,\Omega) = \underline{54.0\,V}$$

c) Apply the loop rule to loop (3):

$$-(2.00\,A)R - \varepsilon_1 + \varepsilon_2 = 0$$

$$R = \frac{\varepsilon_2 - \varepsilon_1}{2.00\,A} = \frac{54.0\,V - 36.0\,V}{2.00\,A} = \underline{9.00\,\Omega}$$

Note: Apply the loop rule to loop (4), as a check of our calculations:

$$-(2.00\,A)R - (3.00\,A)(4.00\,\Omega) + (5.00\,A)(6.00\,\Omega) = 0$$

$$-(2.00\,A)(9.00\,\Omega) - 12.0\,V + 30.0\,V = 0$$

$$-18.0\,V + 18.0\,V = 0 \checkmark$$

27-13

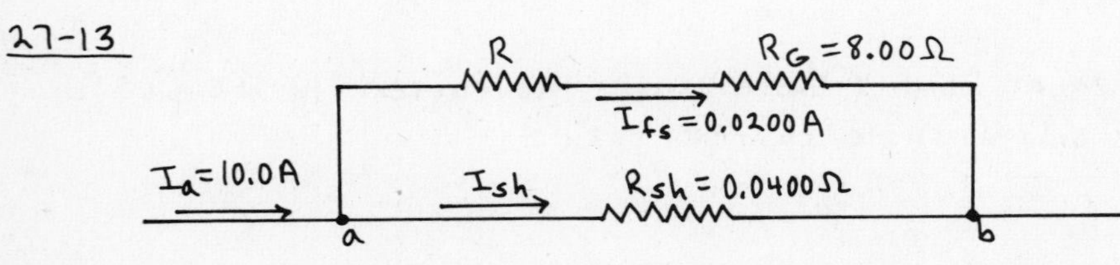

27-13 (cont)

We want that $I_a = 10.0\text{ A}$ in the external circuit produces $I_{fs} = 0.0200\text{ A}$ through the galvanometer coil. Applying the junction rule to point a gives

$$I_a - I_{fs} - I_{sh} = 0$$

$$I_{sh} = I_a - I_{fs} = 10.0\text{ A} - 0.0200\text{ A} = 9.98\text{ A}$$

The potential difference V_{ab} between points a and b must be the same for both paths between these two points

$$\Rightarrow I_{fs}(R + R_G) = I_{sh} R_{sh}$$

$$R = \frac{I_{sh} R_{sh}}{I_{fs}} - R_G = \frac{(9.98\text{ A})(0.0400\,\Omega)}{0.0200\text{ A}} - 8.00\,\Omega = 19.96\,\Omega - 8.00\,\Omega = \underline{12.0\,\Omega}$$

27-15

For each range setting the circuit has the form

$I_{fs} = 1.00\times10^{-3}\text{ A}$ $R_G = 25.0\,\Omega$ R V

3.00 V

For $V = 3.00\text{ V}$, $R = R_1$ and the total meter resistance R_m is $R_m = R_G + R_1$.

$$V = I_{fs} R_m \Rightarrow R_m = \frac{V}{I_{fs}} = \frac{3.00\text{ V}}{1.00\times10^{-3}\text{ A}} = 3.00\times10^{3}\,\Omega$$

$$R_m = R_G + R_1 \Rightarrow R_1 = R_m - R_G = 3.00\times10^{3}\,\Omega - 25.0\,\Omega = \underline{2975\,\Omega}$$

15.0 V

For $V = 15.0\text{ V}$, $R = R_1 + R_2$ and the total meter resistance is $R_m = R_G + R_1 + R_2$.

$$V = I_{fs} R_m \Rightarrow R_m = \frac{V}{I_{fs}} = \frac{15.0\text{ V}}{1.00\times10^{-3}\text{ A}} = 1.50\times10^{4}\,\Omega$$

$$R_2 = R_m - R_G - R_1 = 1.50\times10^{4}\,\Omega - 25.0\,\Omega - 2975\,\Omega = \underline{1.20\times10^{4}\,\Omega}$$

150 V

For $V = 150\text{ V}$, $R = R_1 + R_2 + R_3$ and the total meter resistance R_m is $R_m = R_G + R_1 + R_2 + R_3$.

$$V = I_{fs} R_m \Rightarrow R_m = \frac{V}{I_{fs}} = \frac{150\text{ V}}{1.00\times10^{-3}\text{ A}} = 1.50\times10^{5}\,\Omega$$

$$R_3 = R_m - R_G - R_1 - R_2 = 1.50\times10^{5}\,\Omega - 25.0\,\Omega - 2975\,\Omega - 1.20\times10^{4}\,\Omega = \underline{1.35\times10^{5}\,\Omega}$$

27-21

a)

$+q$ $-q$ C i R

By Kirchhoff's loop rule,

$$v_c - v_R = 0$$

$$\frac{q}{C} - iR = 0 \Rightarrow i = \frac{q}{RC}$$

Just after the connection is made the charge q on the capacitor hasn't had any time to decrease from its initial value of $6.00\times10^{-8}\text{ C}$, so

$$i = \frac{q}{RC} = \frac{6.00\times10^{-8}\text{ C}}{(4.00\times10^{5}\,\Omega)(2.50\times10^{-10}\text{ F})} = \underline{6.00\times10^{-4}\text{ A}}$$

27-21 (cont)

b) Eq. (27-15): $\tau = RC = (4.00\times10^{5}\,\Omega)(2.50\times10^{-10}\,F) = \underline{1.00\times10^{-4}\,s}$

27-25

a) $P = VI \Rightarrow I = \frac{P}{V}$

toaster: $I_t = \frac{P_t}{V} = \frac{1800\,W}{120\,V} = \underline{15.0\,A}$

frying pan: $I_f = \frac{P_f}{V} = \frac{1400\,W}{120\,V} = \underline{11.7\,A}$

lamp: $I_l = \frac{P_l}{V} = \frac{75\,W}{120\,V} = \underline{0.62\,A}$

b) Total current in the circuit is $I_{tot} = I_t + I_f + I_l = 15.0\,A + 11.7\,A + 0.62\,A = 27.3\,A$. This will blow the 20 A fuse.

27-27

a) When the heater element is first turned on it is at room temperature and has resistance $R = 22.0\,\Omega$.

$I = \frac{V}{R} = \frac{120\,V}{22.0\,\Omega} = \underline{5.45\,A}$

$P = \frac{V^2}{R} = \frac{(120\,V)^2}{22.0\,\Omega} = \underline{655\,W}$

b) Find the resistance R_T of the element at the operating temperature of 280 °C:

Eq. (26-14): $R_T = R_0(1+\alpha(T-T_0)) = 22.0\,\Omega(1+(3.00\times10^{-3}(C^\circ)^{-1}))(280\,^\circ C - 23\,^\circ C) = 38.96\,\Omega$

$I = \frac{V}{R_T} = \frac{120\,V}{38.96\,\Omega} = \underline{3.08\,A}$

$P = \frac{V^2}{R} = \frac{(120\,V)^2}{38.96\,\Omega} = \underline{370\,W}$

Problems

27-29

a) Two of the resistors in series would each dissipate one-half the total, or 1.00 W, which is ok. But the series combination would have an equivalent resistance of 2000 Ω, not the 1000 Ω that is required. Resistors in parallel have an equivalent resistance less than that of the individual resistors, so the solution is two in series in parallel with another two in series.

R = 1000 Ω R R R a b V_{ab} = R_s R_s = R_{eq}

R_s is the resistance equivalent to two of the 1000 Ω resistors in series:

$R_s = R + R = 2000\,\Omega$.

27-29 (cont)

R_{eq} is the resistance equivalent to the two $R_s = 2000\ \Omega$ resistors in parallel: $\frac{1}{R_{eq}} = \frac{1}{R_s} + \frac{1}{R_s} = \frac{2}{R_s} \Rightarrow R_{eq} = \frac{R_s}{2} = \frac{2000\ \Omega}{2} = 1000\ \Omega$

This combination does have the required $1000\ \Omega$ equivalent resistance. It will be shown in part (b) that a total of 2.00 W can be dissipated without exceeding the power rating of each individual resistor.

b) For the combination with equivalent resistance $R_{eq} = 1000\ \Omega$ to dissipate 2.00 W the voltage V_{ab} applied to the network must be given by

$$P = \frac{V_{ab}^2}{R_{eq}} \Rightarrow V_{ab} = \sqrt{P R_{eq}} = \sqrt{(2.00\text{ W})(1000\text{ W})} = 44.72\text{ V}$$

This means 44.72 V across each parallel branch and $\frac{1}{2}(44.72\text{ V}) = 22.36$ V across each $1000\ \Omega$ resistor. The power dissipated by each individual resistor is then $P = \frac{V^2}{R} = \frac{(22.36\text{ V})^2}{1000\ \Omega} = 0.500$ W, which is less than the maximum allowed value of 1.00 W.

27-33

a) Break in the circuit between points a and b $\Rightarrow$ no current in the middle branch that contains the $3.00\ \Omega$ resistor and the 10.0 V battery. The circuit therefore has a single current path. Find the current, so that the potential drops across the resistors can be calculated.

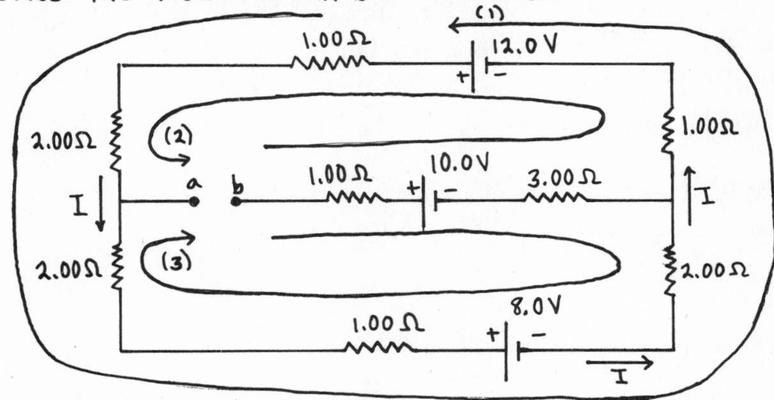

Apply the loop rule to loop (1):

$$+12.0\text{ V} - I(1.00\ \Omega + 2.00\ \Omega + 2.00\ \Omega + 1.00\ \Omega) - 8.0\text{ V} - I(2.00\ \Omega + 1.00\ \Omega) = 0$$

$$I = \frac{12.0\text{ V} - 8.0\text{ V}}{9.00\ \Omega} = 0.4444\text{ A}$$

To find V_{ab} start at point b and travel to a, adding up the potential rises and drops. Travel on path (2) shown on the diagram. The $1.00\ \Omega$ and $3.00\ \Omega$ resistors in the middle branch have no current through them and hence no voltage across them. Therefore, $V_b - 10.0\text{ V} - I(1.00\ \Omega) + 12.0\text{ V} - I(1.00\ \Omega + 2.00\ \Omega) = V_a$

$$V_{ab} = V_a - V_b = 12.0\text{ V} - 10.0\text{ V} - I(4.00\ \Omega) = 2.0\text{ V} - (0.4444\text{ A})(4.00\ \Omega) = +0.22\text{ V}$$

(point a is at higher potential)

As a check on this calculation we can also compute V_{ab} by traveling from b to a on path (3):

27-33 (cont)

$V_b - 10.0\text{ V} + 8.0\text{ V} + I(2.00\,\Omega + 1.00\,\Omega + 2.00\,\Omega) = V_a$

$V_{ab} = -2.00\text{ V} + (0.4444\text{ A})(5.00\,\Omega) = +0.22\text{ V}$, which checks.

b) With points a and b connected by a wire there are three current branches:

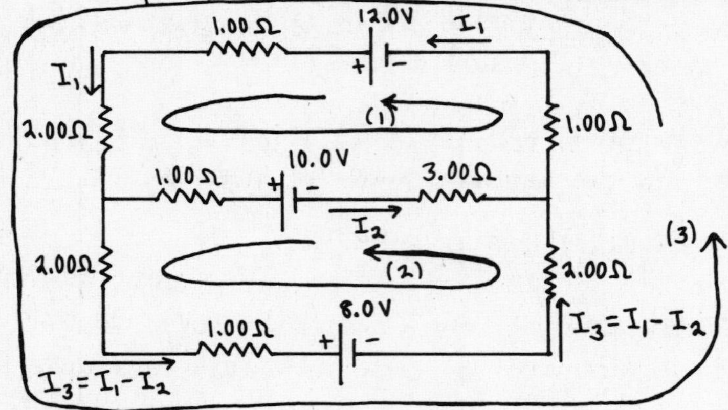

The junction rule has been used to write the third current (in the 8.0 V battery) in terms of the other two.

Apply the loop rule to loop (1):

$12.0\text{ V} - I_1(1.00\,\Omega) - I_1(2.00\,\Omega) - I_2(1.00\,\Omega) - 10.0\text{ V} - I_2(3.00\,\Omega) - I_1(1.00\,\Omega) = 0$

$2.0\text{ V} - I_1(4.00\,\Omega) - I_2(4.00\,\Omega) = 0$

$\boxed{(2.00\,\Omega)I_1 + (2.00\,\Omega)I_2 = 1.0\text{ V}}$ equation (1)

Apply the loop rule to loop (2):

$-(I_1 - I_2)(2.00\,\Omega) - (I_1 - I_2)(1.00\,\Omega) - 8.0\text{ V} - (I_1 - I_2)(2.00\,\Omega) + I_2(3.00\,\Omega) + 10.0\text{ V} + I_2(1.00\,\Omega) = 0$

$\boxed{2.0\text{ V} - (5.00\,\Omega)I_1 + (9.00\,\Omega)I_2 = 0}$ equation (2)

Solve eq. (1) for I_2 and use this to replace I_2 in eq. (2):

$I_2 = 0.50\text{ A} - I_1$

$2.0\text{ V} - (5.00\,\Omega)I_1 + (9.00\,\Omega)(0.50\text{ A} - I_1) = 0$

$(14.0\,\Omega)I_1 = 6.50\text{ V} \Rightarrow I_1 = \frac{6.50\text{ V}}{14.0\,\Omega} = \underline{0.464\text{ A}}$

$I_2 = 0.500\text{ A} - 0.464\text{ A} = 0.036\text{ A}$

The current in the 12.0 V battery is $I_1 = 0.464\text{ A}$.

27-35

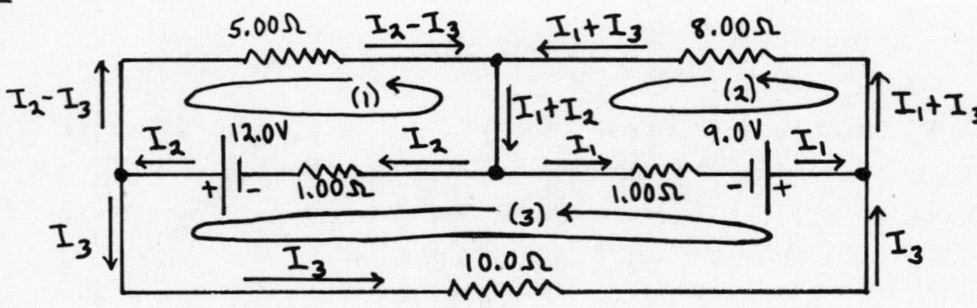

The current in each branch has been written in terms of I_1, I_2, and I_3 such that the junction rule is satisfied at each junction point.

Apply the loop rule to loop (1):

$-12.0\text{ V} + I_2(1.00\,\Omega) + (I_2 - I_3)\,5.00\,\Omega = 0$

27-35 (cont)

$$\boxed{I_2(6.00\,\Omega) - I_3(5.00\,\Omega) = 12.0\text{ V}} \qquad \text{eq.(1)}$$

Apply the loop rule to loop (2):

$$-I_1(1.00\,\Omega) + 9.00\text{ V} - (I_1 + I_3)(8.00\,\Omega) = 0$$

$$\boxed{I_1(9.00\,\Omega) + I_3(8.00\,\Omega) = 9.00\text{ V}} \qquad \text{eq.(2)}$$

Apply the loop rule to loop (3):

$$-I_3(10.0\,\Omega) - 9.00\text{ V} + I_1(1.00\,\Omega) - I_2(1.00\,\Omega) + 12.0\text{ V} = 0$$

$$\boxed{-I_1(1.00\,\Omega) + I_2(1.00\,\Omega) + I_3(10.0\,\Omega) = 3.00\text{ V}} \qquad \text{eq.(3)}$$

eq.(1) $\Rightarrow I_2 = 2.00\text{ A} + \frac{5}{6}I_3$; eq.(2) $\Rightarrow I_1 = 1.00\text{ A} - \frac{8}{9}I_3$

Use these in eq.(3) $\Rightarrow -(1.00\text{A} - \frac{8}{9}I_3)(1.00\,\Omega) + (2.00\text{A} + \frac{5}{6}I_3)(1.00\,\Omega) + I_3(10.0\,\Omega) = 3.00\text{V}$

$$\left(\frac{16+15+180}{18}\right)I_3 = 2.00\text{ A}$$

$$\frac{211}{18}I_3 = 2.00\text{A} \Rightarrow I_3 = \frac{18}{211}(2.00\text{ A}) = \underline{0.171\text{ A}}$$

Then $I_2 = 2.00\text{A} + \frac{5}{6}I_3 = 2.00\text{ A} + \frac{5}{6}(0.171\text{ A}) = \underline{2.14\text{ A}}$

$I_1 = 1.00\text{A} - \frac{8}{9}I_3 = 1.00\text{ A} - \frac{8}{9}(0.171\text{ A}) = \underline{0.848\text{ A}}$

27-37

Two unknown currents I_1 (through the 2.00Ω resistor) and I_2 (through the 5.00 Ω resistor) are labeled on the circuit diagram. The current through the 4.00 Ω resistor has been written as $I_2 - I_1$ by using the junction rule. Apply the loop rule to loops (1) and (2) to get two equations for the unknown currents I_1 and I_2. Loop (3) can then be used to check the results.

loop (1):

$$+20.0\text{ V} - I_1(2.00\,\Omega) - 14.0\text{V} + (I_2 - I_1)(4.00\,\Omega) = 0$$

$$6.00\,I_1 - 4.00\,I_2 = 6.00\text{ A}$$

$$\boxed{3.00\,I_1 - 2.00\,I_2 = 3.00\text{ A}} \qquad \text{eq.(1)}$$

loop (2):

$$+36.0\text{ V} - I_2(5.00\,\Omega) - (I_2 - I_1)(4.00\,\Omega) = 0$$

$$\boxed{-4.00\,I_1 + 9.00\,I_2 = 36.0\text{ A}} \qquad \text{eq.(2)}$$

solve eq.(1) for $I_1 \Rightarrow I_1 = 1.00\text{ A} + \frac{2}{3}I_2$

Use in eq.(2) $\Rightarrow -4.00(1.00\text{A} + \frac{2}{3}I_2) + 9.00\,I_2 = 36.0\text{A}$

$(-\frac{8}{3} + 9.00)\,I_2 = 40.0\text{ A} \Rightarrow I_2 = \underline{6.32\text{ A}}$

Then $I_1 = 1.00\text{A} + \frac{2}{3}I_2 = 1.00\text{ A} + \frac{2}{3}(6.32\text{A}) = \underline{5.21\text{A}}$

Current through the 2.00Ω resistor : $I_1 = 5.21$ A.

Current through the 5.00Ω resistor: $I_2 = 6.32$ A.

Current through the 4.00Ω resistor : $I_2 - I_1 = 6.32\text{A} - 5.21\text{A} = \underline{1.11\text{A}}$

27-37 (cont)

Loop (3) to check: $+20.0\text{ V} - I_1(2.00\,\Omega) - 14.0\text{ V} + 36.0\text{ V} - I_2(5.00\,\Omega) = 0$

$$(5.21\text{ A})(2.00\,\Omega) + (6.32\text{ A})(5.00\,\Omega) = 42.0\text{ V}$$

$$10.4\text{ V} + 31.6\text{ V} = 42.0\text{ V} \checkmark$$

27-39

a)

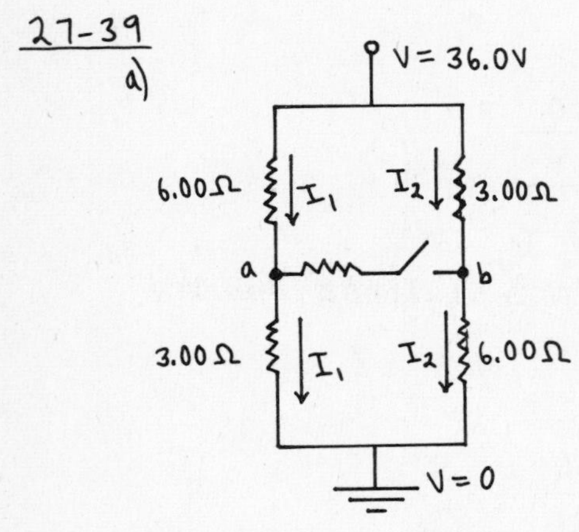

With the switch open there is no current through it and there are only the two currents I_1 and I_2 indicated in the sketch.

The potential drop across each parallel branch is 36.0 V:

$$-I_1(6.00\,\Omega + 3.00\,\Omega) + 36.0\text{ V} = 0$$

$$I_1 = \frac{36.0\text{ V}}{6.00\,\Omega + 3.00\,\Omega} = 4.00\text{ A}$$

$$-I_2(3.00\,\Omega + 6.00\,\Omega) + 36.0\text{ V} = 0$$

$$I_2 = \frac{36.0\text{ V}}{3.00\,\Omega + 6.00\,\Omega} = 4.00\text{ A}$$

To calculate $V_{ab} = V_a - V_b$ start at point b and travel to point a, adding up all the potential rises and drops along the way. We can do this by going from b up through the $3.00\,\Omega$ resistor:

$$V_b + I_2(3.00\,\Omega) - I_1(6.00\,\Omega) = V_a$$

$$V_a - V_b = (4.00\text{ A})(3.00\,\Omega) - (4.00\text{ A})(6.00\,\Omega) = 12.0\text{ V} - 24.0\text{ V} = -12.0\text{ V}$$

$V_{ab} = \underline{-12.0\text{ V}}$ (point a is 12.0 V lower in potential than point b)

Alternatively, we can go from point b down through the $6.00\,\Omega$ resistor:

$$V_b - I_2(6.00\,\Omega) + I_1(3.00\,\Omega) = V_a$$

$$V_a - V_b = -(4.00\text{ A})(6.00\,\Omega) + (4.00\text{ A})(3.00\,\Omega) = -24.0\text{ V} + 12.0\text{ V} = -12.0\text{ V} \checkmark$$

b)

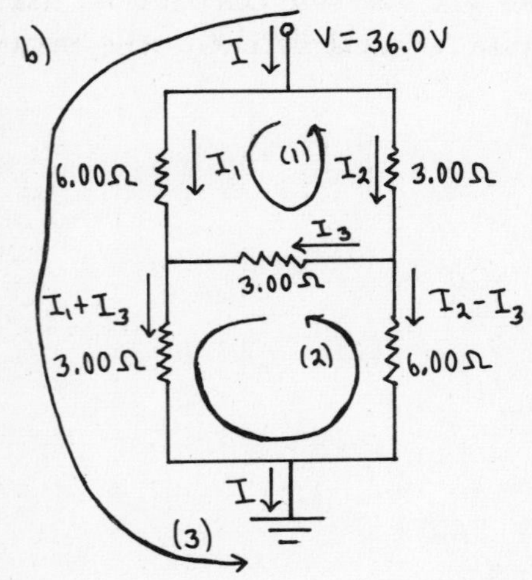

The three unknown currents I_1, I_2, I_3 are labeled on the sketch. Apply the loop rule to loops (1), (2), and (3).

loop (1):

$$-I_1(6.00\,\Omega) + I_3(3.00\,\Omega) + I_2(3.00\,\Omega) = 0$$

$$\boxed{I_2 = 2I_1 - I_3} \quad \text{eq. (1)}$$

loop (2):

$$-(I_1 + I_3)(3.00\,\Omega) + (I_2 - I_3)(6.00\,\Omega) - I_3(3.00\,\Omega) = 0$$

$$6I_2 - 12I_3 - 3I_1 = 0$$

$$2I_2 - 4I_3 - I_1 = 0$$

Use eq. (1) to replace I_2:

$$4I_1 - 2I_3 - 4I_3 - I_1 = 0$$

$$3I_1 = 6I_3 \Rightarrow \boxed{I_1 = 2I_3} \quad \text{eq. (2)}$$

loop (3) (The loop is completed through the battery [not shown], in the direction from the − to the + terminal.):

$$-I_1(6.00\,\Omega) - (I_1 + I_3)(3.00\,\Omega) + 36.0\text{ V} = 0$$

27-39 (cont)

$$9I_1 + 3I_3 = 36.0\text{A}$$

$$\boxed{3I_1 + I_3 = 12.0\text{ A}} \qquad \text{eq. (3)}$$

Use eq.(2) in eq.(3) to replace I_1:

$$3(2I_3) + I_3 = 12.0\text{A}$$

$$I_3 = \frac{12.0\text{A}}{7} = 1.71\text{A}$$

$$I_1 = 2I_3 = 3.42\text{A}$$

$$I_2 = 2I_1 - I_3 = 2(3.42\text{ A}) - 1.71\text{A} = 5.13\text{ A}$$

The current through the switch is $I_3 = \underline{1.71\text{ A}}$.

c) From the results in part (a) the current through the battery is $I = I_1 + I_2 = 3.42\text{ A} + 5.13\text{ A} = 8.55\text{ A}$. The equivalent circuit is a single resistor that produces the same current through the 36.0 V battery.

36.0 V

I=8.55 A

R

$$-IR + 36.0\text{ V} = 0$$

$$R = \frac{36.0\text{ V}}{I} = \frac{36.0\text{ V}}{8.55\text{ A}} = \underline{4.21\,\Omega}$$

27-41

a)

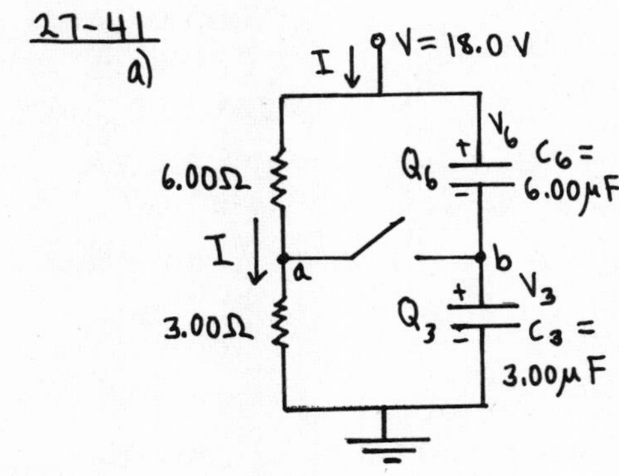

After the capacitors have been charged to their final charge there is no current through them. There is only one current path.

$$-I(6.00\,\Omega + 3.00\,\Omega) + 18.0\text{ V} = 0$$

$$I = \frac{18.0\text{ V}}{9.00\,\Omega} = 2.00\text{ A}$$

There is also a potential difference of 18.0 V applied across the two capacitors in series. The capacitors have the same charge: $Q_3 = Q_6 = Q$.

$$V_1 + V_3 = 18.0\text{ V}$$

$$Q\left(\frac{1}{C_6} + \frac{1}{C_3}\right) = 18.0\text{ V}$$

$$Q\left(\frac{C_3 + C_6}{C_3 C_6}\right) = 18.0\text{ V}$$

$$Q = \left(\frac{C_3 C_6}{C_3 + C_6}\right)(18.0\text{ V}) = \left[\frac{(3.00\times10^{-6}\text{F})(6.00\times10^{-6}\text{F})}{3.00\times10^{-6}\text{F} + 6.00\times10^{-6}\text{F}}\right] 18.0\text{V} = 36.0\,\mu\text{C}$$

And then $V_6 = \frac{Q}{C_6} = \frac{36.0\,\mu\text{C}}{6.00\,\mu\text{F}} = 6.00\text{ V}$, $V_3 = \frac{Q}{C_3} = \frac{36.0\,\mu\text{C}}{3.00\,\mu\text{F}} = 12.0\text{ V}$

$$V_b + V_6 - I(6.00\,\Omega) = V_a$$

$$V_a - V_b = 6.00\text{ V} - (2.00\text{ A})(6.00\,\Omega) = \underline{-6.00\text{ V}}$$

or

$$V_b - V_3 + I(3.00\,\Omega) = V_a$$

$$V_a - V_b = (2.00\text{A})(3.00\,\Omega) - 12.0\text{ V} = -6.00\text{V} \checkmark$$

b) $V_{ab} < 0$; point b is at higher potential

27-41 (cont)

c)

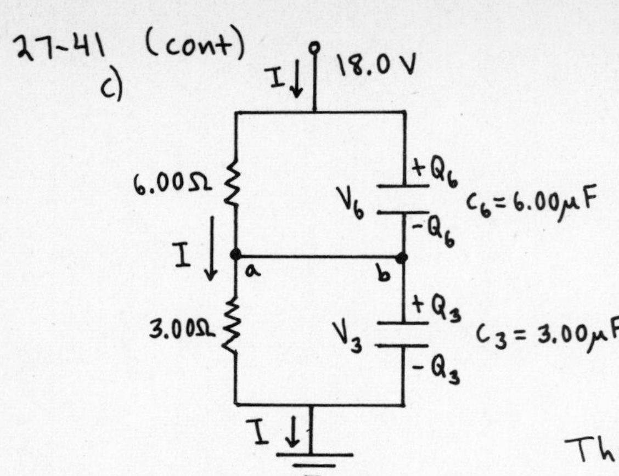

The only current path is still the one through both resistors, so $I = 2.00$ A as calculated in part (a).

The effect of closing the switch is to put points a and b at the same potential, so that $I(6.00\,\Omega) = 12.0$ V is the potential V_6 across the 6.00μF capacitor and $I(3.00\,\Omega) = 6.00$ V is the potential V_3 across the 3.00μF capacitor.

The potential of point b above ground is $V_b = I(3.00\,\Omega) = V_3 = \underline{6.00\text{ V}}$

d) The charges on the capacitors after the switch is closed are

$$Q_6 = C_6 V_6 = (6.00\,\mu F)(12.0\text{ V}) = 72.0\,\mu C$$
$$Q_3 = C_3 V_3 = (3.00\,\mu F)(6.00\text{ V}) = 18.0\,\mu C$$

As calculated in part (a), with the switch open the charge on each capacitor was 36.0μC.

switch open

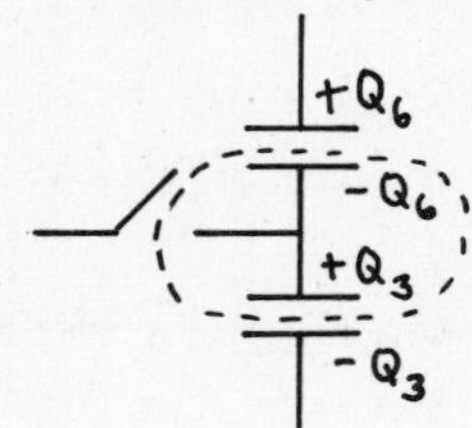

The net charge on the conductor enclosed by the dashed line is zero.

switch closed

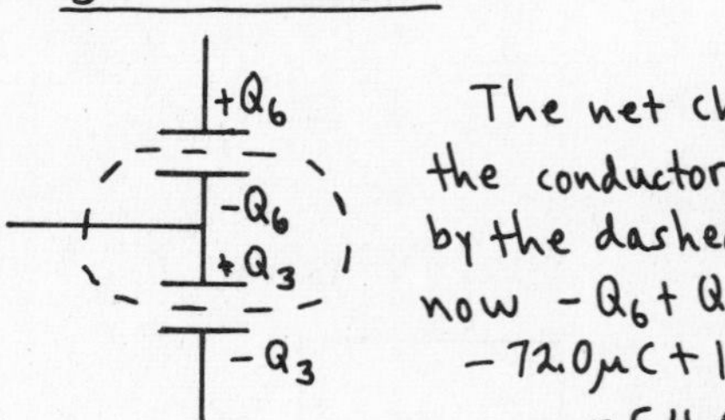

The net charge on the conductor enclosed by the dashed line is now $-Q_6 + Q_3 = -72.0\,\mu C + 18.0\,\mu C = \underline{-54.0\,\mu C}$

This charge must have flowed through the switch when it was closed.

27-43

a)

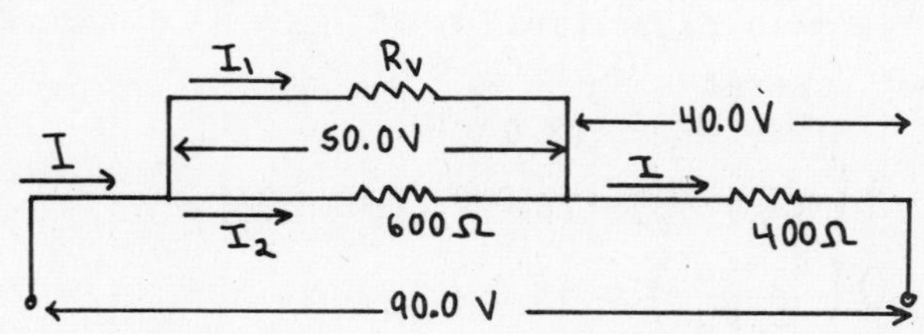

The voltmeter reads the potential difference across its terminals, which is 50.0 V. If we can find the current I_1 through the voltmeter then we can use Ohm's law to find its resistance.

The voltage drop across the 400Ω resistor is 90.0 V − 50.0 V = 40.0 V, so

$$I = \frac{V}{R} = \frac{40.0\text{ V}}{400\,\Omega} = 0.100\text{ A}$$

The voltage drop across the 600Ω resistor is 50.0 V, so $I_2 = \frac{V}{R} = \frac{50.0\text{ V}}{600\,\Omega} = 0.08333$. Then $I = I_1 + I_2 \Rightarrow I_1 = I - I_2 = 0.100\text{ A} - 0.08333\text{ A} = 0.01667$ A.

$$R_V = \frac{V}{I_1} = \frac{50.0\text{ V}}{0.01667\text{ A}} = \underline{3000\,\Omega}$$

27-43 (cont)

b)

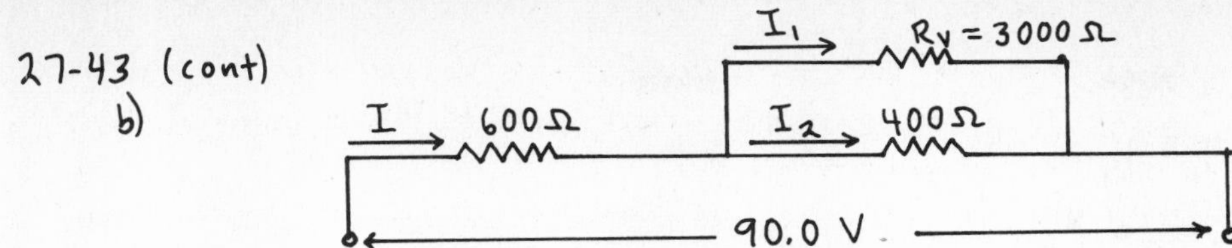

Replace the two resistors in parallel by their equivalent:

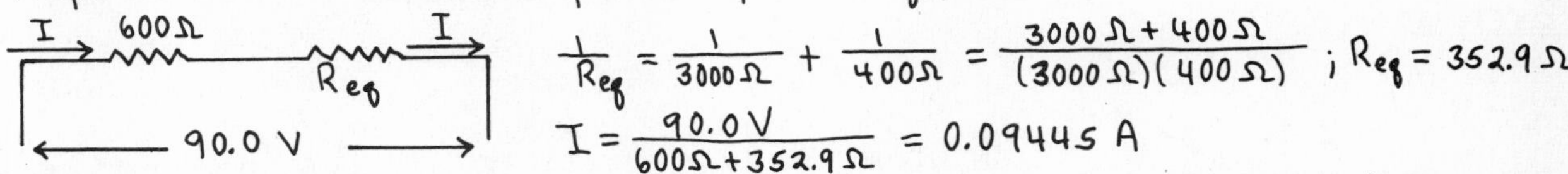

$$\frac{1}{R_{eq}} = \frac{1}{3000\,\Omega} + \frac{1}{400\,\Omega} = \frac{3000\,\Omega + 400\,\Omega}{(3000\,\Omega)(400\,\Omega)}\ ;\ R_{eq} = 352.9\,\Omega$$

$$I = \frac{90.0\text{ V}}{600\,\Omega + 352.9\,\Omega} = 0.09445\text{ A}$$

The potential drop across the $600\,\Omega$ resistor then is $IR = (0.09445\text{ A})(600\,\Omega) = 56.67\text{ V}$, so the potential drop across the $400\,\Omega$ resistor and across the voltmeter (what the voltmeter reads) is $90.0\text{ V} - 56.67\text{ V} = \underline{33.3\text{ V}}$.

27-49

$\mathcal{E} = 200\text{ V}$; i ; R ; C ; $i = 8.00\times10^{-4}\text{ A}$

Initially, the charge on the capacitor is zero, so by $v = \frac{q}{C}$ the voltage across the capacitor is zero. Therefore, the loop rule gives

$$\mathcal{E} - iR = 0 \Rightarrow R = \frac{\mathcal{E}}{i} = \frac{200\text{ V}}{8.00\times10^{-4}\text{ A}} = \underline{2.50\times10^{5}\,\Omega}$$

The time constant is given by $\tau = RC$ (Eq. 27-15), so $C = \frac{\tau}{R} = \frac{6.00\text{ s}}{2.50\times10^{5}\,\Omega} = \underline{24.0\,\mu\text{F}}$

27-51

Eq. (27-14) $i = \frac{\mathcal{E}}{R}e^{-t/RC}$

a) $P = \mathcal{E}i$

The total energy supplied by the battery is $\int_0^\infty P\,dt = \int_0^\infty \mathcal{E}i\,dt = \frac{\mathcal{E}^2}{R}\int_0^\infty e^{-t/RC}\,dt$
$= \frac{\mathcal{E}^2}{R}\left(-RCe^{-t/RC}\Big|_0^\infty\right) = C\mathcal{E}^2$

b) $P = i^2R$

The total energy dissipated in the resistor is $\int_0^\infty P\,dt = \int_0^\infty i^2R\,dt = \frac{\mathcal{E}^2}{R}\int_0^\infty e^{-2t/RC}\,dt$
$= \frac{\mathcal{E}^2}{R}\left(-\frac{RC}{2}e^{-2t/RC}\Big|_0^\infty\right) = \frac{1}{2}C\mathcal{E}^2.$

c) The final charge on the capacitor is $Q = C\mathcal{E}$. The energy stored is $U = \frac{Q^2}{2C} = \frac{1}{2}C\mathcal{E}^2$.

Final energy stored in capacitor $(\frac{1}{2}C\mathcal{E}^2)$ = total energy supplied by battery $(C\mathcal{E}^2)$
− energy dissipated in the resistor $(\frac{1}{2}C\mathcal{E}^2)$.

d) $\frac{1}{2}$ of the energy supplied by the battery is stored in the capacitor. This fraction is independent of R.

CHAPTER 28

Exercises 1, 3, 7, 9, 11, 15, 19, 23, 25, 27, 29

Problems 31, 33, 39, 41, 43, 47, 49

Exercises

28-1

Or, view from above

The force must be in the direction the particle is deflected.

$\vec{F}$ is in the direction of $\vec{v} \times \vec{B}$ as given by the right-hand rule, so q is positive.

28-3

$$\vec{v} = (-3.00 \times 10^4 \text{ m/s})\hat{i} + (5.00 \times 10^4 \text{ m/s})\hat{j}$$

a) $\vec{B} = (1.40\text{T})\hat{i}$

$\vec{F} = q\vec{v} \times \vec{B} = (-2.50 \times 10^{-8}\text{C})(1.40\text{T})[(-3.00 \times 10^4 \text{ m/s})\hat{i} \times \hat{i} + (5.00 \times 10^4 \text{ m/s})\hat{j} \times \hat{i}]$

$\hat{i} \times \hat{i} = 0, \quad \hat{j} \times \hat{i} = -\hat{k}$

$\vec{F} = (-2.50 \times 10^{-8}\text{C})(1.40\text{T})(5.00 \times 10^4 \text{ m/s})(-\hat{k}) = +(1.75 \times 10^{-3} \text{ N})\hat{k}$

The right-hand rule gives that $\vec{v} \times \vec{B}$ is directed into the paper ($-z$-direction). The charge is negative so $\vec{F}$ is opposite to $\vec{v} \times \vec{B}$; $\vec{F}$ is in the $+z$-direction. This agrees with the direction calculated with unit vectors.

b) $\vec{B} = (1.40\text{T})\hat{k}$

$\vec{F} = q\vec{v} \times \vec{B} = (-2.50 \times 10^{-8}\text{C})(1.40\text{T})[(-3.00 \times 10^4 \text{ m/s})\hat{i} \times \hat{k} + (5.00 \times 10^4 \text{ m/s})\hat{j} \times \hat{k}]$

$\hat{i} \times \hat{k} = -\hat{j}; \quad \hat{j} \times \hat{k} = \hat{i}$

$\vec{F} = (+1.05 \times 10^{-3}\text{N})(-\hat{j}) - (1.75 \times 10^{-3}\text{N})(\hat{i}) = -[(1.75 \times 10^{-3}\text{N})\hat{i} + (1.05 \times 10^{-3}\text{N})\hat{j}]$

$\vec{v} \times \vec{B}$ (by right-hand rule)

The direction of $\vec{F}$ is opposite to $\vec{v} \times \vec{B}$ since q is negative. The direction of $\vec{F}$ computed from the right-hand rule agrees qualitatively with the direction calculated with unit vectors.

28-7

$$\Phi_B = \int \vec{B} \cdot d\vec{A}$$

Circular area in the xy-plane $\Rightarrow A = \pi r^2 = \pi(0.400\text{m})^2 = 0.5027 \text{ m}^2$ and $d\vec{A}$ is in the z-direction.

a) $\vec{B} = (1.60\text{T})\hat{k}$; $\vec{B}$ and $d\vec{A}$ are parallel ($\phi = 0°$) so $\vec{B} \cdot d\vec{A} = B\,dA$

B is constant over the circular area so

$\Phi_B = \int \vec{B} \cdot d\vec{A} = \int B\,dA = B\int dA = BA = (1.60\text{T})(0.5027 \text{ m}^2) = \underline{0.804 \text{ Wb}}$

28-7 (cont)

b) $\vec{B}\cdot d\vec{A} = B\cos\phi\, dA$ with $\phi = 30.0°$

B and ϕ are constant over the circular area so

$\Phi_B = \int \vec{B}\cdot d\vec{A} = \int B\cos\phi\, dA = B\cos\phi \int dA = B\cos\phi\, A = (1.60\text{T})(\cos 30.0°)(0.5027\text{m}^2)$

$\Phi_B = \underline{0.697\text{ Wb}}$

c) 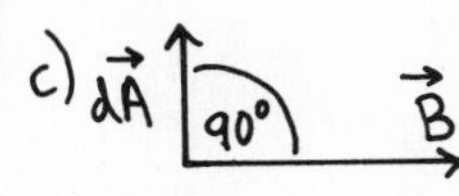

$\vec{B}\cdot d\vec{A} = 0$ since $d\vec{A}$ and $\vec{B}$ are perpendicular ($\phi = 90°$)

$\Phi_B = \int \vec{B}\cdot d\vec{A} = 0$

28-9

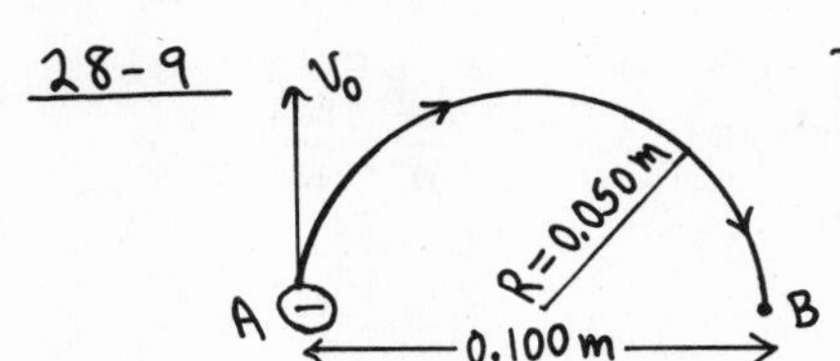

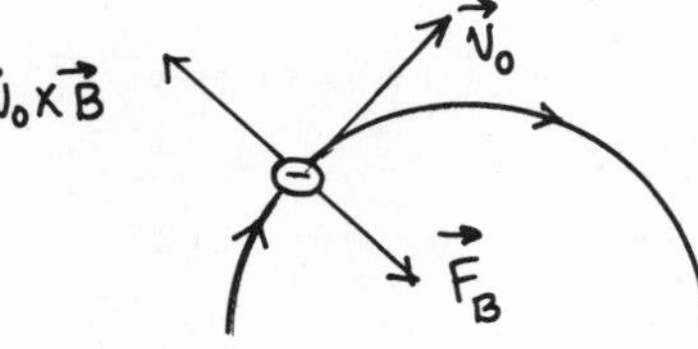

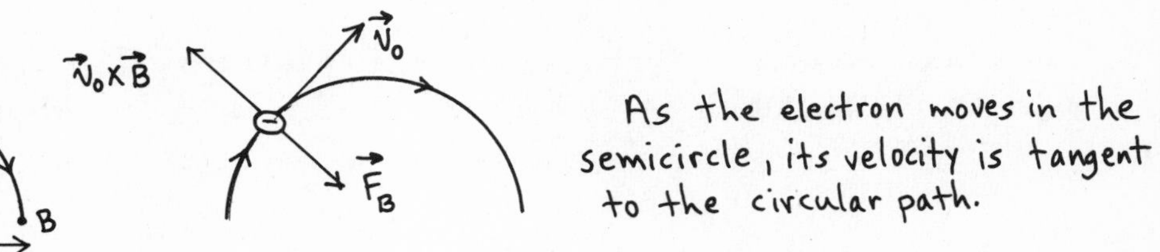

As the electron moves in the semicircle, its velocity is tangent to the circular path.

Circular motion ⇒ the acceleration of the electron $\vec{a}_{rad}$ is directed in toward the center of the circle. Thus the force $\vec{F}_B$ exerted by the magnetic field, since it is the only force on the electron, must be radially inward. Since q is negative, $\vec{F}_B$ is opposite to the direction given by the right-hand rule for $\vec{v}_0 \times \vec{B}$. Thus $\vec{B}$ is <u>into the page</u>.

Apply Newton's 2nd law to calculate the magnitude of B:

$$\Sigma \vec{F} = m\vec{a} \Rightarrow \Sigma F_{rad} = ma_{rad}$$

$$F_B = m\frac{v^2}{R}$$

$$F_B = |q|vB\sin\phi = |q|vB \Rightarrow |q|vB = m\frac{v^2}{R}$$

$$B = \frac{mv}{|q|R} = \frac{(9.109\times10^{-31}\text{kg})(4.00\times10^6\text{m/s})}{(1.602\times10^{-19}\text{C})(0.050\text{ m})} = \underline{4.55\times10^{-4}\text{T}}$$

b) The speed of the electron as it moves along the path is constant. ($\vec{F}_B$ changes the direction of $\vec{v}$ but not its magnitude), so the time is given by the distance divided by v_0. The distance along the semicircular path is πR, so

$$t = \frac{\pi R}{v_0} = \frac{\pi(0.050\text{ m})}{4.00\times10^6\text{ m/s}} = \underline{3.93\times10^{-8}\text{ s}}$$

28-11

a) $\Sigma\vec{F} = m\vec{a} \Rightarrow |q|vB = m\frac{v^2}{R}$

$$v = \frac{|q|BR}{m} = \frac{(1.602\times10^{-19}\text{C})(1.50\text{T})(0.0400\text{m})}{3.34\times10^{-27}\text{kg}} = \underline{2.88\times10^6\text{m/s}}$$

b) $t = \frac{\pi R}{v} = \frac{\pi(0.0400\text{ m})}{2.88\times10^6\text{m/s}} = \underline{4.36\times10^{-8}\text{s}}$

c) Kinetic energy gained = electric potential energy lost

$\frac{1}{2}mv^2 = |q|V$

$$V = \frac{mv^2}{2|q|} = \frac{(3.34\times10^{-27}\text{kg})(2.88\times10^6\text{m/s})^2}{2(1.602\times10^{-19}\text{C})} = 8.65\times10^4\text{ V} = \underline{86.5\text{ kV}}$$

28-15

In the velocity selector $|q|E = |q|vB$

$$v = \frac{E}{B} = \frac{1.20\times10^6\ \text{V/m}}{0.600\ \text{T}} = 2.00\times10^6\ \text{m/s}$$

In the region of the circular path $\Sigma\vec{F} = m\vec{a}$ gives

$$|q|vB = m\frac{v^2}{R} \Rightarrow m = \frac{|q|RB}{v}$$

Singly charged ion $\Rightarrow |q| = +e = 1.602\times10^{-19}\ \text{C}$

$$m = \frac{(1.602\times10^{-19}\ \text{C})(0.728\ \text{m})(0.600\ \text{T})}{2.00\times10^6\ \text{m/s}} = \underline{3.50\times10^{-26}\ \text{kg}}$$

Mass number = # of protons and neutrons. The mass of one proton or neutron is 1.67×10^{-27} kg, so the mass number for this isotope is $\frac{3.50\times10^{-26}\ \text{kg}}{1.67\times10^{-27}\ \text{kg}} = \underline{21}$.

28-19

$$\vec{F} = I\vec{l}\times\vec{B}$$

$I = 7.00\ \text{A}$, $\vec{l} = (0.0100\ \text{m})\hat{\imath}$ (since the current is in the +x-direction)

a) $\vec{B} = -(0.600\ \text{T})\hat{\jmath}$

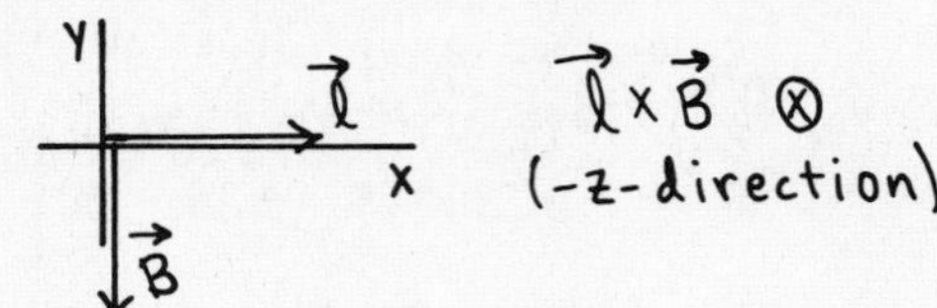

$\vec{F} = (7.00\ \text{A})(0.0100\ \text{m})(-0.600\ \text{T})\,\hat{\imath}\times\hat{\jmath}$

$\hat{\imath}\times\hat{\jmath} = \hat{k}$, so $\vec{F} = -(0.0420\ \text{N})\hat{k}$

b) $\vec{B} = +(0.500\ \text{T})\hat{k}$

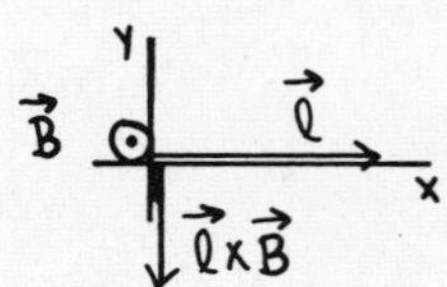

$\vec{F} = (7.00\ \text{A})(0.0100\ \text{m})(0.500\ \text{T})\,\hat{\imath}\times\hat{k}$

$\hat{\imath}\times\hat{k} = -\hat{\jmath}$, so $\vec{F} = -(0.0350\ \text{N})\hat{\jmath}$

c) $\vec{B} = (-0.300\ \text{T})\hat{\imath}$

$\vec{F} = (7.00\ \text{A})(0.0100\ \text{m})(-0.300\ \text{T})\,\hat{\imath}\times\hat{\imath}$

$\hat{\imath}\times\hat{\imath} = 0$, so $F = 0$.

d) $\vec{B} = +(0.200\ \text{T})\hat{\imath} - (0.300\ \text{T})\hat{k}$

$\vec{F} = (7.00\ \text{A})(0.0100\ \text{m})[(0.200\ \text{T})\,\hat{\imath}\times\hat{\imath} - (0.300\ \text{T})\,\hat{\imath}\times\hat{k}]$

$\hat{\imath}\times\hat{\imath} = 0$; $\hat{\imath}\times\hat{k} = -\hat{\jmath}$, so $F = (7.00\ \text{A})(0.0100\ \text{m})(-0.300\ \text{T})(-\hat{\jmath}) = +(0.0210\ \text{N})\hat{\jmath}$

e) $\vec{B} = +(0.900\ \text{T})\hat{\jmath} - (0.400\ \text{T})\hat{k}$

$\vec{F} = (7.00\ \text{A})(0.0100\ \text{m})[+(0.900\ \text{T})(\hat{\imath}\times\hat{\jmath}) - (0.400\ \text{T})(\hat{\imath}\times\hat{k})]$

$\hat{\imath}\times\hat{\jmath} = \hat{k}$; $\hat{\imath}\times\hat{k} = -\hat{\jmath}$

$\vec{F} = (0.0630\ \text{N})\hat{k} - (0.0280\ \text{N})(-\hat{\jmath}) = +(0.0280\ \text{N})\hat{\jmath} + (0.0630\ \text{N})\hat{k}$

28-23

$$U = -\vec{\mu}\cdot\vec{B}$$

$\vec{\mu}$ and $\vec{B}$ parallel $\Rightarrow \phi = 0°$ and $\vec{\mu}\cdot\vec{B} = \mu B\cos\phi = \mu B$

$\vec{\mu}$ and $\vec{B}$ antiparallel $\Rightarrow \phi = 180°$ and $\vec{\mu}\cdot\vec{B} = \mu B\cos\phi = -\mu B$

$U_1 = -\mu B$, $U_2 = +\mu B$

$\Delta U = U_2 - U_1 = 2\mu B = 2(1.30\ \text{A}\cdot\text{m}^2)(0.750\ \text{T}) = \underline{+1.95\ \text{J}}$

28-25

$\vec{\tau} = \vec{\mu} \times \vec{B}$, $U = -\vec{\mu}\cdot\vec{B}$, $\vec{B} = B\vec{j}$

a) $\vec{\mu}\odot \Rightarrow \vec{\mu} = \mu\vec{k} = NIA\vec{k}$

$\vec{\tau} = \vec{\mu}\times\vec{B} = NIAB\,\vec{k}\times\vec{j} = -NIAB\vec{i}$

$U = -\vec{\mu}\cdot\vec{B} = -NIAB\,\vec{k}\cdot\vec{j} = 0$

b) $\vec{\mu}\otimes \Rightarrow \vec{\mu} = \mu\vec{j} = NIA\vec{j}$

$\vec{\tau} = \vec{\mu}\times\vec{B} = NIAB\,\vec{j}\times\vec{j} = 0$

$U = -\vec{\mu}\cdot\vec{B} = -NIAB\,\vec{j}\cdot\vec{j} = -NIAB$

c) $\vec{\mu}\otimes \Rightarrow \vec{\mu} = -\mu\vec{k} = -NIA\vec{k}$

$\vec{\tau} = \vec{\mu}\times\vec{B} = -NIAB\,\vec{k}\times\vec{j} = -NIAB(-\vec{i}) = +NIAB\vec{i}$

$U = -\vec{\mu}\cdot\vec{B} = -(-NIAB)\,\vec{k}\cdot\vec{j} = 0$

d) $\vec{\mu}\odot \Rightarrow \vec{\mu} = -\mu\vec{j} = -NIA\vec{j}$

$\vec{\tau} = \vec{\mu}\times\vec{B} = -NIAB\,(\vec{j}\times\vec{j}) = 0$

$U = -\vec{\mu}\cdot\vec{B} = -(-NIAB)\vec{j}\cdot\vec{j} = +NIAB$

28-27

$I = 4.50\text{A}$, $V = 120\text{V}$, $R_f = 140\Omega$, $R_r = 5.00\Omega$, I_f, I_r, $\mathcal{E}$, a

$\mathcal{E}$ is the induced emf developed by the motor. It is directed so as to oppose the current through the rotor.

a) The field coils and the rotor are in parallel with the applied potential difference V, so $V = I_f R_f \Rightarrow I_f = \frac{V}{R_f} = \frac{120\text{V}}{140\Omega} = \underline{0.857\text{A}}$

b) Apply the junction rule to point a in the circuit diagram

$\Rightarrow I - I_f - I_r = 0$

$I_r = I - I_f = 4.50\text{A} - 0.857\text{A} = \underline{3.64\text{A}}$

c) The potential drop across the rotor, $I_r R_r + \mathcal{E}$, must equal the applied potential difference V: $V = I_r R_r + \mathcal{E}$

$\mathcal{E} = V - I_r R_r = 120\text{V} - (3.64\text{A})(5.00\Omega) = \underline{102\text{V}}$

d) The mechanical power output is the electrical power input minus the rate of dissipation of electrical energy in the resistance of the motor:

electrical power input to the motor

$P_{in} = IV = (4.50\text{A})(120\text{V}) = 540\text{W}$

28-27 (cont)

electrical power loss in the two resistances

$$P_{loss} = I_f^2 R_f + I_r^2 R_r = (0.857\ A)^2(140\ \Omega) + (3.64 A)^2(5.00\ \Omega) = 169\ W$$

mechanical power output

$$P_{out} = P_{in} - P_{loss} = 540\ W - 169\ W = \underline{371\ W}$$

Note: The mechanical power output is the power associated with the induced emf $\mathcal{E}$

$P_{out} = P_{\mathcal{E}} = \mathcal{E} I_r = (102 V)(3.64 A) = 371\ W$, which agrees with the above calculation.

28-29

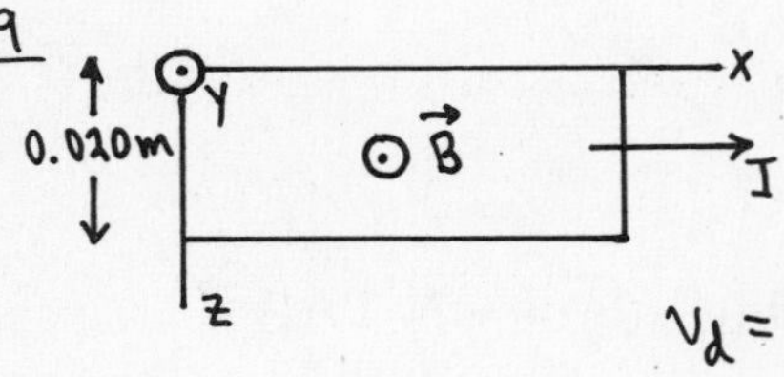

$$J_x = n|q| v_d \Rightarrow v_d = \frac{J_x}{n|q|}$$

$$J_x = \frac{I}{A} = \frac{I}{y_1 z_1} = \frac{140\ A}{(1.00\times10^{-3} m)(0.020 m)} = 7.00\times10^{6}\ A/m^2$$

$$v_d = \frac{J_x}{n|q|} = \frac{7.00\times10^{6}\ A/m^2}{(5.85\times10^{28}/m^3)(1.602\times10^{-19} C)} = \underline{7.47\times10^{-4}\ m/s}$$

b) magnitude of $\vec{E}$

$$|q| E_z = |q| v_d B_y$$

$$E_z = v_d B_y = (7.47\times10^{-4} m/s)(1.50 T) = \underline{1.12\times10^{-3}\ V/m}$$

direction of $\vec{E}$

The drift velocity of the electrons is in the opposite direction to the current.

$v_d \leftarrow \ominus \rightarrow I$ $\quad \vec{v}\times\vec{B}\ \uparrow \quad \vec{F}_B = q\vec{v}\times\vec{B} = -e\,\vec{v}\times\vec{B}\ \downarrow$

$\odot B$

$\uparrow F_E$, $\ominus$, $\downarrow F_B$

$\vec{F}_E$ must oppose $\vec{F}_B \Rightarrow \vec{F}_E$ is in the $-z$-direction

$\vec{F}_E = q\vec{E} = -e\vec{E} \Rightarrow \vec{E}$ is opposite to the direction of $\vec{F}_E$

$\Rightarrow \vec{E}$ is in the $\underline{+z\text{-direction}}$

c) The Hall emf is the potential difference between the two edges of the strip (at $z = 0$ and $z = z_1$) that results from the electric field calculated in part (b).

$$\mathcal{E}_{Hall} = E z_1 = (1.12\times10^{-3} V/m)(0.0200 m) = \underline{2.24\times10^{-5}\ V}$$

Problems

28-31

$$\vec{v}_0 = (4.00\times10^{3} m/s)\,\hat{i} \quad ; \quad \vec{B} = -(0.600 T)\,\hat{j}$$

a) $q = +0.400\times10^{-8}\ C$

$$\vec{F}_B = q\,\vec{v}_0\times\vec{B} = -(q v_0 B)(\hat{i}\times\hat{j}) = -q v_0 B\,\hat{k}$$

$$\vec{F}_E = q\vec{E}$$

undeflected $\Rightarrow \Sigma\vec{F} = 0 \Rightarrow \vec{F}_E = -\vec{F}_B$

28-31 (cont)

$$q\vec{E} = +q v_0 B\vec{k}$$
$$\vec{E} = +v_0 B\vec{k}$$

$\vec{E}$ is in the $+z$-direction and has magnitude $E = v_0 B = (4.00\times10^3\,\text{m/s})(0.600\,\text{T}) = \underline{2.40\times10^3\,\text{V/m}}$

b) $q = -0.400\times10^{-8}\,\text{C}$

$\vec{E} = +v_0 B\vec{k}$, independent of q.

$\vec{E}$ is the same as calculated in part (a): $E = 2.40\times10^3\,\text{V/m}$ and is in the $+z$-direction.

28-33

$v_1 = 3.00\times10^4\,\text{m/s}$, 45°, Y, X, $\vec{F}_1$

$\vec{F} = q\vec{v}\times\vec{B} \Rightarrow \vec{F}$ is perpendicular to $\vec{v}$ and $\vec{B}$

The information given here means that $\vec{B}$ can have no z-component.

Y, X, $\vec{F}_2$, $\vec{v}_2$

$\vec{F}$ is perpendicular to $\vec{v}$ and $\vec{B} \Rightarrow \vec{B}$ can have no x-component

Both pieces of information taken together say that $\vec{B}$ is in the y-direction; $\vec{B} = B_y\vec{j}$.

Use the information given about $\vec{F}_2$ to calculate B_y:

$$\vec{F}_2 = F_2\vec{i};\quad \vec{v}_2 = v_2\vec{k};\quad \vec{B} = B_y\vec{j}$$

$$\vec{F}_2 = q\vec{v}_2\times\vec{B} \Rightarrow F_2\vec{i} = q v_2 B_y\,\vec{k}\times\vec{j} = q v_2 B_y(-\vec{i}) \Rightarrow F_2 = -q v_2 B_y$$

$$B_y = -\frac{F_2}{q v_2} = -\frac{4.00\times10^{-5}\,\text{N}}{(6.00\times10^{-9}\,\text{C})(2.00\times10^4\,\text{m/s})} = -0.333\,\text{T}$$

$\vec{B}$ has magnitude 0.333T and is in the $-y$-direction.

28-39

$$\vec{B} = -(0.220\,\text{T})\vec{k};\quad \vec{v} = (1.00\times10^6\,\text{m/s})(4\vec{i} - 3\vec{j} + 12\vec{k});\quad F = 2.00\,\text{N}$$

a) $\vec{F} = q\vec{v}\times\vec{B}$

$$\vec{F} = q(-0.220\,\text{T})(1.00\times10^6\,\text{m/s})(4\vec{i}\times\vec{k} - 3\vec{j}\times\vec{k} + 12\vec{k}\times\vec{k})$$

$$\vec{i}\times\vec{k} = -\vec{j};\quad \vec{j}\times\vec{k} = \vec{i};\quad \vec{k}\times\vec{k} = 0$$

$$\vec{F} = -q(2.20\times10^5\,\text{N/C})(-4\vec{j} - 3\vec{i}) = +q(2.20\times10^5\,\text{N/C})(3\vec{i} + 4\vec{j})$$

The magnitude of the vector $3\vec{i} + 4\vec{j}$ is $\sqrt{3^2+4^2} = 5$.

$$\Rightarrow F = q(2.20\times10^5\,\text{N/C})(5)$$

$$q = \frac{F}{5(2.20\times10^5\,\text{N/C})} = \frac{2.00\,\text{N}}{5(2.20\times10^5\,\text{N/C})} = \underline{1.82\times10^{-6}\,\text{C}}$$

b) $\sum\vec{F} = m\vec{a} \Rightarrow \vec{a} = \frac{\vec{F}}{m}$

$$\vec{F} = q(2.20\times10^5\,\text{N/C})(3\vec{i}+4\vec{j}) = (1.82\times10^{-6}\,\text{C})(2.20\times10^5\,\text{N/C})(3\vec{i}+4\vec{j}) = 0.4004\,\text{N}\,(3\vec{i}+4\vec{j})$$

28-39 (cont) $\vec{a} = \frac{\vec{F}}{m} = \frac{0.4004\ N\ (3\vec{i}+4\vec{j})}{1.50\times10^{-15}\ kg} = (2.67\ m/s^2)(3\vec{i}+4\vec{j})$

c) $\vec{F}$ is in the xy-plane, so in the z-direction the particle moves with constant speed 12.0×10^{6} m/s.

In the xy-plane the force $\vec{F}$ causes the particle to move in a circle, with $\vec{F}$ directed in toward the center of the circle.

$\vec{F} = m\vec{a} \Rightarrow F = m\frac{v^2}{R} \Rightarrow R = \frac{mv^2}{F}$

$v^2 = v_x^2 + v_y^2 = (4.00\times10^{6}\ m/s)^2 + (-3.00\times10^{6}\ m/s)^2 = 25.00\times10^{12}\ m^2/s^2$

$R = \frac{mv^2}{F} = \frac{(1.50\times10^{-15}\ kg)(25.00\times10^{12}\ m^2/s^2)}{2.00\ N} = \underline{0.0188\ m}$

d) Eq. (28-12): cyclotron frequency $\omega = \frac{v}{R}$

The circular motion is in the xy-plane, so $v = \sqrt{v_x^2 + v_y^2} = 5.00\times10^{6}$ m/s

$\omega = \frac{v}{R} = \frac{5.00\times10^{6}\ m/s}{0.0188\ m} = 2.66\times10^{8}$ rad/s

e) The period of the motion in the xy-plane is given by

$T = \frac{2\pi}{\omega} = \frac{2\pi}{2.66\times10^{8}\ rad/s} = 2.36\times10^{-8}\ s$

In $t = 2T$ the particle has returned to the same x and y coordinates.

The z-component of the motion is motion with a constant velocity of $v_z = 12.0\times10^{6}$ m/s

$\Rightarrow z = z_0 + v_z t = 0 + (12.0\times10^{6}\ m/s)(2)(2.36\times10^{-8}\ s) = 0.566\ m$

The coordinates at $t = 2T$ are $x = R$, $y = 0$, $z = 0.566$ m.

28-41

$\vec{F} = I\vec{l}\times\vec{B}$

$\vec{B} = (0.200\ T)\vec{i}$; $I = 4.00$ A

a) segment ab

$\vec{l} = (0.500\ m)\vec{j}$

$\vec{F}_{ab} = I\vec{l}\times\vec{B} = (4.00\ A)(0.500\ m)(0.200\ T)\vec{j}\times\vec{i} = (0.400\ N)(-\vec{k}) = -(0.400\ N)\vec{k}$

The force has magnitude 0.400 N and is in the -z-direction.

segment bc

$\vec{l} = (0.50\ m)(\vec{i}-\vec{k})$

$\vec{F}_{bc} = I\vec{l}\times\vec{B} = (4.00\ A)(0.500\ m)(0.200\ T)(\vec{i}\times\vec{i} - \vec{k}\times\vec{i})$

$\vec{i}\times\vec{i} = 0$; $\vec{k}\times\vec{i} = \vec{j}$

$\vec{F}_{bc} = -(0.400\ N)\vec{j}$

The force has magnitude 0.400 N and is in the -y-direction.

segment cd

$\vec{l} = (0.500\ m)(-\vec{j}+\vec{k})$

$\vec{F}_{cd} = I\vec{l}\times\vec{B} = (4.00\ A)(0.500\ m)(0.200\ T)(-\vec{j}\times\vec{i} + \vec{k}\times\vec{i})$

$\vec{j}\times\vec{i} = -\vec{k}$; $\vec{k}\times\vec{i} = \vec{j}$

$\vec{F}_{cd} = (0.400\ N)(\vec{j}+\vec{k})$; $F_{cd} = 0.400\ N\sqrt{1^2+1^2} = 0.566\ N$

28-41 (cont)

The force has magnitude 0.566 N and is midway between the +y-axis and the +z-axis.

Segment de

$\vec{l} = -(0.500\text{ m})\vec{k}$

$\vec{F}_{de} = I\vec{l}\times\vec{B} = -(4.00\text{ A})(0.500\text{ m})(0.200\text{ T})\vec{k}\times\vec{i} = -(0.400\text{ N})\vec{j}$

The force has magnitude 0.400 N and is in the -y-direction.

segment ef

$\vec{l} = -(0.500\text{ m})\vec{i}$

$\vec{F}_{ef} = I\vec{l}\times\vec{B} = -(4.00\text{ A})(0.500\text{ m})(0.200\text{ T})(\vec{i}\times\vec{i}) = 0$

b) $\vec{F}_{tot} = \vec{F}_{ab} + \vec{F}_{bc} + \vec{F}_{cd} + \vec{F}_{de} + \vec{F}_{ef}$

$F_{tot} = (0.400\text{ N})(-\vec{k} - \vec{j} + \vec{j} + \vec{k} - \vec{j}) = -(0.400\text{ N})\vec{j}$

The total force has magnitude <u>0.400 N</u> and is in the <u>-y- direction</u>.

28-43

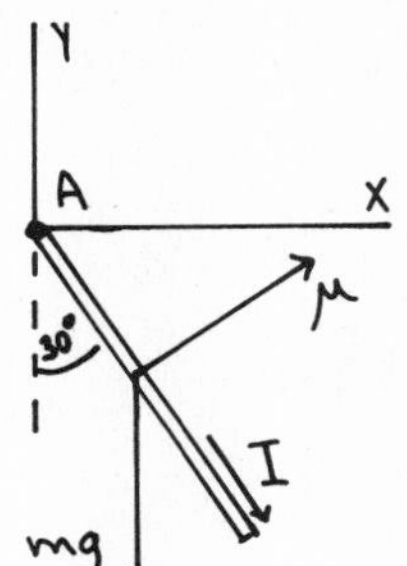

Use $\sum \tau_A = 0$, where point A is along the z-axis.

$\tau_{mg} = mgr\sin\phi = mg(0.0400\text{ m})\sin 30°$

The torque is clockwise; $\vec{\tau}_{mg}$ is directed into the paper.

For the loop to be in equilibrium the torque due to $\vec{B}$ must be counterclockwise (opposite to $\vec{\tau}_{mg}$) and $\tau_B = \tau_{mg}$. $\vec{\tau}_B = \vec{\mu}\times\vec{B}$. For this torque to be counterclockwise ($\vec{\tau}_B$ directed out of the paper) $\vec{B}$ must be in the <u>+y- direction</u>.

$\tau_B = \mu B\sin\phi = IAB\sin 60°$

$\tau_B = \tau_{mg} \Rightarrow IAB\sin 60° = mg(0.0400\text{ m})\sin 30°$

$m = (0.100\text{ g/cm})\,2(8.00\text{ cm} + 6.00\text{ cm}) = 2.80\text{ g} = 2.80\times 10^{-3}\text{ kg}$

$A = (0.0800\text{ m})(0.0600\text{ m}) = 4.80\times 10^{-3}\text{ m}^2$

$$B = \frac{mg(0.0400\text{ m})(\sin 30°)}{IA\sin 60°} = \frac{(2.80\times 10^{-3}\text{ kg})(9.80\text{ m/s}^2)(0.0400\text{ m})\sin 30°}{(8.00\text{ A})(4.80\times 10^{-3}\text{ m}^2)\sin 60°} = \underline{0.0165\text{ T}}$$

28-47

a) After the wire leaves the mercury, its acceleration is g, downward. The wire travels upward a total distance of 0.700 m from its initial position. Its ends lose

28-47 (cont)

contact with the mercury after the wire has traveled 0.050 m, so the wire travels upward 0.650 m after it leaves the mercury. Consider the motion of the wire after it leaves the mercury. Take +y to be upward and take the origin at the position of the wire as it leaves the mercury.

$v_{oy} = ?$
$a_y = -9.80\ m/s^2$
$y - y_0 = +0.650\ m$
$v_y = 0$ (at maximum height)

$$v_y^2 \rightarrow 0 = v_{oy}^2 + 2a_y(y-y_0)$$
$$v_{oy} = \sqrt{-2a_y(y-y_0)} = \sqrt{-2(-9.80\ m/s^2)(+0.650\ m)} = \underline{3.57\ m/s}$$

b) Now consider the motion of the wire while it is in contact with the mercury. Take +y to be upward and the origin at the initial position of the wire. Calculate the acceleration:

$y - y_0 = +0.050\ m$
$v_{oy} = 0$ (starts from rest)
$v_y = +3.57\ m/s$ (from part (a))
$a_y = ?$

$$v_y^2 = v_{oy}^2 \rightarrow 0 + 2a_y(y-y_0)$$
$$a_y = \frac{v_y^2}{2(y-y_0)} = \frac{(3.57\ m/s)^2}{2(0.050\ m)} = 127\ m/s^2$$

Free-body diagram for the wire

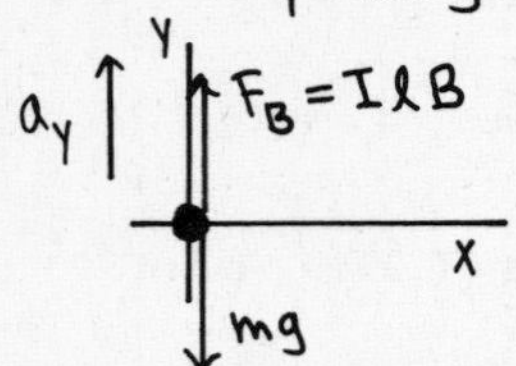

$$\Sigma F_y = ma_y$$
$$F_B - mg = ma_y$$
$$I\ell B = m(g + a_y)$$
$$I = \frac{m(g+a_y)}{\ell B}$$

ℓ is the length of the horizontal section of the wire; $\ell = 0.250\ m$

$$I = \frac{(9.79\times10^{-5}\ kg)(9.80\ m/s^2 + 127\ m/s^2)}{(0.250\ m)(0.0180\ T)} = \underline{2.98\ A}$$

c) $V = IR \Rightarrow R = \frac{V}{I} = \frac{1.50\ V}{2.98\ A} = \underline{0.503\ \Omega}$

28-49

a)

$\vec{\mu}$ is out of the plane of the paper, in the +z-direction $\Rightarrow \vec{\mu} = \mu\vec{k}$.
$\mu = IA = (15.0\ A)(8.00\times10^{-4}\ m^2) = \underline{0.0120\ A\cdot m^2}$

b) $\vec{\tau} = (1.00\times10^{-3}\ N\cdot m)(-6\vec{i} + 8\vec{j})$

$\vec{\tau} = \vec{\mu}\times\vec{B}$; $\vec{\mu} = \mu\vec{k}$; $\vec{B} = B_x\vec{i} + B_y\vec{j} + B_z\vec{k}$

$\vec{\tau} = \vec{\mu}\times\vec{B} = \mu(B_x\vec{k}\times\vec{i} + B_y\vec{k}\times\vec{j} + B_z\vec{k}\times\vec{k}) = -(\mu B_y)\vec{i} + \mu B_x\vec{j}$

Compare this to the expression given for $\vec{\tau} \Rightarrow B_y = \frac{6.00\times10^{-3}\ N\cdot m}{0.0120\ A\cdot m^2} = \underline{+0.500\ T}$; $B_x = \frac{8.00\times10^{-3}\ N\cdot m}{0.0120\ A\cdot m^2} = \underline{+0.667\ T}$

B_z doesn't contribute to the torque since $\vec{\mu}$ is along the z-direction.

But $B = 2.60\ T$ and $B_x^2 + B_y^2 + B_z^2 = B^2$

$\Rightarrow B_z = \pm\sqrt{B^2 - B_x^2 - B_y^2} = \pm\sqrt{(2.60\ T)^2 - (0.500\ T)^2 - (0.667\ T)^2} = \pm 2.46\ T$

$U = -\vec{\mu}\cdot\vec{B}$ is negative determines the sign of B_z:

$U = -\vec{\mu}\cdot\vec{B} = -(\mu\vec{k})\cdot(B_x\vec{i} + B_y\vec{j} + B_z\vec{k}) = -\mu B_z$

U negative $\Rightarrow B_z$ positive, so $B_z = \underline{+2.46\ T}$

CHAPTER 29

Exercises 1, 3, 5, 11, 15, 17, 21, 23, 27, 29, 31

Problems 33, 37, 39, 43, 45, 47, 49, 53

Exercises

29-1

$\vec{B} = \frac{\mu_0}{4\pi}\frac{q\vec{v}\times\hat{r}}{r^2} = \frac{\mu_0}{4\pi}\frac{q\vec{v}\times\vec{r}}{r^3}$, since $\hat{r} = \frac{\vec{r}}{r}$.

$\vec{v} = (8.00\times10^6\ \text{m/s})\vec{i}$; $\vec{r}$ is the vector from the charge to the point where the field is calculated.

a) $\vec{r} = (0.500\ \text{m})\vec{i}$; $r = 0.500\ \text{m}$

$\vec{v}\times\vec{r} = (8.00\times10^6\ \text{m/s})(0.500\ \text{m})\,\vec{i}\times\vec{i} = 0$; $\underline{B = 0}$

b) $\vec{r} = -(0.500\ \text{m})\vec{j}$; $r = 0.500\ \text{m}$

$\vec{v}\times\vec{r} = -(8.00\times10^6\ \text{m/s})(0.500\ \text{m})\,\vec{i}\times\vec{j} = -(4.00\times10^6\ \text{m}^2/\text{s})\vec{k}$

$\vec{B} = (1\times10^{-7}\ \text{T}\cdot\text{m/A})\frac{(3.00\times10^{-6}\ \text{C})(4.00\times10^6\ \text{m}^2/\text{s})}{(0.500\ \text{m})^3}(-\vec{k}) = \underline{-(9.60\times10^{-6}\ \text{T})\vec{k}}$

c) $\vec{r} = +(0.500\ \text{m})\vec{k}$; $r = 0.500\ \text{m}$

$\vec{v}\times\vec{r} = (8.00\times10^6\ \text{m/s})(0.500\ \text{m})\,\vec{i}\times\vec{k} = -(4.00\times10^6\ \text{m}^2/\text{s})\vec{j}$

$\vec{B} = (1\times10^{-7}\ \text{T}\cdot\text{m/A})\frac{(3.00\times10^{-6}\ \text{C})(4.00\times10^6\ \text{m}^2/\text{s})}{(0.500\ \text{m})^3}(-\vec{j}) = -(9.60\times10^{-6}\ \text{T})\vec{j}$

d) $\vec{r} = -(0.500\ \text{m})\vec{j} + (0.500\ \text{m})\vec{k}$; $r = \sqrt{(0.500\text{m})^2 + (0.500\text{m})^2} = 0.7071\ \text{m}$

$\vec{v}\times\vec{r} = (8.00\times10^6\ \text{m/s})(0.500\ \text{m})(-\vec{i}\times\vec{j} + \vec{i}\times\vec{k}) = (4.00\times10^6\ \text{m}^2/\text{s})(-\vec{j} - \vec{k})$

$\vec{B} = -(1\times10^{-7}\ \text{T}\cdot\text{m/A})\frac{(3.00\times10^{-6}\ \text{C})(4.00\times10^6\ \text{m}^2/\text{s})}{(0.7071\ \text{m})^3}(\vec{j}+\vec{k}) = \underline{-(3.39\times10^{-6}\ \text{T})(\vec{j}+\vec{k})}$

29-3

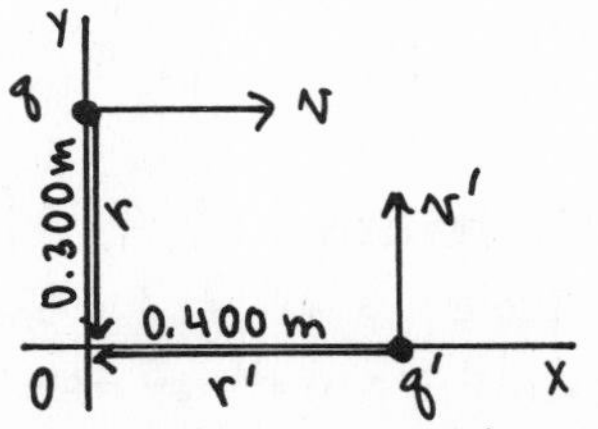

$\vec{B}_{tot} = \vec{B}_q + \vec{B}_{q'}$

$v = v' = 6.00\times10^5\ \text{m/s}$

$q = +5.00\times10^{-6}\ \text{C}$; $q' = -3.00\times10^{-6}\ \text{C}$

$\vec{B} = \left(\frac{\mu_0}{4\pi}\right)\frac{q\vec{v}\times\vec{r}}{r^3}$

Field $\vec{B}_q$ due to q:

$\vec{v} = (6.00\times10^5\ \text{m/s})\vec{i}$; $\vec{r} = -(0.300\ \text{m})\vec{j}$

$\vec{v}\times\vec{r} = -(6.00\times10^5\ \text{m/s})(0.300\ \text{m})(\vec{i}\times\vec{j}) = -(1.80\times10^5\ \text{m}^2/\text{s})\vec{k}$

$\vec{B}_q = (1\times10^{-7}\ \text{T}\cdot\text{m/A})\frac{(5.00\times10^{-6}\ \text{C})(1.80\times10^5\ \text{m}^2/\text{s})}{(0.300\ \text{m})^3}\vec{k} = -(3.333\times10^{-6}\ \text{T})\vec{k}$

29-3 (cont)

Field $\vec{B}_{q'}$ due to q':

$\vec{v} = (6.00\times10^{5}\text{ m/s})\vec{j}\ ;\ \vec{r} = -(0.400\text{ m})\vec{i}$

$\vec{v}\times\vec{r} = -(6.00\times10^{5}\text{ m/s})(0.400\text{ m})(\vec{j}\times\vec{i}) = +(2.40\times10^{5}\text{ m}^2/\text{s})\vec{k}$

$$\vec{B}_{q'} = (1\times10^{-7}\text{ T}\cdot\text{m/A})\frac{(-3.00\times10^{-6}\text{ C})(2.40\times10^{5}\text{ m}^2/\text{s})}{(0.400\text{ m})^3}\vec{k} = -(1.125\times10^{-6}\text{ T})\vec{k}$$

$$\vec{B}_{tot} = \vec{B}_q + \vec{B}_{q'} = -(3.333\times10^{-6}\text{ T} + 1.125\times10^{-6}\text{ T})\vec{k} = -(4.46\times10^{-6}\text{ T})\vec{k}$$

The resultant field at the origin has magnitude 4.46×10^{-6} T and is in the $-z$-direction.

29-5

$$d\vec{B} = \frac{\mu_0 I}{4\pi}\frac{d\vec{l}\times\hat{r}}{r^2} = \frac{\mu_0 I}{4\pi}\frac{d\vec{l}\times\vec{r}}{r^3}$$

Straightline segment on left:

Consider a small segment dl of the wire.
$d\vec{l}\times\vec{r} = 0 \Rightarrow dB = 0 \Rightarrow$ no contribution to B at P

Straightline segment on right:

$d\vec{l}\times\vec{r} = 0 \Rightarrow dB = 0 \Rightarrow$ no contribution to B at P

Semicircular section:

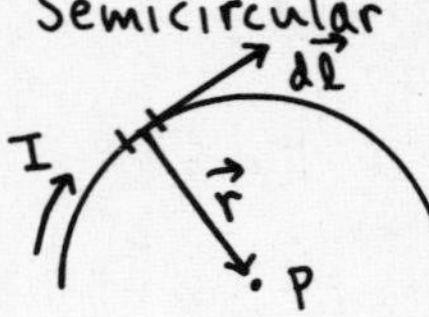

Divide this section up into small segments of length dl.
$d\vec{l}\times\vec{r}$ has direction ⊗ (into the page)
The magnitude of $d\vec{l}\times\vec{r}$ is $|d\vec{l}\times\vec{r}| = r\sin\phi\, dl$
$d\vec{l}$ (tangential) and $\vec{r}$ (radial) are perpendicular so $\phi = 90°$
$r = R \Rightarrow |d\vec{l}\times\vec{r}| = R\,dl$

The magnitude dB of the magnetic field at P due to this infinitesimal segment is

$$dB = \frac{\mu_0 I}{4\pi}\frac{|d\vec{l}\times\vec{r}|}{r^3} = \frac{\mu_0 I}{4\pi}\frac{R\,dl}{R^3} = \frac{\mu_0 I\,dl}{4\pi R^2}$$

The total $\vec{B}$ at P due to the current in the entire semicircle is $\vec{B} = \int d\vec{B}$, where the integral is over all small segments into which the semicircle is divided. But the direction of $d\vec{B}$ is the same (into the page) for each small segment of the semicircle so all the $d\vec{B}$ add and

$$B = \int dB = \int \frac{\mu_0 I\,dl}{4\pi R^2} = \frac{\mu_0 I}{4\pi R^2}\int dl$$

But $\int dl = \pi R$, the length of the semicircle

$$\Rightarrow B = \frac{\mu_0 I}{4\pi R^2}(\pi R) = \frac{\mu_0 I}{4R}$$

(This is half the value for the field at the center of a circular loop (Eq. 29-17).)

29-11

$B = \frac{\mu_0 I}{2\pi r}$ (Eq. 29-9), and the direction of $\vec{B}$ is given by the right-hand rule.

a) At point P midway between the wires, the magnetic field of each current is directed into the paper. $\vec{B}_1$ and $\vec{B}_2$ are in the same direction, so $B = B_1 + B_2$.

$$B_1 = B_2 = \frac{\mu_0 I}{2\pi a} \Rightarrow B = 2\frac{\mu_0 I}{2\pi a} = \frac{\mu_0 I}{\pi a}$$

b)

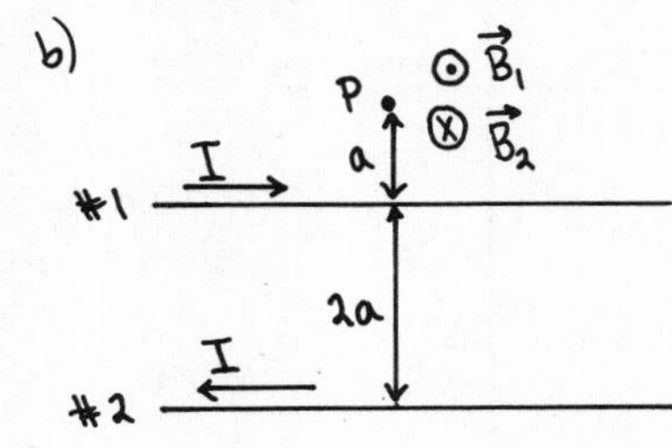

At point P $B_1 > B_2$ since P is closer to wire #1. $\vec{B}_1$ is out of the paper and $\vec{B}_2$ is into the paper. Thus $B = B_1 - B_2$.

$$B_1 = \frac{\mu_0 I}{2\pi a}, \quad B_2 = \frac{\mu_0 I}{2\pi(3a)}$$

$$B = B_1 - B_2 = \frac{\mu_0 I}{2\pi a}\left(1 - \frac{1}{3}\right) = \frac{\mu_0 I}{3\pi a}$$

c) At point P midway between the wires $\vec{B}_1$ and $\vec{B}_2$ are in opposite directions, so $B = B_1 - B_2$.

But $B_1 = B_2 = \frac{\mu_0 I}{2\pi a}$, so $B = 0$.

d)

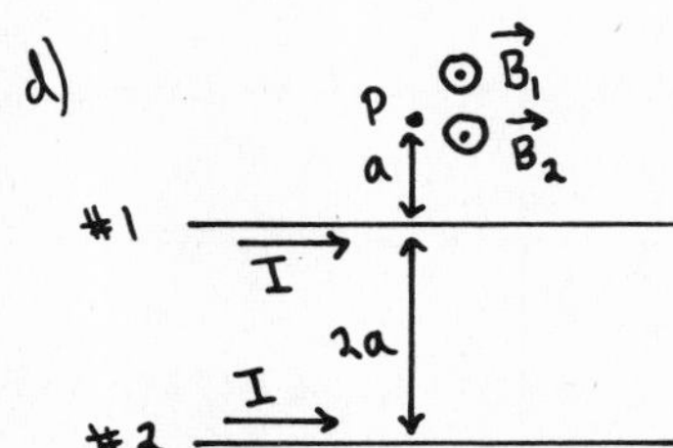

At point P both $\vec{B}_1$ and $\vec{B}_2$ are out of the paper, so $B = B_1 + B_2$.

$$B_1 = \frac{\mu_0 I}{2\pi a}, \quad B_2 = \frac{\mu_0 I}{2\pi(3a)} \Rightarrow B = \frac{\mu_0 I}{2\pi a}\left(1 + \frac{1}{3}\right) = \frac{2\mu_0 I}{3\pi a}$$

29-15

The wire CD rises until the upward force F_I due to the currents balances the downward force of gravity.

Currents in opposite directions $\Rightarrow$ the force is repulsive $\Rightarrow$ F_I is upward, as shown.

Eq. (29-11) $\Rightarrow F_I = \frac{\mu_0 I^2}{2\pi h} L$ where L is the length of wire CD and h is the distance between the wires

$$mg = (5.00 \times 10^{-3}\ \text{kg/m})\, L g$$

$$F_I - mg = 0 \Rightarrow \frac{\mu_0 I^2 L}{2\pi h} = (5.00 \times 10^{-3}\ \text{kg/m})\, L g$$

$$h = \frac{\mu_0 I^2}{2\pi g (5.00 \times 10^{-3}\ \text{kg/m})} = \frac{(2\times 10^{-7}\ \text{T}\cdot\text{m/A})(40.0\ \text{A})^2}{(9.80\ \text{m/s}^2)(5.00 \times 10^{-3}\ \text{kg/m})} = 6.53 \times 10^{-3}\ \text{m} = \underline{6.53\ \text{mm}}$$

29-17

The magnetic field at the center of N circular loops is given by Eq. (29-17): $B = \frac{\mu_0 N I}{2a}$. Thus $N = \frac{2aB}{\mu_0 I} = \frac{2(9.00\times10^{-2}\,m)(4.19\times10^{-4}\,T)}{(4\pi\times10^{-7}\,T\cdot m/A)(2.50\,A)} = \underline{24}$.

29-21

a) The magnetic field near the center of a long solenoid is given by Eq. (29-21): $B = \mu_0 n I$.
turns per unit length $n = \frac{B}{\mu_0 I} = \frac{0.140\,T}{(4\pi\times10^{-7}\,T\cdot m/A)(10.0\,A)} = \underline{1.11\times10^4\ \text{turns/m}}$

b) $N = nL = (1.11\times10^4\ \text{turns/m})(0.500\,m) = 5.55\times10^3$ turns.
Each turn of radius R has a length $2\pi R$ of wire. The total length of wire required is $N(2\pi R) = (5.55\times10^3)(2\pi)(3.00\times10^{-2}\,m) = \underline{1.05\times10^3\,m}$.

29-23

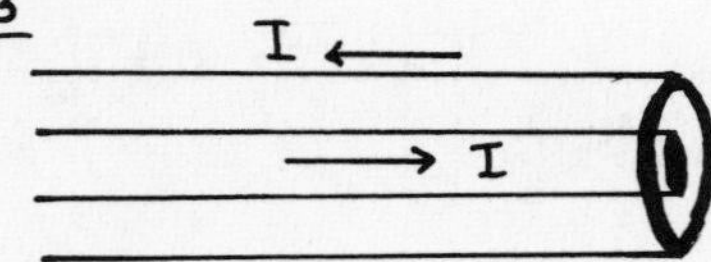

$\oint \vec{B}\cdot d\vec{l} = \mu_0 I_{encl}$

a) end view

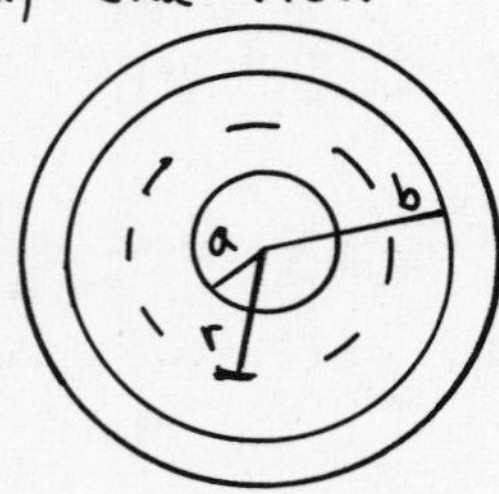

Apply Ampere's law to a path that is a circle of radius r, where $a < r < b$.
By symmetry $\vec{B}$ is tangent to this path and constant around it $\Rightarrow \oint \vec{B}\cdot d\vec{l} = \oint B\,dl = B\oint dl = B(2\pi r)$
$I_{encl} = I$ (All of the current of the inner conductor but none of the outer conductor is enclosed by the path.)
$B(2\pi r) = \mu_0 I \Rightarrow B = \frac{\mu_0 I}{2\pi r}$

b) Apply Ampere's law to a circle of radius r where $r > c$.
As in part (a), $\oint \vec{B}\cdot d\vec{l} = B(2\pi r)$
But now $I_{encl} = 0$, since all the current in each conductor is enclosed by the path and these currents are equal and opposite.
$\Rightarrow B(2\pi r) = 0 \Rightarrow B = 0$.

29-27

$B = K_m \frac{\mu_0 N I}{2\pi r}$ (Eq. 29-22, with μ_0 replaced by $K_m\mu_0$)

a) $K_m = 1400$
$I = \frac{2\pi r B}{\mu_0 K_m N} = \frac{(3.80\times10^{-2}\,m)(0.150\,T)}{(2\times10^{-7}\,T\cdot m/A)(1400)(400)} = \underline{0.0509\,A}$

b) $K_m = 5200$
$I = \frac{2\pi r B}{\mu_0 K_m N} = \frac{(3.80\times10^{-2}\,m)(0.150\,T)}{(2\times10^{-7}\,T\cdot m/A)(5200)(400)} = \underline{0.0137\,A}$

29-29

a) $i_c = 2.00\times10^{-3}$ A

$q = 0$ at $t = 0$

The amount of charge brought to the plates by the charging current in time t is

$q = i_c t = (2.00\times10^{-3}\text{ A})(5.00\times10^{-6}\text{ s}) = \underline{1.00\times10^{-8}\text{ C}}$

$E = \frac{\sigma}{\epsilon_0} = \frac{q}{\epsilon_0 A} = \frac{1.00\times10^{-8}\text{ C}}{(8.854\times10^{-12}\text{ C}^2/\text{N}\cdot\text{m}^2)(4.00\times10^{-4}\text{ m}^2)} = \underline{2.82\times10^{6}\text{ V/m}}$

$V = Ed = (2.82\times10^{6}\text{ V/m})(3.00\times10^{-3}\text{ m}) = \underline{8.46\times10^{3}\text{ V}}$

b) $E = \frac{q}{\epsilon_0 A}$

$\frac{dE}{dt} = \frac{dq/dt}{\epsilon_0 A} = \frac{i_c}{\epsilon_0 A} = \frac{2.00\times10^{-3}\text{ A}}{(8.854\times10^{-12}\text{ C}^2/\text{N}\cdot\text{m}^2)(4.00\times10^{-4}\text{ m}^2)} = \underline{5.65\times10^{11}\text{ V/m}\cdot\text{s}}$

Since i_c is constant $\frac{dE}{dt}$ does not vary in time.

c) $j_D = \epsilon_0 \frac{dE}{dt}$ (Eq. 29-35, with ϵ replaced by ϵ_0 since there is vacuum between the plates.)

$j_D = (8.854\times10^{-12}\text{ C}^2/\text{N}\cdot\text{m}^2)(5.65\times10^{11}\text{ V/m}\cdot\text{s}) = \underline{5.00\text{ A/m}^2}$

$i_D = j_D A = (5.00\text{ A/m}^2)(4.00\times10^{-4}\text{ m}^2) = \underline{2.00\times10^{-3}\text{ A}}$; $i_D = i_c$

29-31

a) $i_c = i_D$, so $j_D = \frac{i_D}{A} = \frac{i_c}{A} = \frac{0.600\text{ A}}{\pi r^2} = \frac{0.600\text{ A}}{\pi(0.0500\text{ m})^2} = \underline{76.4\text{ A/m}^2}$

b) $j_D = \epsilon_0 \frac{dE}{dt} \Rightarrow \frac{dE}{dt} = \frac{j_D}{\epsilon_0} = \frac{76.4\text{ A/m}^2}{8.854\times10^{-12}\text{ C}^2/\text{N}\cdot\text{m}^2} = \underline{8.63\times10^{12}\text{ V/m}\cdot\text{s}}$

c) Apply Ampere's law $\oint \vec{B}\cdot d\vec{l} = \mu_0(i_c + i_D)$ (Eq. 29-34) to a circular path with radius $r = 0.0250$ m.

end view

By symmetry the magnetic field is tangent to the path and constant around it $\Rightarrow \oint \vec{B}\cdot d\vec{l} = \oint B\,dl = B\oint dl = B(2\pi r)$

$i_c = 0$ (no conduction current flows through the air space between the plates)

The displacement current enclosed by the path is $j_D \pi r^2$.

Thus $B(2\pi r) = \mu_0(j_D \pi r^2)$

$B = \frac{1}{2}\mu_0 j_D r = \frac{1}{2}(4\pi\times10^{-7}\text{ T}\cdot\text{m/A})(76.4\text{ A/m}^2)(0.0250\text{ m}) = \underline{1.20\times10^{-6}\text{ T}}$

d) Apply Ampere's law $\oint \vec{B}\cdot d\vec{l} = \mu_0(i_c + i_D)$ to a circular path with radius $r = 0.100$ m.

$\oint \vec{B}\cdot d\vec{l} = B(2\pi r)$ as in part (c)

$i_c = 0$

The displacement current enclosed by the path is $j_D \pi R^2 = i_c$.

Thus $B(2\pi r) = \mu_0 i_c$

$B = \frac{\mu_0 i_c}{2\pi r} = (2\times10^{-7}\text{ T}\cdot\text{m/A})\frac{0.600\text{ A}}{0.100\text{ m}} = \underline{1.20\times10^{-6}\text{ T}}$

Problems

29-33

r ⊖ → v

I

At the electron's position the magnetic field $\vec{B}$ due to the current in the wire has magnitude

$$B = \frac{\mu_0 I}{2\pi r} = \frac{(2\times10^{-7}\,\text{T}\cdot\text{m/A})(1.50\,\text{A})}{0.0800\,\text{m}} = 3.75\times10^{-6}\,\text{T}$$

By the right-hand rule, $\vec{B}$ is ⊙ (out of page) at this point.

$\vec{B}$ ⊙ ⊖ → $\vec{v}$ $\vec{v}\times\vec{B}$ ↓   But $\vec{F} = q\vec{v}\times\vec{B} = -e\,\vec{v}\times\vec{B}$ has the direction ↑ (away from the wire)

$$F = |q|vB\sin\phi = evB = (1.602\times10^{-19}\,\text{C})(5.00\times10^{4}\,\text{m/s})(3.75\times10^{-6}\,\text{T}) = 3.00\times10^{-20}\,\text{N}$$

The force has magnitude $\underline{3.00\times10^{-20}\,\text{N}}$ and is directed <u>away from the wire</u>.

29-37

a)

⊗ $I_1 = 6.00\,\text{A}$

1.00 m

⊙ I_2

0.50 m

B_1 ← P → B_2

$\vec{B}_1$ and $\vec{B}_2$ must be equal and opposite for the resultant field at P to be zero ⇒ $\vec{B}_2$ to the right ⇒ I_2 <u>out of page</u>.

$$B_1 = \frac{\mu_0 I_1}{2\pi r_1} = \frac{\mu_0}{2\pi}\frac{6.00\,\text{A}}{1.50\,\text{m}}$$

$$B_2 = \frac{\mu_0 I_2}{2\pi r_2} = \frac{\mu_0}{2\pi}\frac{I_2}{0.50\,\text{m}}$$

$$B_1 = B_2 \Rightarrow \frac{\mu_0}{2\pi}\left(\frac{6.00\,\text{A}}{1.50\,\text{m}}\right) = \frac{\mu_0}{2\pi}\left(\frac{I_2}{0.50\,\text{m}}\right)$$

$$I_2 = \left(\frac{0.50\,\text{m}}{1.50\,\text{m}}\right)(6.00\,\text{A}) = \underline{2.00\,\text{A}}$$

b)

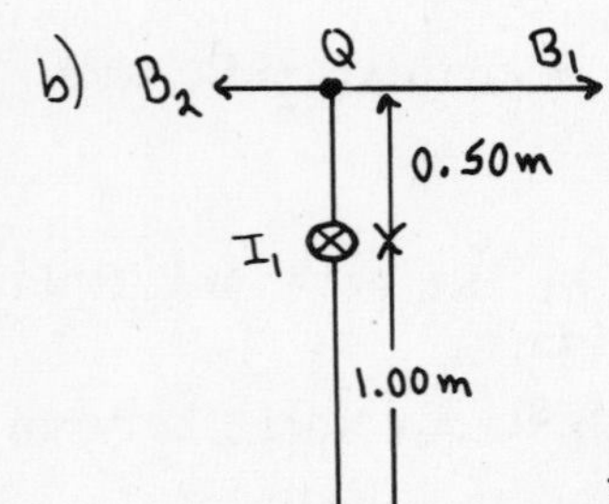

$$B_1 = \frac{\mu_0 I_1}{2\pi r_1} = (2\times10^{-7}\,\text{T}\cdot\text{m/A})\left(\frac{6.00\,\text{A}}{0.50\,\text{m}}\right) = 2.40\times10^{-6}\,\text{T}$$

$$B_2 = \frac{\mu_0 I_2}{2\pi r_2} = (2\times10^{-7}\,\text{T}\cdot\text{m/A})\left(\frac{2.00\,\text{A}}{1.50\,\text{m}}\right) = 2.67\times10^{-7}\,\text{T}$$

$\vec{B}_1$ and $\vec{B}_2$ are in opposite directions and $B_1 > B_2$

⇒ $B = B_1 - B_2 = 2.40\times10^{-6}\,\text{T} - 2.67\times10^{-7}\,\text{T} = \underline{2.13\times10^{-6}\,\text{T}}$, and $\vec{B}$ is <u>to the right</u>.

c)

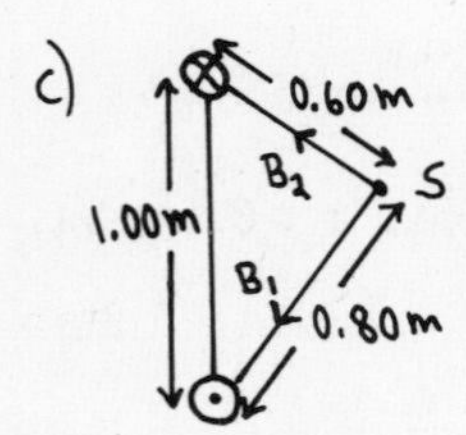

$$B_1 = \frac{\mu_0 I}{2\pi r_1} = (2\times10^{-7}\,\text{T}\cdot\text{m/A})\frac{6.00\,\text{A}}{0.60\,\text{m}} = 2.00\times10^{-6}\,\text{T}$$

$$B_2 = \frac{\mu_0 I}{2\pi r_2} = (2\times10^{-7}\,\text{T}\cdot\text{m/A})\frac{2.00\,\text{A}}{0.80\,\text{m}} = 5.00\times10^{-7}\,\text{T}$$

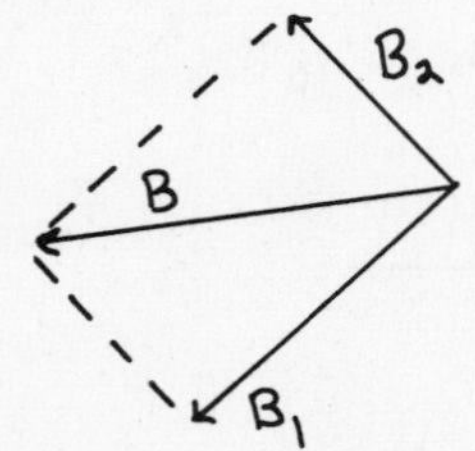

B_1 and B_2 are at right angles to each other, so the magnitude of their resultant is given by

$$B = \sqrt{B_1^2 + B_2^2} = \sqrt{(2.00\times10^{-6}\,\text{T})^2 + (5.00\times10^{-7}\,\text{T})^2} = \underline{2.06\times10^{-6}\,\text{T}}$$

29-39

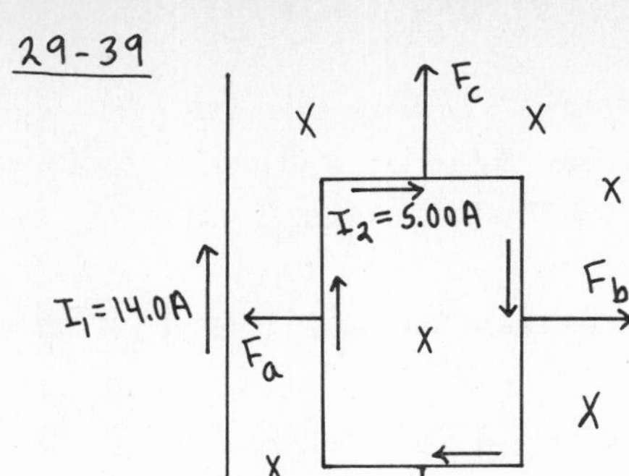

At the location of the loop the magnetic field $\vec{B}_1$ due to the current in the long, straight wire is directed into the page. The direction of the force on each side of the rectangle given by the right-hand rule is shown in the sketch.

$\vec{F}_c$ and $\vec{F}_d$ are equal and opposite, so $\vec{F}_c + \vec{F}_d = 0$.

$F_a = I_2 L B_1$ where $L = 0.200\,m$ and B_1 is the magnetic field due to I_1 at a distance of 0.010 m from the wire.

$$B_1 = \frac{\mu_0 I}{2\pi r} = \frac{\mu_0 I_1}{2\pi(0.010m)} = (2\times10^{-7}\,T\cdot m/A)\left(\frac{14.0A}{0.010\,m}\right) = 2.80\times10^{-4}\,T$$

$$F_a = I_2 L B_1 = (5.00A)(0.200\,m)(2.80\times10^{-4}T) = 2.80\times10^{-4}N$$

$F_b = I_2 L B_1$ where now B_1 is the magnetic field due to I_1 at a distance of 0.100 m from the wire.

$$B_1 = \frac{\mu_0 I}{2\pi r} = \frac{\mu_0 I_1}{2\pi(0.100m)} = (2\times10^{-7}T\cdot m/A)\frac{14.0A}{0.100m} = 2.80\times10^{-5}\,T$$

$$F_b = I_2 L B_1 = (5.00A)(0.200\,m)(2.80\times10^{-5}T) = 2.80\times10^{-5}N$$

$F_a > F_b$ and $\vec{F}_a$ and $\vec{F}_b$ are in opposite directions, so the resultant force $\vec{F}$ has magnitude $F = F_a - F_b = 2.80\times10^{-4}N - 2.80\times10^{-5}N = \underline{2.52\times10^{-4}N}$ and is directed <u>toward the wire</u>.

29-43

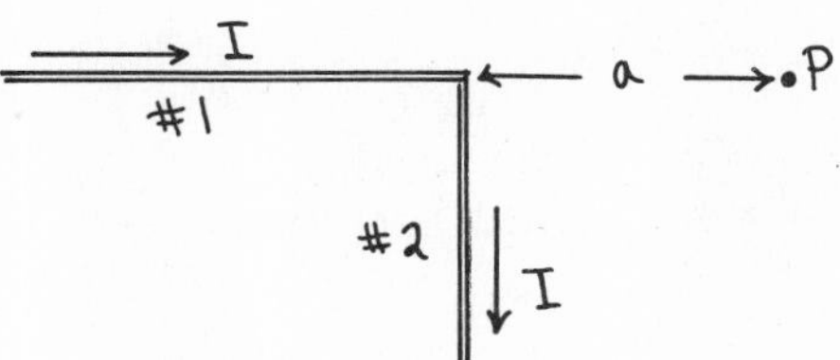

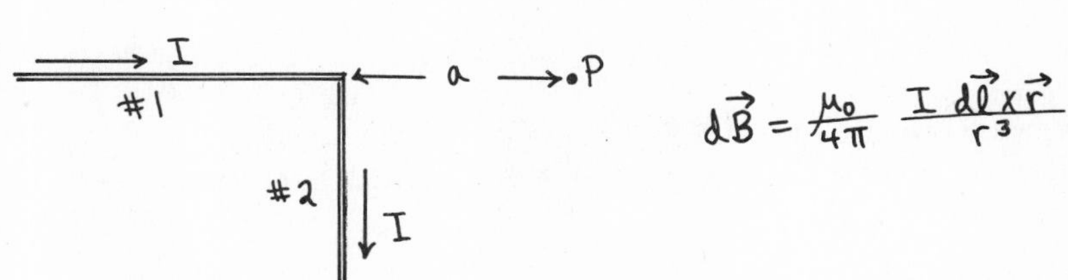

$$d\vec{B} = \frac{\mu_0}{4\pi}\frac{I\,d\vec{l}\times\vec{r}}{r^3}$$

Section #1

dl
r
P

$d\vec{l}$ and $\vec{r}$ are parallel, so $d\vec{l}\times\vec{r} = 0$, $dB = 0$, and this section of the wire produces no magnetic field at P.

Section #2

a
x
$r = \sqrt{a^2+y^2}$
$\pi-\phi$
ϕ
y
dl

Divide this section of the wire up into short segments of length $dl = dy$. For each segment $d\vec{l}\times\vec{r}$ is directed out of the page ($\odot$) so the fields due to each segment all add and give a resultant field $\vec{B}$ directed <u>out of the page</u>.

$$|d\vec{l}\times\vec{r}| = r\sin\phi\,dl = r\sin(\pi-\phi)\,dl = \sqrt{a^2+y^2}\left(\frac{a}{\sqrt{a^2+y^2}}\right)dy = a\,dy$$

$$dB = \frac{\mu_0 I}{4\pi}\frac{|d\vec{l}\times\vec{r}|}{r^3} = \frac{\mu_0 I}{4\pi}\frac{a\,dy}{(a^2+y^2)^{3/2}}$$

$$B = \int dB = \frac{\mu_0 I a}{4\pi}\int_0^\infty \frac{dy}{(a^2+y^2)^{3/2}} = \frac{\mu_0 I a}{4\pi}\left[\frac{1}{a^2}\frac{y}{(a^2+y^2)^{1/2}}\Big|_0^\infty\right] = \frac{\mu_0 I}{4\pi a}$$

29-45

a)

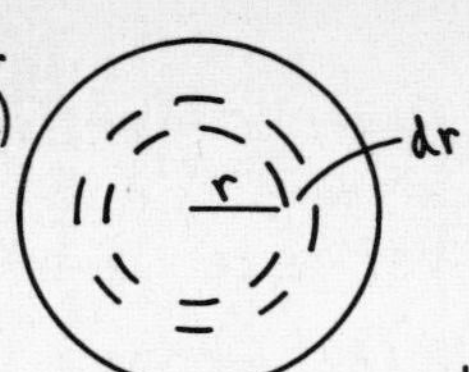

To integrate J over the cross section of the wire divide the wire cross section up into thin concentric rings of radius r and width dr. (We can't just take $I = JA = J\pi R^2$, since J varies across the cross section.)

The area of such a ring is dA, and the current through it is $dI = J\,dA$:

$$dA = 2\pi r\,dr \Rightarrow dI = J\,dA = \alpha r(2\pi r\,dr) = 2\pi\alpha r^2\,dr$$

$$I = \int dI = 2\pi\alpha\int_0^R r^2\,dr = 2\pi\alpha\left(\tfrac{1}{3}R^3\right) = \frac{2\pi\alpha R^3}{3} \Rightarrow \alpha = \frac{3I}{2\pi R^3}$$

b) (i) $r \le R$

Apply Ampere's law to a circle of radius $r < R$:

$\oint \vec{B}\cdot d\vec{l} = \oint B\,dl = B\oint dl = B(2\pi r)$, by the symmetry and direction of $\vec{B}$.

The current passing through the path is $I_{encl} = \int dI$, where in the integral r goes from 0 to r.

$$I_{encl} = 2\pi\alpha\int_0^r r^2\,dr = \frac{2\pi\alpha r^3}{3} = \frac{2\pi}{3}\left(\frac{3I}{2\pi R^3}\right)r^3 = \frac{Ir^3}{R^3}$$

Thus $\oint \vec{B}\cdot d\vec{l} = \mu_0 I_{encl} \Rightarrow B(2\pi r) = \mu_0\dfrac{Ir^3}{R^3} \Rightarrow B = \dfrac{\mu_0 I r^2}{2\pi R^3}$

(ii) $r \ge R$

Apply Ampere's law to a circle of radius $r > R$:

$\oint \vec{B}\cdot d\vec{l} = \oint B\,dl = B\oint dl = B(2\pi r)$

$I_{encl} = I$; all the current in the wire passes through the path

Thus $\oint \vec{B}\cdot d\vec{l} = \mu_0 I_{encl} \Rightarrow B(2\pi r) = \mu_0 I$

$$B = \frac{\mu_0 I}{2\pi r}$$

Note: At $r = R$ the expression in (i) $(r \le R)$ gives $B = \dfrac{\mu_0 I}{2\pi R}$.
At $r = R$ the expression in (ii) $(r \ge R)$ gives $B = \dfrac{\mu_0 I}{2\pi R}$, which is the same.

29-47

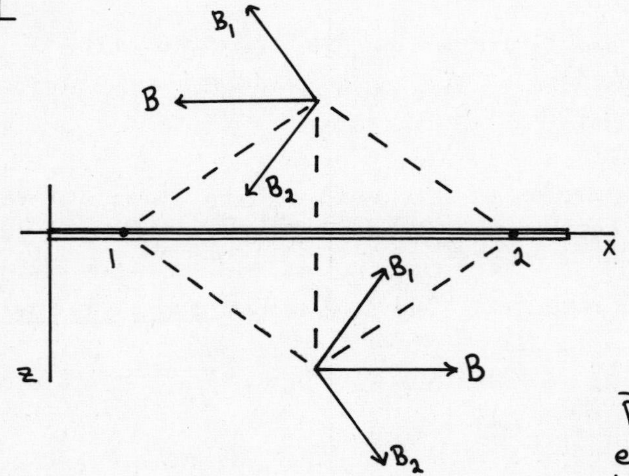

Consider the individual currents in pairs, where the currents in each pair are equidistant on either side of the point where $\vec{B}$ is being calculated.

The sketch shows that for each pair the z-components cancel, and that above the sheet the field is in the -x-direction and that below the sheet it is in the +x-direction.

Also, by symmetry the magnitude of $\vec{B}$ a distance a above the sheet must equal the magnitude of $\vec{B}$ a distance a below the sheet.

Now that we have deduced the symmetry of $\vec{B}$, apply Ampere's law.

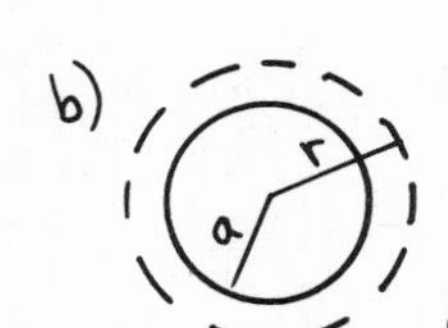

29-47 (cont)

Use a path that is a rectangle:

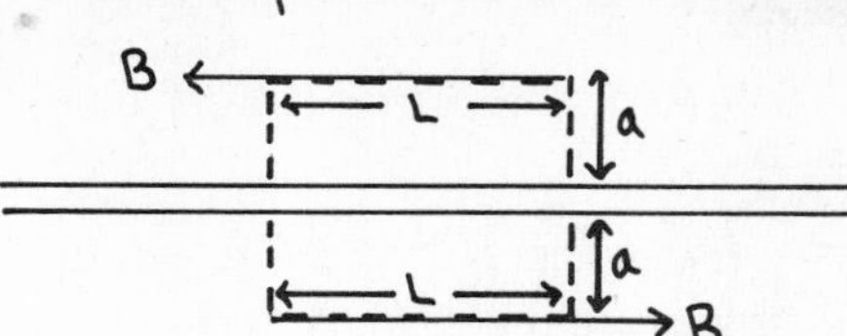

$\oint \vec{B}\cdot d\vec{l} = \mu_0 I_{encl}$

I is directed out of the page, so for I to be positive the integral around the path is taken in the counterclockwise direction.

Since $\vec{B}$ is parallel to the sheet, on the sides of the rectangle that have length 2a $\oint \vec{B}\cdot d\vec{l} = 0$. Thus $\oint \vec{B}\cdot d\vec{l} = 2BL$.

n conductors per unit length and current I out of the page in each gives $I_{encl} = InL$.

Ampere's law then gives $2BL = \mu_0 InL$

$$B = \tfrac{1}{2}\mu_0 In$$

Note that B is independent of the distance a from the sheet. Compare this result to Eq. (23-11) for the electric field due to an infinite sheet of charge.

29-49

a) Divide the cross section of the cylinder into thin concentric rings of radius r and width dr. The current through each ring is $dI = J\,dA = J\,2\pi r\,dr$.

$$dI = \frac{2I_0}{\pi a^2}\left[1-(r/a)^2\right]2\pi r\,dr = \frac{4I_0}{a^2}\left[1-\left(\frac{r}{a}\right)^2\right]r\,dr$$

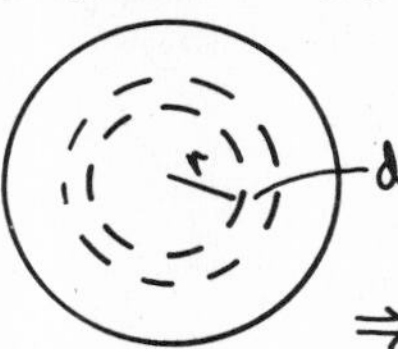

The total current I is obtained by integrating dI over the cross section

$$\Rightarrow I = \int_0^a dI = \frac{4I_0}{a^2}\int_0^a\left(1-\frac{r^2}{a^2}\right)r\,dr = \frac{4I_0}{a^2}\left(\frac{1}{2}r^2 - \frac{1}{4}\frac{r^4}{a^2}\right)\Big|_0^a = I_0,$$ as was to be shown.

b)

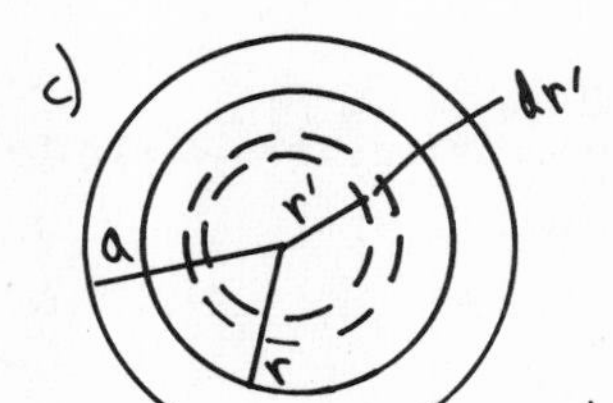

Apply Ampere's law to a path that is a circle of radius $r>a$.

$\oint \vec{B}\cdot d\vec{l} = B(2\pi r)$

$I_{encl} = I_0$ (the path encloses the entire cylinder)

$\oint \vec{B}\cdot d\vec{l} = \mu_0 I_{encl} \Rightarrow B(2\pi r) = \mu_0 I_0$

$$B = \frac{\mu_0 I_0}{2\pi r}$$

c)

Divide the cross section of the cylinder up into concentric rings of radius r′ and width dr′, as was done in part (a). The current dI through each ring is

$$dI = \frac{4I_0}{a^2}\left[1-\left(\frac{r'}{a}\right)^2\right]r'\,dr'$$

The current, I is obtained by integrating dI from $r'=0$ to $r'=r$:

$$I = \int dI = \frac{4I_0}{a^2}\int_0^r\left[1-\left(\frac{r'}{a}\right)^2\right]r'\,dr' = \frac{4I_0}{a^2}\left[\left(\frac{1}{2}r'^2 - \frac{1}{4}\frac{r'^4}{a^2}\right)\Big|_0^r\right]$$

$$I = \frac{4I_0}{a^2}\left(\frac{r^2}{2} - \frac{r^4}{4a^2}\right) = \frac{I_0 r^2}{a^2}\left(2-\frac{r^2}{a^2}\right)$$

d)

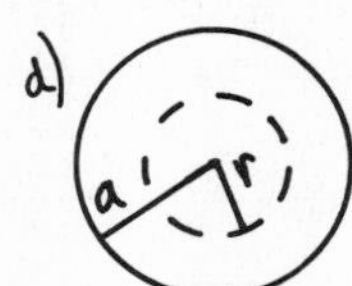

Apply Ampere's law to a path that is a circle of radius $r<a$.

$\oint \vec{B}\cdot d\vec{l} = B(2\pi r)$

$I_{encl} = \frac{I_0 r^2}{a^2}\left(2-\frac{r^2}{a^2}\right)$ (from part (c))

29-49 (cont)

$$\oint \vec{B}\cdot d\vec{l} = \mu_0 I_{encl} \Rightarrow B(2\pi r) = \mu_0 \frac{I_0 r^2}{a^2}\left(2 - \frac{r^2}{a^2}\right)$$

$$B = \frac{\mu_0 I_0}{2\pi}\frac{r}{a^2}\left(2 - r^2/a^2\right)$$

result in part (b) evaluated at $r=a$: $B = \frac{\mu_0 I_0}{2\pi a}$

result in part (d) evaluated at $r=a$: $B = \frac{\mu_0 I}{2\pi}\frac{a}{a^2}\left(2 - \frac{a^2}{a^2}\right) = \frac{\mu_0 I}{2\pi a}$;

the two results, one for $r>a$ and the other for $r<a$, agree at $r=a$.

29-53

a)

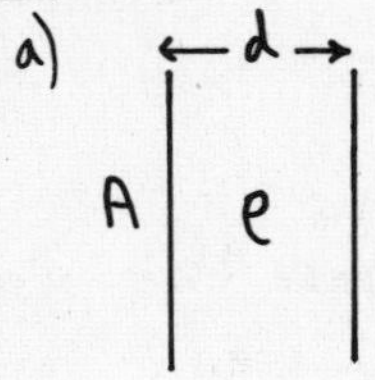

Apply Ohm's law to the dielectric:

$$i(t) = \frac{v(t)}{R}$$

$$v(t) = \frac{q(t)}{C} \text{ and } C = K\frac{\epsilon_0 A}{d} \Rightarrow v = \frac{d}{K\epsilon_0 A}q(t)$$

The resistance R of the dielectric slab is $R = \frac{\rho d}{A}$.

Thus $i(t) = \frac{v(t)}{R} = \frac{q(t)\,d}{K\epsilon_0 A}\left(\frac{A}{\rho d}\right) = \frac{q(t)}{K\epsilon_0\rho}$.

But the current $i(t)$ in the dielectric is related to the rate of change $\frac{dq}{dt}$ of the charge $q(t)$ on the plates by $i(t) = -\frac{dq}{dt}$ (a positive i in the direction from the + to the − plate of the capacitor corresponds to a decrease in the charge).

Use this in the above $\Rightarrow -\frac{dq}{dt} = \frac{1}{K\rho\epsilon_0}q(t)$

$$\frac{dq}{q} = -\frac{dt}{K\rho\epsilon_0}$$

Integrate both sides of this equation from $t=0$, when $q = Q_0$, to a later time t when the charge is $q(t)$

$$\Rightarrow \int_{Q_0}^{q}\frac{dq}{q} = -\frac{1}{K\rho\epsilon_0}\int_0^t dt$$

$$\ln(q/Q_0) = -\frac{t}{K\rho\epsilon_0} \Rightarrow q(t) = Q_0 e^{-t/K\rho\epsilon_0}$$

Then $i(t) = -\frac{dq}{dt} = \frac{Q_0}{K\rho\epsilon_0}e^{-t/K\rho\epsilon_0}$.

$$j_c(t) = \frac{i(t)}{A} = \frac{Q}{KA\rho\epsilon_0}e^{-t/K\rho\epsilon_0}$$

The conduction current flows from the positive to the negative plate of the capacitor.

b) $E(t) = \frac{q(t)}{\epsilon A} = \frac{q(t)}{K\epsilon_0 A}$

$$j_D(t) = \epsilon\frac{dE}{dt} = K\epsilon_0\frac{dE}{dt} = K\epsilon_0\frac{dq(t)/dt}{K\epsilon_0 A} = -\frac{i_c(t)}{A} = -j_c(t)$$

The minus sign means that $j_D(t)$ is directed from the negative towards the positive plate. $\vec{E}$ is from + to − but $\frac{dE}{dt}$ is negative (E decreases) so $j_D(t)$ is from − to +.

CHAPTER 30

Exercises 1, 3, 7, 9, 13, 15, 17, 21, 25

Problems 27, 29, 35, 37, 39

Exercises

30-1

The magnitude of the induced emf is $|\mathcal{E}| = \left|\frac{d\Phi_B}{dt}\right|$

$\Phi_B = BA$, so $\frac{d\Phi_B}{dt} = A\frac{dB}{dt}$

$|\mathcal{E}| = A\left|\frac{dB}{dt}\right| = (0.0900\ m^2)(0.190\ T/s) = 0.0171\ V$

$I = \frac{|\mathcal{E}|}{R} = \frac{0.0171\ V}{0.400\ \Omega} = \underline{0.0428\ A}$

Note that the induced emf does not depend on B, just on $\frac{dB}{dt}$.

30-3

The magnitude of the average induced emf is

$|\mathcal{E}_{av}| = N\left|\frac{\Delta\Phi_B}{\Delta t}\right|$

$\Delta\Phi_B = \Phi_{Bf} - \Phi_{Bi}$

initial

$\vec{B}$ 45° 45° $\vec{A}$, plane of coil

$\Phi_{Bi} = BA\cos\phi = (1.40\ T)(0.120\ m)(0.250\ m)\cos 45°$

$\Phi_{Bi} = 0.0297\ Wb$

final

$\vec{B}$ $\vec{A}$

$\Phi_{Bf} = BA\cos\phi = (1.40\ T)(0.120\ m)(0.250\ m)\cos 0° = 0.0420\ Wb$

Then $|\mathcal{E}_{av}| = N\left|\frac{\Delta\Phi_B}{\Delta t}\right| = 50\left(\frac{0.0420\ Wb - 0.0297\ Wb}{0.0800\ s}\right) = \underline{7.69\ V}$

30-7

The magnitude of the average emf induced in the coil is

$|\mathcal{E}_{av}| = N\left|\frac{\Delta\Phi_B}{\Delta t}\right|$

Initially, $\Phi_{Bi} = BA\cos\phi = BA$

The final flux is zero, so

$|\mathcal{E}_{av}| = N\frac{|\Phi_{Bf} - \Phi_{Bi}|}{\Delta t} = \frac{NBA}{\Delta t}$

The average induced current is $I = \frac{|\mathcal{E}_{av}|}{R} = \frac{NBA}{R\Delta t}$

The total charge that flows through the coil is $Q = I\Delta t = \left(\frac{NBA}{R\Delta t}\right)\Delta t = \frac{NBA}{R}$.

30-9

Exercise 30-7 derived the result $Q = \frac{NBA}{R}$.
In this problem the flux changes from its maximum value of $\Phi_B = BA$ to zero, so this equation applies. R is the total resistance so here is $50.0\,\Omega + 30.0\,\Omega = 80.0\,\Omega$.

$$Q = \frac{NBA}{R} \Rightarrow B = \frac{QR}{NA} = \frac{(6.00\times10^{-5}\,\text{C})(80.0\,\Omega)}{160(4.00\times10^{-4}\,\text{m}^2)} = \underline{0.0750\,\text{T}}$$

30-13

a)

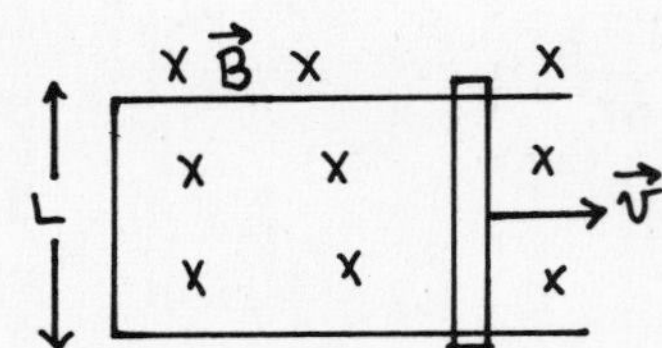

The magnitude of the induced emf is given by
$$|\mathcal{E}| = vLB \Rightarrow v = \frac{|\mathcal{E}|}{LB} = \frac{3.20\,\text{V}}{(0.400\,\text{m})(1.20\,\text{T})} = \underline{6.67\,\text{m/s}}$$

b) $I = \frac{|\mathcal{E}|}{R} = \frac{3.20\,\text{V}}{0.800\,\Omega} = \underline{4.00\,\text{A}}$

c) The direction of the induced emf and induced currents can be determined as discussed in Sect. 30-2:

Let positive $\vec{A}$ be into the page (in the same direction as $\vec{B}$).
$\vec{B}$ is in the positive direction (the direction of $\vec{A}$) so Φ_B is positive.
The magnitude of the flux through the circuit is increasing as the area of the circuit is getting larger, so $d\Phi_B/dt$ is positive.
Then by $\mathcal{E} = -\frac{d\Phi_B}{dt}$, $\mathcal{E}$ is negative. With our chosen direction of $\vec{A}$ clockwise is positive, so a negative $\mathcal{E}$ is counterclockwise and I is counterclockwise.

Force on the current in the rod: $\vec{F} = I\vec{l}\times\vec{B}$

The force on the rod due to the induced currents is to the left.
$$F = IlB\sin\phi = ILB = (4.00\,\text{A})(0.400\,\text{m})(1.20\,\text{T}) = 1.92\,\text{N}$$
The force on the rod has magnitude 1.92 N and is directed to the left (opposite to $\vec{v}$).

30-15

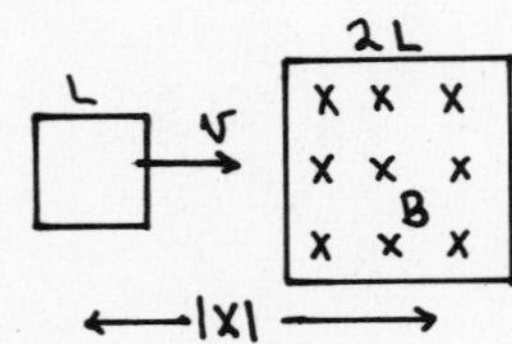

For $x < -\frac{3L}{2}$ or $x > \frac{3L}{2}$ the loop is completely outside the field region, $\Phi_B = 0$, and $\frac{d\Phi_B}{dt} = 0$. Thus $\mathcal{E} = 0$ and $I = 0$, so there is no force from the magnetic field and the external force F necessary to maintain constant velocity is zero.

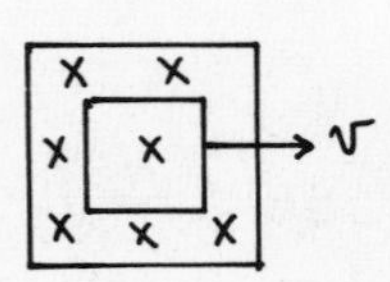

For $-\frac{L}{2} < x < \frac{L}{2}$ the loop is completely inside the field region and $\Phi_B = BL^2$. But $\frac{d\Phi_B}{dt} = 0$ so $\mathcal{E} = 0$ and $I = 0$. There is no force $\vec{F} = I\vec{l}\times\vec{B}$ from the magnetic field and the external force F necessary to maintain constant velocity is zero.

30-15 (cont)

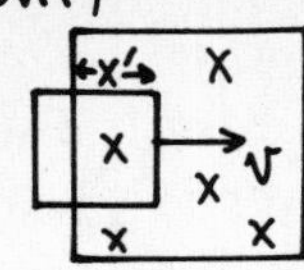

For $-\frac{3L}{2} < x < -\frac{L}{2}$ the loop is entering the field region. Let x' be the length of the loop that is within the field. Then $|\Phi_B| = BLx'$ and $\left|\frac{d\Phi_B}{dt}\right| = BLv$. The magnitude of the induced emf is $|\mathcal{E}| = \left|\frac{d\Phi_B}{dt}\right| = BLv$ and the induced current is $I = \frac{|\mathcal{E}|}{R} = \frac{BLv}{R}$.

Let $\vec{A}$ be directed into the plane of the figure. Then Φ_B is positive. The flux is positive and increasing in magnitude, so $\frac{d\Phi_B}{dt}$ is positive. Then by Faraday's law $\mathcal{E}$ is negative, and with our choice for direction of $\vec{A}$ a negative $\mathcal{E}$ is counterclockwise. The current induced in the loop is counterclockwise. Then $\vec{F} = I\vec{l} \times \vec{B}$ gives that $\vec{F}$ is to the left and has magnitude

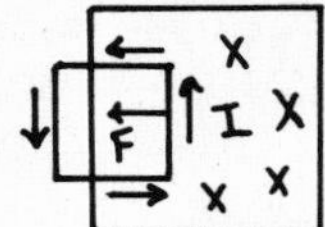

$$F = ILB = \left(\frac{BLv}{R}\right)LB = \frac{B^2L^2v}{R}$$

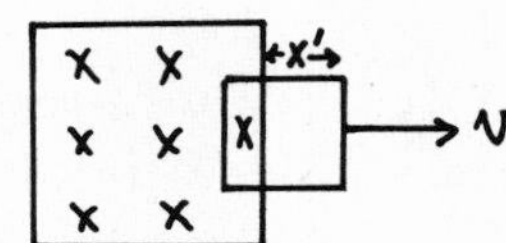

For $\frac{L}{2} < x < \frac{3L}{2}$ the loop is leaving the field region. Let x' be the length of the loop that is outside the field. Then $|\Phi_B| = BL(L-x')$ and $\left|\frac{d\Phi_B}{dt}\right| = BLv$. The magnitude of the induced emf is $|\mathcal{E}| = \left|\frac{d\Phi_B}{dt}\right| = BLv$ and the induced current is $I = \frac{|\mathcal{E}|}{R} = \frac{BLv}{R}$.

Again let $\vec{A}$ be directed into the plane of the figure. Then Φ_B is positive and decreasing in magnitude, so $\frac{d\Phi_B}{dt}$ is negative. Then by Faraday's law $\mathcal{E}$ is positive, and with our choice for direction of $\vec{A}$ a positive $\mathcal{E}$ is clockwise. The current induced in the loop is clockwise.

Then $\vec{F} = I\vec{l} \times \vec{B}$ gives that $\vec{F}$ is to the left and has magnitude $F = ILB = \left(\frac{BLv}{R}\right)LB = \frac{B^2L^2v}{R}$.

a)

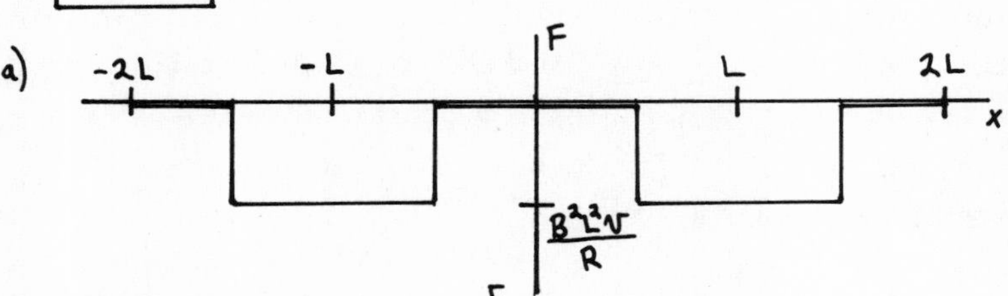

b)

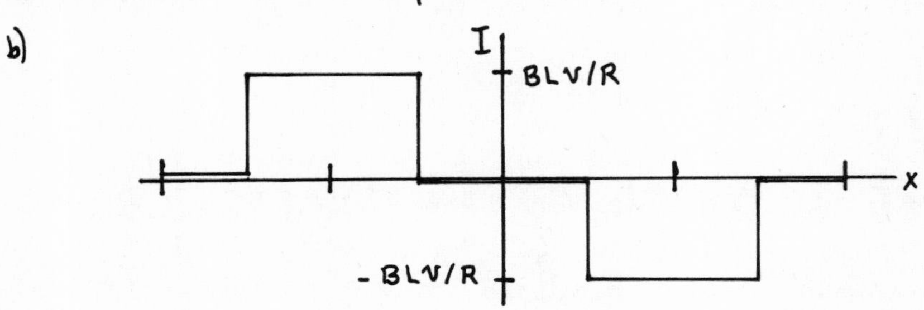

30-17

a)

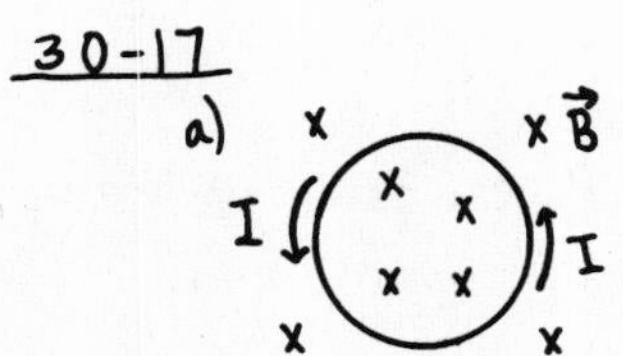

B increasing. The flux through the coil is directed into the paper and is increasing. By Lenz's law the field of the induced current, $\vec{B}_{induced}$, is directed out of the page (⊙) inside the loop, to oppose the increase in flux. This $\vec{B}_{induced}$ is produced by a <u>counterclockwise</u> current.

30-17 (cont)

The sign convention for Faraday's law is an alternative means of determing the direction of I. Let $\vec{A}$ be directed into the page. Then Φ_B is positive. Since B is increasing the flux is increasing in magnitude and Φ_B positive gives $\frac{d\Phi_B}{dt} > 0$. $\mathcal{E} = -\frac{d\Phi_B}{dt}$ then gives $\mathcal{E} < 0$. By our choice of $\vec{A}$, $\mathcal{E} < 0$ means $\mathcal{E}$ is counterclockwise $\Rightarrow$ I is counterclockwise. This agrees with the direction we deduced from Lenz's law.

b) B decreasing. The flux through the coil is directed into the paper and is decreasing. By Lenz's law the field of the induced current, $\vec{B}_{induced}$, is directed into the paper (⊗) inside the loop, to oppose the decrease in flux. This $\vec{B}_{induced}$ is produced by a clockwise current.

Using Faraday's law: Let $\vec{A}$ be directed into the page. Then Φ_B is positive. Since B is decreasing in magnitude and with Φ_B positive, $\frac{d\Phi_B}{dt} < 0$. $\mathcal{E} = -\frac{d\Phi_B}{dt}$ then gives $\mathcal{E} > 0$. By our choice of $\vec{A}$, $\mathcal{E} > 0$ means $\mathcal{E}$ is clockwise $\Rightarrow$ I is clockwise. This agrees with the direction we deduced from Lenz's law.

c) B constant means $\frac{d\Phi_B}{dt} = 0$ so $\mathcal{E} = 0$ and $I = 0$. There is no induced current.

30-21

a) Because of the axial symmetry and the absence of any electric charge, the field lines are concentric circles.

b) $\vec{E}$ is tangent to the ring. The direction of $\vec{E}$ (clockwise or counterclockwise) is the direction in which current will be induced in the ring. Use the sign convention for Faraday's law to deduce this direction. Let $\vec{A}$ be into the paper. Then Φ_B is positive. B decreasing then means $\frac{d\Phi_B}{dt}$ is negative, so by $\mathcal{E} = -\frac{d\Phi_B}{dt}$, $\mathcal{E}$ is positive and therefore clockwise. Thus $\vec{E}$ is clockwise around the ring.

To calculate E apply $\oint \vec{E} \cdot d\vec{l} = -\frac{d\Phi_B}{dt}$ to a circular path that coincides with the ring.

$$\oint \vec{E} \cdot d\vec{l} = E(2\pi r)$$

$$\Phi_B = B\pi r^2 \; ; \; \left|\frac{d\Phi_B}{dt}\right| = \pi r^2 \left|\frac{dB}{dt}\right|$$

$$E(2\pi r) = \pi r^2 \left|\frac{dB}{dt}\right| \Rightarrow E = \tfrac{1}{2} r \left|\frac{dB}{dt}\right| = \tfrac{1}{2}(0.100\text{ m})(0.0600\text{ T/s}) = \underline{3.00\times10^{-3}\text{ V/m}}$$

c) The induced emf has magnitude $\mathcal{E} = \oint \vec{E} \cdot d\vec{l} = E(2\pi r) = (3.00\times10^{-3}\text{ V/m})(2\pi)(0.100\text{ m})$

$\mathcal{E} = 1.885\times10^{-3}$ V.

Then $I = \frac{\mathcal{E}}{R} = \frac{1.884\times10^{-3}\text{ V}}{2.00\ \Omega} = \underline{9.42\times10^{-4}\text{ A}}$.

d) $\mathcal{E} = IR$ for any segment of the ring. Therefore, the potential difference between points a and b, or between any other points of the ring, is zero.

30-21 (cont)

e) The induced emf $\mathcal{E}$ remains the same, but now there is no current and no IR potential drop. the potential difference will be $\mathcal{E} = 1.88\times10^{-3}$ V (from part (c)).

30-25

Apply Ampere's law to a circular path of radius $r < R$ where R is the radius of the wire.

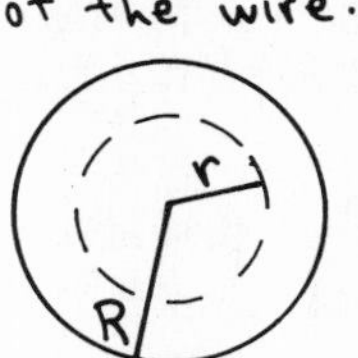

$$\oint \vec{B}\cdot d\vec{l} = \mu_0 \left(I_c + \epsilon_0 \frac{d\Phi_E}{dt}\right)$$

There is no displacement current, so $\oint \vec{B}\cdot d\vec{l} = \mu_0 I_c$.

The magnetic field inside the superconducting material is zero, so $\oint \vec{B}\cdot d\vec{l} = 0$. But then Ampere's law says that $I_c = 0$; there can be no conduction current through the path. This same argument applies to any circular path with $r < R$, so all the current must be at the surface of the wire.

Problems

30-27

a) $|\mathcal{E}| = \left|\frac{d\Phi_B}{dt}\right| = BLv.$

$I = \frac{BLv}{R}.$

Use $\mathcal{E} = -\frac{d\Phi_B}{dt}$ to find the direction of I:

Let $\vec{A}$ be into the page. Then $\Phi_B > 0$. The area of the circuit is increasing, so $\frac{d\Phi_B}{dt} > 0$. Then $\mathcal{E} < 0$ and with our direction for $\vec{A}$ this means that $\mathcal{E}$ and I are counterclockwise, as shown on the sketch. The force F_I on the rod due to the induced current is given by $\vec{F}_I = I\vec{l}\times\vec{B}$. This gives $\vec{F}_I$ to the left with magnitude $F_I = ILB = \left(\frac{BLv}{R}\right)LB = \frac{B^2L^2v}{R}$. Note that $\vec{F}_I$ is directed to oppose the motion of the rod, as required by Lenz's law.

The net force on the rod is $F - F_I$, so its acceleration is $a = \frac{F - F_I}{m} = \frac{F - B^2L^2v/R}{m}$. The rod starts with $v = 0$ and $a = \frac{F}{m}$. As the speed v increases the acceleration a decreases. When $a = 0$ the rod has reached its terminal speed v_t.

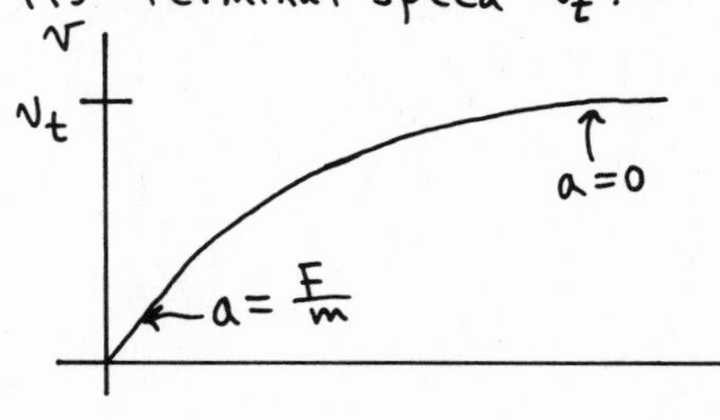

(Recall that a is the slope of the tangent to the v versus t curve.)

b) $v = v_t$ when $a = 0 \Rightarrow \frac{F - B^2L^2v_t/R}{m} = 0$

$$v_t = \frac{RF}{B^2L^2}$$

30-29

a) The emf induced in a thin slice is

$$d\mathcal{E} = \vec{v} \times \vec{B} \cdot d\vec{l}$$

Assume that $\vec{B}$ is directed out of the page. Then $\vec{v} \times \vec{B}$ is directed radially outward and $dl = dr$, so

$$\vec{v} \times \vec{B} \cdot d\vec{l} = vB\,dr$$

$$v = r\omega \Rightarrow d\mathcal{E} = \omega B r\,dr$$

The $d\mathcal{E}$ for all the thin slices that make up the rod are in series so they add:

$$\mathcal{E} = \int d\mathcal{E} = \int_0^L \omega B r\,dr = \tfrac{1}{2}\omega B L^2 = \tfrac{1}{2}(8.00\,\text{rad/s})(0.500\,\text{T})(0.620\,\text{m})^2 = \underline{0.769\,\text{V}}$$

b) No current flows so there is no IR potential drop. Thus the potential difference between the ends equals the emf of 0.769 V calculated in part (a).

30-35

Calculate the induced electric field at each point and then use $\vec{F} = q\vec{E}$.

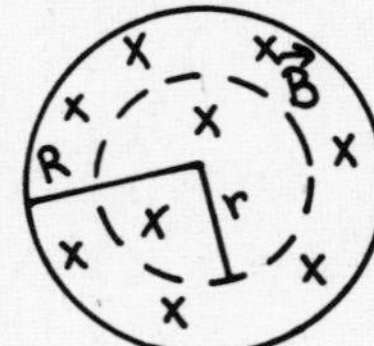

Apply $\oint \vec{E} \cdot d\vec{l} = -\frac{d\Phi_B}{dt}$ to a concentric circle of radius r. Take $\vec{A}$ to be into the page, in the direction of $\vec{B}$. B increasing then gives $\frac{d\Phi_B}{dt} > 0$, so $\oint \vec{E} \cdot d\vec{l}$ is negative. This means that $\vec{E}$ is tangent to the circle in the counterclockwise direction.

$$\oint \vec{E} \cdot d\vec{l} = -E(2\pi r)$$

$$-\frac{d\Phi_B}{dt} = -\pi r^2 \frac{dB}{dt}$$

$$\Rightarrow -E(2\pi r) = -\pi r^2 \frac{dB}{dt} \Rightarrow E = \tfrac{1}{2} r \frac{dB}{dt}$$

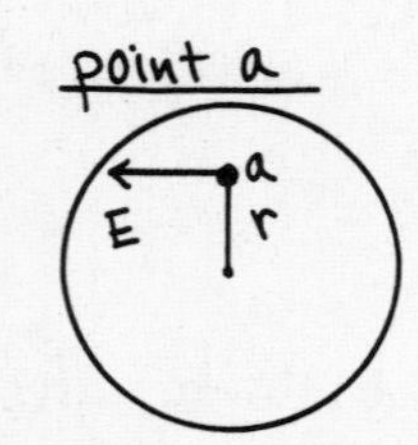

point a

$$F = qE = \tfrac{1}{2} q r \frac{dB}{dt}$$

$\vec{F}$ is to the left.

($\vec{F}$ is in the same direction as $\vec{E}$ since q is positive.)

point b

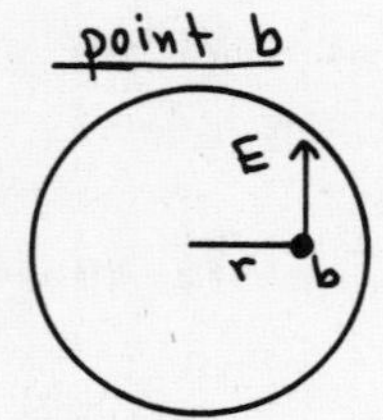

$$F = qE = \tfrac{1}{2} q r \frac{dB}{dt}$$

$\vec{F}$ is toward the top of the page.

point c

$r = 0$ here, so $E = 0$ and $F = 0$.

30-37

a) and b)

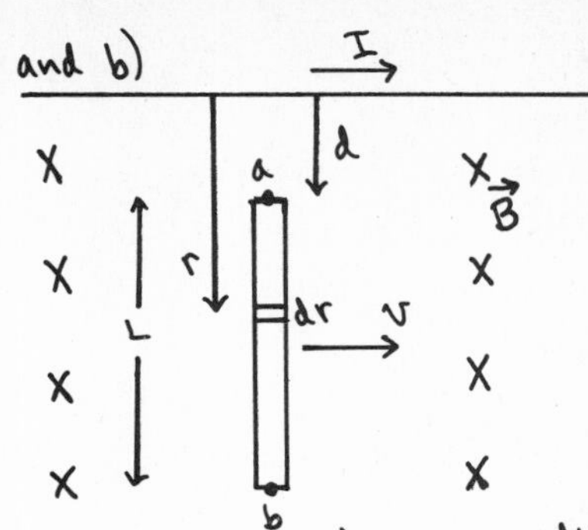

The magnetic field of the wire is given by $B = \frac{\mu_0 I}{2\pi r}$ and varies along the length of the bar. At every point along the bar $\vec{B}$ has direction into the page. Divide the bar up into thin slices.

The emf $d\mathcal{E}$ induced in each slice is given by $d\mathcal{E} = \vec{v} \times \vec{B} \cdot d\vec{l}$. $\vec{v} \times \vec{B}$ is directed toward the wire, so $d\mathcal{E} = -vB\,dr = -v\left(\frac{\mu_0 I}{2\pi r}\right) dr$. The total emf induced in the bar is

$$V_{ab} = \int_a^b d\mathcal{E} = -\int_d^{d+L} \frac{\mu_0 I v}{2\pi r} dr = -\frac{\mu_0 I v}{2\pi}\int_d^{d+L} \frac{dr}{r} = -\frac{\mu_0 I v}{2\pi}\left[\ln(r)\Big|_d^{d+L}\right]$$

$$V_{ab} = -\frac{\mu_0 I v}{2\pi}\left(\ln(d+L) - \ln(d)\right) = -\frac{\mu_0 I v}{2\pi}\ln\left(1 + \frac{L}{d}\right)$$

The minus sign means that V_{ab} is negative; point a is at higher potential than point b. (The force $\vec{F} = q\vec{v} \times \vec{B}$ on positive charge carriers in the bar is toward a, so a is at higher potential.)

c)

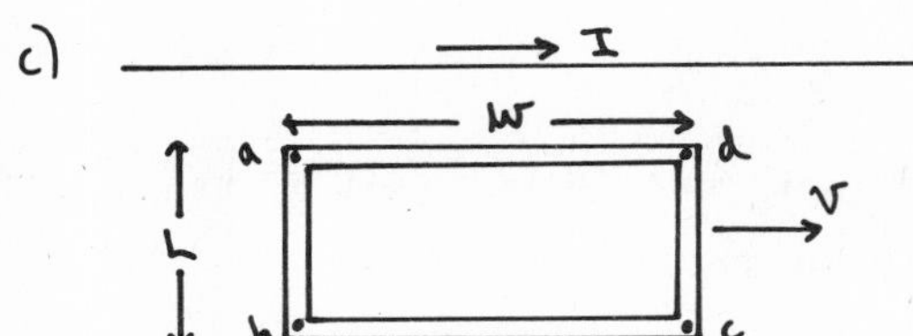

As the loop moves to the right the magnetic flux through it doesn't change. Thus $\mathcal{E} = -\frac{d\Phi_B}{dt} = 0$ and $I = 0$.

This result can also be understood as follows. The induced emf in section ab puts point a at higher potential; the induced emf in section dc puts point d at higher potential. If you travel around the loop then these two induced emf's sum to zero. There is no net emf in the loop and hence no current.

30-39

a)

When the wire has speed v the induced emf is $\mathcal{E} = Bva$ and the induced current is $I = \frac{\mathcal{E}}{R} = \frac{Bva}{R}$. The induced current flows upward in the wire as shown, so the force $\vec{F} = I\vec{l} \times \vec{B}$ exerted by the magnetic field on the induced current is to the left. $\vec{F}$ opposes the motion of the wire, as it must by Lenz's law. The magnitude of the force is $F = IaB = \frac{B^2a^2v}{R}$.

b) Apply $\Sigma\vec{F} = m\vec{a}$ to the wire. Take +x to be toward the right and let the origin be at the location of the wire at $t = 0$, so $x_0 = 0$.

$$\Sigma F_x = ma_x$$

$$-F = ma_x \Rightarrow a_x = -\frac{F}{m} = -\frac{B^2a^2v}{mR}$$

$$a = \frac{dv}{dt} = -\frac{B^2a^2v}{mR}$$

30-29 (cont) $\frac{dv}{v} = -\frac{B^2a^2}{mR}\,dt$

$$\int_{v_0}^{v} \frac{dv}{v} = -\frac{B^2a^2}{mR}\int_0^t dt$$

$$\ln v - \ln v_0 = -\frac{B^2a^2t}{mR}$$

$$\ln\left(\frac{v}{v_0}\right) = -\frac{B^2a^2t}{mR} \Rightarrow v = v_0\, e^{-B^2a^2t/mR}$$

Note: $t = 0 \Rightarrow v = v_0$ and $v \to 0$ when $t \to \infty$

$$v = \frac{dx}{dt} = v_0\, e^{-B^2a^2t/mR}$$

$$dx = v_0\, e^{-B^2a^2t/mR}\, dt$$

$$\int_0^x dx = \int_0^t v_0\, e^{-B^2a^2t/mR}\, dt$$

$$x = v_0\left(-\frac{mR}{B^2a^2}\right)\left[e^{-B^2a^2t/mR}\right]\Big|_0^t = \frac{mRv_0}{B^2a^2}\left(1 - e^{-B^2a^2t/mR}\right)$$

Comes to rest $\Rightarrow v \to 0$. This happens when $t \to \infty$.
$t \to \infty$ gives $x = \frac{mRv_0}{B^2a^2}$. Thus the distance the wire travels before coming to rest is given by $\frac{mRv_0}{B^2a^2}$, as was to be shown.

CHAPTER 31

Exercises 3, 5, 9, 11, 13, 19, 21, 25, 27

Problems 29, 35, 37, 39, 41, 43, 45

Exercises

31-3

a) $|\mathcal{E}_2| = M\left|\frac{di_1}{dt}\right| = (0.0300\ \text{H})(0.0500\ \text{A/s}) = \underline{0.00150\ \text{V}}$; yes, it is constant.

b) $|\mathcal{E}_1| = M\left|\frac{di_2}{dt}\right|$; M is a property of the pair of coils so it is the same as in part (a)

$\Rightarrow |\mathcal{E}_1| = \underline{0.00150\ \text{V}}$

31-5

a) $|\mathcal{E}| = L\left|\frac{di}{dt}\right| \Rightarrow L = \frac{|\mathcal{E}|}{|di/dt|} = \frac{0.0500\ \text{V}}{0.0600\ \text{A/s}} = \underline{0.833\ \text{H}}$

b) $N\Phi_B = Li \Rightarrow \Phi_B = \frac{Li}{N} = \frac{(0.833\ \text{H})(0.800\ \text{A})}{300} = \underline{2.22\times10^{-3}\ \text{Wb}}$

31-9

$L = \frac{N\Phi_B}{i}$

If the magnetic field is uniform inside the solenoid $\Phi_B = BA$.

Eq. (29-21): $B = \mu_0 n i = \mu_0\left(\frac{N}{\ell}\right)i \Rightarrow \Phi_B = \frac{\mu_0 N i A}{\ell}$

$L = \frac{N}{i}\left(\frac{\mu_0 N i A}{\ell}\right) = \frac{\mu_0 N^2 A}{\ell}$

31-11

$U = \frac{1}{2}LI^2 \Rightarrow L = \frac{2U}{I^2} = \frac{2(0.350\ \text{J})}{(25.0\ \text{A})^2} = 0.00112\ \text{H}$

Example 31-3 gives the inductance of a toroidal solenoid to be $L = \frac{\mu_0 N^2 A}{2\pi r}$

$N = \sqrt{\frac{2\pi r L}{\mu_0 A}} = \sqrt{\frac{2\pi(0.120\ \text{m})(0.00112\ \text{H})}{(4\pi\times10^{-7}\ \text{T}\cdot\text{m/A})(4.00\times10^{-4}\ \text{m}^2)}} = \underline{1296}$

31-13

a) The energy density (energy per unit volume) in a magnetic field (in vacuum) is given by $u = \frac{U}{V} = \frac{B^2}{2\mu_0}$ (Eq. 31-10).

$\Rightarrow V = \frac{2\mu_0 U}{B^2} = \frac{2(4\pi\times10^{-7}\ \text{T}\cdot\text{m/A})(3.60\times10^6\ \text{J})}{(0.500\ \text{T})^2} = \underline{36.2\ \text{m}^3}$

b) $u = \frac{U}{V} = \frac{B^2}{2\mu_0}$

$\Rightarrow B = \sqrt{\frac{2\mu_0 U}{V}} = \sqrt{\frac{2(4\pi\times10^{-7}\ \text{T}\cdot\text{m/A})(3.60\times10^6\ \text{J})}{0.125\ \text{m}^3}} = \underline{8.51\ \text{T}}$

31-19

$\frac{di}{dt}$ is positive as the current increases from its initial value of zero.

$$\mathcal{E} - v_R - v_L = 0$$

$$\mathcal{E} - iR - L\frac{di}{dt} = 0 \Rightarrow i = \frac{\mathcal{E}}{R}\left(1 - e^{-(R/L)t}\right)$$

a) Initially ($t=0$) $i = 0 \Rightarrow \mathcal{E} - L\frac{di}{dt} = 0$

$$\frac{di}{dt} = \frac{\mathcal{E}}{L} = \frac{12.0\ \text{V}}{3.00\ \text{H}} = \underline{4.00\ \text{A/s}}$$

b) $\mathcal{E} - iR - L\frac{di}{dt} = 0$ (Use this equation rather than Eq. (31-14) since i rather than t is given.)

$$\frac{di}{dt} = \frac{\mathcal{E} - iR}{L} = \frac{12.0\ \text{V} - (1.00\ \text{A})(9.00\ \Omega)}{3.00\ \text{H}} = \underline{1.00\ \text{A/s}}$$

c) $$i = \frac{\mathcal{E}}{R}\left(1 - e^{-(R/L)t}\right) = \frac{12.0\ \text{V}}{9.00\ \Omega}\left(1 - e^{-\left(\frac{9.00\ \Omega}{3.00\ \text{H}}\right)(0.200\ \text{s})}\right) = 1.333\ \text{A}\left(1 - e^{-0.6}\right) = \underline{0.601\ \text{A}}$$

d) Final steady state $\Rightarrow t \to \infty$ and $\frac{di}{dt} \to 0$, so $\mathcal{E} - iR = 0$.

$$i = \frac{\mathcal{E}}{R} = \frac{12.0\ \text{V}}{9.00\ \Omega} = \underline{1.33\ \text{A}}$$

31-21

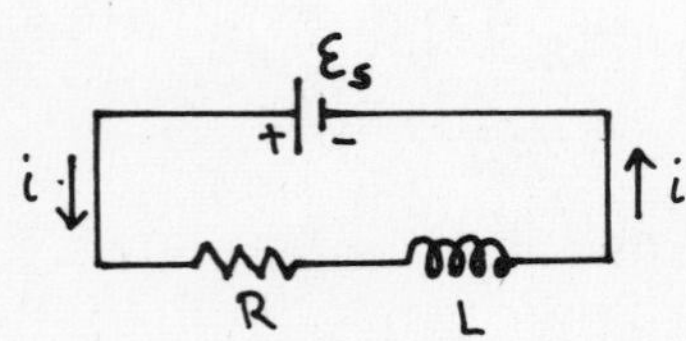

$$\mathcal{E}_s - iR - L\frac{di}{dt} = 0$$

constant current is established $\Rightarrow \frac{di}{dt} = 0$

$$i = \frac{\mathcal{E}_s}{R} = \frac{120\ \text{V}}{500\ \Omega} = 0.240\ \text{A}$$

a)

$$i = I_0 e^{-(R/L)t}$$

$$t = 0 \Rightarrow i = I_0 = \underline{0.240\ \text{A}}$$

(The inductor prevents an instantaneous change in the current; the current in the inductor just after S_2 is closed and S_1 is opened equals the current in the inductor just before this done.)

b) $$i = I_0 e^{-(R/L)t} = (0.240\ \text{A})e^{-\left(\frac{500\ \Omega}{0.200\ \text{H}}\right)(2.00 \times 10^{-4}\ \text{s})} = (0.240\ \text{A})e^{-0.5} = \underline{0.146\ \text{A}}$$

c) If we trace around the loop in the direction of the current the potential falls as we travel through the resistor so it must rise as we pass through the inductor: $V_{ab} > 0$ and $V_{bc} < 0$. Point c is at higher potential. $V_{ab} + V_{bc} = 0 \Rightarrow V_{bc} = -V_{ab}$, or

$$V_{cb} = +V_{ab} = iR = (0.146\ \text{A})(500\ \Omega) = \underline{73.0\ \text{V}}$$

d) $i = I_0 e^{-(R/L)t}$

$$i = \tfrac{1}{2}I_0 \Rightarrow \tfrac{1}{2}I_0 = I_0 e^{-(R/L)t} \Rightarrow \tfrac{1}{2} = e^{-(R/L)t}$$

31-21 (cont)

Take natural logs of both sides of this equation $\Rightarrow \ln\left(\frac{1}{2}\right) = -\frac{R}{L}t$

$$t = \frac{L}{R}\ln 2 \quad \left(\text{since } \ln\left(\tfrac{1}{2}\right) = -\ln 2\right)$$

$$t = \left(\frac{0.200\text{ H}}{500\ \Omega}\right)\ln 2 = \underline{2.77\times10^{-4}\text{ s}}$$

31-25

a) Eq. (31-22) $\omega = \frac{1}{\sqrt{LC}} = \frac{1}{\sqrt{(3.00\text{H})(6.00\times10^{-4}\text{ F})}} = \underline{23.6\text{ rad/s}}$

The period is given by $T = \frac{2\pi}{\omega} = \frac{2\pi}{23.6\text{ rad/s}} = \underline{0.266\text{ s}}$

b)

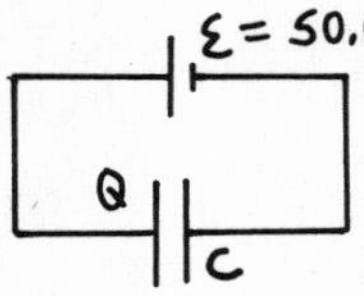

$$\mathcal{E} - \frac{Q}{C} = 0$$

$$Q = \mathcal{E}C = (50.0\text{ V})(6.00\times10^{-4}\text{F}) = \underline{0.0300\text{ C}}$$

c) $U = \frac{1}{2}CV^2 = \frac{1}{2}(6.00\times10^{-4}\text{F})(50.0\text{V})^2 = \underline{0.750\text{ J}}$

d) $q = Q\cos(\omega t + \phi)$ (Eq. 31-21)

$q = Q$ at $t = 0 \Rightarrow \phi = 0$

$q = Q\cos(\omega t) = (0.0300\text{C})\cos([23.6\text{ rad/s}][0.0444\text{ s}]) = \underline{0.0150\text{ C}}$

e) $i = -\omega Q\sin(\omega t + \phi)$ (Eq. 31-23)

$i = -(23.6\text{ rad/s})(0.0300\text{ C})\sin([23.6\text{ rad/s}][0.0444\text{s}]) = \underline{-0.613\text{ A}}$

or $\frac{1}{2}Li^2 + \frac{q^2}{2C} = \frac{Q^2}{2C} \Rightarrow i = \pm\sqrt{\frac{1}{LC}}\sqrt{Q^2 - q^2}$ (Eq. 31-25)

$i = \pm(23.6\text{ rad/s})\sqrt{(0.0300\text{C})^2 - (0.0150\text{C})^2} = \pm0.613\text{A}$, which checks

f) $U_C = \frac{q^2}{2C} = \frac{1}{2}\frac{(0.0150\text{ C})^2}{6.00\times10^{-4}\text{F}} = \underline{0.187\text{ J}}$

$U_L = \frac{1}{2}Li^2 = \frac{1}{2}(3.00\text{ H})(0.613\text{ A})^2 = \underline{0.563\text{ J}}$

Note: $U_C + U_L = 0.187\text{J} + 0.563\text{J} = 0.750\text{J}$. This agrees with the total energy initially stored in the capacitor, $U = \frac{Q^2}{2C} = \frac{(0.0300\text{C})^2}{2(6.00\times10^{-4}\text{F})} = 0.750\text{ J}$.

31-27

a) $\omega = \frac{1}{\sqrt{LC}} = \frac{1}{\sqrt{(0.500\text{H})(2.00\times10^{-4}\text{F})}} = \underline{100\text{ rad/s}}$

b) $\omega' = \sqrt{\frac{1}{LC} - \frac{R^2}{4L^2}}$

10% decrease $\Rightarrow \omega' = 0.90\,\omega = 0.90\frac{1}{\sqrt{LC}}$

$$\sqrt{\frac{1}{LC} - \frac{R^2}{4L^2}} = (0.90)\frac{1}{\sqrt{LC}}$$

Square both sides of the equation: $\frac{1}{LC} - \frac{R^2}{4L^2} = 0.81\frac{1}{LC}$

$$\frac{R^2}{4L^2} = \frac{0.19}{LC} \Rightarrow R = \sqrt{0.76\frac{L}{C}} = \sqrt{0.76\frac{0.500\text{H}}{2.00\times10^{-4}\text{F}}} = \underline{43.6\ \Omega}$$

Problems

31-29

a) $|\mathcal{E}| = L\left|\frac{di}{dt}\right| \Rightarrow L = \frac{|\mathcal{E}|}{|di/dt|}$

$\frac{di}{dt} = \frac{\Delta i}{\Delta t}$ (since rate of increase is constant) $= \frac{50.0\text{A}}{10.0\text{s}} = 5.00\text{A/s}$

$L = \frac{40.0\text{V}}{5.00\text{A/s}} = \underline{8.00\text{ H}}$

b) $N\Phi_B = Li = (8.00\text{H})(50.0\text{A}) = \underline{400\text{ Wb}}$

c) Rate at which electrical energy is being dissipated by the resistance is $P_R = i^2R = (50.0\text{A})^2(25.0\Omega) = 6.25\times10^4\text{ W}$.

Rate at which electrical energy is being stored in the magnetic field of the inductor is $P_L = |\mathcal{E}_L|i = Li\left|\frac{di}{dt}\right| = (8.00\text{ H})(50.0\text{A})(5.00\text{A/s}) = 2.00\times10^3\text{ W}$.

The ratio is $\frac{P_L}{P_R} = \frac{2.00\times10^3\text{W}}{6.25\times10^4\text{ W}} = \underline{0.0320}$

31-35

a) end view

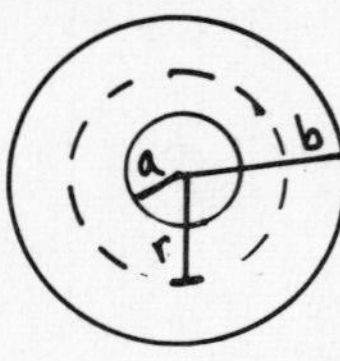

Apply Ampere's law to a circular path of radius r.

$$\oint \vec{B}\cdot d\vec{l} = \mu_0 I_{encl}$$

$$\oint \vec{B}\cdot d\vec{l} = B(2\pi r)$$

$I_{encl} = i$, the current in the inner conductor

Thus $B(2\pi r) = \mu_0 i \Rightarrow B = \frac{\mu_0 i}{2\pi r}$

b) $u = \frac{B^2}{2\mu_0}$

$dU = u\,dV$, where $dV = 2\pi r l\,dr$

$$dU = \frac{1}{2\mu_0}\left(\frac{\mu_0 i}{2\pi r}\right)^2(2\pi r l)\,dr = \frac{\mu_0 i^2 l}{4\pi r}\,dr$$

c) $U = \int dU = \frac{\mu_0 i^2 l}{4\pi}\int_a^b \frac{dr}{r} = \frac{\mu_0 i^2 l}{4\pi}\ln r\Big|_a^b = \frac{\mu_0 i^2 l}{4\pi}(\ln b - \ln a) = \frac{\mu_0 i^2 l}{4\pi}\ln\left(\frac{b}{a}\right)$

d) Eq. (31-8): $U = \frac{1}{2}Li^2$

Part (c): $U = \frac{\mu_0 i^2 l}{4\pi}\ln\left(\frac{b}{a}\right)$

$\Rightarrow \frac{1}{2}Li^2 = \frac{\mu_0 i^2 l}{4\pi}\ln\left(\frac{b}{a}\right)$

$L = \frac{\mu_0 l}{2\pi}\ln\left(\frac{b}{a}\right)$, the same result as calculated in part (d) of Problem 31-34.

31-37

Energy density in the electric field: $u_E = \frac{1}{2}\epsilon_0 E^2$ (Eq. 25-11)

Energy density in the magnetic field: $u_B = \frac{B^2}{2\mu_0}$ (Eq. 31-9)

$u_E = u_B \Rightarrow \frac{1}{2}\epsilon_0 E^2 = \frac{B^2}{2\mu_0}$

$B = \sqrt{\mu_0\epsilon_0}\;E = \sqrt{(4\pi\times10^{-7}\text{ T}\cdot\text{m/A})(8.854\times10^{-12}\text{ C}^2/\text{N}\cdot\text{m}^2)}\,(500\text{ V/m}) = \underline{1.67\times10^{-6}\text{ T}}$

31-39

$L = 3.00\ \text{H},\ R = 9.00\ \Omega,\ \mathcal{E} = 12.0\ \text{V}$

$i = \frac{\mathcal{E}}{R}(1 - e^{-t/\tau}),\ \tau = \frac{L}{R}$

a) Eq. (31-8): $U_L = \frac{1}{2}Li^2$

$t = \tau \Rightarrow i = \frac{\mathcal{E}}{R}(1 - e^{-1}) = \frac{12.0\ \text{V}}{9.00\ \Omega}(1 - e^{-1}) = 0.8428\ \text{A}$

$U_L = \frac{1}{2}Li^2 = \frac{1}{2}(3.00\ \text{H})(0.8428\ \text{A})^2 = \underline{1.06\ \text{J}}$

Exercise 31-22(c): $P_L = \frac{dU_L}{dt} = Li\frac{di}{dt}$

$i = \frac{\mathcal{E}}{R}(1 - e^{-t/\tau});\ \frac{di}{dt} = \frac{\mathcal{E}}{L}e^{-(R/L)t} = \frac{\mathcal{E}}{L}e^{-t/\tau}$

$\Rightarrow P_L = L\left(\frac{\mathcal{E}}{R}(1 - e^{-t/\tau})\right)\left(\frac{\mathcal{E}}{L}e^{-t/\tau}\right) = \frac{\mathcal{E}^2}{R}(e^{-t/\tau} - e^{-2t/\tau})$

$U_L = \int_0^\tau P_L\,dt = \frac{\mathcal{E}^2}{R}\int_0^\tau (e^{-t/\tau} - e^{-2t/\tau})\,dt = \frac{\mathcal{E}^2}{R}\left[-\tau e^{-t/\tau} + \frac{\tau}{2}e^{-t/\tau}\right]\Big|_0^\tau$

$U_L = -\frac{\mathcal{E}^2}{R}\tau\left[e^{-t/\tau} - \frac{1}{2}e^{-2t/\tau}\right]\Big|_0^\tau = \frac{\mathcal{E}^2}{R}\tau\left[1 - \frac{1}{2} - e^{-1} + \frac{1}{2}e^{-2}\right] = \frac{\mathcal{E}^2}{2R}\left(\frac{L}{R}\right)(1 - 2e^{-1} + e^{-2})$

$U_L = \frac{1}{2}\left(\frac{\mathcal{E}}{R}\right)^2 L(1 - 2e^{-1} + e^{-2}) = \frac{1}{2}\left(\frac{12.0\ \text{V}}{9.00\ \Omega}\right)^2(3.00\ \text{H})(0.3996) = 1.06\ \text{J}$, which checks.

b) Exercise 31-22(a): The rate at which the battery supplies energy is

$P_\mathcal{E} = \mathcal{E}i = \mathcal{E}\left(\frac{\mathcal{E}}{R}(1 - e^{-t/\tau})\right) = \frac{\mathcal{E}^2}{R}(1 - e^{-t/\tau})$

$U_\mathcal{E} = \int_0^\tau P_\mathcal{E}\,dt = \frac{\mathcal{E}^2}{R}\int_0^\tau (1 - e^{-t/\tau})\,dt = \frac{\mathcal{E}^2}{R}\left[t + \tau e^{-t/\tau}\right]\Big|_0^\tau = \frac{\mathcal{E}^2}{R}(\tau + \tau e^{-1} - \tau)$

$U_\mathcal{E} = \frac{\mathcal{E}^2}{R}\tau e^{-1} = \frac{\mathcal{E}^2}{R}\left(\frac{L}{R}\right)e^{-1} = \left(\frac{\mathcal{E}}{R}\right)^2 Le^{-1} = \left(\frac{12.0\ \text{V}}{9.00\ \Omega}\right)^2(3.00\ \text{H})(0.3679) = \underline{1.96\ \text{J}}$

c) $P_R = i^2R = \frac{\mathcal{E}^2}{R}(1 - e^{-t/\tau})^2 = \frac{\mathcal{E}^2}{R}(1 - 2e^{-t/\tau} + e^{-2t/\tau})$

$U_R = \int_0^\tau P_R\,dt = \frac{\mathcal{E}^2}{R}\int_0^\tau (1 - 2e^{-t/\tau} + e^{-2t/\tau})\,dt = \frac{\mathcal{E}^2}{R}\left[t + 2\tau e^{-t/\tau} - \frac{\tau}{2}e^{-2t/\tau}\right]\Big|_0^\tau$

$U_R = \frac{\mathcal{E}^2}{R}\left[\tau + 2\tau e^{-1} - \frac{\tau}{2}e^{-2} - 2\tau + \frac{\tau}{2}\right] = \frac{\mathcal{E}^2}{R}\left[-\frac{\tau}{2} + 2\tau e^{-1} - \frac{\tau}{2}e^{-2}\right] = \frac{\mathcal{E}^2}{2R}\tau\left[-1 + 4e^{-1} - e^{-2}\right]$

$U_R = \frac{\mathcal{E}^2}{2R}\left(\frac{L}{R}\right)\left[-1 + 4e^{-1} - e^{-2}\right] = \left(\frac{\mathcal{E}}{R}\right)^2\frac{1}{2}L\left[-1 + 4e^{-1} - e^{-2}\right] = \left(\frac{12.0\ \text{V}}{9.00\ \Omega}\right)^2\frac{1}{2}(3.00\ \text{H})(0.3362) = \underline{0.90\ \text{J}}$

d) $U_\mathcal{E} = U_R + U_L$ $(1.96\ \text{J} = 0.90\ \text{J} + 1.06\ \text{J})$

The energy supplied by the battery equals the sum of the energy stored in the magnetic field of the inductor and the energy dissipated in the resistance of the inductor.

31-41

The equation is $-iR - L\frac{di}{dt} - \frac{q}{C} = 0$.

Multiply by $-i \Rightarrow i^2R + Li\frac{di}{dt} + \frac{iq}{C} = 0$.

$\frac{d}{dt}U_L = \frac{d}{dt}\left(\frac{1}{2}Li^2\right) = \frac{1}{2}L\frac{d}{dt}(i^2) = \frac{1}{2}L\left(2i\frac{di}{dt}\right) = Li\frac{di}{dt}$, the second term

$\frac{d}{dt}U_C = \frac{d}{dt}\left(\frac{q^2}{2C}\right) = \frac{1}{2C}\frac{d}{dt}(q^2) = \frac{1}{2C}2q\frac{dq}{dt} = \frac{qi}{C}$, the third term

$i^2R = P_R$, the rate at which electrical energy is dissipated in the resistance.

$\frac{d}{dt}U_L = P_L$, the rate at which the amount of energy stored in the inductor is changing.

$\frac{d}{dt}U_C = P_C$, the rate at which the amount of energy stored in the capacitor is changing.

The equation says that $P_R + P_L + P_C = 0$; the net rate of change of energy in the circuit is zero.

31-41 (cont)

Note: At any given time one of P_C or P_L is negative. If the current and U_L are increasing the charge on the capacitor and U_C are decreasing, and vice versa.

31-43

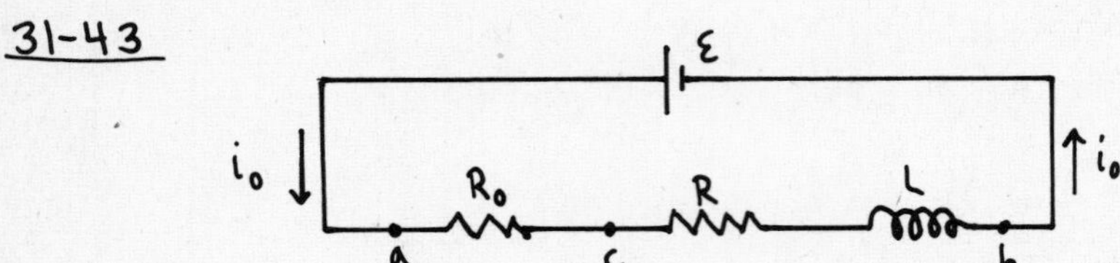

a) At time $t=0$, $i=0$ so $V_{ac} = iR_0 = 0$. By the loop rule $\mathcal{E} - V_{ac} - V_{cb} = 0$, so $V_{cb} = \mathcal{E} - V_{ac} = \mathcal{E} = 60.0\ \text{V}$. ($iR = 0$ so this potential of 60.0 V is across the inductor and is an induced emf produced by the changing current.)

b) After a long time $\frac{di}{dt} \to 0$ so the potential $-L\frac{di}{dt}$ across the inductor becomes zero. The loop rule gives $\mathcal{E} - i_0(R_0 + R) = 0$.

$$i_0 = \frac{\mathcal{E}}{R_0 + R} = \frac{60.0\ \text{V}}{50.0\ \Omega + 150\ \Omega} = 0.300\ \text{A}$$

$$V_{ac} = i_0 R_0 = (0.300\ \text{A})(50.0\ \Omega) = 15.0\ \text{V}$$

$$V_{cb} = i_0 R + L\frac{di}{dt}\!\!\to 0 = (0.300\ \text{A})(150\ \Omega) = 45.0\ \text{V}$$ (Note that $V_{ac} + V_{cb} = \mathcal{E}$.)

c) $\mathcal{E} - V_{ac} - V_{cb} = 0$

$$\mathcal{E} - iR_0 - iR - L\frac{di}{dt} = 0$$

$$L\frac{di}{dt} = \mathcal{E} - i(R_0 + R)$$

$$-\left(\frac{L}{R+R_0}\right)\frac{di}{dt} = i - \frac{\mathcal{E}}{R+R_0}$$

$$\frac{di}{i - \frac{\mathcal{E}}{R+R_0}} = -\left(\frac{R+R_0}{L}\right)dt$$

Integrate from $t=0$, when $i=0$, to t, when $i = i_0$

$$\int_0^{i_0} \frac{di}{i - \frac{\mathcal{E}}{R+R_0}} = -\frac{R+R_0}{L}\int_0^t dt$$

$$\ln\left(i - \frac{\mathcal{E}}{R+R_0}\right)\Big|_0^{i_0} = -\left(\frac{R+R_0}{L}\right)t \Rightarrow \ln\left(i_0 - \frac{\mathcal{E}}{R+R_0}\right) - \ln\left(-\frac{\mathcal{E}}{R+R_0}\right) = -\left(\frac{R+R_0}{L}\right)t$$

$$\ln\left(\frac{i_0 - \frac{\mathcal{E}}{R+R_0}}{-\mathcal{E}/R+R_0}\right) = -\left(\frac{R+R_0}{L}\right)t$$

Take exponentials of both sides $\Rightarrow$ $\dfrac{i_0 - \frac{\mathcal{E}}{R+R_0}}{-\frac{\mathcal{E}}{R+R_0}} = e^{-\left(\frac{R+R_0}{L}\right)t}$

$$i_0 = \frac{\mathcal{E}}{R+R_0}\left(1 - e^{-\left(\frac{R+R_0}{L}\right)t}\right)$$

Substituting in the numerical values gives

$$i_0 = \frac{60.0\ \text{V}}{50\ \Omega + 150\ \Omega}\left(1 - e^{-\left(\frac{200\ \Omega}{5.00\ \text{H}}\right)t}\right) = (0.300\ \text{A})\left(1 - e^{-t/0.025\ \text{s}}\right)$$

At $t \to 0$, $i_0 = (0.300\ \text{A})(1-1) = 0$ (agrees with part (a))

At $t \to \infty$, $i_0 = (0.300\ \text{A})(1-0) = 0.300\ \text{A}$ (agrees with part (b))

31-43 (cont)

$V_{ac} = i_0 R_0 = \frac{\mathcal{E} R_0}{R+R_0}\left(1-e^{-\left(\frac{R+R_0}{L}\right)t}\right) = 15.0\text{ V}(1-e^{-t/0.025\text{ s}})$

$V_{cb} = \mathcal{E} - V_{ac} = 60.0\text{ V} - (15.0\text{ V})(1-e^{-t/0.025\text{ s}}) = 15.0\text{ V}(3.00+e^{-t/0.025\text{ s}})$

At $t \to 0$, $V_{ac} = 0$, $V_{cb} = 60.0$ V (agrees with part (a)).

At $t \to \infty$, $V_{ac} = 15.0$ V, $V_{cb} = 45.0$ V (agrees with part (b)).

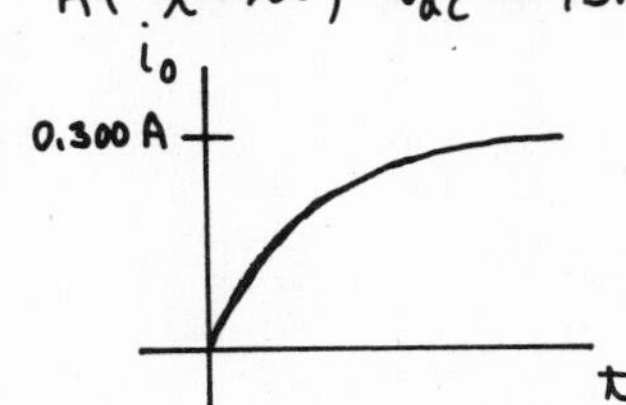

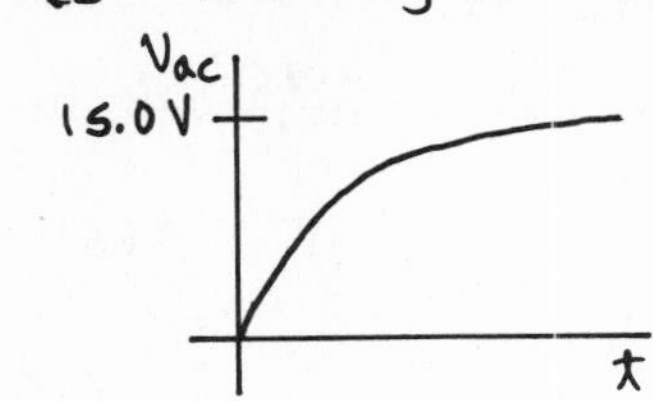

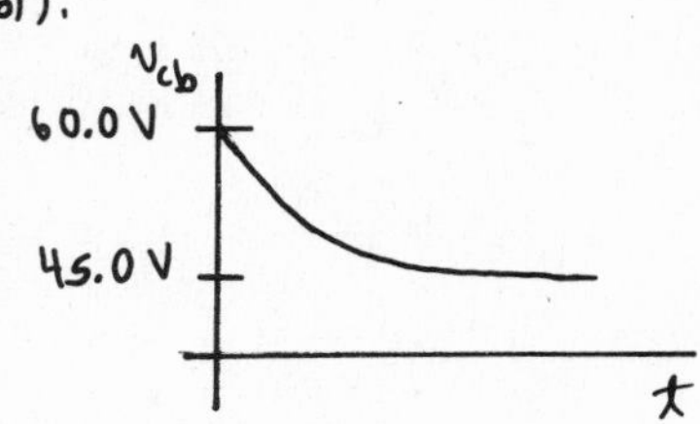

31-45

a) With switch S closed the circuit is

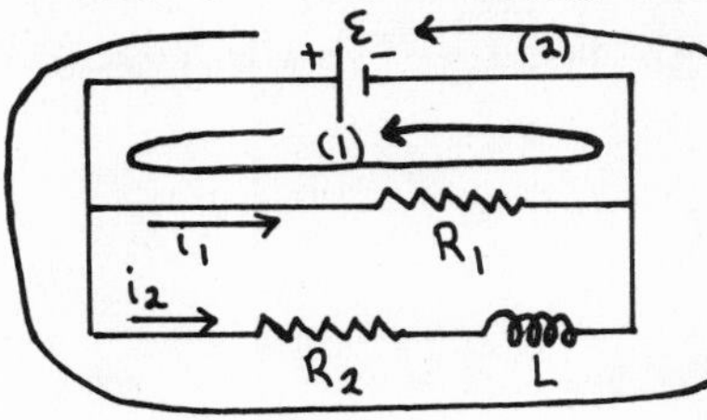

loop (1)

$\mathcal{E} - i_1 R_1 = 0$

$i_1 = \frac{\mathcal{E}}{R_1}$ (independent of t)

loop (2)

$\mathcal{E} - i_2 R_2 - L\frac{di_2}{dt} = 0$

This in the form of equation (31-11), so the solution is analagous to Eq. (31-13): $i_2 = \frac{\mathcal{E}}{R_2}(1-e^{-R_2 t/L})$

b) The expressions derived in part (a) give that as $t \to \infty$,

$i_1 = \frac{\mathcal{E}}{R_1}$, $i_2 = \frac{\mathcal{E}}{R_2}$

(Since $\frac{di}{dt} \to 0$ at steady-state the inductance has no effect on the circuit.)

c) The circuit now becomes

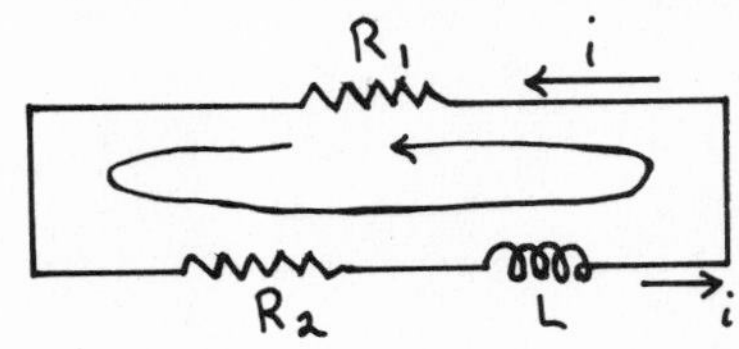

Let $t = 0$ now be when S is opened.

At $t = 0$, $i = \frac{\mathcal{E}}{R_2}$.

The loop rule applied to the single current loop gives

$-i(R_1+R_2) - L\frac{di}{dt} = 0$ (Now $\frac{di}{dt}$ is negative.)

$L\frac{di}{dt} = -i(R_1+R_2) \Rightarrow \frac{di}{i} = -\left(\frac{R_1+R_2}{L}\right)dt$

Integrate from $t=0$, when $i = I_0 = \frac{\mathcal{E}}{R_2}$, to t.

$\int_{I_0}^{i} \frac{di}{i} = -\left(\frac{R_1+R_2}{L}\right)\int_0^t dt$

$\ln\frac{i}{I_0} = -\left(\frac{R_1+R_2}{L}\right)t$

Take exponentials of both sides of this equation

$\Rightarrow i = I_0 e^{-(R_1+R_2)t/L} = \frac{\mathcal{E}}{R_2} e^{-(R_1+R_2)t/L}$

d) $L = 48.0$ H

$P_{R_1} = \frac{V^2}{R_1} = 60.0\text{ W} \Rightarrow R_1 = \frac{V^2}{P_{R_1}} = \frac{(120\text{ V})^2}{60.0\text{ W}} = 240\,\Omega$

31-45 (cont)

We are asked to find R_2 and $\mathcal{E}$. Use the expression derived in part (c).

$I_0 = 0.900\text{ A} \Rightarrow \frac{\mathcal{E}}{R_2} = 0.900\text{ A}$

$i = 0.300\text{A}$ when $t = 0.200\text{ s}$, so $i = \frac{\mathcal{E}}{R_2} e^{-(R_1+R_2)t/L}$ gives

$0.300\text{ A} = 0.900\text{ A}\, e^{-(R_1+R_2)t/L}$

$\frac{1}{3} = e^{-(R_1+R_2)t/L} \Rightarrow \ln 3 = \frac{(R_1+R_2)t}{L}$

$R_2 = \frac{L \ln 3}{t} - R_1 = \frac{(48.0\text{ H})\ln 3}{0.200\text{ s}} - 240\,\Omega = 263.7\,\Omega - 240\,\Omega = \underline{23.7\,\Omega}$

Then $\frac{\mathcal{E}}{R_2} = 0.900\text{ A} \Rightarrow \mathcal{E} = (0.900\text{A})\,R_2 = (0.900\text{A})(23.7\,\Omega) = \underline{21.3\text{ V}}$.

e) Use the expressions derived in part (a). The current through the light bulb before the switch is opened is $i_1 = \frac{\mathcal{E}}{R_1} = \frac{21.3\text{ V}}{240\,\Omega} = \underline{0.0888\text{ A}}$

When the switch is opened the current through the light bulb jumps from 0.0888 A to 0.900 A.

CHAPTER 32

Exercises 3, 7, 9, 11, 13, 17, 21, 23, 25

Problems 29, 31, 33, 35, 39, 41, 45

Exercises

32-3

a) $X_L = \omega L = 2\pi f L = 2\pi(50.0\text{ Hz})(1.00\text{ H}) = \underline{314\ \Omega}$

b) $X_L = 2\pi f L \Rightarrow L = \dfrac{X_L}{2\pi f} = \dfrac{1.00\ \Omega}{2\pi(50.0\text{ Hz})} = \underline{3.18\times10^{-3}\text{ H}}$

c) $X_c = \dfrac{1}{\omega C} = \dfrac{1}{2\pi f C} = \dfrac{1}{2\pi(50.0\text{Hz})(1.00\times10^{-6}\text{F})} = \underline{3.18\times10^{3}\ \Omega}$

d) $X_c = \dfrac{1}{2\pi f C} \Rightarrow C = \dfrac{1}{2\pi f\ X_c} = \dfrac{1}{2\pi(50.0\text{Hz})(1.00\ \Omega)} = \underline{3.18\times10^{-3}\text{F}}$

32-7

$V = IX_c \Rightarrow X_c = \dfrac{V}{I} = \dfrac{170\text{V}}{1.20\text{A}} = 141.7\ \Omega$

$X_c = \dfrac{1}{\omega C} = \dfrac{1}{2\pi f\ C} \Rightarrow C = \dfrac{1}{2\pi f X_c} = \dfrac{1}{2\pi(60.0\text{Hz})(141.7\ \Omega)} = 1.87\times10^{-5}\text{F} = \underline{18.7\ \mu\text{F}}$

32-9

[Circuit: source V in series with R and C]

No inductor $\Rightarrow X_L = 0$

$R = 300\ \Omega,\ C = 4.00\times10^{-6}\text{F},\ V = 50.0\text{ V},\ \omega = 100\text{ rad/s}$

a) $X_c = \dfrac{1}{\omega C} = \dfrac{1}{(100\text{rad/s})(4.00\times10^{-6}\text{F})} = 2500\ \Omega$

$Z = \sqrt{R^2 + (X_L - X_c)^2} = \sqrt{R^2 + X_c^2} = \sqrt{(300\ \Omega)^2 + (2500\ \Omega)^2} = \underline{2518\ \Omega}$

b) $I = \dfrac{V}{Z} = \dfrac{50.0\text{V}}{2518\ \Omega} = \underline{0.0199\text{ A}}$

c) The voltage amplitude across the resistor is
$V_R = IR = (0.0199\text{A})(300\ \Omega) = \underline{5.97\text{ V}}$

The voltage amplitude across the capacitor is
$V_c = IX_c = (0.0199\text{ A})(2500\ \Omega) = \underline{49.8\text{ V}}$

d) $\tan\phi = \dfrac{X_L - X_c}{R} = \dfrac{0 - 2500\ \Omega}{300\ \Omega} = -8.333 \Rightarrow \phi = \underline{-83.2^\circ}$

The phase angle is negative, so the source voltage lags behind the current.

32-9 (cont)

e) The phasor diagram is qualitatively

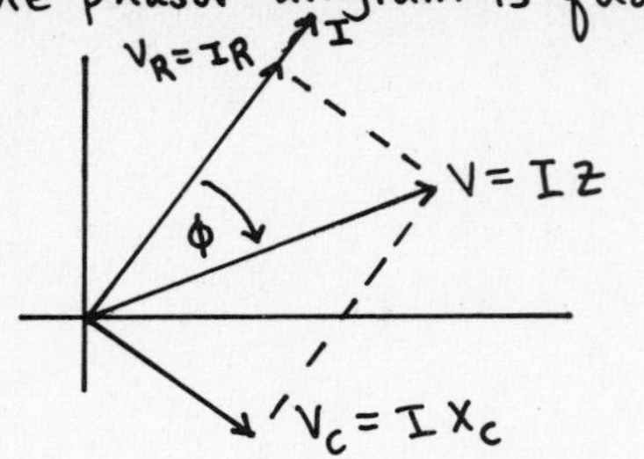

32-11

$R = 300\,\Omega$, $L = 0.100\,H$, $C = 0.500 \times 10^{-6}\,F$

a) $f = 500\,Hz$

$X_L = \omega L = 2\pi f L = 2\pi(500\,Hz)(0.100\,H) = 314.2\,\Omega$

$X_C = \frac{1}{\omega C} = \frac{1}{2\pi f C} = \frac{1}{2\pi(500\,Hz)(0.500\times10^{-6}F)} = 636.6\,\Omega$

$Z = \sqrt{R^2 + (X_L - X_C)^2} = \sqrt{(300\,\Omega)^2 + (314.2\,\Omega - 636.6\,\Omega)^2} = \underline{440\,\Omega}$

$\tan\phi = \frac{X_L - X_C}{R} = \frac{314.2\,\Omega - 636.6\,\Omega}{300\,\Omega} = -1.075 \Rightarrow \phi = \underline{-47.1^\circ}$

The phase angle is negative, so the source voltage lags behind the current. Since $X_C > X_L$ the phasor diagram is qualitatively

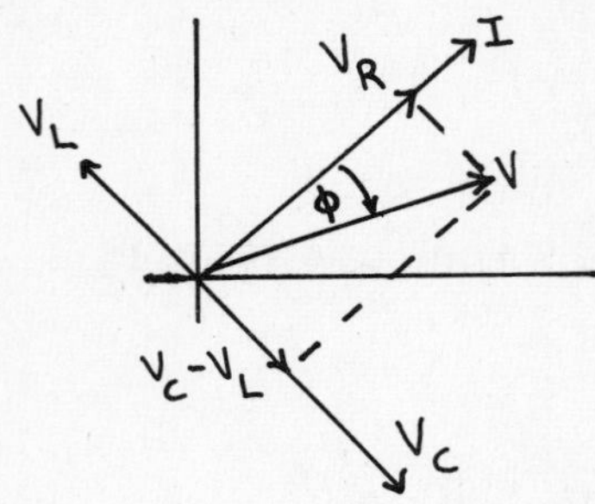

b) $f = 1000\,Hz$

$X_L = \omega L = 2\pi f L = 2\pi(1000\,Hz)(0.100\,H) = 628.3\,\Omega$

$X_C = \frac{1}{\omega C} = \frac{1}{2\pi f C} = \frac{1}{2\pi(1000\,Hz)(0.500\times10^{-6}F)} = 318.3\,\Omega$

$Z = \sqrt{R^2 + (X_L - X_C)^2} = \sqrt{(300\,\Omega)^2 + (628.3\,\Omega - 318.3\,\Omega)^2} = \underline{431\,\Omega}$

$\tan\phi = \frac{X_L - X_C}{R} = \frac{628.3\,\Omega - 318.3\,\Omega}{300\,\Omega} = +1.033 \Rightarrow \phi = \underline{45.9^\circ}$

The phase angle is positive, so the source voltage leads the current. Since $X_L > X_C$ the phasor diagram is qualitatively

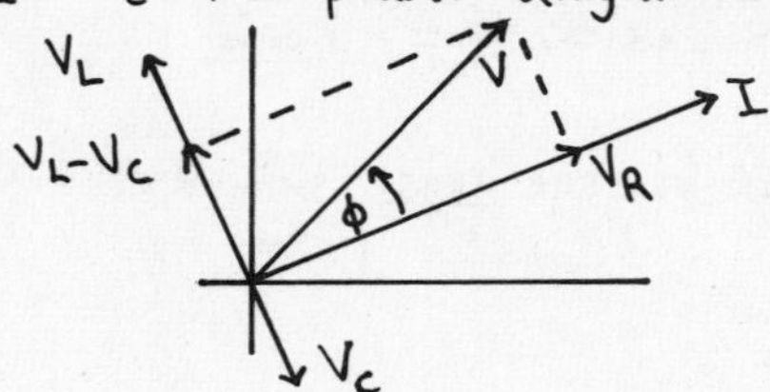

32-13

$R = 300\,\Omega,\ L = 0.100\text{ H},\ C = 0.500\times10^{-6}\text{ F},\ f = 1000\text{ Hz},\ I_{rms} = 0.250\text{ A}$

a) $P = V_{rms} I_{rms} \cos\phi$ (Eq. 32-31)

$X_L = \omega L = 2\pi f L = 2\pi(1000\text{ Hz})(0.100\text{ H}) = 628.3\,\Omega$

$X_C = \frac{1}{\omega C} = \frac{1}{2\pi f C} = \frac{1}{2\pi(1000\text{ Hz})(0.500\times10^{-6}\text{ F})} = 318.3\,\Omega$

$Z = \sqrt{R^2 + (X_L - X_C)^2} = \sqrt{(300\,\Omega)^2 + (628.3\,\Omega - 318.3\,\Omega)^2} = 431.3\,\Omega$

$V_{rms} = I_{rms} Z = (0.250\text{ A})(431.3\,\Omega) = 107.8\text{ V}$

$\tan\phi = \frac{X_L - X_C}{R} = \frac{628.3\,\Omega - 318.3\,\Omega}{300\,\Omega} = 1.033 \Rightarrow \phi = 45.94°$

Then $P = V_{rms} I_{rms} \cos\phi = (107.8\text{ V})(0.250\text{ A})\cos(45.94°) = \underline{18.7\text{ W}}$

b) In the resistor the current and voltage are in phase, so
$P = V_{R,rms} I_{rms} = (I_{rms} R) I_{rms} = I_{rms}^2 R = (0.250\text{ A})^2(300\,\Omega) = \underline{18.7\text{ W}}$

c) For the capacitor the instantaneous voltage and current are 90° out of phase, and the average power consumed is zero. Electrical energy is periodically stored and released in the capacitor, but not converted to other forms and thereby lost to the circuit.

d) For the inductor the instantaneous voltage and current are 90° out of phase, and the average power consumed is zero. Electrical energy is periodically stored and released in the magnetic field of the inductor, but not converted to other forms and thereby lost to the circuit.

e) part (a): $P_{source} = 18.7\text{ W}$
part (b): $P_R = 18.7\text{ W}$
part (c): $P_C = 0$
part (d): $P_L = 0$

$P_{source} = P_R + P_C + P_L$; the average power delivered by the source equals the sum of the average power consumed in the resistor, capacitor, and inductor.

32-17

$R = 400\,\Omega,\ L = 0.900\text{ H},\ C = 2.00\,\mu\text{F}$

a) The resonant angular frequency is given by Eq. (32-32)

$\omega_0 = \frac{1}{\sqrt{LC}} = \frac{1}{\sqrt{(0.900\text{ H})(2.00\times10^{-6}\text{ F})}} = 745.4\text{ rad/s}\ ;\ 745\text{ rad/s}$

b) At resonance $X_L = X_C$, so $\tan\phi = \frac{X_L - X_C}{R} = 0$ and $\phi = 0$. The source voltage and the current in the circuit are in phase.

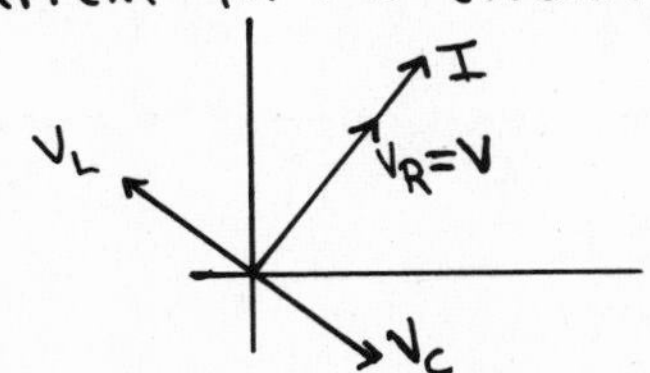

32-17 (cont)

c) $X_L = \omega L = (745.4 \text{ rad/s})(0.900 \text{ H}) = 671\,\Omega$

$X_C = \frac{1}{\omega C} = \frac{1}{(745.4 \text{ rad/s})(2.00\times10^{-6}\text{ F})} = 671\,\Omega$

$X_L = X_C \Rightarrow Z = \sqrt{R^2 + (X_L - X_C)^2} = R = 400\,\Omega$ (with $(X_L - X_C)$ crossed out, $\to 0$)

$I = \frac{V}{Z} = \frac{110 \text{ V}}{400\,\Omega} = 0.275 \text{ A}\ ;\ I_{rms} = \frac{I}{\sqrt{2}} = 0.1945 \text{ A}$

V_1 reads $V_{R,rms} = I_{rms} R = (0.1945 \text{ A})(400\,\Omega) = 77.8 \text{ V}$

V_2 reads $V_{L,rms} = I_{rms} X_L = (0.1945 \text{ A})(671\,\Omega) = 130 \text{ V}$

V_3 reads $V_{C,rms} = I_{rms} X_C = (0.1945 \text{ A})(671\,\Omega) = 130 \text{ V}$

V_4 reads $(V_L - V_C)_{rms} = I_{rms}(X_L - X_C) = 0$. At any instant the voltages across the inductor and capacitor are equal and opposite and sum to zero.

V_5 reads $V_{rms} = \frac{110 \text{ V}}{\sqrt{2}} = 77.8 \text{ V}$, the same as V_1.

d) $\omega_0 = \frac{1}{\sqrt{LC}}$ is independent of R, so $\omega_0 = 745 \text{ rad/s}$, the same as when $R = 400\,\Omega$.

e) At resonance $X_L = X_C$ and $Z = R$.

$I_{rms} = \frac{V_{rms}}{Z} = \frac{V_{rms}}{R} = \frac{77.8 \text{ V}}{100\,\Omega} = \underline{0.778 \text{ A}}.$

The current at resonance is inversely proportional to the resistance so increases by a factor of 4 when the resistance is decreased by a factor of 4.

32-21

$R = 400\,\Omega,\ L = 0.500 \text{ H},\ C = 0.600\times10^{-6}\text{ F},\ V = 120 \text{ V}$

a) The resonant angular frequency of the parallel L-R-C circuit is the same as for the series L-R-C circuit:

$\omega_0 = \frac{1}{\sqrt{LC}} = \frac{1}{\sqrt{(0.500 \text{ H})(0.600\times10^{-6}\text{ F})}} = 1826 \text{ rad/s}$

$f_0 = \frac{\omega_0}{2\pi} = \frac{1826 \text{ rad/s}}{2\pi \text{ rad}} = 291 \text{ Hz}$

b) The phasor diagram is like Fig. 32-15b, but with $I_C = I_L$ so $\phi = 0$.

V

I_C $\quad I = I_R$

I_L

c) At resonance $X_L = X_C$ so $\frac{1}{Z} = \sqrt{\frac{1}{R^2} + \left(\frac{1}{X_C} - \frac{1}{X_L}\right)^2} = \frac{1}{R}$ (with $\left(\frac{1}{X_C} - \frac{1}{X_L}\right)$ crossed out, $\to 0$)

$Z = R = 400\,\Omega$

$I = \frac{V}{Z} = \frac{120 \text{ V}}{400\,\Omega} = \underline{0.300 \text{ A}}$

d) For the parallel circuit the voltage across each circuit element equals the source voltage.

$X_L = \omega L = (1826 \text{ rad/s})(0.500 \text{ H}) = 913\,\Omega$

$X_C = \frac{1}{\omega C} = \frac{1}{(1826 \text{ rad/s})(0.600\times10^{-6}\text{ F})} = 913\,\Omega$

32-21 (cont)

The amplitude of the current through the resistor is $I_R = \frac{V}{R} = \frac{120\ V}{400\ \Omega} = \underline{0.300\ A}$.
The amplitude of the current through the inductor is $I_L = \frac{V}{X_L} = \frac{120\ V}{913\ \Omega} = \underline{0.131\ A}$.
The amplitude of the current through the capacitor branch is $I_C = \frac{V}{X_C} = \frac{120\ V}{913\ \Omega} = \underline{0.131\ A}$.

(At any instant in time the inductor and capacitor currents are equal in magnitude and opposite in direction, so their sum is zero. At resonance then the current through the source equals the current through the resistor.)

32-23

$R = 500\ \Omega,\ L = 0.200\ H,\ C = 0.200\ \mu F,\ V = 300\ V$

a) $\omega = 500$ rad/s

$X_L = \omega L = (500\ \text{rad/s})(0.200\ H) = 100\ \Omega$

$X_C = \frac{1}{\omega C} = \frac{1}{(500\ \text{rad/s})(0.200\times10^{-6}F)} = 1.00\times10^4\ \Omega$

$$\frac{1}{Z} = \sqrt{\frac{1}{R^2} + \left(\frac{1}{X_C} - \frac{1}{X_L}\right)^2} = \sqrt{\left(\frac{1}{500\ \Omega}\right)^2 + \left(\frac{1}{1.00\times10^4\ \Omega} - \frac{1}{100\ \Omega}\right)^2} = \underline{99.0\ \Omega}$$

$I = \frac{V}{Z} = \frac{300\ V}{99.0\ \Omega} = \underline{3.03\ A}$

b) $\omega = \omega_0 \Rightarrow X_L = X_C$ and $Z = R = \underline{500\ \Omega}$

$I = \frac{V}{Z} = \frac{V}{R} = \frac{300\ V}{500\ \Omega} = \underline{0.600\ A}$

(At resonance the impedance is a maximum and the current through the source is a minimum.)

32-25

a) Eq. (32-37): $\frac{V_2}{V_1} = \frac{N_2}{N_1} \Rightarrow \frac{N_1}{N_2} = \frac{V_1}{V_2} = \frac{120\ V}{8.00\ V} = \underline{15}$

b) $I_2 = \frac{V_2}{R} = \frac{8.00\ V}{4.00\ \Omega} = \underline{2.00\ A}$

c) $P_2 = I_2^2 R = (2.00\ A)^2(4.00\ \Omega) = \underline{16.0\ W}$

d) The power drawn from the line by the transformer is the 16.0 W that is delivered to the load.

$P = \frac{V^2}{R} \Rightarrow R = \frac{V^2}{P} = \frac{(120\ V)^2}{16.0\ W} = \underline{900\ \Omega}$

And $\left(\frac{N_1}{N_2}\right)^2(4.00\ \Omega) = (15)^2(4.00\ \Omega) = 900\ \Omega$, as was to be shown.

Problems

32-29

The voltage across the coil leads the current in it by $30.0° \Rightarrow \phi = +30.0°$.
$\tan\phi = \frac{X_L - X_C}{R}$. But there is no capacitance in the circuit so $X_C = 0$.

32-29 (cont)

$\tan\phi = \frac{X_L}{R} \Rightarrow X_L = R\tan\phi = (40.0\,\Omega)\tan 30.0° = 23.09\,\Omega$

$X_L = \omega L = 2\pi f L \Rightarrow L = \frac{X_L}{2\pi f} = \frac{23.09\,\Omega}{2\pi(100\text{ Hz})} = \underline{0.0367\text{ H}}$

32-31

a) $V_C = IX_C \Rightarrow I = \frac{V_C}{X_C} = \frac{720\text{ V}}{600\,\Omega} = \underline{1.20\text{ A}}$

b) $V = IZ \Rightarrow Z = \frac{V}{I} = \frac{240\text{ V}}{1.20\text{ A}} = \underline{200\,\Omega}$

c) $Z^2 = R^2 + (X_L - X_C)^2$

$X_L - X_C = \pm\sqrt{Z^2 - R^2} \Rightarrow X_L = X_C \pm \sqrt{Z^2 - R^2}$

$X_L = 600\,\Omega \pm \sqrt{(200\,\Omega)^2 - (150\,\Omega)^2} = 600\,\Omega \pm 132.3\,\Omega$

$X_L = \underline{732\,\Omega \text{ or } 468\,\Omega}$

32-33

$P = V_{rms} I_{rms} \cos\phi$

But from the phasor diagram (Fig. 32-8) $V_{rms}\cos\phi = V_R = I_{rms}R$

$\Rightarrow P = I_{rms}^2 R$

$R = \frac{P}{I_{rms}^2} = \frac{12.0\text{ J/s}}{(0.500\text{ A})^2} = 48.0\,\Omega$

Then $Z = \sqrt{R^2 + (X_L - X_C)^2} = \sqrt{(48.0\,\Omega)^2 + (25.0\,\Omega - 0)^2} = \underline{54.1\,\Omega}$

32-35

a) Source voltage lags current $\Rightarrow X_C > X_L \Rightarrow$ add an inductor in series with the circuit.

b) power factor $\cos\phi$ equals $1 \Rightarrow \phi = 0 \Rightarrow X_C = X_L$.

Calculate the present value of $X_C - X_L$ to see how much more X_L is needed:

From the phasor diagram $\cos\phi = \frac{R}{Z}$ (Exercise 32-15).

$R = Z\cos\phi = (50.0\,\Omega)(0.400) = 20.0\,\Omega$

$\tan\phi = \frac{X_L - X_C}{R} \Rightarrow X_L - X_C = R\tan\phi$

$\cos\phi = 0.400 \Rightarrow \phi = -66.42°$ (ϕ is negative since the voltage lags the current.)

$X_L - X_C = R\tan\phi = (20.0\,\Omega)\tan(-66.42°) = -45.82\,\Omega$

Therefore need to add $45.8\,\Omega$ of X_L.

$X_L = \omega L = 2\pi f L \Rightarrow L = \frac{X_L}{2\pi f} = \frac{45.8\,\Omega}{2\pi(60.0\text{ Hz})} = \underline{0.121\text{ H}}$, amount of inductance to add

32-39

V_s C R L V_{out}

$X_L = \omega L,\ X_C = \frac{1}{\omega C}$

$Z = \sqrt{R^2 + (\omega L - \frac{1}{\omega C})^2}$

32-39 (cont)

$$I = \frac{V_s}{Z} = \frac{V_s}{\sqrt{R^2+(\omega L-\frac{1}{\omega C})^2}}$$

The voltages across the resistor and the inductor are 90° out of phase, so

$$V_{out} = \sqrt{V_R^2+V_L^2} = I\sqrt{R^2+X_L^2} = I\sqrt{R^2+\omega^2L^2} = V_s\sqrt{\frac{R^2+\omega^2L^2}{R^2+(\omega L-\frac{1}{\omega C})^2}}$$

$$\frac{V_{out}}{V_s} = \sqrt{\frac{R^2+\omega^2L^2}{R^2+(\omega L-\frac{1}{\omega C})^2}}$$

ω small

$$R^2+(\omega L-\frac{1}{\omega C})^2 \to \frac{1}{\omega^2C^2}$$

$$R^2+\omega^2L^2 \to R^2$$

Therefore $\frac{V_{out}}{V_s} \xrightarrow{\omega\,small} \sqrt{\frac{R^2}{\frac{1}{\omega^2C^2}}} = \omega RC.$

ω large

$$R^2+\omega^2L^2 \to \omega^2L^2$$

$$R^2+(\omega L-\frac{1}{\omega C})^2 \to R^2+\omega^2L^2 \to \omega^2L^2$$

Therefore $\frac{V_{out}}{V_s} \xrightarrow{\omega\,large} \sqrt{\frac{\omega^2L^2}{\omega^2L^2}} = 1.$

32-41

$R = 100\,\Omega$, $C = 0.100\times10^{-6}$ F, $L = 0.100$ H, $V = 160$ V

a) resonance $\Rightarrow X_C = X_L \Rightarrow \frac{1}{\omega_0 C} = \omega_0 L \Rightarrow \omega_0 = \frac{1}{\sqrt{LC}}$

$$\omega_0 = \frac{1}{\sqrt{(0.100\,\text{H})(0.100\times10^{-6}\,\text{F})}} = \underline{1.00\times10^4\ \text{rad/s}}$$

b) At resonance $Z = R$

$$I = \frac{V}{Z} = \frac{160\,\text{V}}{100\,\Omega} = \underline{1.60\ \text{A}}$$

c) $I_R = \frac{V}{R} = \frac{160\,\text{V}}{100\,\Omega} = \underline{1.60\ \text{A}}$

d) $I_L = \frac{V}{X_L}$

$$X_L = \omega L = (1.00\times10^4\,\text{rad/s})(0.100\,\text{H}) = 1.00\times10^3\,\Omega$$

$$I_L = \frac{160\,\text{V}}{1.00\times10^3\,\Omega} = \underline{0.160\ \text{A}}$$

e) $X_C = \frac{1}{\omega C} = \frac{1}{(1.00\times10^4\,\text{rad/s})(0.100\times10^{-6}\,\text{F})} = 1.00\times10^3\,\Omega$

$$I_C = \frac{V}{X_C} = \frac{160\,\text{V}}{1.00\times10^3\,\Omega} = \underline{0.160\ \text{A}}$$

f) inductor: $U_{max} = \frac{1}{2}LI_L^2 = \frac{1}{2}(0.100\,\text{H})(0.160\,\text{A})^2 = \underline{1.28\times10^{-3}\ \text{J}}$

capacitor: $U_{max} = \frac{1}{2}CV_C^2$; parallel circuit $\Rightarrow V_C = V = 160$ V

$U_{max} = \frac{1}{2}(0.100\times10^{-6}\,\text{F})(160\,\text{V})^2 = \underline{1.28\times10^{-3}\ \text{J}}$, the same as the maximum energy stored in the inductor.

32-45

Consider the cycle of the repeating current from $t_1 = \frac{T}{2}$ to $t_2 = \frac{3T}{2}$. In this interval $i = \frac{2I_0}{T}(t - T)$.

$$I_{av} = \frac{1}{t_2 - t_1}\int_{t_1}^{t_2} i\,dt = \frac{1}{T}\int_{T/2}^{3T/2} \frac{2I_0}{T}(t - T)\,dt = \frac{2I_0}{T^2}\left(\frac{1}{2}t^2 - Tt\right)\Big|_{T/2}^{3T/2}$$

$$I_{av} = \frac{2I_0}{T^2}\left(\frac{9T^2}{8} - \frac{3T^2}{2} - \frac{T^2}{8} + \frac{T^2}{2}\right) = (2I_0)\frac{1}{8}(9 - 12 - 1 + 4) = \frac{I_0}{4}(13 - 13) = 0.$$

$$I_{rms}^2 = (I^2)_{av} = \frac{1}{t_2 - t_1}\int_{t_1}^{t_2} i^2\,dt = \frac{1}{T}\int_{T/2}^{3T/2} \frac{4I_0^2}{T^2}(t - T)^2\,dt$$

$$I_{rms}^2 = \frac{4I_0^2}{T^3}\int_{T/2}^{3T/2}(t - T)^2\,dt = \frac{4I_0^2}{T^3}\,\frac{1}{3}(t - T)^3\Big|_{T/2}^{3T/2} = \frac{4I_0^2}{3T^3}\left[\left(\frac{T}{2}\right)^3 - \left(-\frac{T}{2}\right)^3\right]$$

$$I_{rms}^2 = \frac{I_0^2}{6}[1 + 1] = \frac{1}{3}I_0^2$$

$$I_{rms} = \sqrt{I_{rms}^2} = \frac{I_0}{\sqrt{3}}$$

CHAPTER 33

Exercises 3, 5, 9, 13, 15, 17

Problems 21, 25, 27, 29, 31, 33, 35

Exercises

33-3

a) $c = f\lambda \Rightarrow \lambda = \frac{c}{f} = \frac{3.00\times10^8\text{ m/s}}{90.9\times10^6\text{ Hz}} = \underline{3.30\text{ m}}$

b) Eq. (33-4): $E = cB = (3.00\times10^8\text{ m/s})(2.40\times10^{-11}\text{ T}) = \underline{7.20\times10^{-3}\text{ V/m}}$

33-5

$\omega = 2\pi f = 2\pi(6.00\times10^{13}\text{ Hz}) = 3.77\times10^{14}\text{ rad/s}$

$k = \frac{2\pi}{\lambda} = \frac{2\pi f}{c} = \frac{\omega}{c} = \frac{3.77\times10^{14}\text{ rad/s}}{3.00\times10^8\text{ m/s}} = 1.26\times10^6\text{ rad/m}$

$B_{max} = 4.00\times10^{-4}\text{ T}$

$E_{max} = c B_{max} = (3.00\times10^8\text{ m/s})(4.00\times10^{-4}\text{ T}) = 1.20\times10^5\text{ V/m}$

$\vec{B}$ is along the x-axis. $\vec{E}\times\vec{B}$ is in the direction of propagation (the +y-direction); from this we can deduce the direction of $\vec{E}$.

Y, $\vec{E}\times\vec{B}$, $\vec{B}$, X, $\vec{E}$, Z

$\vec{E}$ is along the z-axis.

The equations are in the form of Eqs. (33-17):

$\vec{E} = E_{max}\vec{k}\sin(\omega t - ky) = (1.20\times10^5\text{ V/m})\vec{k}\sin((3.77\times10^{14}\text{ rad/s})t - (1.26\times10^6\text{ rad/m})y)$

$\vec{B} = B_{max}\vec{i}\sin(\omega t - ky) = (4.00\times10^{-4}\text{ T})\vec{i}\sin((3.77\times10^{14}\text{ rad/s})t - (1.26\times10^6\text{ rad/m})y)$

33-9

a) $E_{max} = c B_{max} \Rightarrow B_{max} = \frac{E_{max}}{c} = \frac{0.0600\text{ V/m}}{3.00\times10^8\text{ m/s}} = 2.00\times10^{-10}\text{ T}$

b) $I = S_{av} = \frac{E_{max}B_{max}}{2\mu_0} = \frac{(0.0600\text{ V/m})(2.00\times10^{-10}\text{ T})}{2(4\pi\times10^{-7}\text{ T}\cdot\text{m/A})} = 4.775\times10^{-6}\text{ W/m}^2$

I gives the power per unit area. As in Example 33-3 surround the antenna with a sphere of radius $R = 50.0\times10^3\text{ m}$. The upper half of this sphere has area $A = 2\pi R^2$. All the power radiated passes through this surface, so the power per unit area at this surface is given by $I = \frac{P}{2\pi R^2}$.

Thus $P = 2\pi R^2 I = 2\pi(50.0\times10^3\text{ m})^2(4.775\times10^{-6}\text{ W/m}^2) = 7.50\times10^4\text{ W} = \underline{75.0\text{ kW}}$

c) The discussion in part (b) shows that since $P = I\,2\pi R^2$, IR^2 is constant

33-9 (cont)

$$\Rightarrow I_1 R_1^2 = I_2 R_2^2. \text{ But } I = \frac{E_{max}^2}{2\mu_0 c}, \text{ so}$$

$$E_{max,1}^2 R_1^2 = E_{max,2}^2 R_2^2 \Rightarrow E_{max,1} R_1 = E_{max,2} R_2$$

$$R_2 = R_1 \frac{E_{max,1}}{E_{max,2}} = 50.0\text{km}\left(\frac{0.0600 \text{ V/m}}{0.0300 \text{ V/m}}\right) = \underline{100.0 \text{ km}}$$

33-13

a) By Eq. (33-28) the average momentum density is $\frac{p}{V} = \frac{S_{av}}{c^2} = \frac{I}{c^2}$

$$\frac{p}{V} = \frac{0.60\times10^3 \text{ W/m}^2}{(3.00\times10^8 \text{ m/s})^2} = 6.67\times10^{-15} \frac{\text{kg}\cdot\text{m/s}}{\text{m}^3} = \underline{6.7\times10^{-15} \text{ kg/m}^2\cdot\text{s}}$$

b) By Eq. (33-29) the average momentum flow rate per unit area is

$$\frac{S_{av}}{c} = \frac{I}{c} = \frac{0.60\times10^3 \text{ W/m}^2}{3.00\times10^8 \text{ m/s}} = \underline{2.0\times10^{-6} \text{ Pa}}$$

33-15

a) $\lambda = \frac{v}{f} = \frac{2.20\times10^8 \text{ m/s}}{20.0\times10^6 \text{ Hz}} = \underline{11.0 \text{ m}}$

b) $\lambda = \frac{c}{f} = \frac{3.00\times10^8 \text{ m/s}}{20.0\times10^6 \text{ Hz}} = \underline{15.0 \text{ m}}$

c) $n = \frac{c}{v} = \frac{3.00\times10^8 \text{ m/s}}{2.20\times10^8 \text{ m/s}} = \underline{1.36}$

d) $n = \sqrt{KK_m} \cong \sqrt{K} \Rightarrow K = n^2 = (1.36)^2 = \underline{1.85}$

33-17

a) By Eq. (33-39) we see that the nodal planes of the $\vec{B}$ field are a distance $\frac{\lambda}{2}$ apart, so $\frac{\lambda}{2} = 5.00$ mm $\Rightarrow \lambda = \underline{10.0 \text{ mm}}$.

b) By Eq. (33-38) we see that the nodal planes of the $\vec{E}$ field are also a distance $\frac{\lambda}{2} = \underline{5.00 \text{ mm}}$ apart.

c) $v = f\lambda = (2.00\times10^{10} \text{ Hz})(10.0\times10^{-3} \text{ m}) = \underline{2.00\times10^8 \text{ m/s}}$

Problems

33-21

Eq. (33-12): $\frac{\partial E_y}{\partial x} = -\frac{\partial B_z}{\partial t}$

Take $\frac{\partial}{\partial t} \Rightarrow \frac{\partial^2 E_y}{\partial x \partial t} = -\frac{\partial^2 B_z}{\partial t^2}$

Eq. (33-14): $-\frac{\partial B_z}{\partial x} = \epsilon_0 \mu_0 \frac{\partial E_y}{\partial t}$

Take $\frac{\partial}{\partial x} \Rightarrow -\frac{\partial^2 B_z}{\partial x^2} = \epsilon_0 \mu_0 \frac{\partial^2 E_y}{\partial t \partial x} \Rightarrow \frac{\partial^2 E_y}{\partial t \partial x} = -\frac{1}{\epsilon_0 \mu_0} \frac{\partial^2 B_z}{\partial x^2}$

33-21 (cont)

But $\frac{\partial^2 E_y}{\partial x \partial t} = \frac{\partial^2 E_y}{\partial t \partial x}$ (the order in which the partial derivatives are taken doesn't change the result)

So $-\frac{\partial^2 B_z}{\partial t^2} = -\frac{1}{\epsilon_0 \mu_0} \frac{\partial^2 B_z}{\partial x^2}$

$\frac{\partial^2 B_z}{\partial x^2} = \epsilon_0 \mu_0 \frac{\partial^2 B_z}{\partial t^2}$, as was to be shown.

33-25

a) I gives the energy flow per unit time per unit area

$I = \frac{1}{A}\frac{dU}{dt} \Rightarrow \frac{dU}{dt} = AI$

$I = \frac{E_{max}^2}{2\mu_0 c} = \frac{(15.0\ \text{V/m})^2}{2(4\pi \times 10^{-7}\ \text{T}\cdot\text{m/A})(3.00\times 10^8\ \text{m/s})} = 0.298\ \text{W/m}^2$

$\frac{dU}{dt} = AI = (6.00\times 10^{-4}\ \text{m}^2)(0.298\ \text{W/m}^2) = \underline{1.79\times 10^{-4}\ \text{W}}$

b) The light is reflected by the mirror, so the average pressure is

$\frac{2I}{c} = \frac{2(0.298\ \text{W/m}^2)}{3.00\times 10^8\ \text{m/s}} = \underline{1.99\times 10^{-9}\ \text{Pa}}$

c) Surround the light bulb with an imaginary sphere with radius $R = 4.00\text{m}$ and surface area $A = 4\pi R^2$. All the power radiated by the bulb passes through this surface, so $I = \frac{P}{A} = \frac{P}{4\pi R^2}$.

$P = 4\pi R^2 I = 4\pi(4.00\text{m})^2(0.298\ \text{W/m}^2) = \underline{59.9\ \text{W}}$

33-27

a) The intensity is the power per unit area:

$I = \frac{P}{A} = \frac{6.00\times 10^{-3}\ \text{W}}{\pi(2.00\times 10^{-3}\text{m})^2} = 477\ \text{W/m}^2$

$I = \frac{E_{max}^2}{2\mu_0 c} \Rightarrow E_{max} = \sqrt{2\mu_0 c I} = \sqrt{2(4\pi\times 10^{-7}\ \text{T}\cdot\text{m/A})(3.00\times 10^8\ \text{m/s})(477\ \text{W/m}^2)} = \underline{600\ \text{V/m}}$

$B_{max} = \frac{E_{max}}{c} = \frac{600\ \text{V/m}}{3.00\times 10^8\ \text{m/s}} = \underline{2.00\times 10^{-6}\ \text{T}}$

b) The energy density in the electric field is (Eq. 25-11) $u_E = \frac{1}{2}\epsilon_0 E^2$. $E = E_{max}\sin(\omega t - kx)$ and the average value of $\sin^2(\omega t - kx)$ is $\frac{1}{2}$. The average energy density in the electric field then is $u_{E,av} = \frac{1}{4}\epsilon_0 E_{max}^2$ $= \frac{1}{4}(8.85\times 10^{-12}\ \text{C}^2/\text{N}\cdot\text{m}^2)(600\ \text{V/m})^2 = 7.96\times 10^{-7}\ \text{J/m}^3$

The energy density in the magnetic field is (Eq. 31-9) $u_B = \frac{B^2}{2\mu_0}$. The average value is $u_{B,av} = \frac{B_{max}^2}{4\mu_0} = \frac{(2.00\times 10^{-6}\text{T})^2}{4(4\pi\times 10^{-7}\ \text{T}\cdot\text{m/A})} = 7.96\times 10^{-7}\ \text{J/m}^3$.

c) The total energy in this length of beam is the total energy density $u_{av} = u_{E,av} + u_{B,av} = 1.59\times 10^{-6}\ \text{W/m}^3$ times the volume of this part of the beam

33-27 (cont)

$\Rightarrow U = u_{av} LA = (1.59\times10^{-6}\ \text{W/m}^3)(0.500\ \text{m})(\pi)(2.00\times10^{-3}\ \text{m})^2 = \underline{1.00\times10^{-11}\ \text{J}}$

This quantity can also be calculated as the power output times the time it takes the light to travel $L = 0.500$ m:

$U = P\left(\frac{L}{c}\right) = (6.00\times10^{-3}\ \text{W})\left(\frac{0.500\ \text{m}}{3.00\times10^{8}\ \text{m/s}}\right) = 1.0\times10^{-11}\ \text{J}$, which checks.

33-29

a) For a long solenoid $B = \mu_0 n i$ (Eq. 29-21) and is uniform over the cross section of the solenoid. i changing $\Rightarrow \frac{dB}{dt} = \mu_0 n \frac{di}{dt}$

Calculate the induced electric field from Faraday's law: $\oint \vec{E}\cdot d\vec{l} = -\frac{d\Phi_B}{dt}$.

Assume that $\vec{B}$ is directed into the page, and take this to be the direction of $\vec{A}$.

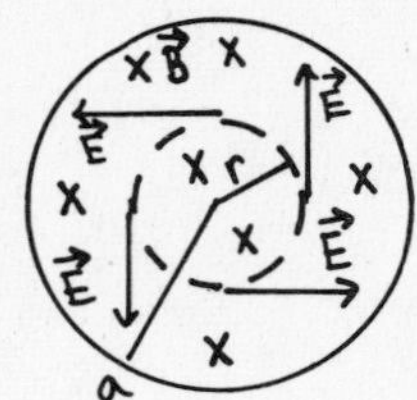

Apply Faraday's law to the circular path of radius r shown in the sketch. Φ_B is positive and increasing $\Rightarrow \frac{d\Phi_B}{dt}$ is positive

$\Rightarrow \oint \vec{E}\cdot d\vec{l}$ is negative; $\vec{E}$ is counterclockwise around the path.

$\oint \vec{E}\cdot d\vec{l} = -E(2\pi r)$

$\frac{d\Phi_B}{dt} = \frac{d}{dt}(B\pi r^2) = \pi r^2 \frac{dB}{dt}$

$\Rightarrow -E(2\pi r) = -\pi r^2 \frac{dB}{dt} \Rightarrow E = \frac{1}{2} r \frac{dB}{dt}$

$\vec{S} = \frac{1}{\mu_0}\vec{E}\times\vec{B}$ is radially inward.

$S = \frac{1}{\mu_0} EB$

$E = \frac{1}{2} r \frac{dB}{dt} = \frac{1}{2} r \mu_0 n \frac{di}{dt}$

$B = \mu_0 n i$

$\Rightarrow S = \frac{1}{\mu_0}\left(\frac{1}{2} r \mu_0 n \frac{di}{dt}\right)(\mu_0 n i) = \frac{1}{2}\mu_0 n^2 r i \frac{di}{dt}$

33-31

a) The direction of $\vec{E}$ is parallel to the axis of the cylinder, in the direction of the current. From Eq. 26-7, $E = \rho J = \rho \frac{I}{\pi a^2}$. ($E$ is uniform across the cross section of the conductor.)

b) Cross-sectional view of the conductor; take the current to be coming out of the page.

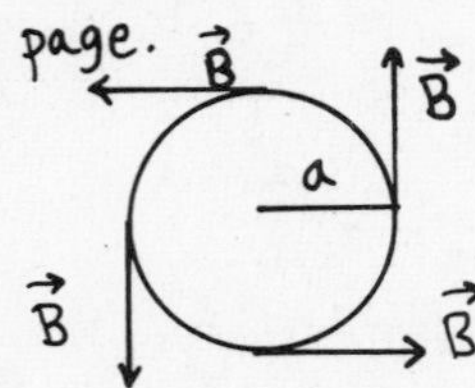

Apply Ampere's law to a circle of radius a.

$\oint \vec{B}\cdot d\vec{l} = B(2\pi a)$

$I_{encl} = I$

$\oint \vec{B}\cdot d\vec{l} = \mu_0 I_{encl} \Rightarrow B(2\pi a) = \mu_0 I$

$B = \frac{\mu_0 I}{2\pi a}$

The direction of $\vec{B}$ is counterclockwise around the circle.

33-31 (cont)

c) The direction of $\vec{S} = \frac{1}{\mu_0}\vec{E}\times\vec{B}$ is radially inward.

$$S = \frac{1}{\mu_0}EB = \frac{1}{\mu_0}\left(\frac{\rho I}{\pi a^2}\right)\left(\frac{\mu_0 I}{2\pi a}\right) = \frac{\rho I^2}{2\pi^2 a^3}$$

d) The rate of energy flow P is given by S times the surface area of a length l of the conductor: $P = SA = S(2\pi a l) = \frac{\rho I^2}{2\pi^2 a^3}(2\pi a l) = \frac{\rho l I^2}{\pi a^2}$.

But $R = \frac{\rho l}{\pi a^2}$, so the result from the Poynting vector is $P = RI^2$. This agrees with $P_R = I^2 R$, the rate at which electrical energy is being dissipated by the resistance of the wire.

33-33

The magnitude of the induced emf is given by Faraday's law: $|\mathcal{E}| = \left|\frac{d\Phi_B}{dt}\right|$.

$\Phi_B = B\pi R^2$, where $R = 0.200$ m is the radius of the loop. (This assumes that the magnetic field is uniform across the loop, an excellent approximation.)

$\Rightarrow |\mathcal{E}| = \pi R^2 \left|\frac{dB}{dt}\right|$

$B = B_{max}\sin(\omega t - kx) \Rightarrow \frac{dB}{dt} = B_{max}\,\omega\cos(\omega t - kx)$

The maximum value of $\frac{dB}{dt}$ is $B_{max}\omega$, so $\boxed{|\mathcal{E}|_{max} = \pi R^2 B_{max}\,\omega}$.

$R = 0.200$ m, $\omega = 2\pi f = 2\pi(10.0\times10^6\text{ Hz}) = 6.283\times10^7$ rad/s

Calculate the intensity I at this distance from the source, and from that the magnetic field amplitude B_{max}: $I = \frac{P}{4\pi r^2} = \frac{1.00\times10^6\text{ W}}{4\pi(500\text{ m})^2} = 0.3183\text{ W/m}^2$.

$$I = \frac{E_{max}^2}{2\mu_0 c} = \frac{(cB_{max})^2}{2\mu_0 c} = \frac{c}{2\mu_0}B_{max}^2$$

$$\Rightarrow B_{max} = \sqrt{\frac{2\mu_0 I}{c}} = \sqrt{\frac{2(4\pi\times10^{-7}\text{ T}\cdot\text{m/A})(0.3183\text{ W/m}^2)}{3.00\times10^8\text{ m/s}}} = 5.164\times10^{-8}\text{ T}.$$

Then $|\mathcal{E}|_{max} = \pi R^2 B_{max}\,\omega = \pi(0.200\text{m})^2(5.164\times10^{-8}\text{T})(6.283\times10^7\text{rad/s}) = \underline{0.408\text{ V}}$

33-35

The time rate of change of the momentum of the light emitted by the flashlight (the momentum flow rate in the beam) is $\frac{dp}{dt} = \frac{AS}{c}$, where A is the area of the beam and S is its intensity. But $SA = P$, the power output of the flashlight, so $\frac{dp}{dt} = \frac{P}{c}$.

By Newton's second law $\frac{dp}{dt}$ equals the force F exerted by the light on the astronaut, so $F = \frac{P}{c}$. Then $F = ma$ gives $ma = \frac{P}{c}$.

$$a = \frac{P}{mc} = \frac{100\text{ W}}{(200\text{kg})(3.00\times10^8\text{m/s})} = 1.67\times10^{-9}\text{m/s}^2$$

Now that we have the acceleration we can use a constant-acceleration kinematic equation to find the time: $x - x_0 = v_0 t^{\,0} + \frac{1}{2}at^2$

$$t = \sqrt{\frac{2(x-x_0)}{a}} = \sqrt{\frac{2(16.0\text{ m})}{1.67\times10^{-9}\text{m/s}^2}} = (1.384\times10^5\text{s})\left(\frac{1\text{ h}}{3600\text{ s}}\right) = \underline{38.5\text{ h}}$$

CHAPTER 34

Exercises 3, 7, 9, 11, 17, 19

Problems 27, 29, 31, 33, 35, 37, 39

Exercises

34-3

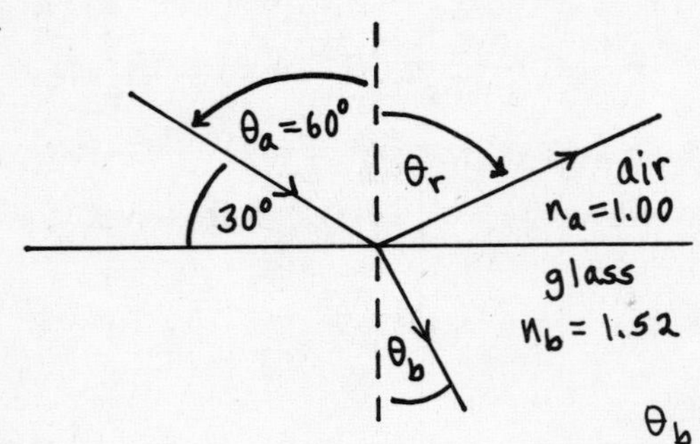

a) $\theta_r = \theta_a = 60.0°$
$\Rightarrow$ The reflected ray makes an angle of $90.0° - \theta_r = \underline{30.0°}$ with the surface of the glass.

b) $n_a \sin\theta_a = n_b \sin\theta_b$, where the angles are measured from the normal to the interface.

$$\sin\theta_b = \frac{n_a \sin\theta_a}{n_b} = \frac{(1.00)(\sin 60.0°)}{1.52} = 0.5698$$

$\theta_b = 34.73° \Rightarrow$ The refracted ray makes an angle of $90.0° - \theta_b = \underline{55.3°}$ with the surface of the glass.

34-7

a) Eq. (34-1): $n = \frac{c}{v} \Rightarrow v = \frac{c}{n} = \frac{2.998 \times 10^8\ \text{m/s}}{1.70} = \underline{1.76 \times 10^8\ \text{m/s}}$

b) Eq. (34-5): $\lambda = \frac{\lambda_0}{n} = \frac{500\ \text{nm}}{1.70} = \underline{294\ \text{nm}}$

34-9

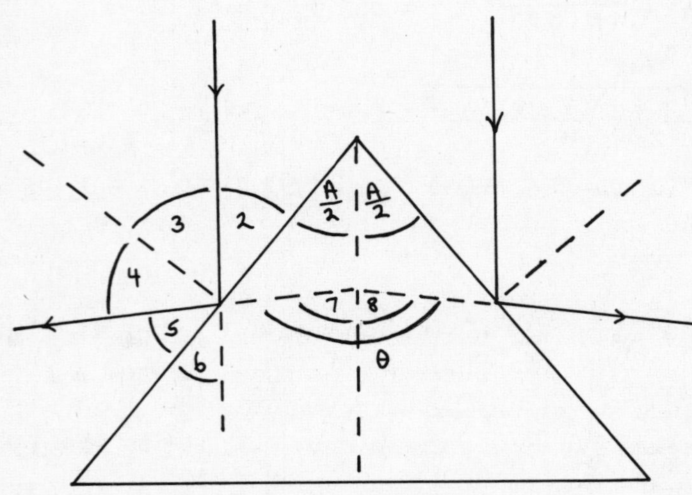

Angle θ is angle 7 + angle 8. We are asked to show that $\theta = 2A$. By symmetry angle 7 = angle 8, so we need to show that angle 7 = A.

From the sketch we see that angle 5 + angle 6 = angle 7 and that angle 2 = $\frac{A}{2}$.
But angle 2 = angle 6, so **angle 6 = $\frac{A}{2}$**.

The law of reflection $\theta_a = \theta_r$ says that angle 3 = angle 4. But then angle 2 = angle 5. Thus **angle 5 = $\frac{A}{2}$**.

Then angle 7 = angle 5 + angle 6 = $\frac{A}{2} + \frac{A}{2} = A$, which completes the proof.

34-11

a) Define the index of refraction of air for sound waves to be 1.00. Then $n_{water} = \frac{v_{air}}{v_{water}}$ (Eq. 34-1) $\Rightarrow n_{water} = \frac{344\ \text{m/s}}{1320\ \text{m/s}} = 0.2606$. $n_{water} < n_{air}$; air has the larger index of refraction for sound waves since sound travels slower in air than it does in water.

b) $n_a \sin\theta_a = n_b \sin\theta_b$

$\theta_a = \theta_{crit}$ when $\sin\theta_b = 1 \Rightarrow \sin\theta_{crit} = \frac{n_b \sin 90^\circ}{n_a} = \frac{n_{water}}{n_{air}} = \frac{0.2606}{1.00} = 0.2606$

$\Rightarrow \theta_{crit} = \underline{15.1^\circ}$

34-17

a)

#1 #2 #3 60° 30°

$I = I_{max} \cos^2\phi$ (Eq. 34-7)

After the first filter the intensity is $I_1 = \frac{1}{2} I_0$ and the light is linearly polarized along the axis of the first polarizer.

After the second filter the intensity is

$I_2 = I_1 \cos^2\phi = (\frac{1}{2} I_0)(\cos 60.0^\circ)^2 = 0.125\ I_0$

and the light is linearly polarized along the axis of the second polarizer.

After the third filter the intensity is

$I_3 = I_2 \cos^2\phi = (0.125\ I_0)(\cos 30.0^\circ)^2 = 0.0938\ I_0$

and the light is linearly polarized along the axis of the third polarizer.

b)

#1 90° #3

After the first filter the intensity is $I_1 = \frac{1}{2} I_0$ and the light is linearly polarized along the axis of the first polarizer.
After the next filter the intensity is $I_3 = I_1 \cos^2\phi = \frac{1}{2} I_0 (\cos 90^\circ)^2 = 0$. No light is passed.

34-19

a) Reflected beam completely linearly polarized $\Rightarrow$ the angle of incidence equals the polarizing angle $\Rightarrow \theta_p = 58.6^\circ$.

$\tan\theta_p = \frac{n_b}{n_a} \Rightarrow n_{glass} = n_{air} \tan\theta_p = (1.00)\tan 58.6^\circ = \underline{1.64}$.

b) $n_a \sin\theta_a = n_b \sin\theta_b$

$\sin\theta_b = \frac{n_a \sin\theta_a}{n_b} = \frac{1.00 (\sin 58.6^\circ)}{1.64} = 0.5205 \Rightarrow \theta_b = 31.4^\circ$.

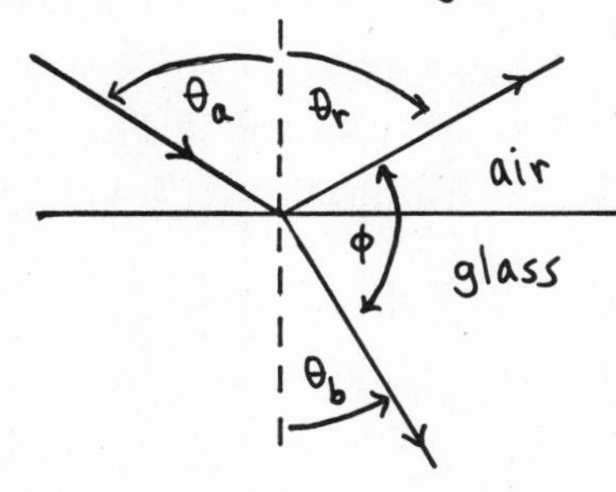

Note: $\phi = 180.0^\circ - \theta_r - \theta_b$ and $\theta_r = \theta_a$
$\Rightarrow \phi = 180.0^\circ - 58.6^\circ - 31.4^\circ = 90.0^\circ$; the reflected ray and the refracted ray are perpendicular to each other.

Problems

34-27

Before the liquid is poured in:

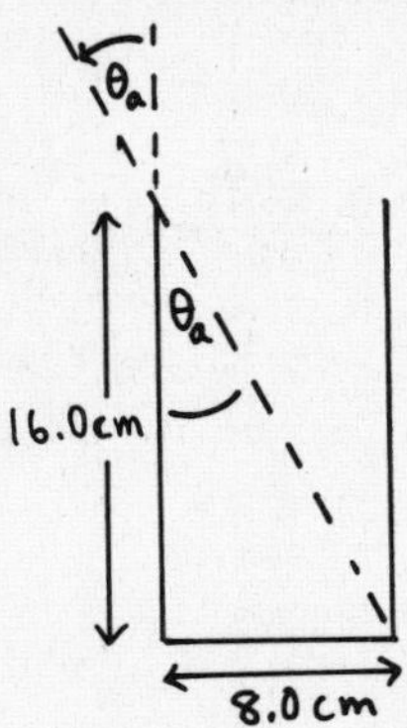

$$\tan\theta_a = \frac{8.0\ \text{cm}}{16.0\ \text{cm}} = 0.500 \Rightarrow \theta_a = 26.57^\circ$$

After the liquid is poured in, θ_a is the same and the refracted ray passes through the center of the bottom of the glass:

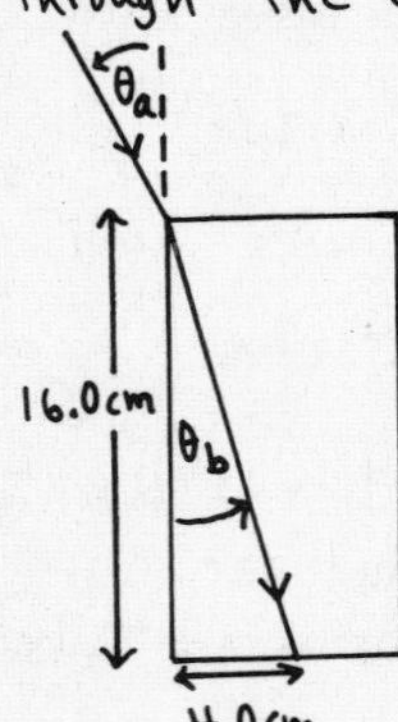

$$\tan\theta_b = \frac{4.0\ \text{cm}}{16.0\ \text{cm}} = 0.250 \Rightarrow \theta_b = 14.04^\circ$$

Then can use Snell's law to find n_b, the refractive index of the liquid:

$$n_a \sin\theta_a = n_b \sin\theta_b$$

$$n_b = \frac{n_a \sin\theta_a}{\sin\theta_b} = \frac{(1.00)\sin(26.57^\circ)}{\sin(14.04^\circ)} = \underline{1.84}$$

34-29

air | 2.62mm | $n_b = 1.00$ | refracted ray
glass | θ_{crit} | θ_{crit} | θ_r | totally internally reflected ray, at inner edge of halo | 3.00mm
air | focused spot at front of plate | 5.24 mm

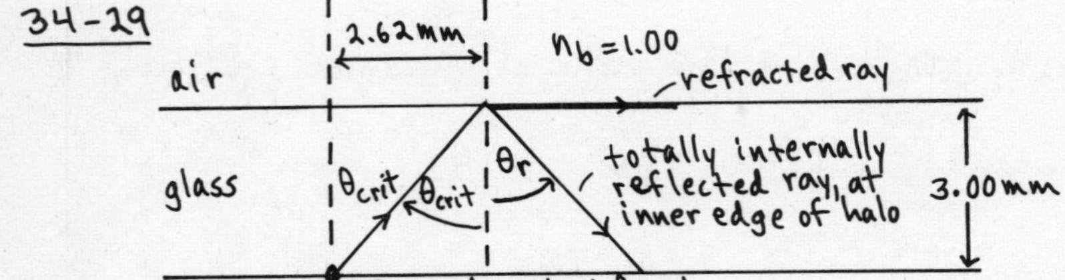

From the distances given in the sketch

$$\tan\theta_{crit} = \frac{2.62\ \text{mm}}{3.00\ \text{mm}} = 0.8733$$

$$\Rightarrow \theta_{crit} = 41.13^\circ$$

Apply Snell's law to the total internal reflection to find the refractive index of the glass:

$$n_a \sin\theta_a = n_b \sin\theta_b$$

$$n_{glass} \sin\theta_{crit} = 1.00 \sin 90^\circ$$

$$n_{glass} = \frac{1}{\sin\theta_{crit}} = \frac{1}{\sin 41.13^\circ} = \underline{1.52}$$

34-31

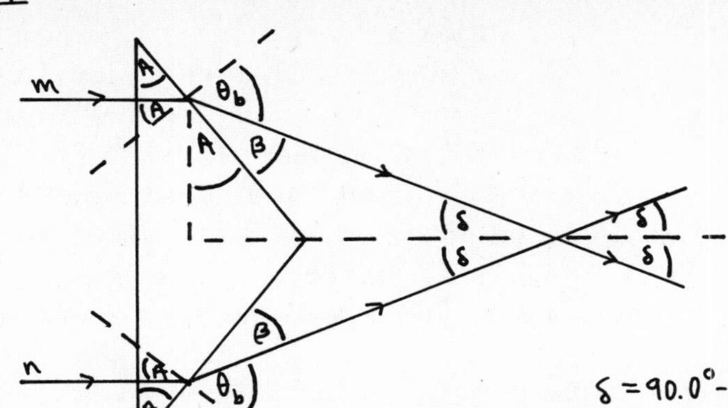

Apply Snell's law to the refraction of each ray as it emerges from the glass. The angle of incidence equals the angle $A = 30.0°$.

$n_a \sin\theta_a = n_b \sin\theta_b$

$n_{glass} \sin 30.0° = 1.00 \sin\theta_b$

$\sin\theta_b = n_{glass} \sin 30.0° = (1.48)\sin 30.0° = 0.740$

$\Rightarrow \theta_b = 47.73°$

$\beta = 90.0° - \theta_b = 42.27°$

$\delta = 90.0° - A - \beta = 90.0° - 30.0° - 42.27° = 17.7°$

The angle between the two rays is $2\delta = \underline{35.4°}$.

34-33

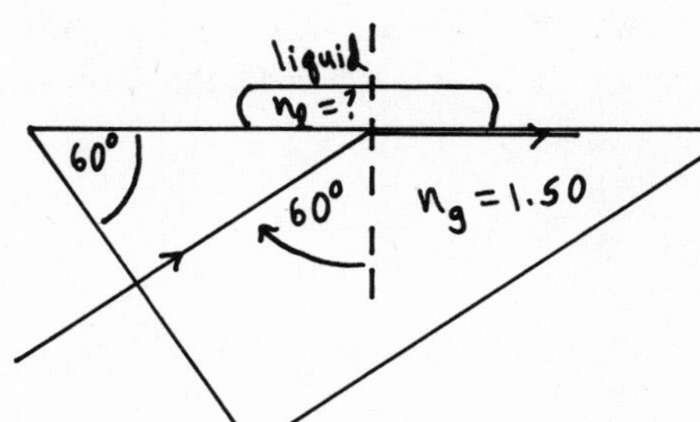

$n_a \sin\theta_a = n_b \sin\theta_b$

$n_a = n_g = 1.50$

$n_b = n_\ell = ?$

$\theta_a = 60°$

Maximum n_ℓ for total reflection $\Rightarrow \theta_b = 90°$

$(1.50)\sin 60° = n_\ell \sin 90°$

$n_\ell = (1.50)\sin 60° = \underline{1.30}$

34-35

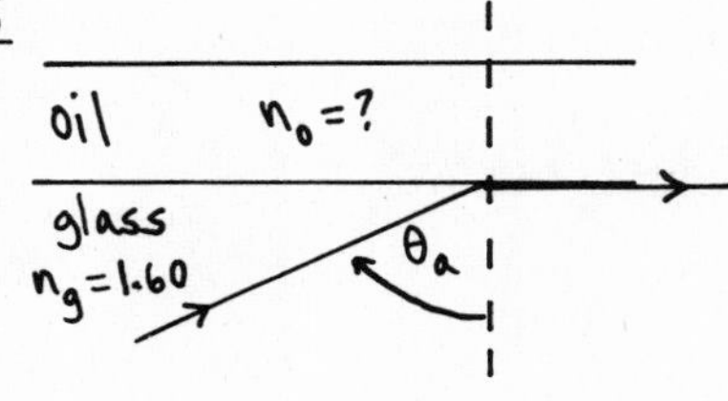

$n_a \sin\theta_a = n_b \sin\theta_b$

$n_a = n_g = 1.60$

$n_b = n_o = ?$

$\theta_a = 55.0°$

Maximum n_o for total reflection $\Rightarrow \theta_b = 90°$

$(1.60)\sin 55.0° = n_o \sin 90°$

$n_o = (1.60)\sin 55.0° = \underline{1.31}$

34-37

a)

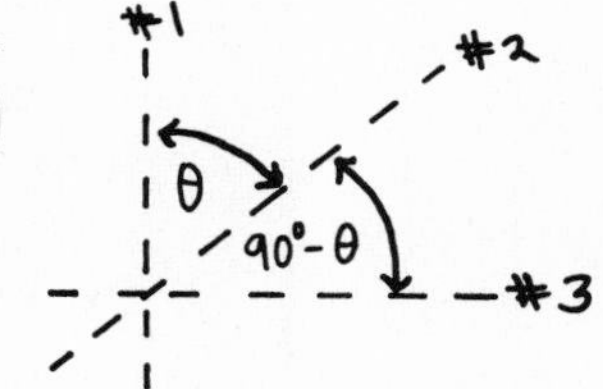

1st filter: $I_1 = \frac{1}{2} I_0$

2nd filter: $I_2 = I_1 (\cos\theta)^2 = \frac{1}{2} I_0 \cos^2\theta$

3rd filter: $I_3 = I_2 (\cos(90° - \theta))^2 = \frac{1}{2} I_0 \cos^2\theta \cos^2(90° - \theta)$

$\cos(90° - \theta) = \cos(\theta - 90°) = \sin\theta$ (Using the trig identities in Appendix B.)

Therefore, $I_3 = \frac{1}{2} I_0 \cos^2\theta \sin^2\theta$

But $\cos\theta \sin\theta = \frac{1}{2}\sin 2\theta$ (Appendix B again)

$\Rightarrow I_3 = \frac{1}{8} I_0 (\sin 2\theta)^2$

b) I_3 maximum $\Rightarrow \sin 2\theta = 1 \Rightarrow \underline{\theta = 45°}$

34-39

a)

polarization axis of polarized component

$\phi-\theta$

axis of polarizer

Y, X, θ, ϕ

The polarizer passes $\frac{1}{2}$ of the intensity of the polarized component, independent of ϕ. Out of the intensity I_p of the polarized component the polarizer passes intensity $I_p \cos^2(\phi-\theta)$, where $(\phi-\theta)$ is the angle between the plane of polarization and the axis of the polarizer.

The total transmitted intensity is $I = \frac{1}{2} I_0 + I_p \cos^2(\phi-\theta)$. This is maximum when $\theta=\phi$ and from the table of data this occurs for ϕ between $30°$ and $40°$, say at $35° \Rightarrow \underline{\theta = 35°}$.

Alternatively, the total transmitted intensity is minimum when $\phi-\theta = 90°$ and from the data this occurs for $\phi = 125° \Rightarrow \theta = \phi - 90° = 125° - 90° = 35°$, in agreement with the above.

b) $I = \frac{1}{2} I_0 + I_p \cos^2(\phi-\theta)$

Use data at two values of ϕ to determine the two constants I_0 and I_p. Use data where the I_p term is large ($\phi = 30°$) and where it is small ($\phi = 130°$) to have the greatest sensitivity to both I_0 and I_p:

$\phi = 30° \Rightarrow 24.8\ \text{W/m}^2 = \frac{1}{2} I_0 + I_p \cos^2(30° - 35°)$

$$\boxed{24.8\ \text{W/m}^2 = 0.500\, I_0 + 0.9924\, I_p}$$

$\phi = 130° \Rightarrow 5.2\ \text{W/m}^2 = \frac{1}{2} I_0 + I_p \cos^2(130° - 35°)$

$$\boxed{5.2\ \text{W/m}^2 = 0.500\, I_0 + 0.0076\, I_p}$$

Subtract the second equation from the first

$\Rightarrow 19.6\ \text{W/m}^2 = 0.9848\, I_p$

$I_p = \underline{19.9\ \text{W/m}^2}$

And then $I_0 = 2(5.2\ \text{W/m}^2 - 0.0076(19.9\ \text{W/m}^2)) = \underline{10.1\ \text{W/m}^2}$

CHAPTER 35

Exercises 1, 5, 9, 13, 15, 17, 21, 23, 25, 29

Problems 33, 35, 37, 39, 41, 43, 47, 49, 51, 53, 57, 61, 63, 65

Exercises

35-1

Plane mirror: $s = -s'$ (Eq. 35-1) and $m = \frac{y'}{y} = -\frac{s'}{s} = +1$ (Eq. 35-2).

4.0 cm — 60.0 cm — 60.0 cm — 4.0 cm

$s' = -s = -60\text{ cm}$

$|y'| = |m|\,|y| = (+1)(4.0\text{ cm}) = 4.0\text{ cm}$

The image is 60.0 cm to the right of the mirror and is 4.0 cm tall.

35-5

concave $\Rightarrow R = +20.0\text{ cm}$; $f = \frac{R}{2} = +10.0\text{ cm}$

a)

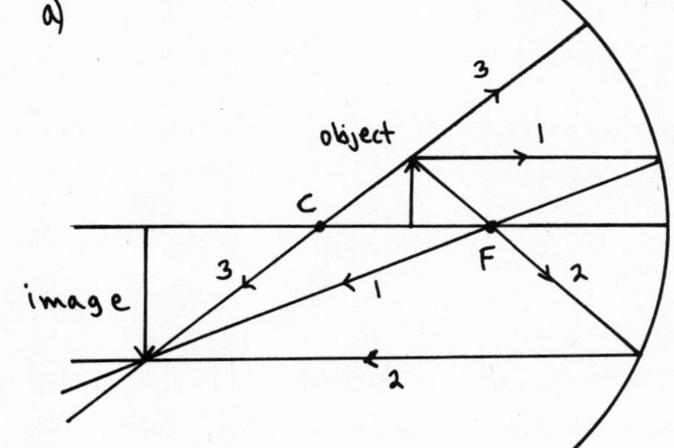

Three principal rays, numbered as in Sect. 35-3 are shown.

The principal ray diagram shows that the image is real, inverted, enlarged, and lies outside the center of curvature.

b) $\frac{1}{s} + \frac{1}{s'} = \frac{1}{f}$

$\frac{1}{s'} = \frac{1}{f} - \frac{1}{s} = \frac{s-f}{sf} \Rightarrow s' = \frac{sf}{s-f} = \frac{(15.0\text{ cm})(10.0\text{ cm})}{15.0\text{ cm} - 10.0\text{ cm}} = +30.0\text{ cm}$

$s' > 0 \Rightarrow$ real image, 30.0 cm to left of mirror vertex

$m = -\frac{s'}{s} = -\frac{30\text{ cm}}{15\text{ cm}} = -2.00$ ($m < 0 \Rightarrow$ inverted image)

$|y'| = |m|\,|y| = 2.00\,(0.400\text{ cm}) = 0.800\text{ cm}$

The image is 30.0 cm to the left of the mirror vertex. It is real, inverted, and 0.800 cm tall (enlarged). The calculation agrees with the image characterization from the principal ray diagram.

35-9

convex $\Rightarrow R = -24.0$ cm ; $f = \frac{R}{2} = -12.0$ cm

$s = 8.00$ cm, so that $|s| < |f|$

a)

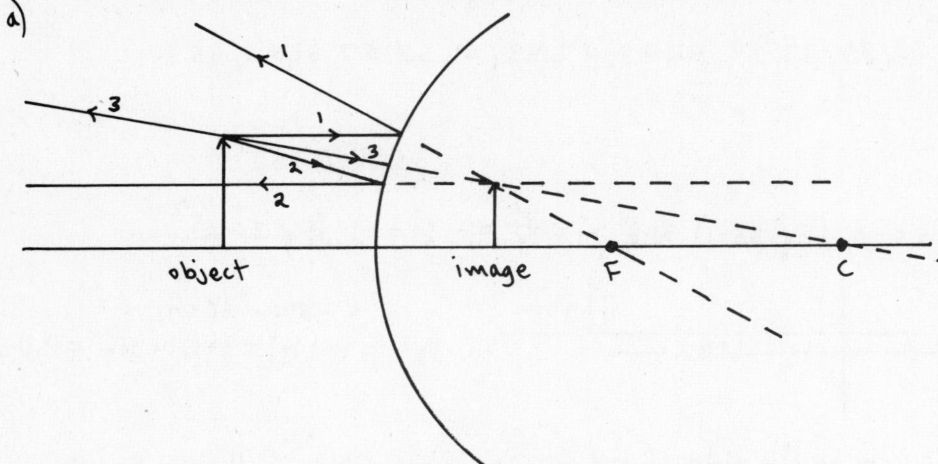

Three principal rays, numbered as in Sect. 35-3, are shown.

The principal ray diagram shows that the image is virtual, erect, reduced in size, and located closer to the mirror than the focal point.

b) $\frac{1}{s} + \frac{1}{s'} = \frac{1}{f}$

$\frac{1}{s'} = \frac{1}{f} - \frac{1}{s} = \frac{s-f}{sf} \Rightarrow s' = \frac{sf}{s-f} = \frac{(8.00\text{ cm})(-12.0\text{ cm})}{8.00\text{ cm} - (-12.0\text{ cm})} = -4.80$ cm

$s' < 0 \Rightarrow$ virtual image 4.80 cm to right of mirror vertex

$m = -\frac{s'}{s} = -\frac{-4.80\text{ cm}}{8.00\text{ cm}} = +0.600$ ($m > 0 \Rightarrow$ erect image)

$|y'| = |m||y| = (+0.600)(1.50\text{ cm}) = 0.90$ cm

The image is 4.80 cm to the right of the mirror vertex. It is 0.90 cm tall (reduced), erect, and virtual. The calculation agrees with the image characterization from the principal ray diagram.

35-13

$n_a = n_l$ $R = +3.00$ cm $n_b = n_g = 1.50$

object   image

$\leftarrow s = +60.0$ cm $\rightarrow \leftarrow s' = +90.0$ cm $\rightarrow$

Let n_l be the refractive index of the liquid. s, s', and R are all positive.

$\frac{n_a}{s} + \frac{n_b}{s'} = \frac{n_b - n_a}{R}$ (Eq. 35-11)

$\frac{n_l}{60.0\text{cm}} + \frac{1.50}{90.0\text{ cm}} = \frac{1.50 - n_l}{3.00\text{ cm}}$

Multiply the equation by 180.0 cm

$\Rightarrow 3n_l + 3 = 90 - 60n_l$

$63n_l = 87 \Rightarrow n_l = \frac{87}{63} = \underline{1.38}$

35-15

$y = 2.00$ mm $n_a = 1.00$ $R = -5.0$ cm $n_b = 1.50$

$\leftarrow s = +25.0$ cm $\rightarrow$

35-15 (cont)

$$\frac{n_a}{s}+\frac{n_b}{s'}=\frac{n_b-n_a}{R}$$

$$\frac{1.00}{25.0\text{ cm}}+\frac{1.50}{s'}=\frac{1.50-1.00}{-5.0\text{ cm}}$$

Multiply by 25.0 cm

$$\Rightarrow \quad 1.00+\frac{37.5\text{ cm}}{s'}=-2.50$$

$$\frac{37.5\text{cm}}{s'}=-3.50 \Rightarrow s'=-\frac{37.5\text{ cm}}{3.50}=-10.7\text{ cm}$$

Eq. (35-12): $m=-\frac{n_a s'}{n_b s}=-\frac{(1.00)(-10.7\text{cm})}{(1.50)(+25.0\text{ cm})}=+0.285$

$|y'|=|m||y|=(0.285)(2.00\text{mm})=\underline{0.57\text{mm}}$

The image is virtual ($s'<0$) and is 10.7 cm to the left of the vertex. The image is erect ($m>0$) and is 0.57 mm tall.

35-17

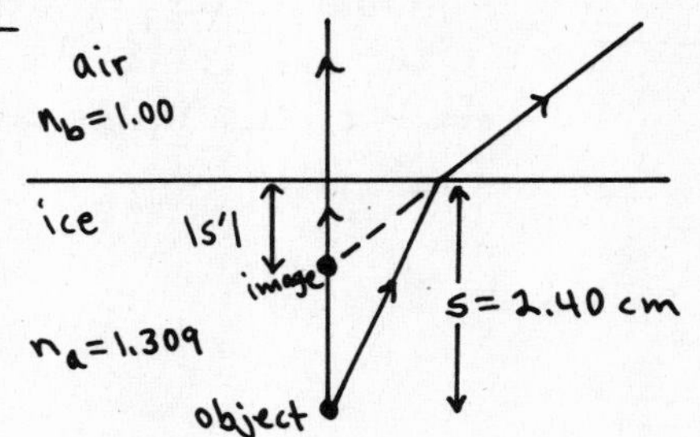

$$\frac{n_a}{s}+\frac{n_b}{s'}=\frac{n_b-n_a}{R}\ ; \quad R\to\infty \Rightarrow \frac{n_b-n_a}{R}\to 0$$

$$\frac{n_a}{s}+\frac{n_b}{s'}=0$$

$$s'=-\frac{n_b s}{n_a}=-\frac{(1.00)(2.40\text{ cm})}{1.309}=-1.83\text{ cm}$$

The apparent depth is $\underline{1.83\text{ cm}}$.

35-21

$f=-10.0$ cm

real image 15.0 cm to right of lens $\Rightarrow s'=+15.0$ cm

$$\frac{1}{s}+\frac{1}{s'}=\frac{1}{f} \Rightarrow \frac{1}{s}=\frac{1}{f}-\frac{1}{s'}=\frac{s'-f}{s'f}$$

$$s=\frac{s'f}{s'-f}=\frac{(+15.0\text{cm})(-10.0\text{cm})}{+15.0\text{ cm}-(-10.0\text{ cm})}=-\frac{150\text{ cm}}{25}=-6.00\text{ cm}$$

$$m=-\frac{s'}{s}=-\frac{+15.0\text{cm}}{-6.00\text{ cm}}=+2.50$$

$$m=\frac{y'}{y} \Rightarrow |y|=\frac{|y'|}{|m|}=\frac{1.00\text{ cm}}{2.50}=0.400\text{ cm}$$

$s<0 \Rightarrow$ virtual object 6.00 cm to right of lens

$m>0 \Rightarrow$ image is erect with respect to the object

The height of the object is 0.400 cm.

35-23

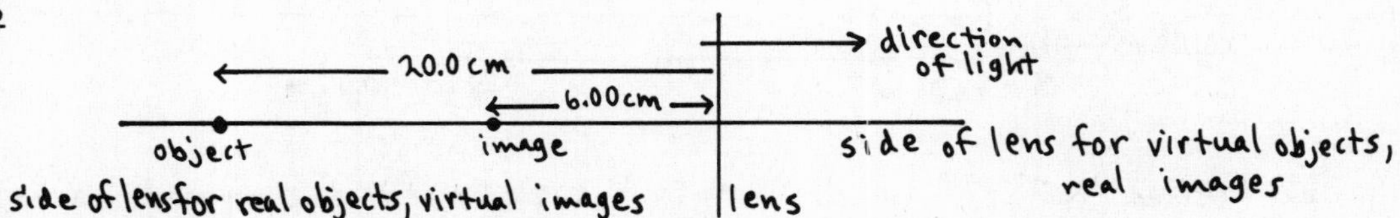

35-23 (cont)

$s = +20.0\text{ cm}$

$s' = -6.00\text{ cm}$ (virtual image)

a) $\frac{1}{s} + \frac{1}{s'} = \frac{1}{f}$

$\frac{1}{f} = \frac{s+s'}{ss'} \Rightarrow f = \frac{ss'}{s+s'} = \frac{(+20.0\text{ cm})(-6.00\text{ cm})}{20.0\text{ cm} - 6.00\text{ cm}} = \underline{-8.57\text{ cm}}$

$f < 0 \Rightarrow$ lens is <u>diverging</u>

b) $m = -\frac{s'}{s} = -\frac{-6.00\text{cm}}{+20.0\text{cm}} = +0.300$; $m > 0 \Rightarrow$ image is <u>erect</u>

$m = \frac{y'}{y} \Rightarrow |y'| = |m||y| = (+0.300)(0.400\text{cm}) = \underline{0.120\text{ cm}}$

c)

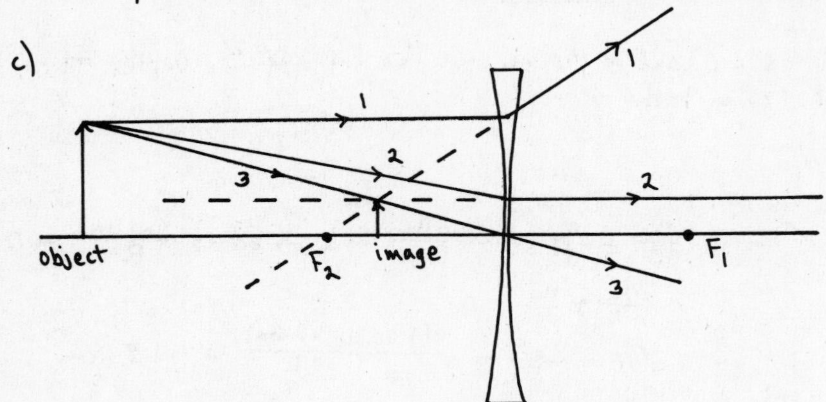

The three principal rays are numbered as in Sect 35-6. The principal ray diagram shows that the image is virtual, erect, reduced in size, and is to the left of the lens inside the focal point.

35-25

$f = +6.00\text{ cm}$ ($f > 0$ since the lens is converging)

$m = -\frac{s'}{s}$; image erect $\Rightarrow m > 0$, so must have $s' < 0$ (image is virtual)

$m = \frac{y'}{y} \Rightarrow |m| = \frac{|y|}{|y'|} = \frac{40.0\text{cm}}{5.00\text{ cm}} = 8.00$

$s' = -ms = -(8.00)s$

Use this in $\frac{1}{s} + \frac{1}{s'} = \frac{1}{f} \Rightarrow \frac{1}{s} + \frac{1}{-8.00s} = \frac{1}{6.00\text{cm}}$

$\frac{7}{8.00s} = \frac{1}{6.00\text{cm}} \Rightarrow s = \frac{7(6.00\text{cm})}{8.00} = 5.25\text{cm}$

$s' = -(8.00)s = -(8.00)(5.25\text{ cm}) = -42.0\text{ cm}$

The object is 5.25 cm to the left of the lens.
$s' < 0$, so the image is 42.0 cm to the left of the lens and is virtual.

35-29

object

1.00 mm

n=1.00 (air)

20.0 cm

50.0cm

n=1.50

n=1.00 (air)

$R_1 = +5.00\text{cm}$

$R_2 = -10.0\text{ cm}$

a) The image formed by the first surface is the object for the second surface.

35-29 (cont)

b) Consider the image formed by the first surface:

$n_a = 1.00$ $\qquad \frac{n_a}{s} + \frac{n_b}{s'} = \frac{n_b - n_a}{R}$

$n_b = 1.50$

$s = +20.0\text{ cm}$ $\qquad \frac{1.00}{20.0\text{cm}} + \frac{1.50}{s'} = \frac{1.50 - 1.00}{+5.00\text{ cm}}$

$R = +5.00\text{ cm}$ $\qquad$ Multiply by 20.0 cm $\Rightarrow 1.00 + \frac{30.0\text{cm}}{s'} = 2.00$

$$\frac{30.0\text{cm}}{s'} = 1.00 \Rightarrow s' = +30.0\text{cm}$$

The image is 30.0 cm to the right of the first surface. Thus it is 50.0 cm − 30.0 cm = 20.0 cm to the left of the second surface. The object distance for the second surface is $\underline{+20.0\text{ cm}}$.

c) For the second surface $s > 0$; real object. This "object" is on the side of the surface from which the rays actually come.

d) Consider the image formed by the second surface:

$n_a = 1.50$ $\qquad \frac{n_a}{s} + \frac{n_b}{s'} = \frac{n_b - n_a}{R}$

$n_b = 1.00$

$R = -10.0\text{ cm}$ $\qquad \frac{1.50}{20.0\text{cm}} + \frac{1.00}{s'} = \frac{1.00 - 1.50}{-10.0\text{ cm}}$

$s = +20.0\text{ cm}$ $\qquad \frac{1.50}{20.0\text{cm}} + \frac{1.00}{s'} = +\frac{1}{20.0\text{cm}}$

$$1.50 + \frac{20.0\text{cm}}{s'} = 1.00 \Rightarrow \frac{20.0\text{ cm}}{s'} = -0.50 \Rightarrow s' = -40.0\text{cm}$$

$s' < 0 \Rightarrow$ the image formed by the second surface is 40.0 cm to the left of the second vertex.

Problems

35-33

For a plane mirror $s' = -s$; $v = \frac{ds}{dt}$, $v' = \frac{ds'}{dt} \Rightarrow v' = -v$. The velocities of the object and image are equal in magnitude and opposite in direction. Thus both you and your image are approaching the mirror surface at 3.00 m/s, from opposite directions. Your image is therefore moving at 6.00 m/s relative to you.

35-35

a) Image is to be formed on screen $\Rightarrow$ real image; $s' > 0$. Mirror to screen distance is 4.00 m, so $s' = +4.00\text{ m}$.

$m = -\frac{s'}{s} < 0$ since both s and s' are positive.

$$|m| = \frac{|y'|}{|y|} = \frac{35.0\text{ cm}}{0.500\text{ cm}} = 70.0 \Rightarrow m = -70.0$$

$$m = -\frac{s'}{s} \Rightarrow s = -\frac{s'}{m} = -\frac{4.00\text{m}}{-70.0} = +0.0571\text{ m} = \underline{+5.71\text{ cm}}$$

b) $\frac{1}{s} + \frac{1}{s'} = \frac{2}{R} \Rightarrow \frac{2}{R} = \frac{s+s'}{ss'}$

$R = 2\left(\frac{ss'}{s+s'}\right) = 2\left(\frac{(5.71\text{ cm})(4.00\text{ m})}{4.00\text{ m} + 5.71\text{ cm}}\right) = \underline{11.3\text{ cm}}$ $\quad$ (Note that R is calculated to be positive, which is the correct sign for a concave mirror.)

35-37

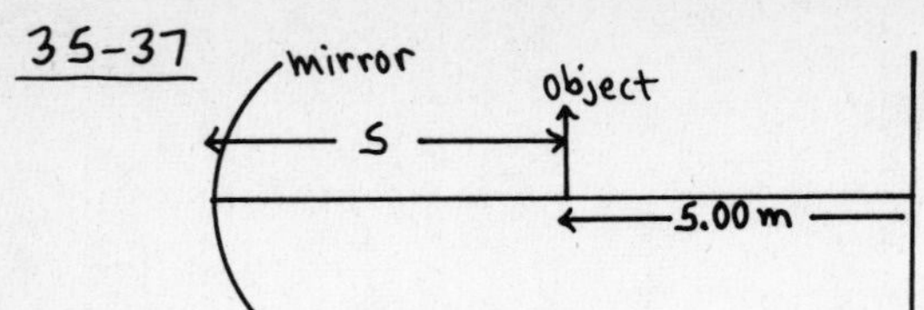

$s>0$; image projected on wall $\Rightarrow s'>0$

From the sketch $\boxed{s' = s + 5.00\text{ m}}$.

$m = -\frac{s'}{s}$ and $s>0,\ s'>0 \Rightarrow m<0$

$|m| = \frac{|y'|}{|y|} = 3.00 \Rightarrow m = -3.00$

$m = -\frac{s'}{s} \Rightarrow -3.00 = -\frac{s'}{s} \Rightarrow \boxed{s' = 3.00s}$

Use this in the first equation $\Rightarrow 3.00s = s + 5.00\text{ m} \Rightarrow s = \frac{5.00\text{ m}}{2.00} = 2.50\text{ m}$

$s' = 3.00s = 3.00(2.50\text{ m}) = 7.50\text{ m}$

The mirror should be 7.50 m from the wall.

$\frac{1}{s} + \frac{1}{s'} = \frac{2}{R} \Rightarrow \frac{2}{R} = \frac{s+s'}{ss'}$

$R = 2\left(\frac{ss'}{s+s'}\right) = 2\frac{(2.50\text{ m})(7.50\text{ m})}{2.50\text{ m} + 7.50\text{ m}} = \underline{3.75\text{ m}}$

Note that R is calculated to be positive, which is the correct sign for a concave mirror.

35-39

a) convex $\Rightarrow R<0$; $R = -12.0\text{ cm}$; $f = \frac{R}{2} = -6.00\text{ cm}$

$\frac{1}{s} + \frac{1}{s'} = \frac{1}{f} \Rightarrow \frac{1}{s'} = \frac{1}{f} - \frac{1}{s} = \frac{s-f}{sf} \Rightarrow s' = \frac{sf}{s-f} = \frac{(-6.00\text{ cm})s}{s + 6.00\text{ cm}}$

s is negative, so write $s = -|s| \Rightarrow s' = \frac{+(6.00\text{ cm})|s|}{6.00\text{ cm} - |s|}$

Thus $s'>0$ (real image) for $|s| < 6.00\text{ cm}$. (A real image is formed if the virtual object is closer to the mirror vertex than the focus.)

b) $m = -\frac{s'}{s}$; real image $\Rightarrow s'>0$; virtual object $\Rightarrow s<0 \Rightarrow m>0$; image is erect.

c)

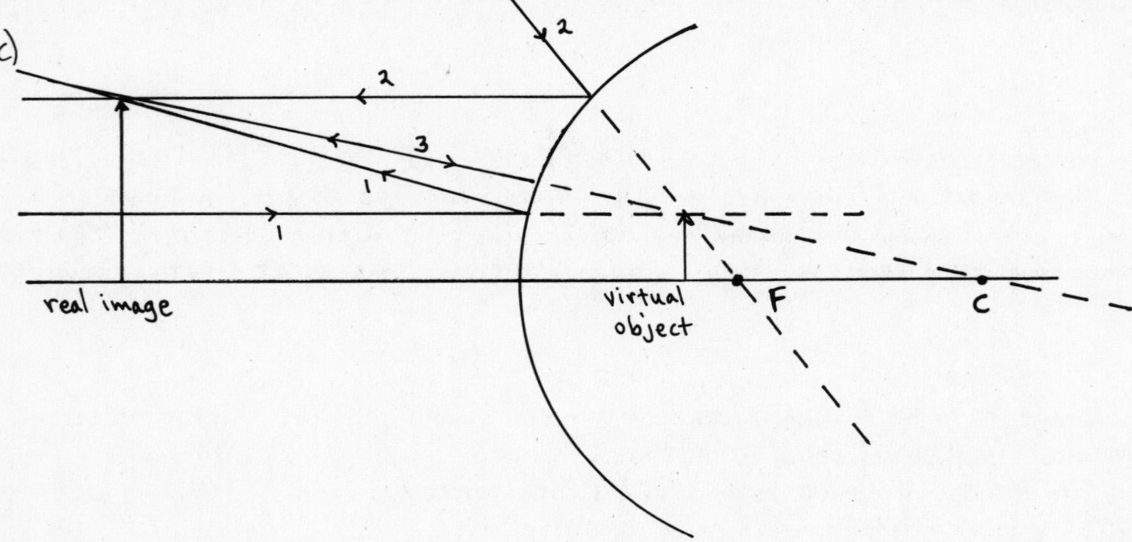

35-41

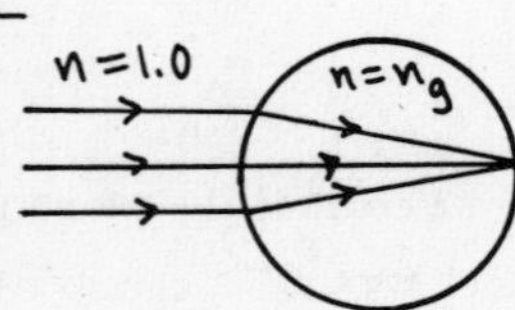

Image is formed by refraction at the front surface of the sphere. Let n_g be the index of refraction of the glass.

$s = \infty$

$s' = +2r$, where r is the radius of the sphere

35-41 (cont)

$n_a = 1.00, \quad n_b = n_g, \quad R = +r$

$$\frac{n_a}{s} + \frac{n_b}{s'} = \frac{n_b - n_a}{R}$$

$$\frac{1}{\infty} + \frac{n_g}{2r} = \frac{n_g - 1.00}{r}$$

$$\frac{n_g}{2r} = \frac{n_g}{r} - \frac{1}{r} \Rightarrow \frac{n_g}{2r} = \frac{1}{r} \Rightarrow \underline{n_g = 2.00}$$

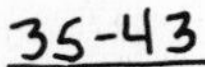

35-43

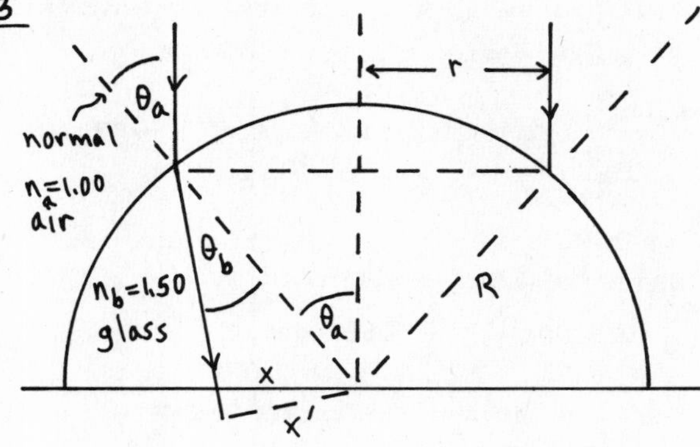

The width of the incident beam is exaggerated in the sketch, to make it easier to draw.

The diameter of the circle of light formed on the table is $2x$. Note the two right triangles containing the angles θ_a and θ_b.

$r = 0.200$ cm is the radius of the incident beam.

$R = 10.0$ cm is the radius of the glass hemisphere

θ_a and θ_b small $\Rightarrow x \simeq x'$; $\sin\theta_a = \frac{r}{R}$, $\sin\theta_b = \frac{x'}{R} \simeq \frac{x}{R}$

Snell's law: $n_a \sin\theta_a = n_b \sin\theta_b$

Use the above expressions for $\sin\theta_a$, $\sin\theta_b \Rightarrow n_a \frac{r}{R} = n_b \frac{x}{R}$

$$n_a r = n_b x \Rightarrow x = \frac{n_a r}{n_b} = \frac{1.00\,(0.200\text{ cm})}{1.50} = 0.133\text{ cm}$$

The diameter of the circle on the table is $2x = 2(0.133\text{ cm}) = \underline{0.266\text{ cm}}$.

b) R divides out of the expression; the result for the diameter of the spot is independent of the radius R of the hemisphere.

35-47

convex mirror $\Rightarrow R = -2.00$ m; $f = \frac{R}{2} = -1.00$ m

Find the relation between s and s':

$$\frac{1}{s} + \frac{1}{s'} = \frac{1}{f}$$

$$\frac{1}{s'} = \frac{1}{f} - \frac{1}{s} = \frac{s-f}{sf} \Rightarrow s' = \frac{sf}{s-f}$$

The speed of the image is $v' = \frac{ds'}{dt} = \frac{ds'}{ds}\frac{ds}{dt}$

$$\frac{ds'}{ds} = \frac{f}{s-f} - \frac{sf}{(s-f)^2} = \frac{(s-f)f - sf}{(s-f)^2} = -\frac{f^2}{(s-f)^2} = -\left(\frac{f}{s-f}\right)^2$$

$\frac{ds}{dt} = v$, the speed of the object

Thus $v' = -\left(\frac{f}{s-f}\right)^2 v$. (The quantity $\left(\frac{f}{s-f}\right)^2$ is always positive, so the minus sign means that the object and image move in opposite directions.)

a) $s = 10.0$ m

$$v' = -\left(\frac{1.00\text{ m}}{10.0\text{ m} - (-1.00\text{ m})}\right)^2 (3.00\text{ m/s}) = \underline{-0.0248\text{ m/s}}$$

35-47 (cont)

b) $s = 2.0\,m$

$$v' = -\left(\frac{1.00\,m}{2.0\,m + 1.00\,m}\right)^2 (3.00\,m/s) = \underline{-0.333\,m/s}$$

35-49

air $n=1.00$

benzene $n=1.50$ — 2.00 cm

water $n=1.33$ — 5.00 cm

The bottom of the water serves as the object. An image is formed by the water/benzene interface. This image serves as an object for the benzene/air interface, which forms the final image.

$$\frac{n_a}{s} + \frac{n_b}{s'} = \frac{n_b - n_a}{R}; \text{ flat (plane) surface} \Rightarrow R \to \infty$$

$$\Rightarrow \frac{n_a}{s} + \frac{n_b}{s'} = 0 \Rightarrow s' = -\left(\frac{n_b}{n_a}\right)s$$

Image formed by refraction at the water/benzene interface:

$n_a = 1.33$ $\qquad s' = -\left(\frac{n_b}{n_a}\right)s = -\left(\frac{1.50}{1.33}\right)(+5.00\,cm) = -5.639\,cm$

$n_b = 1.50$

$s = +5.00\,cm$ $\qquad$ The first image is 5.639 cm below the water/benzene interface and thus 5.639 cm + 2.00 cm = 7.639 cm below the benzene/air interface.

Image formation by refraction at the benzene/air interface:

$n_a = 1.50$ $\qquad s' = -\left(\frac{n_b}{n_a}\right)s = -\frac{1.00}{1.50}(+7.639\,cm) = -5.13\,cm$

$n_b = 1.00$

$s = +7.639\,cm$ $\qquad$ Thus this final image of the bottom of the water layer is 5.13 cm below the top surface of the benzene.

35-51

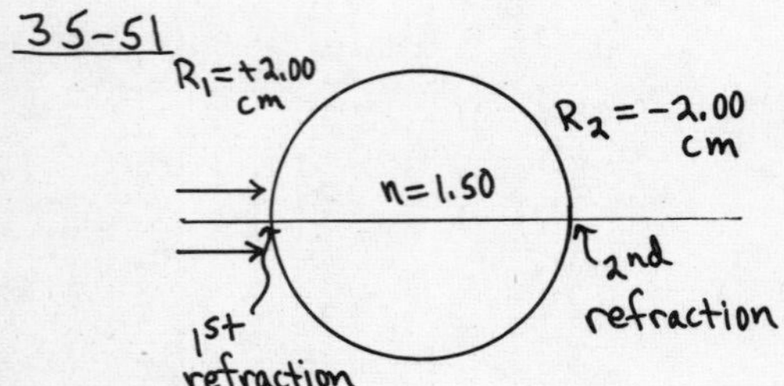

The image formed by the first refraction serves as the object for the second refraction.

The two vertexes are a distance 2R = 4.00 cm apart.

1st refraction

$n_a = 1.00$

$n_b = 1.50$

$R = +2.00\,cm$

$s = \infty$ (parallel rays)

$$\frac{n_a}{s} + \frac{n_b}{s'} = \frac{n_b - n_a}{R}$$

$$\frac{1.0}{\infty} + \frac{1.50}{s'} = \frac{1.50 - 1.00}{+2.00\,cm}$$

$$\frac{1.50}{s'} = \frac{1}{4.00\,cm} \Rightarrow s' = (1.50)(4.00\,cm) = 6.00\,cm$$

$s' > 0 \Rightarrow$ the first image is 6.00 cm to the right of the first vertex, so is 2.00 cm to the right of the second vertex. This image serves as a virtual object for the second refraction, with $s = -2.00\,cm$.

35-51 (cont)

2nd refraction

$n_a = 1.50$

$n_b = 1.00$

$R = -2.00$ cm

$s = -2.00$ cm

$$\frac{n_a}{s} + \frac{n_b}{s'} = \frac{n_b - n_a}{R}$$

$$\frac{1.50}{-2.00\text{ cm}} + \frac{1.00}{s'} = \frac{1.00 - 1.50}{-2.00\text{ cm}}$$

$$\frac{1}{s'} = \frac{3}{4.00\text{cm}} + \frac{1}{4.00\text{cm}} = \frac{1}{1.00\text{cm}} \Rightarrow s' = 1.00\text{cm}$$

$s' > 0$, so this final image is 1.00 cm to the right of the second vertex, or 3.00 cm from the center of the sphere.

35-53

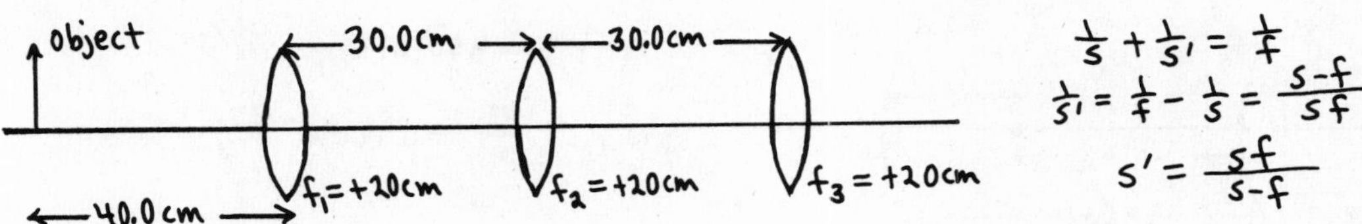

$$\frac{1}{s} + \frac{1}{s'} = \frac{1}{f}$$

$$\frac{1}{s'} = \frac{1}{f} - \frac{1}{s} = \frac{s-f}{sf}$$

$$s' = \frac{sf}{s-f}$$

lens #1

$s = +40.0$ cm

$f = +20.0$ cm

$$s' = \frac{sf}{s-f} = \frac{(+40.0\text{ cm})(+20.0\text{ cm})}{+40.0\text{ cm} - 20.0\text{ cm}} = +40.0\text{ cm}$$

The image formed by the first lens is 40.0 cm to the right of the first lens so it is 40.0 cm − 30.0 cm = 10.0 cm to the right of the second lens.

lens #2

$s = -10.0$ cm

$f = +20.0$ cm

$$s' = \frac{sf}{s-f} = \frac{(-10.0\text{cm})(+20.0\text{ cm})}{-10.0\text{cm} - 20.0\text{ cm}} = +6.667\text{ cm}$$

The image formed by the second lens is 6.667 cm to the right of the second lens, so it is 30.0cm − 6.667cm = 23.33 cm to the left of the third lens.

lens #3

$s = +23.33$ cm

$f = +20.0$ cm

$$s' = \frac{sf}{s-f} = \frac{(23.33\text{ cm})(+20.0\text{cm})}{23.33\text{ cm} - 20.0\text{ cm}} = +140.0\text{ cm}$$

The final image is 140 cm to the right of the third lens.

35-57

x

14.0 cm

screen (image)

object

lens

$$s = x$$

$$s' = 14.0\text{ cm}$$

$$\frac{1}{s} + \frac{1}{s'} = \frac{1}{f} \Rightarrow \frac{1}{x} + \frac{1}{14.0\text{cm}} = \frac{1}{f}$$

x+2.00cm

2.00cm

10.0cm

2.00 cm

$$s = x + 2.00\text{cm}$$

$$s' = 10.0\text{ cm}$$

$$\frac{1}{s} + \frac{1}{s'} = \frac{1}{f} \Rightarrow \frac{1}{x+2.00\text{cm}} + \frac{1}{10.0\text{cm}} = \frac{1}{f}$$

Equate these two expressions for $\frac{1}{f}$

$$\Rightarrow \frac{1}{x} + \frac{1}{14.0\text{cm}} = \frac{1}{x+2.00\text{ cm}} + \frac{1}{10.0\text{cm}}$$

$$\frac{1}{x} - \frac{1}{x+2.00\text{cm}} = \frac{1}{10.0\text{cm}} - \frac{1}{14.0\text{cm}}$$

35-57 (cont)

$$\frac{x+2.00\text{ cm}-x}{x(x+2.00\text{ cm})} = \frac{7-5}{70.0\text{cm}} \Rightarrow \frac{2.00\text{ cm}}{x(x+2.00\text{cm})} = \frac{2}{70.0\text{cm}}$$

$$x^2+(2.00\text{ cm})x - 70.0\text{ cm}^2 = 0$$

$$x = \tfrac{1}{2}\left(-2.00 \pm \sqrt{4.00+4(70.0)}\right)\text{ cm}$$

x must be positive $\Rightarrow x = \tfrac{1}{2}(-2.00+16.85)\text{ cm} = 7.425\text{ cm}$

Then $\frac{1}{x}+\frac{1}{14.0\text{cm}} = \frac{1}{f} \Rightarrow \frac{1}{f} = \frac{1}{7.425\text{ cm}} + \frac{1}{14.0\text{cm}} \Rightarrow \underline{f = +4.85\text{cm}}$

($f>0$; the lens is converging)

35-61

a)

$|R| = 0.400$ m

$|f| = \frac{|R|}{2} = 0.200\text{m}$

#1 #2 ← x → ← L − x → ← L=0.600 m →

Image formed by convex mirror (mirror #1)

convex $\Rightarrow f_1 = -0.200$ m

$$s_1 = L - x$$

$$s_1' = \frac{s_1 f_1}{s_1 - f_1} = \frac{(L-x)(-0.200\text{m})}{L-x+0.200\text{ m}} = -(0.200\text{ m})\left(\frac{0.600\text{ m}-x}{0.800\text{ m}-x}\right) < 0$$

The image is $(0.200\text{ m})\left(\frac{0.600\text{ m}-x}{0.800\text{m}-x}\right)$ to the left of mirror #1, so is

$0.600\text{m}+(0.200\text{m})\left(\frac{0.600\text{m}-x}{0.800\text{m}-x}\right) = \frac{0.480\text{m}^2 - 0.600\text{m}\,x + 0.120\text{m}^2 - 0.200\text{m}\,x}{0.800\text{ m}-x} = \frac{0.600\text{ m}^2 - 0.800\text{m}\,x}{0.800\text{m}-x}$ to the left of mirror #2

Image formed by concave mirror (mirror #2)

concave $\Rightarrow f_2 = +0.200$ m

$$s_2 = \frac{0.600\text{ m}^2 - 0.800\text{ m}\,x}{0.800\text{ m}-x}$$

Rays return to the source $\Rightarrow s_2' = x$

$$\frac{1}{s}+\frac{1}{s'} = \frac{1}{f} \Rightarrow \frac{0.800\text{ m}-x}{0.600\text{ m}^2-0.800\text{m}\,x} + \frac{1}{x} = \frac{1}{0.200\text{ m}} \Rightarrow \frac{0.800\text{ m}-x}{0.600\text{m}^2-0.800\text{m}\,x} = \frac{x-0.200\text{ m}}{0.200\text{m}\,x}$$

$$\cancel{0.160\text{m}^2 x} - 0.200\text{m}\,x^2 = 0.600\text{ m}^2 x - 0.120\text{ m}^3 - 0.800\text{m}\,x^2 + \cancel{0.160\text{ m}^2 x}$$

$$(0.600\text{m})x^2 - (0.600\text{m}^2)x + 0.120\text{ m}^3 = 0$$

$$x^2 - (1.00\text{ m})x + 0.20\text{ m}^2 = 0 \Rightarrow x = \tfrac{1}{2}\left(1.00 \pm \sqrt{1.00-4(0.20)}\right)\text{ m}$$

$x = (0.500 \pm 0.224)$ m ; $x = 0.724$ m (impossible; can't have $x > L = 0.600$m) or $\underline{x = 0.276\text{m}}$

b)

#2 #1 ← x → ← L−x → ← L=0.600 m →

Image formed by concave mirror (mirror #1)

concave $\Rightarrow f_1 = +0.200$ m

$$s_1 = x$$

$$s_1' = \frac{s_1 f_1}{s_1 - f_1} = \frac{(0.200\text{m})x}{x-0.200\text{m}}$$

35-61 (cont)

The image is $\frac{(0.200\text{m})x}{x-0.200\text{m}}$ to left of mirror #1, so $s_2 = 0.600\text{m} - \frac{(0.200\text{m})x}{x-0.200\text{m}} = \frac{(0.400\text{m})x - 0.120\text{m}^2}{x-0.200\text{m}}$

Image formed by convex mirror (mirror #2)

convex $\Rightarrow f_2 = -0.200$ m

Rays return to the source $\Rightarrow s_2' = L - x = 0.600\text{ m} - x$

$$\frac{1}{s}+\frac{1}{s'}=\frac{1}{f} \Rightarrow \frac{x-0.200\text{ m}}{(0.400\text{m})x-0.120\text{m}^2}+\frac{1}{0.600\text{m}-x}=-\frac{1}{0.200\text{m}}$$

$$\frac{x-0.200\text{m}}{(0.400\text{m})x-0.120\text{m}^2}=-\left(\frac{0.800\text{ m}-x}{0.120\text{m}-0.200\text{m }x}\right)$$

$$\Rightarrow -(0.600\text{ m})x^2+(0.600\text{ m}^2)x-(0.120\text{ m}^3)=0$$

$$x^2-(1.00\text{ m})x+0.20\text{ m}^2=0$$

This is the same quadratic equation as obtained in part (a), so again $x = \underline{0.276\text{ m}}$.

35-63

Thin-walled glass $\Rightarrow$ the glass has no effect on the light rays. The problem is that of refraction by a sphere of water surrounded by air.

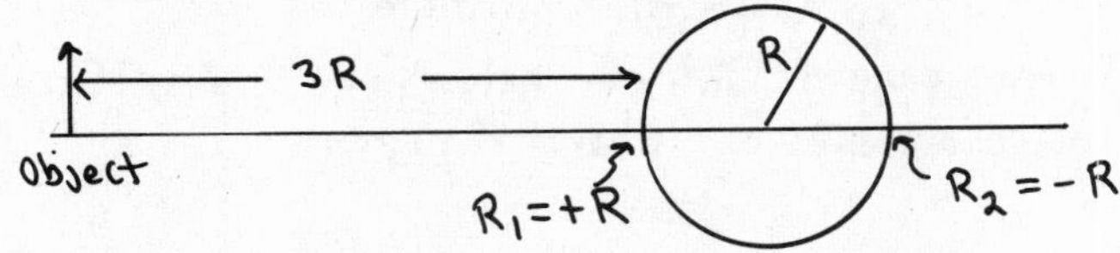

First refraction (air $\rightarrow$ water):

$n_a = 1.00$ (air)

$n_b = 1.333$ (water)

$s_1 = 3R$

$R_1 = +R$

$$\frac{n_a}{s_1}+\frac{n_b}{s_1'}=\frac{n_b-n_a}{R_1}$$

$$\frac{1}{3R}+\frac{1.333}{s_1'}=\frac{1.333-1}{R}$$

$$\frac{1.333}{s_1'}=\frac{0.333}{R}-\frac{0.333}{R}=0 \Rightarrow s_1'=\infty \text{ (parallel rays)}$$

Second refraction (water $\rightarrow$ air):

$n_a = 1.333$ (water)

$n_b = 1.00$

$s_2 = -\infty$

$R_2 = -R$

$$\frac{n_a}{s_2}+\frac{n_b}{s_2'}=\frac{n_b-n_a}{R_2}$$

$$\frac{1.333}{-\infty}+\frac{1.00}{s_2'}=\frac{1.00-1.333}{-R}$$

$$\frac{1}{s_2'}=+\frac{1}{3R} \Rightarrow s_2'=+3R$$

The final image is 3R to the right of the second surface $\Rightarrow$ 4R from the center of the sphere, on the opposite side from the object.

35-65

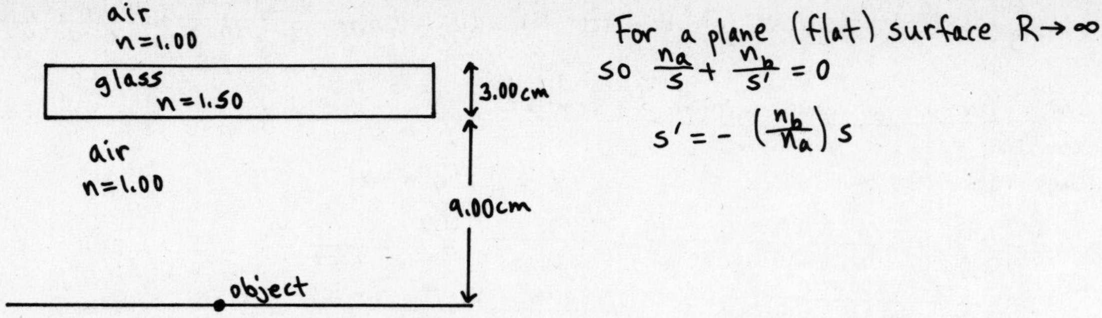

For a plane (flat) surface $R \to \infty$,

so $\frac{n_a}{s} + \frac{n_b}{s'} = 0$

$$s' = -\left(\frac{n_b}{n_a}\right) s$$

First refraction (air → glass)

$n_a = 1.00$

$n_b = 1.50$

$s = 9.00\text{ cm}$

$$s' = -\left(\frac{n_b}{n_a}\right) s = -\frac{1.50}{1.00}(9.00\text{ cm}) = -13.5\text{ cm}$$

The image is 13.5 cm below the lower surface of the glass, so is 13.5 cm + 3.00 cm = 16.5 cm below the upper surface.

Second refraction (glass → air)

$n_a = 1.50$

$n_b = 1.00$

$s = +16.5\text{ cm}$

$$s' = -\left(\frac{n_b}{n_a}\right) s = -\frac{1.00}{1.50}(16.5\text{ cm}) = -11.0\text{ cm}$$

The image of the page is 11.0 cm below the top surface of the glass plate and therefore 1.00 cm above the page.

CHAPTER 36

Exercises 3, 5, 7, 9, 15, 17, 19, 23

Problems 25, 29, 31

Exercises

36-3

a) We need $m = -\dfrac{24\times10^{-3}\,\text{m}}{100\,\text{m}} = -2.4\times10^{-4}$, or $m = -\dfrac{36\times10^{-3}\,\text{m}}{150\,\text{m}} = -2.4\times10^{-4}$.

$s \gg f$ so $s' \approx f$

Then $m = -\dfrac{s'}{s} = -\dfrac{f}{s} \Rightarrow \dfrac{f}{s} = 2.4\times10^{-4} \Rightarrow f = 2.4\times10^{-4}(150\,\text{m}) = 0.036\,\text{m} = 36\,\text{mm}$

A smaller f means a smaller s' and a smaller m, so with $f = 35$ mm the object's image nearly fills the picture area.

b) We need $m = -\dfrac{36\times10^{-3}\,\text{m}}{2.0\,\text{m}} = -1.80\times10^{-2}$.

Then, as in part (a), $\dfrac{f}{s} = 1.8\times10^{-2} \Rightarrow f = (12.0\,\text{m})(1.8\times10^{-2}) = 0.216\,\text{m} = 216\,\text{mm}$.

Therefore use the 200 mm lens.

36-5

a) $f/2.8$ lens $\Rightarrow$ f-number $= 2.8$

$\text{f-number} = \dfrac{f}{D}$ (Eq. 36-1) $\Rightarrow D = \dfrac{f}{\text{f-number}} = \dfrac{90.0\,\text{mm}}{2.8} = \underline{32.1\,\text{mm}}$

b) $f/5.6$ lens $\Rightarrow$ f-number $= 5.6$; $D = \dfrac{f}{\text{f-number}} = \dfrac{90.0\,\text{mm}}{5.6} = 16.1\,\text{mm}$

D is smaller by a factor of 2. The aperture area is smaller by a factor of $2^2 = 4$; so need an exposure time larger by a factor of $4 \Rightarrow 4\left(\dfrac{1}{60\,\text{s}}\right) = \dfrac{1}{15\,\text{s}}$.

36-7

a) $f = 0.100$ m

$s' = 5.00$ m

$s = ?$

$\dfrac{1}{s} + \dfrac{1}{s'} = \dfrac{1}{f} \Rightarrow \dfrac{1}{s} = \dfrac{1}{f} - \dfrac{1}{s'} = \dfrac{s'-f}{s'f}$

$s = \dfrac{s'f}{s'-f} = \dfrac{(5.00\,\text{m})(0.100\,\text{m})}{5.00\,\text{m} - 0.100\,\text{m}} = 0.1020\,\text{m} = \underline{10.2\,\text{cm}}$

b) $f = 0.100$ m

$s = 0.100$ m

$\dfrac{1}{s} + \dfrac{1}{s'} = \dfrac{1}{f}$

$s' = \dfrac{sf}{s-f} = \dfrac{(0.100\,\text{m})(0.100\,\text{m})}{0.100\,\text{m} - 0.100\,\text{m}} \rightarrow \infty$; no, the screen cannot be moved to achieve focus

36-9

a) $\dfrac{1}{f} = +3.00$ diopters $\Rightarrow f = +\dfrac{1}{3.00}\,\text{m} = +0.3333\,\text{m}$ (converging lens)

The purpose of the corrective lens is to take an object at 25 cm from the eye and form a virtual image at the eye's near point.

36-9 (cont)

$s = 25\text{ cm}$ $\quad \frac{1}{s}+\frac{1}{s'}=\frac{1}{f} \Rightarrow s'=\frac{sf}{s-f}=\frac{(25\text{cm})(33.33\text{cm})}{25\text{cm}-33.33\text{cm}}=-100\text{ cm}$

$f = 33.33\text{ cm}$

The eye's near point is 100 cm from the eye.

b) $\frac{1}{f}=-0.600$ diopters $\Rightarrow f=-\frac{1}{0.600}\text{ m}=-1.667\text{ m}$ (diverging lens)

The purpose of this corrective lens is to take an object at infinity and form a virtual image of it at the eye's far point.

$f=-166.7\text{ cm}$ $\quad \frac{1}{s}+\frac{1}{s'}=\frac{1}{f}$

$s=\infty$ $\quad s=\infty \Rightarrow \frac{1}{s'}=\frac{1}{f} \Rightarrow s=f=-166.7\text{ cm}$

The eye's far point is 167 cm from the eye.

36-15

a) $s'=-25.0\text{ cm}$ $\quad \frac{1}{s}+\frac{1}{s'}=\frac{1}{f} \Rightarrow \frac{1}{s}=\frac{1}{f}-\frac{1}{s'}=\frac{s'-f}{s'f}$

$f=+12.0\text{ cm}$

$s=?$ $\quad s=\frac{s'f}{s'-f}=\frac{(-25.0\text{cm})(+12.0\text{cm})}{-25.0\text{ cm}-12.0\text{ cm}}=\underline{+8.11\text{ cm}}$

b) $m=-\frac{s'}{s}=-\frac{-25.0\text{ cm}}{8.11\text{ cm}}=+3.083$

$|m|=\frac{|y'|}{|y|} \Rightarrow |y'|=|m||y|=(3.083)(1.00\text{ mm})=\underline{3.08\text{ mm}}$

36-17

a) $f_1=1.60\text{cm}$ $\quad f_2=2.50\text{ cm}$

21.4 cm

objective $\quad$ eyepiece

Final image at $\infty \Rightarrow$ object for eyepiece is at its focal point. But the object for the eyepiece is the image of the objective, so the image formed by the objective is $21.4\text{ cm}-2.50\text{ cm}=18.9\text{ cm}$ to the right of the lens.

Image formation by the objective:

$s'=+18.9\text{ cm}$ $\quad \frac{1}{s}+\frac{1}{s'}=\frac{1}{f} \Rightarrow \frac{1}{s}=\frac{1}{f}-\frac{1}{s'}=\frac{s'-f}{s'f}$

$f=+1.60\text{ cm}$

$s=?$ $\quad s=\frac{s'f}{s'-f}=\frac{(18.9\text{cm})(1.60\text{ cm})}{18.9\text{ cm}-1.6\text{ cm}}=1.75\text{cm}=\underline{17.5\text{ mm}}$

b) $m_1=-\frac{s'}{s}=-\frac{18.9\text{ cm}}{1.75\text{ cm}}=-10.8$

The magnitude of this linear magnification is $\underline{10.8}$.

c) $M=m_1M_2$

$M_2=\frac{25\text{ cm}}{f_2}=\frac{25\text{ cm}}{2.50\text{ cm}}=10.0$

$M=m_1M_2=-(10.8)(10.0)=\underline{-108}$

36-19

$f_1 = 90.0$ cm (objective) ; $f_2 = 20.0$ cm (eyepiece)

a) Eq. (36-5) $M = -\frac{f_1}{f_2} = -\frac{90.0\text{ cm}}{20.0\text{ cm}} = \underline{-4.50}$

b) $s = 2.00 \times 10^3$ m

$s' = f_1 = 90.0$ cm (since s is very large $s' \simeq f$)

$m = -\frac{s'}{s} = -\frac{0.900\text{ m}}{2.00\times10^3\text{ m}} = -4.50\times10^{-4}$

$|y'| = |m||y| = (4.50\times10^{-4})(80.0\text{ m}) = 0.0360\text{ m} = \underline{3.60\text{ cm}}$

c) The angular size of the object is $\theta = \frac{80.0\text{ m}}{2.00\times10^3\text{ m}} = 0.0400$ rad

$M = \frac{\theta'}{\theta}$ (Eq. 36-5) $\Rightarrow$ angular size of image is $\theta' = M\theta = -4.50(0.0400\text{ rad}) = \underline{-0.180\text{ rad}}$
(The minus sign shows that the final image is inverted.)

36-23

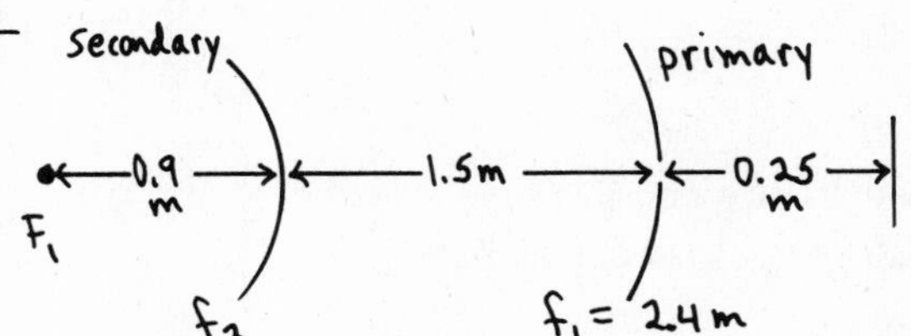

Far distant object $\Rightarrow$ image of primary mirror is formed at its focus, which is $2.4\text{ m} - 1.5\text{ m} = 0.9\text{ m}$ behind the secondary mirror. The image formed by the secondary mirror is located at the detector.

Image formation by the secondary mirror:

$s = -0.9$ m

$s' = +1.75$ m

$f = ?$

$\frac{1}{s} + \frac{1}{s'} = \frac{1}{f} \Rightarrow \frac{1}{f} = \frac{s+s'}{ss'}$

$f = \frac{ss'}{s+s'} = \frac{(-0.9\text{ m})(1.75\text{ m})}{-0.9\text{ m} + 1.75\text{ m}} = -1.85\text{ m}$

$R = 2f = 2(-1.85\text{ m}) = -3.7\text{ m}$

The secondary mirror should be convex ($R < 0$) with radius of curvature 3.7m.

Problems

36-25

a) $f = 50.0\times10^{-3}$ m

$|m| = \frac{|y'|}{|y|} = \frac{\frac{2}{3}(30.0\times10^{-3}\text{ m})}{3.20\text{ m}} = 6.25\times10^{-3}$

Image on film $\Rightarrow$ image is real. $s>0, s'>0 \Rightarrow m = -\frac{s'}{s} < 0 \Rightarrow m = -6.25\times10^{-3}$

$s' = -ms = -(-6.25\times10^{-3})s = +(6.25\times10^{-3})s$

Use this in $\frac{1}{s} + \frac{1}{s'} = \frac{1}{f} \Rightarrow \frac{1}{s} + \frac{1}{(6.25\times10^{-3})s} = \frac{1}{f}$

$\frac{161}{s} = \frac{1}{f} \Rightarrow s = 161f = 161(50.0\times10^{-3}\text{ m}) = \underline{8.05\text{ m}}$

b) Fill the viewfinder frame $\Rightarrow |m| = \frac{30.0\times10^{-3}\text{ m}}{3.20\text{ m}} = 9.375\times10^{-3}$.

Thus $s' = (9.375\times10^{-3})s$.

36-25 (cont)

$$\frac{1}{s}+\frac{1}{s'}=\frac{1}{f} \Rightarrow \frac{1}{s}+\frac{1}{(9.375\times10^{-3})s}=\frac{1}{f}$$

$$\frac{107.7}{s}=\frac{1}{f} \Rightarrow s=107.7f=107.7(50.0\times10^{-3}\,\text{m})=\underline{5.38\,\text{m}}$$

36-29

air $n_a=1.00$, $n_b=1.40$; object, $s=30.0$ cm, $s'=?$, retina; cornea $R=+0.75$ cm

$$\frac{n_a}{s}+\frac{n_b}{s'}=\frac{n_b-n_a}{R}$$

$$\frac{1.00}{30.0\,\text{cm}}+\frac{1.40}{s'}=\frac{1.40-1.00}{0.75\,\text{cm}}$$

$$\frac{1.40}{s'}=0.5333\,\text{cm}^{-1}-0.0333\,\text{cm}^{-1}=0.5000\,\text{cm}^{-1}$$

$$s'=\underline{2.80\,\text{cm}}$$

The distance from the cornea vertex to retina for this eye is 2.80 cm. Exercise 36-10 says that this distance for a normal eye is 2.60 cm, so the nearsighted eye is elongated.

36-31

$f_1=9.00$ mm (objective), $f_2=6.00$ cm (eyepiece)

a) The total magnification of the final image is $m_{tot}=m_1m_2$.

object; objective; image of objective; eyepiece; $s_1=?$; $s_1'=18.0$ cm; $s_2=?$; $s_2'=100$ cm; screen; $f_1=0.900$ cm; $f_2=6.00$ cm

Find the object distance s_1 for the objective:

$s_1'=+18.0$ cm
$f_1=0.900$ cm
$s_1=?$

$$\frac{1}{s_1}+\frac{1}{s_1'}=\frac{1}{f_1} \Rightarrow \frac{1}{s_1}=\frac{1}{f_1}-\frac{1}{s_1'}=\frac{s_1'-f_1}{s_1'f_1}$$

$$s_1=\frac{s_1'f_1}{s_1'-f_1}=\frac{(18.0\,\text{cm})(0.900\,\text{cm})}{18.0\,\text{cm}-0.900\,\text{cm}}=0.9474\,\text{cm}$$

Find the object distance s_2 for the eyepiece:

$s_2'=100$ cm
$f_2=6.00$ cm
$s_2=?$

$$\frac{1}{s_2}+\frac{1}{s_2'}=\frac{1}{f_2} \Rightarrow s_2=\frac{s_2'f_2}{s_2'-f_2}=\frac{(100\,\text{cm})(6.00\,\text{cm})}{100\,\text{cm}-6.00\,\text{cm}}=6.383\,\text{cm}$$

Now we can calculate the magnification for each lens:

$$m_1=-\frac{s_1'}{s_1}=-\frac{18.0\,\text{cm}}{0.9474\,\text{cm}}=-19.0$$

$$m_2=-\frac{s_2'}{s_2}=-\frac{100\,\text{cm}}{6.383\,\text{cm}}=-15.67$$

$$m_{tot}=m_1m_2=(-19.0)(-15.67)=\underline{298}$$

b) From the sketch we see that the distance between the two lenses is $s_1'+s_2=18.0\,\text{cm}+6.383\,\text{cm}=\underline{24.4\,\text{cm}}$

CHAPTER 37

Exercises 1, 3, 5, 9, 13, 17, 19, 25

Problems 27, 29, 33, 35, 37

Exercises

37-1

For constructive interference the path difference $r_2 - r_1 = 1.80 \times 10^{-6}$ m is an integer number of wavelengths: $r_2 - r_1 = m\lambda$, $m = 1, 2, 3, \ldots$ (Since $r_2 - r_1$ is specified as nonzero and positive, $m = 0$ and negative m values are excluded.)

$$\Rightarrow \lambda = \frac{r_2 - r_1}{m} = \frac{1.80 \times 10^{-6}\text{ m}}{m}, \quad m = 1, 2, 3, \ldots$$

$m = 3 \Rightarrow 0.600 \times 10^{-6}\text{ m} = \underline{600\text{ nm}}$

$m = 4 \Rightarrow 0.450 \times 10^{-6}\text{ m} = \underline{450\text{ nm}}$

Other values of λ are outside the range of visible wavelengths.

37-3

A ← x → P ... B; ←110 m→

The distance of point P from each coherent source is $r_A = x$ and $r_B = 110\text{ m} - x$.

The path difference is $r_B - r_A = 110\text{ m} - 2x$. For constructive interference this path difference is an integer multiple of the wavelength:

$$r_B - r_A = m\lambda, \quad m = 0, \pm 1, \pm 2, \ldots$$

$$\lambda = \frac{c}{f} = \frac{2.998 \times 10^8\text{ m/s}}{7.50 \times 10^6\text{ Hz}} = 40.0\text{ m}$$

$$\Rightarrow 110\text{ m} - 2x = m(40.0\text{ m}) \Rightarrow x = \frac{110\text{ m} - m(40.0\text{ m})}{2} = 55.0\text{ m} - (20.0\text{ m})m$$

x must lie in the range 0 to 110 m, since P is said to be between the two antennas.

$m = 0 \Rightarrow x = 55.0\text{ m}$

$m = +1 \Rightarrow x = 55.0\text{ m} - 20.0\text{ m} = 35.0\text{ m}$

$m = +2 \Rightarrow x = 55.0\text{ m} - 40.0\text{ m} = 15.0\text{ m}$

$m = -1 \Rightarrow x = 55.0\text{ m} + 20.0\text{ m} = 75.0\text{ m}$

$m = -2 \Rightarrow x = 55.0\text{ m} + 40.0\text{ m} = 95.0\text{ m}$

All other values of m give values of x out of the allowed range.

Constructive interference will occur for $x = 15.0$ m, 35.0 m, 55.0 m, 75.0 m, and 95.0 m.

37-5

The dark lines correspond to destructive interference and hence are located by Eq. (37-5): $d\sin\theta = (m + \frac{1}{2})\lambda \Rightarrow \sin\theta = \frac{(m+\frac{1}{2})\lambda}{d}$, $m = 0, \pm 1, \pm 2, \ldots$

1st dark line $\Rightarrow m = 0$

2nd dark line $\Rightarrow m = 1 \Rightarrow \sin\theta_1 = \frac{3\lambda}{2d} = \frac{3(600 \times 10^{-9}\text{ m})}{2(0.300 \times 10^{-3}\text{ m})} = 3.00 \times 10^{-3} \Rightarrow \theta_1 = 3.00 \times 10^{-3}$ rad

3rd dark line $\Rightarrow m = 2 \Rightarrow \sin\theta_2 = \frac{5\lambda}{2} = \frac{5(600 \times 10^{-9}\text{ m})}{2(0.300 \times 10^{-3}\text{ m})} = 5.00 \times 10^{-3} \Rightarrow \theta_2 = 5.00 \times 10^{-3}$ rad

(Note that θ_1 and θ_2 are small so that the approximation $\theta \simeq \sin\theta \simeq \tan\theta$ is valid.)

37-5 (cont)

The distance of each dark line from the center of the central bright band is given by $y_m = R\tan\theta$, where $R = 0.600\,m$ is the distance to the screen.

$\tan\theta \simeq \theta \Rightarrow y_m = R\theta_m$

$y_1 = R\theta_1 = (0.600\,m)(3.00\times10^{-3}\,rad) = 1.80\times10^{-3}\,m$

$y_2 = R\theta_2 = (0.600\,m)(5.00\times10^{-3}\,rad) = 3.00\times10^{-3}\,m$

$\Delta y = y_2 - y_1 = 3.00\times10^{-3}\,m - 1.80\times10^{-3}\,m = 1.20\times10^{-3}\,m = \underline{1.20\,mm}$

37-9

a) The minima are located by $d\sin\theta = (m+\frac{1}{2})\lambda$.

The first minimum is for $m=0 \Rightarrow \sin\theta = \frac{\lambda}{2d} = \frac{500\times10^{-9}\,m}{2(0.200\times10^{-3}\,m)} = 1.25\times10^{-3}$

$\Rightarrow \theta = 1.25\times10^{-3}\,rad$ (θ is small so $\theta \simeq \sin\theta \simeq \tan\theta$)

The distance on the screen is $y_1 = R\tan\theta \simeq R\theta$, where R is the distance to the screen. $y_1 = R\theta_1 = (0.800\,m)(1.25\times10^{-3}\,rad) = 1.00\times10^{-3}\,m = \underline{1.00\,mm}$.

b) Since θ is small we can use Eq. (37-15):

$$I = I_0\cos^2\left(\frac{\pi d y}{\lambda R}\right)$$

We want I for $y = \frac{1}{2}y_1 = 0.500\times10^{-3}\,m$.

$$I = (6.00\times10^{-6}\,W/m^2)\left(\cos\left(\frac{\pi(0.200\times10^{-3}\,m)(0.500\times10^{-3}\,m)}{(500\times10^{-9}\,m)(0.800\,m)}\right)\right)^2 = \underline{3.00\times10^{-6}\,W/m^2}$$

37-13

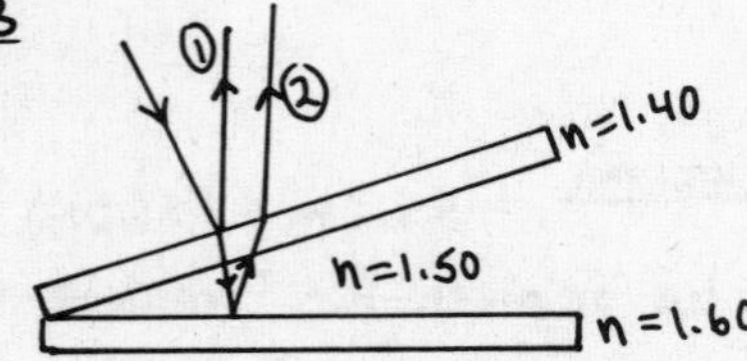

When ray ① reflects off the top of the wedge of silicone grease it undergoes a 180° phase change ($n=1.40 < n=1.50$).

When ray ② reflects off the top of the lower plate it undergoes a 180° phase change ($n=1.50 < n=1.60$).

There is no net phase difference introduced by the phase changes on reflection, and the condition for an interference minimum is $2t = (m+\frac{1}{2})\lambda$, where λ is the wavelength is the silicone grease, $\lambda = \frac{\lambda_0}{n} = \frac{500\times10^{-9}\,m}{1.50} = 333\times10^{-9}\,m$.

As in Example 37-4 (Fig. 37-9), $\frac{t}{x} = \frac{h}{l} \Rightarrow t = \frac{hx}{l}$. Then $\frac{2hx}{l} = (m+\frac{1}{2})\lambda$.

$x_m = (m+\frac{1}{2})\frac{\lambda l}{2h}$; $x_{m+1} = (m+\frac{3}{2})\frac{\lambda l}{2h}$

The spacing between adjacent dark fringes is $\Delta x = x_{m+1} - x_m = \frac{\lambda l}{2h} = \frac{(333\times10^{-9}\,m)(0.10\,m)}{2(0.020\times10^{-3}\,m)}$

$\Delta x = 0.833\times10^{-3}\,m = \underline{0.833\,mm}$.

37-17

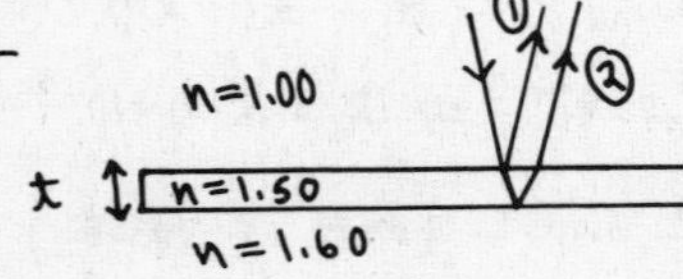

Both rays ① and ② undergo a 180° phase change on reflection, so there is no net phase difference introduced and the condition for destructive interference is $2t = (m+\frac{1}{2})\lambda$.

37-17 (cont) $t = \frac{(m+\frac{1}{2})\lambda}{2}$; thinnest film $\Rightarrow m=0 \Rightarrow t = \frac{\lambda}{4}$.

$\lambda = \frac{\lambda_0}{1.50} \Rightarrow t = \frac{\lambda_0}{4(1.50)} = \frac{500\times10^{-9}\,m}{4(1.50)} = 8.33\times10^{-8}\,m = \underline{83.3\,nm}$

37-19

Eq. (37-19) $\Rightarrow x = m\frac{\lambda}{2} = 2000\left(\frac{606\times10^{-9}\,m}{2}\right) = 606\times10^{-6}\,m = \underline{0.606\,mm}$

37-25

By conservation of linear momentum, since the initial momentum is zero the nucleus and the photon must have equal and opposite momenta.

Find the recoil speed of the nucleus:

$K = \frac{1}{2}mv^2 \Rightarrow v = \sqrt{\frac{2K}{m}} = \sqrt{\frac{2(3.132\times10^{-22}\,J)}{9.454\times10^{-26}\,kg}} = 81.40\,m/s$

The momentum of the recoiling nucleus is

$p = mv = (9.454\times10^{-26}\,kg)(81.40\,m/s) = \underline{7.70\times10^{-24}\,kg\cdot m/s}$

The gamma ray photon must have $p = 7.70\times10^{-24}\,kg\cdot m/s$.

$E = pc = (7.70\times10^{-24}\,kg\cdot m/s)(2.998\times10^{8}\,m/s) = \underline{2.31\times10^{-15}\,J}$

$p = \frac{h}{\lambda} \Rightarrow \lambda = \frac{h}{p} = \frac{6.626\times10^{-34}\,J\cdot s}{7.70\times10^{-24}\,kg\cdot m/s} = 8.61\times10^{-11}\,m = \underline{0.0861\,nm}$

$c = f\lambda \Rightarrow f = \frac{c}{\lambda} = \frac{2.998\times10^{8}\,m/s}{8.61\times10^{-11}\,m} = \underline{3.48\times10^{18}\,Hz}$

Problems

37-27

The only effect of the water is to change the wavelength to

$\lambda = \frac{\lambda_0}{n} = \frac{600\times10^{-9}\,m}{1.333} = 450\times10^{-9}\,m$

$\theta_1 \simeq \sin\theta_1 = \frac{3\lambda}{2d} = \frac{3(450\times10^{-9}\,m)}{2(0.300\times10^{-3}\,m)} = 2.25\times10^{-3}\,rad$

$\theta_2 \simeq \sin\theta_2 = \frac{5\lambda}{2d} = \frac{5(450\times10^{-9}\,m)}{2(0.300\times10^{-3}\,m)} = 3.75\times10^{-3}\,rad$

$y_1 \simeq R\theta_1 = (0.600\,m)(2.25\times10^{-3}\,rad) = 1.35\times10^{-3}\,m$

$y_2 \simeq R\theta_2 = (0.600\,m)(3.75\times10^{-3}\,rad) = 2.25\times10^{-3}\,m$

$\Delta y = y_2 - y_1 = 2.25\times10^{-3}\,m - 1.35\times10^{-3}\,m = 0.900\times10^{-3}\,m = \underline{0.900\,nm}$

37-29

a) There must be destructive interference between the sound waves from the two speakers.

b) The change in path length must be $\frac{\lambda}{2}$, so $\frac{\lambda}{2} = 0.34\,m$ and $\lambda = 0.68\,m$.

$v = f\lambda \Rightarrow f = \frac{v}{\lambda} = \frac{340\,m/s}{0.68\,m} = \underline{500\,Hz}$

37-29 (cont)

c) The change in path length must equal λ to go from one point of constructive interference to the next, so the speakers must be moved 0.68 m.

<u>37-33</u>

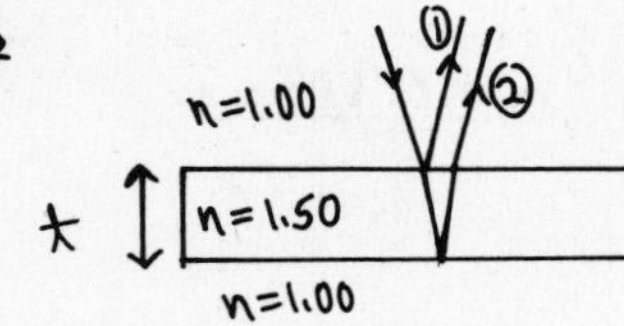

Ray ① undergoes a 180° phase change on reflection at the top surface of the glass.

Ray ② has no phase change on reflection from the lower surface of the glass.

The reflections produce a net phase difference of 180°, so the condition for constructive interference is $2t = (m+\frac{1}{2})\lambda$, $m = 0, 1, 2, \ldots \Rightarrow \lambda = \frac{2t}{m+\frac{1}{2}}$.

λ is the wavelength in the glass plate; $\lambda = \frac{\lambda_0}{n}$

$$\frac{\lambda_0}{n} = \frac{2t}{m+\frac{1}{2}} \Rightarrow \lambda_0 = \frac{2tn}{m+\frac{1}{2}} = \frac{2(0.375\times10^{-6}\text{m})(1.50)}{m+\frac{1}{2}} = \frac{1125\text{ nm}}{m+\frac{1}{2}}$$

m	λ
0	2250 nm
1	750 nm
2	450 nm
3	321 nm
⋮	⋮

Only <u>$\lambda = 450$ nm</u> is within the limits of the visible spectrum.

<u>37-35</u>

This problem deals with Newton's rings (Sect. 37-4). The interference is between rays reflecting from the top and bottom edges of the air that is between the lens and the plate.

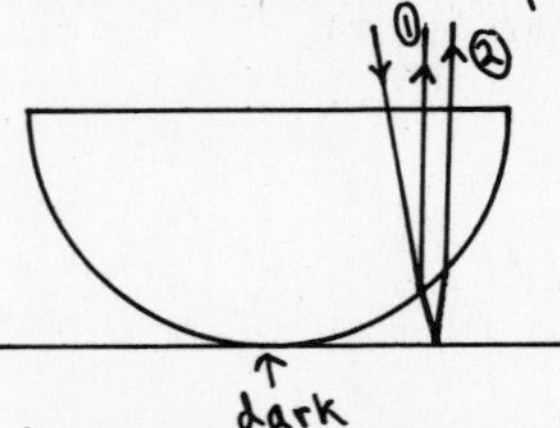

Ray ① does not undergo any phase change on reflection.
Ray ② does undergo a 180° phase change on reflection.

Hence the path difference $2t$, where t is the thickness of the air wedge, must satisfy
$2t = (m+\frac{1}{2})\lambda$, $m = 0, 1, 2, \ldots$ for constructive interference.

Third bright ring $\Rightarrow m = 2$ and $t = \frac{5\lambda}{4} = \frac{5(650\times10^{-9}\text{m})}{4} = 8.125\times10^{-7}\text{ m}$

Now must relate this to the diameter of the ring:

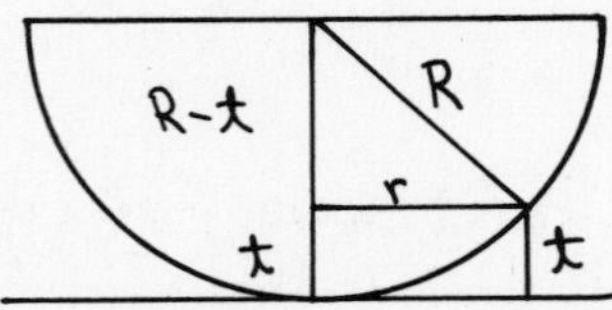

The radius of the ring is r.

$r^2 + (R-t)^2 = R^2$

$\Rightarrow r = \sqrt{R^2 - (R-t)^2} = \sqrt{R^2 - R^2 + 2Rt - t^2}$

$r = \sqrt{2Rt - t^2}$

But $R = 1.20\text{ m} \gg t$, so can neglect t^2 relative to $2Rt$

$\Rightarrow r = \sqrt{2Rt} = \sqrt{2(1.30\text{m})(8.125\times10^{-7}\text{m})} = 1.46\times10^{-3}\text{ m}$.

This is the radius. The diameter of the ring is $2r = 2.91\times10^{-3}\text{ m} =$ <u>2.91 mm</u>

37-37

a) The wavelength in the glass is decreased by a factor of $\frac{1}{n}$, so for the light through the upper slit a shorter path is needed to produce the same phase at the screen. Therefore, the interference pattern is shifted downward on the screen.

b) At a point on the screen located by the angle θ the difference in path length is $d\sin\theta$. This introduces a phase difference of $\phi = \frac{2\pi}{\lambda_0}(d\sin\theta)$, where λ_0 is the wavelength of the light in air or vacuum.

In the thickness L of glass the number of wavelengths is $\frac{L}{\lambda} = \frac{nL}{\lambda_0}$. A corresponding length L of the path of the ray through the lower slit, in air, contains $\frac{L}{\lambda_0}$ wavelengths. The phase difference this introduces is $\phi = 2\pi\left(\frac{nL}{\lambda_0} - \frac{L}{\lambda_0}\right)$
$\Rightarrow \phi = 2\pi(n-1)\frac{L}{\lambda_0}$.

The total phase difference is the sum of these two,
$$\phi = \frac{2\pi}{\lambda_0}(d\sin\theta) + 2\pi(n-1)\frac{L}{\lambda_0} = \frac{2\pi}{\lambda_0}\left(d\sin\theta + L(n-1)\right).$$

Eq. (37-10) then gives $I = I_0\cos^2\left(\frac{\pi}{\lambda_0}\left(d\sin\theta + L(n-1)\right)\right)$.

c) Maxima $\Rightarrow \cos\frac{\phi}{2} = \pm 1 \Rightarrow \frac{\phi}{2} = m\pi, \quad m = 0, \pm 1, \pm 2, \ldots$

$$\frac{\pi}{\lambda_0}\left(d\sin\theta + L(n-1)\right) = m\pi$$
$$d\sin\theta + L(n-1) = m\lambda_0$$
$$\sin\theta = \frac{m\lambda_0 - L(n-1)}{d}$$

CHAPTER 38

Exercises 1, 3, 5, 7, 9, 11, 15, 17, 21

Problems 29, 33, 37, 39

Exercises

38-1

The minima are located by Eq. (38-2):

$$\sin\theta = \frac{m\lambda}{a}, \quad m = \pm 1, \pm 2 \ldots$$

First minimum $\Rightarrow m = 1 \Rightarrow \sin\theta_1 = \frac{\lambda}{a} \Rightarrow \lambda = a\sin\theta_1$

$y_1 = 1.80$ mm

$$y_1 = R\tan\theta_1$$

$$\tan\theta_1 = \frac{y_1}{R} = \frac{1.80\times10^{-3}\text{ m}}{3.00\text{ m}} = 0.600\times10^{-3}$$

$$\Rightarrow \theta_1 = 0.600\times10^{-3}\text{ rad}$$

$$\lambda = a\sin\theta_1 = (0.800\times10^{-3}\text{ m})\sin(0.600\times10^{-3}\text{ rad}) = \underline{480\text{ nm}}$$

38-3

a) $w = 2y_1$

The first minimum is located by

$$\sin\theta_1 = \frac{\lambda}{a} = \frac{633\times10^{-9}\text{ m}}{0.300\times10^{-3}\text{ m}} = 2.11\times10^{-3}$$

$$\theta_1 = 2.11\times10^{-3}\text{ rad}$$

$$y_1 = R\tan\theta_1 = (4.0\text{ m})\tan(2.11\times10^{-3}\text{ rad}) = 8.44\times10^{-3}\text{ m}$$

$$w = 2y_1 = 2(8.44\times10^{-3}\text{ m}) = 1.69\times10^{-2}\text{ m} = \underline{16.9\text{ mm}}$$

b)

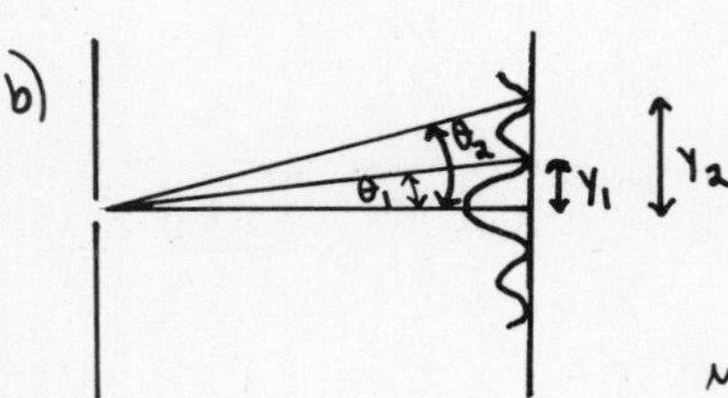

$$w = y_2 - y_1$$

$$y_1 = 8.44\times10^{-3}\text{ m} \quad \text{(part (a))}$$

$$\sin\theta_2 = \frac{2\lambda}{a} = 4.22\times10^{-3} \Rightarrow \theta_2 = 4.22\times10^{-3}\text{ rad}$$

$$y_2 = R\tan\theta_2 = 16.88\times10^{-3}\text{ m}$$

$$w = y_2 - y_1 = 16.88\times10^{-3}\text{ m} - 8.44\times10^{-3}\text{ m} = 8.44\times10^{-3}\text{ m} = \underline{8.44\text{ mm}}$$

38-5

a)

$$\sin\theta_1 = \frac{\lambda}{a} = \frac{500\times10^{-9}\text{ m}}{0.200\times10^{-3}\text{ m}} = 2.50\times10^{-3}; \quad \theta_1 = 2.50\times10^{-3}\text{ rad}$$

$$y_1 = R\tan\theta_1 = (4.00\text{ m})\tan(2.50\times10^{-3}\text{ rad}) = 1.00\text{ m} = \underline{10.0\text{ mm}}$$

b) Midway between the center of the central maximum and the first minimum

$$\Rightarrow y = \tfrac{1}{2}(10.0\text{ mm}) = 5.00\times10^{-3}\text{ m}$$

$$\tan\theta = \frac{y}{R} = \frac{5.00\times10^{-3}\text{ m}}{4.00\text{ m}} = 1.25\times10^{-3}; \quad \theta = 1.25\times10^{-3}\text{ rad}$$

The phase angle β at this point on the screen is

38-5 (cont)

$$\beta = \frac{2\pi}{\lambda} a \sin\theta = \frac{2\pi}{500\times10^{-9}\,\text{m}}(0.200\times10^{-3}\,\text{m})\sin(1.25\times10^{-3}\,\text{rad}) = \pi$$

Then $I = I_0\left[\frac{\sin(\beta/2)}{\beta/2}\right]^2 = (5.00\times10^{-6}\,\text{W/m}^2)\left[\frac{\sin(\frac{\pi}{2})}{\frac{\pi}{2}}\right]^2 = \frac{4}{\pi^2}(5.00\times10^{-6}\,\text{W/m}^2)$

$$I = \underline{2.03\times10^{-6}\,\text{W/m}^2}$$

38-7

Eq. (38-6) : $\beta = \frac{2\pi}{\lambda} a \sin\theta$

$$\lambda = \frac{2\pi}{\beta} a\sin\theta = \frac{2\pi}{\frac{\pi}{2}\,\text{rad}}(0.400\times10^{-3}\,\text{m})\sin 4.0^\circ = 1.12\times10^{-4}\,\text{m} = \underline{112\,\mu\text{m}}$$

38-9

(i) $\phi = \frac{\pi}{2}$

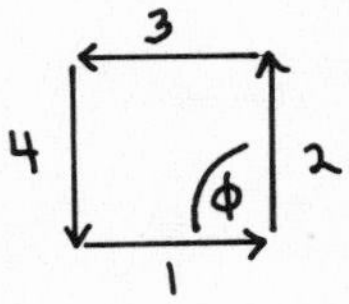

There is destructive interference between the light through slits 1 and 3 and between 2 and 4.

(ii) $\phi = \pi$

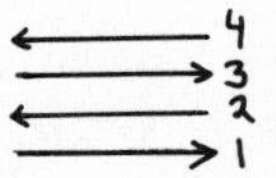

There is destructive interference between the light through slits 1 and 2 and between 3 and 4.

(iii) $\phi = \frac{3\pi}{2}$

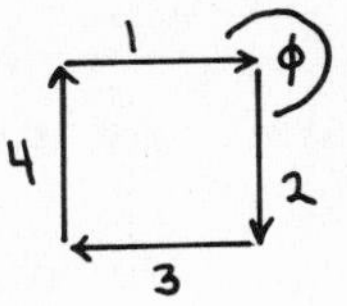

There is destructive interference between the light through slits 1 and 3 and between 2 and 4.

38-11

a) The interference fringes (maxima) are located by $d\sin\theta = m\lambda$, with $m = 0, \pm1, \pm2, \ldots$. The intensity I in the diffraction pattern is given by $I = I_0\left(\frac{\sin(\beta/2)}{\beta/2}\right)^2$, with $\beta = \frac{2\pi}{\lambda} a\sin\theta$.

We want $m = \pm3$ in the first equation to give θ that makes $I = 0$ in the second equation. $d\sin\theta = 3\lambda \Rightarrow \beta = \frac{2\pi}{\lambda} a\left(\frac{3\lambda}{d}\right) = 2\pi\left(3\frac{a}{d}\right)$

$$I = 0 \Rightarrow \frac{\sin(\beta/2)}{\beta/2} = 0 \Rightarrow \beta = 2\pi \Rightarrow 2\pi = 2\pi\left(3\frac{a}{d}\right) \Rightarrow \frac{d}{a} = 3.$$

b) Fringes $m = 0, \pm1, \pm2$ are within the central diffraction maximum and the $m = \pm3$ fringes coincide with the first diffraction minimum.

Find the value of m for the fringe that coincides with the second diffraction minimum: second minimum $\Rightarrow \beta = 4\pi$

$$\beta = \frac{2\pi}{\lambda} a\sin\theta = \frac{2\pi}{\lambda} a \frac{m\lambda}{d} = 2\pi m\left(\frac{a}{d}\right) = 2\pi\frac{m}{3}$$

$$\beta = 4\pi \Rightarrow 4\pi = 2\pi\frac{m}{3} \Rightarrow m = 6$$

38-11 (cont)

Therefore, the $m=+4$ and $m=+5$ fringes are contained within the first diffraction maximum on one side of the central maximum; two fringes.

38-15

5000 lines/cm $\Rightarrow$ 5.00×10^5 lines/m

The slit spacing is $d = \frac{1}{5.00\times10^5}$ m $= 2.00\times10^{-6}$ m

a) The line positions are given by $\sin\theta = m\frac{\lambda}{d}$

first-order $\Rightarrow m=1$

$$\sin\theta_\alpha = \frac{\lambda_\alpha}{d} = \frac{656\times10^{-9}\text{ m}}{2.00\times10^{-6}\text{ m}} = 0.328 \Rightarrow \theta_\alpha = 19.15°$$

$$\sin\theta_\delta = \frac{\lambda_\delta}{d} = \frac{410\times10^{-9}\text{ m}}{2.00\times10^{-6}\text{ m}} = 0.205 \Rightarrow \theta_\delta = 11.83°$$

The angular separation is $\theta_\alpha - \theta_\delta = 19.15° - 11.83° = \underline{7.32°}$.

b) Second-order $\Rightarrow m=2$

$$\sin\theta_\alpha = \frac{2\lambda_\alpha}{d} = 0.656 \Rightarrow \theta_\alpha = 41.00°$$

$$\sin\theta_\delta = \frac{2\lambda_\delta}{d} = 0.410 \Rightarrow \theta_\delta = 24.20°$$

The angular separation is $\theta_\alpha - \theta_\delta = 41.00° - 24.20° = \underline{16.8°}$

Note that the separation in angle is larger in higher orders.

38-17

The maxima occur at angles θ given by Eq. (38-16):

$2d\sin\theta = m\lambda$, where d is the spacing between adjacent atomic planes.

Second-order $\Rightarrow m=2$

$$d = \frac{m\lambda}{2\sin\theta} = \frac{2(0.0820\times10^{-9}\text{ m})}{2\sin 21.4°} = 2.25\times10^{-10}\text{ m} = \underline{0.225\text{ nm}}$$

38-21

Resolved by Rayleigh's criterion $\Rightarrow$ angular separation θ of the objects equals $1.22\frac{\lambda}{D}$.

The angular separation θ of the objects is $\theta = \frac{20.0\times10^3\text{ m}}{1.49\times10^{11}\text{ m}}$, where 1.49×10^{11} m is the distance from the earth to the sun. Thus $\theta = 1.34\times10^{-7}$.

$$\theta = 1.22\frac{\lambda}{D} \Rightarrow D = \frac{1.22\lambda}{\theta} = \frac{1.22(500\times10^{-9}\text{ m})}{1.34\times10^{-7}} = \underline{4.55\text{ m}}$$

Problems

38-29

a) $I = I_0\left[\frac{\sin(\beta/2)}{\beta/2}\right]^2$, where $\beta = \frac{2\pi}{\lambda}a\sin\theta$

$I = \frac{1}{2}I_0 \Rightarrow \frac{\sin(\beta/2)}{\beta/2} = \frac{1}{\sqrt{2}}$

Let $x = \frac{\beta}{2}$; the equation for x is $\frac{\sin x}{x} = \frac{1}{\sqrt{2}} = 0.7071$

Use trial and error to find the value of x that is a solution to this equation.

x	$\frac{\sin x}{x}$	x	$\frac{\sin x}{x}$
1.0 rad	0.841	1.4 rad	0.7039
1.5 rad	0.665	1.39 rad	0.7077
1.2 rad	0.777		

$\Rightarrow x = 1.39$ rad; $\beta = 2x = 2.78$ rad

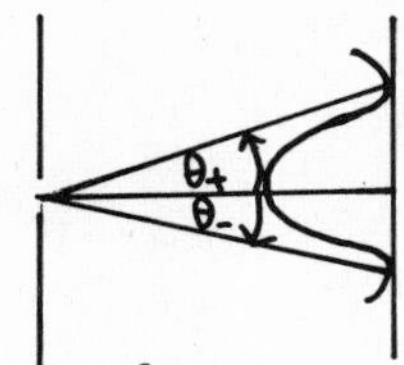

$\Delta\theta = |\theta_+ - \theta_-| = 2\theta_+$

$\sin\theta_+ = \frac{\lambda\beta}{2\pi a} = \frac{\lambda}{a}\left(\frac{2.78\text{ rad}}{2\pi\text{ rad}}\right) = 0.4425\frac{\lambda}{a}$

(i) $\frac{a}{\lambda} = 2 \Rightarrow \sin\theta_+ = 0.4425\left(\frac{1}{2}\right) = 0.2212 \Rightarrow \theta_+ = 12.78°$; $\Delta\theta = 2\theta_+ = \underline{25.6°}$

(ii) $\frac{a}{\lambda} = 5 \Rightarrow \sin\theta_+ = 0.4425\left(\frac{1}{5}\right) = 0.0885 \Rightarrow \theta_+ = 5.077°$; $\Delta\theta = 2\theta_+ = \underline{10.2°}$

(iii) $\frac{a}{\lambda} = 10 \Rightarrow \sin\theta_+ = 0.4425\left(\frac{1}{10}\right) = 0.04425 \Rightarrow \theta_+ = 2.536°$; $\Delta\theta = \underline{5.07°}$

b) $\sin\theta_0 = \frac{\lambda}{a}$ locates the first minimum

(i) $\frac{a}{\lambda} = 2 \Rightarrow \sin\theta_0 = \frac{1}{2} \Rightarrow \theta_0 = 30.0°$; $2\theta_0 = \underline{60.0°}$

(ii) $\frac{a}{\lambda} = 5 \Rightarrow \sin\theta_0 = \frac{1}{5} \Rightarrow \theta_0 = 11.54°$; $2\theta_0 = \underline{23.1°}$

(iii) $\frac{a}{\lambda} = 10 \Rightarrow \sin\theta_0 = \frac{1}{10} \Rightarrow \theta_0 = 5.74°$; $2\theta_0 = \underline{11.5°}$

Note: either definition of the width shows that the central maximum gets narrower as the slit gets wider.

38-33

The condition for an intensity maximum is

$$d\sin\theta = m\lambda,\quad m = 0, \pm1, \pm2, \ldots$$

fourth order $\Rightarrow m = 4$

6000 lines/cm $\Rightarrow 6.00\times10^5$ lines/m, so $d = \frac{1}{6.00\times10^5}\text{ m} = 1.667\times10^{-6}\text{ m}$

The longest observable wavelength is the one that gives $\theta = 90°$ and hence $\sin\theta = 1$.

$$\lambda = \frac{d\sin\theta}{m} = \frac{(1.667\times10^{-6}\text{ m})(1)}{4} = 4.17\times10^{-7}\text{ m} = \underline{417\text{ nm}}$$

38-37

Rayleigh's criterion says that the two objects are resolved if the center of one diffraction pattern coincides with the first minimum of the other.

By Eq. (38-2) the angular position of the first minimum relative to the center of the central maximum is $\sin\theta = \frac{\lambda}{a}$, where a is the slit width. Hence if the objects are resolved according to Rayleigh's criterion, the angular separation between the centers of the images of the two objects must be at least $\frac{\lambda}{a}$.

But as discussed in Example 38-7, the angular separation of the image points equals the angular separation of the object points. The angular separation of the object points is $\frac{y}{s}$, where $y = 1.00\text{ m}$ is the linear separation of the two points and s is their distance from the observer.

Thus $\frac{y}{s} = \frac{\lambda}{a} \Rightarrow s = \frac{ya}{\lambda} = \frac{(1.00\text{m})(0.400\times10^{-3}\text{m})}{500\times10^{-9}\text{m}} = \underline{800\text{ m}}$.

38-39

Resolved by Rayleigh's criterion $\Rightarrow$ angular separation θ of the objects is given by

$$\theta = 1.22\frac{\lambda}{D}$$

$\theta = \frac{y}{s}$ where $y = 45.0\text{ m}$ is the distance between the two objects and s is their distance from the astronaut (her altitude)

$$\frac{y}{s} = 1.22\frac{\lambda}{D}$$

$$s = \frac{yD}{1.22\lambda} = \frac{(45.0\text{m})(4.00\times10^{-3}\text{m})}{1.22(550\times10^{-9}\text{m})} = 2.68\times10^{5}\text{m} = \underline{268\text{ km}}$$

CHAPTER 39

Exercises 1, 3, 7, 9, 11, 15, 17, 21, 25, 31, 33

Problems 37, 39, 43, 45, 47, 49

Exercises

39-1

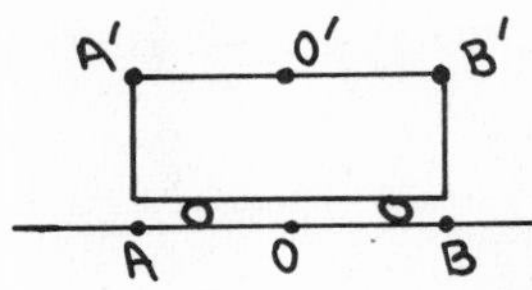

Simultaneous to observer on train ⇒ light pulses from A' and B' arrive at O' at the same time. To observer at O light from A' has a shorter distance to travel than light from B', so O will conclude that the pulse from A(A') started before the pulse at B(B'). To observer at O bolt A appears to strike first.

39-3

a) $\Delta t_0 = 2.2\times10^{-6}\ s$; $\Delta t = 1.9\times10^{-5}\ s$

$\Delta t = \dfrac{\Delta t_0}{\sqrt{1-u^2/c^2}} \Rightarrow 1-\dfrac{u^2}{c^2} = \left(\dfrac{\Delta t_0}{\Delta t}\right)^2$

$\dfrac{u}{c} = \sqrt{1-\left(\dfrac{\Delta t_0}{\Delta t}\right)^2} = \sqrt{1-\left(\dfrac{2.2\times10^{-6}\ s}{1.9\times10^{-5}\ s}\right)^2} = 0.993$; $u = \underline{0.993\ c}$

b) The speed in the laboratory frame is $u = 0.993\ c$; the time measured in this frame is Δt, so the distance as measured in this frame is
$d = u\Delta t = (0.993)(2.998\times10^{8}\ m/s)(1.9\times10^{-5}\ s) = 5.7\times10^{3}\ m = \underline{5.7\ km}$

39-7

$\Delta t = 5.00\ h = 5.00\ h\left(\dfrac{3600\ s}{1\ h}\right) = 1.8\times10^{4}\ s$

The elapsed time for the clock on the plane is Δt_0.

$\Delta t = \dfrac{\Delta t_0}{\sqrt{1-u^2/c^2}} \Rightarrow \Delta t_0 = \Delta t\sqrt{1-u^2/c^2}$

$\dfrac{u}{c}$ small $\Rightarrow \sqrt{1-\dfrac{u^2}{c^2}} = \left(1-\dfrac{u^2}{c^2}\right)^{1/2} \simeq 1-\dfrac{1}{2}\dfrac{u^2}{c^2} \Rightarrow \Delta t_0 = \Delta t\left(1-\dfrac{1}{2}\dfrac{u^2}{c^2}\right)$

The difference in the clock readings is $\Delta t - \Delta t_0 = \dfrac{1}{2}\dfrac{u^2}{c^2}\Delta t = \dfrac{1}{2}\left(\dfrac{400\ m/s}{2.998\times10^{8}\ m/s}\right)^2(1.8\times10^{4}\ s)$
$= 1.6\times10^{-8}\ s.$

The clock on the plane shows the shorter elapsed time.

39-9

$l = l_0\sqrt{1-\dfrac{u^2}{c^2}}$. The length measured when the spacecraft is moving is $l = 160\ m$; l_0 is the length measured in a frame at rest relative to the spacecraft.

$l_0 = \dfrac{l}{\sqrt{1-u^2/c^2}} = \dfrac{160\ m}{\sqrt{1-\left(\dfrac{0.800\ c}{c}\right)^2}} = \underline{267\ m}$

39-11

a) The distance measured in the earth's frame is the proper length $l_0 = 20.0\times10^3\,\text{m}$.

$$l = l_0\sqrt{1-\frac{u^2}{c^2}} = (20.0\times10^3\,\text{m})\sqrt{1-\left(\frac{0.9954c}{c}\right)^2} = 1.92\times10^3\,\text{m} = \underline{1.92\text{ km}}$$

b) Use the lifetime measured in the muon's frame as the time of travel to calculate the distance traveled as measured in that frame.

$$d = u\Delta t = (0.9954)(2.998\times10^8\,\text{m/s})(2.2\times10^{-6}\,\text{s}) = 657\,\text{m} = \underline{0.657\text{ km}}$$

The muon's original height as measured in the muon's frame (part (a)) is 1.92 km, so the fraction is $\frac{0.657\text{ km}}{1.92\text{ km}} = 0.342 = \underline{34.2\%}$.

c) $\Delta t_0 = 2.2\times10^{-6}\,\text{s}$; $\Delta t = ?$

$$\Delta t = \frac{\Delta t_0}{\sqrt{1-u^2/c^2}} = \frac{2.2\times10^{-6}\,\text{s}}{\sqrt{1-\left(\frac{0.9954c}{c}\right)^2}} = \underline{2.3\times10^{-5}\,\text{s}}$$

Use the lifetime in the earth's frame to find the distance traveled in that frame:

$$d = u\Delta t = (0.9954)(2.998\times10^8\,\text{m/s})(2.3\times10^{-5}\,\text{s}) = 6.86\times10^3\,\text{m} = \underline{6.86\text{ km}}$$

The fraction is $\frac{6.86\text{ km}}{20.0\text{ km}} = 0.343 = 34.3\%$, the same fraction as in the muon's frame.

39-15

Use the Lorentz velocity transformation equation, Eq. (39-24): $v' = \frac{v-u}{1-uv/c^2}$.

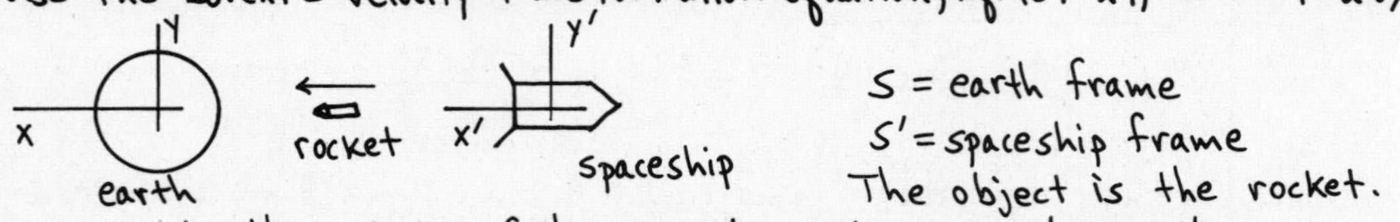

S = earth frame
S' = spaceship frame
The object is the rocket.

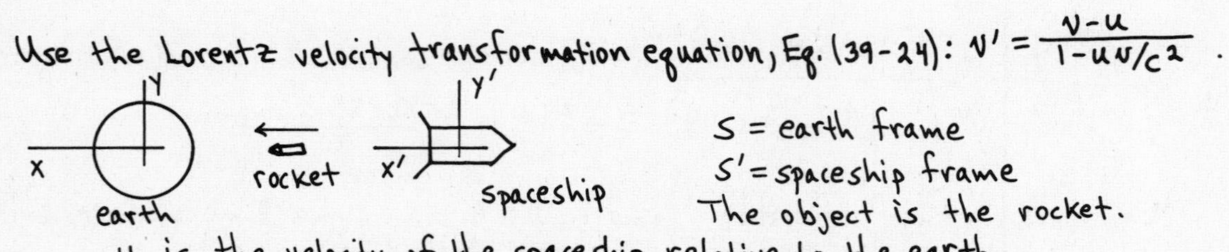

u is the velocity of the spaceship relative to the earth.

$v = +0.360c$
$v' = +0.840c$
(In each frame the rocket is moving in the positive coordinate direction.)

$$v' = \frac{v-u}{1-uv/c^2} \Rightarrow v' - u\frac{vv'}{c^2} = v - u \Rightarrow u\left(1-\frac{vv'}{c^2}\right) = v - v'$$

$$u = \frac{v-v'}{1-\frac{vv'}{c^2}} = \frac{0.360c - 0.840c}{1-\frac{(0.360c)(0.840c)}{c^2}} = -\frac{0.480c}{0.6976} = -0.688c$$

The speed of the spaceship relative to the earth is 0.688c. The minus sign in our result for u means that the spaceship is moving in the $-x$-direction, so it is moving away from the earth.

39-17

Use the Lorentz velocity transformation equation, Eq. (39-24): $v' = \frac{v-u}{1-uv/c^2}$.

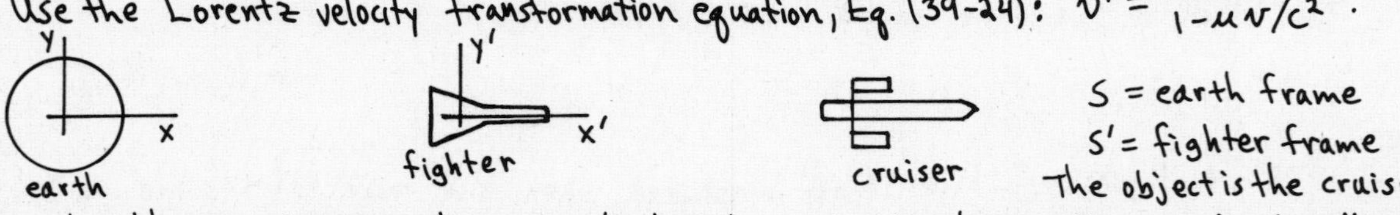

S = earth frame
S' = fighter frame
The object is the cruiser.

With the coordinates shown, each ship is moving in the positive coordinate direction in the earth frame.

u is the velocity of the fighter relative to the earth; $u = +0.900c$

39-17 (cont)

$v = +0.700c$

$v' = ?$ (velocity of the cruiser relative to the fighter)

$$v' = \frac{v-u}{1-uv/c^2} = \frac{0.700c - 0.900c}{1-(0.900c)(0.700c)/c^2} = \frac{-0.200c}{0.37} = -0.541c$$

The cruiser is moving toward the fighter with a speed of 0.541 c.

39-21

Source and observer approaching ⇒ use Eq. (39-27): $f = \sqrt{\frac{c+u}{c-u}}\, f_0$

Solve for u:

$$f^2 = \left(\frac{c+u}{c-u}\right) f_0^2$$

$$(c-u)f^2 = (c+u)f_0^2 \Rightarrow u = \frac{c(f^2 - f_0^2)}{f^2 + f_0^2} = c\left(\frac{(\frac{f}{f_0})^2 - 1}{(\frac{f}{f_0})^2 + 1}\right)$$

Convert to an equation in terms of λ, λ_0 rather than f, f_0:

$$f = \frac{c}{\lambda},\ f_0 = \frac{c}{\lambda_0} \Rightarrow \frac{f}{f_0} = \frac{\lambda_0}{\lambda} \Rightarrow u = c\left(\frac{(\lambda_0/\lambda)^2 - 1}{(\lambda_0/\lambda)^2 + 1}\right)$$

$\lambda_0 = 675\text{ nm}$, $\lambda = 525\text{ nm}$

$$u = \left[\frac{(675/525)^2 - 1}{(675/525)^2 + 1}\right]c = 0.246c = (0.246)(2.998 \times 10^8 \text{ m/s}) = 7.38 \times 10^7 \text{ m/s}; \text{ definitely speeding}$$

39-25

$$\text{Eq. (39-31): } F = \frac{dp}{dt} = \frac{d}{dt}\left(\frac{mv}{\sqrt{1 - v^2/c^2}}\right) = \frac{m}{(1-v^2/c^2)^{1/2}}\frac{dv}{dt} + \frac{mv}{(1-v^2/c^2)^{3/2}}\left(-\frac{1}{2}\right)\left(-\frac{2v}{c^2}\right)\frac{dv}{dt}$$

$$F = \frac{dv}{dt}\frac{m}{(1-v^2/c^2)^{3/2}}\left[1 - \frac{v^2}{c^2} + \frac{v^2}{c^2}\right] = \frac{dv}{dt}\frac{m}{(1-v^2/c^2)^{3/2}}$$

But $\frac{dv}{dt} = a$, so $F = \frac{ma}{(1-v^2/c^2)^{3/2}}$, which is Eq. (39-32).

39-31

a) $K = q\Delta V = (1.602 \times 10^{-19}\text{ C})(1.40 \times 10^5\text{ V}) = 2.243 \times 10^{-14}\text{ J}\left(\frac{1\text{ eV}}{1.602 \times 10^{-19}\text{ J}}\right) = 1.40 \times 10^5\text{ eV} = \underline{0.140\text{ MeV}}$

b) $E = K + mc^2$ (Eq. 39-40)

For an electron $mc^2 = (9.109 \times 10^{-31}\text{ kg})(2.998 \times 10^8\text{ m/s})^2 = 8.187 \times 10^{-14}\text{ J}\left(\frac{1\text{ eV}}{1.602 \times 10^{-19}\text{ J}}\right)$

$= 5.110 \times 10^5\text{ eV} = 0.511\text{ MeV}$

Then $E = K + mc^2 = 0.140\text{ MeV} + 0.511\text{ MeV} = \underline{0.651\text{ MeV}}$

c) $E = \frac{mc^2}{\sqrt{1-v^2/c^2}} \Rightarrow 1 - \frac{v^2}{c^2} = \left(\frac{mc^2}{E}\right)^2$

$v = c\sqrt{1 - \left(\frac{mc^2}{E}\right)^2} = (2.998 \times 10^8\text{ m/s})\sqrt{1 - \left(\frac{0.511\text{ MeV}}{0.651\text{ MeV}}\right)^2} = \underline{1.86 \times 10^8\text{ m/s}}$

d) The classical relation between speed and kinetic energy is

$K = \frac{1}{2}mv^2 \Rightarrow v = \sqrt{\frac{2K}{m}} = \sqrt{\frac{2(2.243 \times 10^{-14}\text{ J})}{9.109 \times 10^{-31}\text{ kg}}} = \underline{2.22 \times 10^8\text{ m/s}}$

(The classical result is too large by about 20%.)

39-33

a) The total energy is given in terms of the momentum by Eq. (39-41)

$E = \sqrt{(mc^2)^2 + (pc)^2} = \sqrt{[(3.32\times10^{-27}\,kg)(2.998\times10^8\,m/s)^2]^2 + [(9.65\times10^{-19}\,kg\cdot m/s)(2.998\times10^8\,m/s)]^2}$

$E = \underline{4.16\times10^{-10}\,J}$

b) $E = K + mc^2 \Rightarrow K = E - mc^2$

$mc^2 = 3.32\times10^{-27}\,kg\,(2.998\times10^8\,m/s)^2 = 2.98\times10^{-10}\,J$

$K = E - mc^2 = 4.16\times10^{-10}\,J - 2.98\times10^{-10}\,J = \underline{1.18\times10^{-10}\,J}$

c) $\dfrac{K}{mc^2} = \dfrac{1.18\times10^{-10}\,J}{2.98\times10^{-10}\,J} = \underline{0.396}$

Problems

39-37

$\Delta t_0 = 2.6\times10^{-8}\,s$

The time measured in the lab must satisfy $d = c\,\Delta t \Rightarrow \Delta t = \dfrac{d}{c} = \dfrac{3.00\times10^3\,m}{2.998\times10^8\,m/s} = 1.00\times10^{-5}\,s$

$\Delta t = \dfrac{\Delta t_0}{\sqrt{1-u^2/c^2}} \Rightarrow \left(1-\dfrac{u^2}{c^2}\right)^{1/2} = \dfrac{\Delta t_0}{\Delta t} \Rightarrow \left(1-\dfrac{u^2}{c^2}\right) = \left(\dfrac{\Delta t_0}{\Delta t}\right)^2$

Write $u = (1-\Delta)c \Rightarrow \left(\dfrac{u}{c}\right)^2 = (1-\Delta)^2 = 1 - 2\Delta + \Delta^2 \simeq 1 - 2\Delta$ since Δ is small.

Use this in the above $\Rightarrow 1 - (1-2\Delta) = \left(\dfrac{\Delta t_0}{\Delta t}\right)^2$

$\Delta = \dfrac{1}{2}\left(\dfrac{\Delta t_0}{\Delta t}\right)^2 = \dfrac{1}{2}\left(\dfrac{2.6\times10^{-8}\,s}{1.00\times10^{-5}\,s}\right)^2 = \underline{3.38\times10^{-6}}$

An alternative calculation is to say that the length of the tube must contract relative to the moving meson so that the meson travels that length before decaying. The contracted length must be $l = c\,\Delta t_0 = (2.998\times10^8\,m/s)(2.6\times10^{-8}\,s) = 7.8\,m$.

$l = l_0\sqrt{1-\dfrac{u^2}{c^2}} \Rightarrow 1 - \dfrac{u^2}{c^2} = \left(\dfrac{l}{l_0}\right)^2$

Then $u = (1-\Delta)c \Rightarrow \Delta = \dfrac{1}{2}\left(\dfrac{l}{l_0}\right)^2 = \dfrac{1}{2}\left(\dfrac{7.8\,m}{3.00\times10^3\,m}\right)^2 = 3.38\times10^{-6}$, which checks.

b) $E = \gamma mc^2$

$\gamma = \dfrac{1}{\sqrt{1-u^2/c^2}} = \dfrac{1}{\sqrt{2\Delta}} = 385$

$E = 385\,(139.6\,MeV) = 5.37\times10^4\,MeV = \underline{53.7\,GeV}$

39-39

$E = mc^2$; the mass increase is due to the heat flow into the ice to melt it.

$E = Q = mL_f = (1.00\,kg)(334\times10^3\,J/kg) = 3.34\times10^5\,J$

$m = \dfrac{E}{c^2} = \dfrac{3.34\times10^5\,J}{(2.998\times10^8\,m/s)^2} = \underline{3.72\times10^{-12}\,kg}$

39-43

In crown glass the speed of light is $v = \frac{c}{n} = \frac{2.998\times10^{8}\,\text{m/s}}{1.52} = 1.972\times10^{8}\,\text{m/s}$.
Calculate the kinetic energy of an electron that has this speed:

$K = mc^2(\gamma - 1)$

$mc^2 = (9.109\times10^{-31}\,\text{kg})(2.998\times10^{8}\,\text{m/s})^2 = 8.187\times10^{-14}\,\text{J}\left(\frac{1\,\text{eV}}{1.602\times10^{-19}\,\text{J}}\right) = 0.5111\,\text{MeV}$

$\gamma = \frac{1}{\sqrt{1 - v^2/c^2}} = \frac{1}{\sqrt{1 - \left(\frac{1.972\times10^{8}\,\text{m/s}}{2.998\times10^{8}\,\text{m/s}}\right)^2}} = 1.328$

$K = mc^2(\gamma - 1) = 0.5111\,\text{MeV}\,(1.328 - 1) = \underline{0.168\,\text{MeV}}$

39-45

An increase in wavelength corresponds to a decrease in frequency $\left(f = \frac{c}{\lambda}\right)$, so the atoms are moving away from the earth.

Receding $\Rightarrow$ use Eq. (39-28): $f = \sqrt{\frac{c-u}{c+u}}\, f_0$

Solve for u: $\left(\frac{f}{f_0}\right)^2 (c+u) = c - u \Rightarrow u = c\left(\frac{1 - (f/f_0)^2}{1 + (f/f_0)^2}\right)$

$f = \frac{c}{\lambda},\ f_0 = \frac{c}{\lambda_0} \Rightarrow \frac{f}{f_0} = \frac{\lambda_0}{\lambda}$

$u = c\left(\frac{1 - (\lambda_0/\lambda)^2}{1 + (\lambda_0/\lambda)^2}\right) = c\left(\frac{1 - (121.6/396.5)^2}{1 + (121.6/396.5)^2}\right) = 0.828c = \underline{2.48\times10^{8}\,\text{m/s}}$

39-47

With a force in the y-direction, Eq. (39-31) gives

$F_y = \frac{dp_y}{dt} = \frac{d}{dt}\frac{mv_y}{\sqrt{1 - v^2/c^2}}$

$v^2 = v_x^2 + v_y^2$; v_x is constant since there is no force in that direction

$\frac{d}{dt}(v^2) = \frac{d}{dt}(v_x^2 + v_y^2) = 2v_y\frac{dv_y}{dt}$

$F_y = \frac{m}{(1 - v^2/c^2)^{1/2}}\frac{dv_y}{dt} + \frac{mv_y}{(1 - v^2/c^2)^{3/2}}\left(-\frac{1}{2}\right)\left(-\frac{1}{c^2}\frac{d}{dt}(v^2)\right)$

$F_y = \frac{m}{(1 - v^2/c^2)^{1/2}}\frac{dv_y}{dt} + \frac{mv_y^2}{(1 - v^2/c^2)^{3/2}}\frac{1}{c^2}\frac{dv_y}{dt}$

Initially (at $t = 0$) the particle is moving in the x-direction, so $v_y = 0$.

Thus at $t = 0$, $F_y = \frac{m}{\sqrt{1 - v^2/c^2}}\frac{dv_y}{dt}$.

But $\frac{dv_y}{dt} = a_y \Rightarrow a_y = \frac{F_y}{m}\sqrt{1 - v^2/c^2}$, as was to be shown.

39-49

According to Eq. (39-32), $a = \frac{dv}{dt} = \frac{F}{m}(1 - v^2/c^2)^{3/2}$.
(One-dimensional motion is assumed, and all the F, v, and a refer to x-components.)

$\Rightarrow \frac{dv}{(1 - v^2/c^2)^{3/2}} = \left(\frac{F}{m}\right)dt$

39-49 (cont)

Integrate from $t=0$ when $v=0$ to time t when the velocity is v.

$$\int_0^v \frac{dv}{(1-v^2/c^2)^{3/2}} = \int_0^t \frac{F}{m}\,dt$$

Since F is constant, $\int_0^t \frac{F}{m}\,dt = \frac{Ft}{m}$

In the velocity integral make the change of variable $y = \frac{v}{c}$; $dy = \frac{dv}{c}$

$$c\int_0^{v/c} \frac{dy}{(1-y^2)^{3/2}} = c\left(\frac{y}{(1-y^2)^{1/2}}\Big|_0^{v/c}\right) = c\,\frac{\frac{v}{c}}{\sqrt{1-(v/c)^2}} = \frac{v}{\sqrt{1-v^2/c^2}}$$

Thus $\frac{v}{\sqrt{1-v^2/c^2}} = \frac{Ft}{m}$.

Solve this equation for v:

$$\frac{v^2}{1-v^2/c^2} = \left(\frac{Ft}{m}\right)^2$$

$$v^2 = \left(\frac{Ft}{m}\right)^2\left(1-\frac{v^2}{c^2}\right)$$

$$v^2\left(1+\left(\frac{Ft}{mc}\right)^2\right) = \left(\frac{Ft}{m}\right)^2$$

$$v = \frac{Ft/m}{\sqrt{1+\left(\frac{Ft}{mc}\right)^2}} = c\,\frac{Ft}{\sqrt{m^2c^2+F^2t^2}}$$

As $t \to \infty$, $\frac{Ft}{\sqrt{m^2c^2+F^2t^2}} \to \frac{Ft}{\sqrt{F^2t^2}} \to 1$, so $v \to c$

Note also that $\frac{Ft}{\sqrt{m^2c^2+F^2t^2}}$ is always less than 1, so $v < c$ always and approaches c only when $t \to \infty$.

CHAPTER 40

Exercises 1, 5, 7, 9, 11, 13, 15, 19, 21, 23, 25, 29, 35

Problems 39, 41, 45, 47, 51, 53, 55, 59

Exercises

40-1

a) $E = hf \Rightarrow f = \frac{E}{h} = \frac{(2.50\times10^{6}\,\text{eV})\left(\frac{1.602\times10^{-19}\,\text{J}}{1\,\text{eV}}\right)}{6.626\times10^{-34}\,\text{J}\cdot\text{s}} = \underline{6.04\times10^{20}\,\text{Hz}}$

b) $c = f\lambda \Rightarrow \lambda = \frac{c}{f} = \frac{2.998\times10^{8}\,\text{m/s}}{6.04\times10^{20}\,\text{Hz}} = \underline{4.96\times10^{-13}\,\text{m}}$

c) λ is about a factor of 100 times larger than a nuclear radius

40-5

a) Eq. (39-42): $E = pc = (2.40\times10^{-27}\,\text{kg}\cdot\text{m/s})(2.998\times10^{8}\,\text{m/s}) = 7.19\times10^{-19}\,\text{J}$

$E = (7.19\times10^{-19}\,\text{J})\left(\frac{1\,\text{eV}}{1.602\times10^{-19}\,\text{J}}\right) = \underline{4.49\,\text{eV}}$

b) Eq. (40-6): $p = \frac{h}{\lambda} \Rightarrow \lambda = \frac{h}{p} = \frac{6.626\times10^{-34}\,\text{J}\cdot\text{s}}{2.40\times10^{-27}\,\text{kg}\cdot\text{m/s}} = 2.76\times10^{-9}\,\text{m} = \underline{276\,\text{nm}}$

This wavelength is shorter than visible wavelengths; it is in the ultraviolet region of the electromagnetic spectrum.

40-7

Eq. (40-3): $\frac{1}{2}mv_{max}^2 = hf - \phi = (6.626\times10^{-34}\,\text{J}\cdot\text{s})(3.40\times10^{15}\,\text{Hz}) - (4.00\,\text{eV})\left(\frac{1.602\times10^{-19}\,\text{J}}{1\,\text{eV}}\right)$

$\frac{1}{2}mv_{max}^2 = 2.253\times10^{-18}\,\text{J} - 6.408\times10^{-19}\,\text{J} = 1.612\times10^{-18}\,\text{J}$

$v_{max} = \sqrt{\frac{2(1.612\times10^{-18}\,\text{J})}{m}} = \sqrt{\frac{2(1.612\times10^{-18}\,\text{J})}{9.109\times10^{-31}\,\text{kg}}} = \underline{1.88\times10^{6}\,\text{m/s}}$

40-9

First find the work function ϕ:

$eV_0 = hf - \phi \Rightarrow \phi = hf - eV_0 = \frac{hc}{\lambda} - eV_0$

$\phi = \frac{(6.626\times10^{-34}\,\text{J}\cdot\text{s})(2.998\times10^{8}\,\text{m/s})}{254\times10^{-9}\,\text{m}} - (1.602\times10^{-19}\,\text{C})(0.181\,\text{V})$

$\phi = 7.821\times10^{-19}\,\text{J} - 2.900\times10^{-20}\,\text{J} = 7.531\times10^{-19}\,\text{J}\left(\frac{1\,\text{eV}}{1.602\times10^{-19}\,\text{J}}\right) = 4.70\,\text{eV}$

The threshold frequency f_{th} is the smallest frequency that still produces photoelectrons. It corresponds to $K_{max} = 0$ in Eq. (40-3), so $hf_{th} = \phi$.

$f = \frac{c}{\lambda} \Rightarrow \frac{hc}{\lambda_{th}} = \phi$

$\lambda_{th} = \frac{hc}{\phi} = \frac{(6.626\times10^{-34}\,\text{J}\cdot\text{s})(2.998\times10^{8}\,\text{m/s})}{7.531\times10^{-19}\,\text{J}} = 2.64\times10^{-7}\,\text{m} = \underline{264\,\text{nm}}$

40-11

a) Eq. (40-8): $\frac{1}{\lambda} = R\left(\frac{1}{2^2} - \frac{1}{n^2}\right)$

$n = 4 \Rightarrow \frac{1}{\lambda} = R\left(\frac{1}{2^2} - \frac{1}{4^2}\right) = R\left(\frac{4-1}{16}\right) = R\,\frac{3}{16}$

$\lambda = \frac{16}{3R} = \frac{16}{3(1.097\times10^7)}\text{ m} = 4.86\times10^{-7}\text{ m} = 486\text{ nm}$

$f = \frac{c}{\lambda} = \frac{2.998\times10^8\text{ m/s}}{4.86\times10^{-7}\text{ m}} = \underline{6.17\times10^{14}\text{ Hz}}$

b) $\lambda = \underline{486\text{ nm}}$, as calculated in part (a).

40-13

$hf = \Delta E\,;\ f = \frac{c}{\lambda} \Rightarrow \frac{hc}{\lambda} = \Delta E \Rightarrow \lambda = \frac{hc}{\Delta E}$

$\Delta E = 3.60\text{ eV}\left(\frac{1.602\times10^{-19}\text{ J}}{1\text{ eV}}\right) = 5.767\times10^{-19}\text{ J}$

$\lambda = \frac{hc}{\Delta E} = \frac{(6.626\times10^{-34}\text{ J}\cdot\text{s})(2.998\times10^8\text{ m/s})}{5.767\times10^{-19}\text{ J}} = 3.44\times10^{-7}\text{ m} = \underline{344\text{ nm}}$

40-15

a) If the particles are treated as point charges, $U = \frac{1}{4\pi\epsilon_0}\frac{q_1 q_2}{r}$.

$q_1 = 2e$ (alpha particle)

$q_2 = 79e$ (gold nucleus)

$U = (8.987\times10^9\text{ N}\cdot\text{m}^2/\text{C}^2)\frac{(2)(79)(1.602\times10^{-19}\text{ C})^2}{8.00\times10^{-14}\text{ m}} = \underline{4.56\times10^{-13}\text{ J}}$

$U = 4.56\times10^{-13}\text{ J}\left(\frac{1\text{ eV}}{1.602\times10^{-19}\text{ J}}\right) = 2.84\times10^6\text{ eV} = \underline{2.84\text{ MeV}}$

b) $K_1 + U_1 = K_2 + U_2$

Alpha particle is initially far from the gold nucleus $\Rightarrow r_1 \simeq \infty$ and $U_1 = 0$.

Alpha particle stops $\Rightarrow K_2 = 0$.

Thus $K_1 = U_2 = 4.56\times10^{-13}\text{ J} = 2.84\text{ MeV}$

c) $K = \frac{1}{2}mv^2 \Rightarrow v = \sqrt{\frac{2K}{m}} = \sqrt{\frac{2(4.56\times10^{-13}\text{ J})}{6.65\times10^{-27}\text{ kg}}} = \underline{1.17\times10^7\text{ m/s}}$

40-19

The force between the electron and the nucleus in He^+ is $F = \frac{1}{4\pi\epsilon_0}\frac{Ze^2}{r^2}$, where $Z = 2$ is the nuclear charge.

All the equations for the hydrogen atom apply to He^+ if we replace e^2 by Ze^2.

a) $E_n = -\frac{1}{\epsilon_0^2}\frac{me^4}{8n^2h^2}$ (hydrogen) becomes

$E_n = -\frac{1}{\epsilon_0^2}\frac{m(Ze^2)^2}{8n^2h^2} = Z^2\left(-\frac{1}{\epsilon_0^2}\frac{me^4}{8n^2h^2}\right)$ (He^+)

The energy levels are all larger in magnitude by a factor of $Z^2 = 2^2 = 4$.

b) $\frac{1}{\lambda} = R\left(\frac{1}{n_1^2} - \frac{1}{n_2^2}\right)$ just as for hydrogen but now R has a different value.

40-19 (cont) $R = \frac{me^4}{8\epsilon_0^2 h^3 c} = 1.097\times10^7\ m^{-1}$ for hydrogen becomes

$R = Z^2\left(\frac{me^4}{8\epsilon_0^2 h^3 c}\right) = 4(1.097\times10^7\ m^{-1}) = 4.388\times10^7\ m^{-1}$ for He^+

Find the longest and shortest wavelengths for each series. Visible wavelengths are between 400 nm and 700 nm.

Lyman $n_1 = 1$

Longest wavelength is for $n_2 = 2$: $\frac{1}{\lambda} = R(1-\frac{1}{4}) = \frac{3R}{4} \Rightarrow \lambda = \frac{4}{3R} = \frac{4}{3(4.388\times10^7 m^{-1})}$

$\lambda = 30.4$ nm; all the Lyman series wavelengths are shorter than visible wavelengths.

Balmer $n_1 = 2$

Longest wavelength is for $n_2 = 3$: $\frac{1}{\lambda} = R(\frac{1}{4}-\frac{1}{9}) = \frac{5R}{36} \Rightarrow \lambda = \frac{36}{5R} = 164$ nm; all the Balmer series wavelengths are shorter than visible wavelengths.

Paschen $n_1 = 3$

Longest wavelength is for $n_2 = 4$: $\frac{1}{\lambda} = R(\frac{1}{9}-\frac{1}{16}) = \frac{7R}{144} \Rightarrow \lambda = \frac{144}{7R} = 469$ nm.

The Paschen series has lines in the visible.

Brackett $n_1 = 4$

Shortest wavelength is for $n_2 \to \infty$: $\frac{1}{\lambda} = R(\frac{1}{16}) \Rightarrow \lambda = \frac{16}{R} = 365$ nm. The Brackett series has lines in the visible.

Pfund $n_1 = 5$

Shortest wavelength is for $n_2 \to \infty$: $\frac{1}{\lambda} = R(\frac{1}{25}) \Rightarrow \lambda = \frac{25}{R} = 570$ nm. The Pfund series has lines in the visible.

$n_1 = 6$ series

Shortest wavelength is for $n_2 \to \infty$: $\frac{1}{\lambda} = R(\frac{1}{36}) \Rightarrow \lambda = \frac{36}{R} = 820$ nm. All wavelengths of this series are longer than the visible region; the same is also true of all higher series.

c) Eq. (40-13): $r_n = \epsilon_0 \frac{n^2 h^2}{\pi m e^2}$ (for hydrogen)

$e^2 \to Ze^2 \Rightarrow r_n = \frac{1}{Z}\left(\epsilon_0 \frac{n^2 h^2}{\pi m e^2}\right)$ for He^+; for a given n the radius of an orbit in He^+ is smaller by a factor of $Z = 2$.

40-21

Power is energy per unit time, so in 1.00 s the energy emitted by the laser is $(1.50\times10^{-3}\ W)(1.00\ s) = 1.50\times10^{-3}$ J.

The energy of each photon is $E = \frac{hc}{\lambda} = \frac{(6.626\times10^{-34} J\cdot s)(2.998\times10^8 m/s)}{633\times10^{-9} m} = 3.138\times10^{-19}$ J

The number of photons emitted each second is the total energy emitted divided by the energy of one photon $\frac{1.50\times10^{-3} J/s}{3.138\times10^{-19} J/photon} = \underline{4.78\times10^{15}\ photons/s}$

40-23 $\frac{n_{5s}}{n_{3p}} = e^{-(E_{5s}-E_{3p})/kT}$

From Fig. 20-21, $E_{5s} = 20.66\,\text{eV}$ and $E_{3p} = 18.70\,\text{eV}$

$E_{5s} - E_{3p} = 20.66\,\text{eV} - 18.70\,\text{eV} = 1.96\,\text{eV}\left(\frac{1.602\times10^{-19}\,\text{J}}{1\,\text{eV}}\right) = 3.140\times10^{-19}\,\text{J}$

$\frac{n_{5s}}{n_{3p}} = e^{-(3.140\times10^{-19}\,\text{J})/[(1.381\times10^{-23}\,\text{J/K})(300\,\text{K})]} = e^{-75.79} = \underline{1.2\times10^{-33}}$

40-25

The kinetic energy K of an electron after acceleration is

$K = q\,\Delta V = (1.602\times10^{-19}\,\text{C})(20.0\times10^{3}\,\text{V}) = 3.204\times10^{-15}\,\text{J}$

The shortest wavelength x rays are those for which the kinetic energy of an electron is converted entirely to the photon energy $hf = \frac{hc}{\lambda}$.

$\frac{hc}{\lambda} = K \Rightarrow \lambda = \frac{hc}{K} = \frac{(6.626\times10^{-34}\,\text{J}\cdot\text{s})(2.998\times10^{8}\,\text{m/s})}{3.204\times10^{-15}\,\text{J}} = 6.20\times10^{-11}\,\text{m} = \underline{0.0620\,\text{nm}}$

40-29

Eq. (40-21): $\lambda' - \lambda = \frac{h}{mc}(1-\cos\phi) = \lambda_c(1-\cos\phi)$

$\lambda' = \lambda + \lambda_c(1-\cos\phi)$

The largest λ' corresponds to $\phi = 180°$, so $\cos\phi = -1$.

Then $\lambda' = \lambda + 2\lambda_c = 0.0600\times10^{-9}\,\text{m} + 2(2.426\times10^{-12}\,\text{m}) = 6.485\times10^{-11}\,\text{m} = \underline{0.0648\,\text{nm}}$

40-35

The wavelength λ_m where the Planck distribution peaks is given by Eq. (40-29):

$\lambda_m = \frac{2.90\times10^{-3}\,\text{m}\cdot\text{K}}{T} = \frac{2.90\times10^{-3}\,\text{m}\cdot\text{K}}{2.7\,\text{K}} = 1.07\times10^{-3}\,\text{m} = \underline{1.07\,\text{mm}}$

(This wavelength is in the microwave portion of the electromagnetic spectrum.)

Problems

40-39

a) One photon dissociates one AgBr molecule, so we need to find the energy required to dissociate a single molecule. The problem states that it requires $1.00\times10^{5}\,\text{J}$ to dissociate one mole of AgBr, and one mole contains Avogadro's number (6.02×10^{23}) of molecules, so the energy required to dissociate one AgBr is

$\frac{1.00\times10^{5}\,\text{J/mol}}{6.02\times10^{23}\,\text{molecules/mol}} = 1.66\times10^{-19}\,\text{J/molecule}$

The photon is to have this energy, so $E = 1.66\times10^{-19}\,\text{J}\left(\frac{1\,\text{eV}}{1.602\times10^{-19}\,\text{J}}\right) = \underline{1.04\,\text{eV}}$

b) $E = \frac{hc}{\lambda} \Rightarrow \lambda = \frac{hc}{E} = \frac{(6.626\times10^{-34}\,\text{J}\cdot\text{s})(3.00\times10^{8}\,\text{m/s})}{1.66\times10^{-19}\,\text{J}} = 1.20\times10^{-6}\,\text{m} = \underline{1200\,\text{nm}}$

c) $c = f\lambda \Rightarrow f = \frac{c}{\lambda} = \frac{3.00\times10^{8}\,\text{m/s}}{1.20\times10^{-6}\,\text{m}} = \underline{2.50\times10^{14}\,\text{Hz}}$

40-39 (cont)

d) $E = hf = (6.626\times10^{-34}\ \text{J}\cdot\text{s})(100\times10^{6}\ \text{Hz}) = 6.63\times10^{-26}\ \text{J}$

$E = 6.63\times10^{-26}\ \text{J}\left(\frac{1\ \text{eV}}{1.602\times10^{-19}\ \text{J}}\right) = \underline{4.14\times10^{-7}\ \text{eV}}$

e) A photon with frequency $f = 100$ MHz has too little energy, by a large factor, to dissociate a AgBr molecule. The photons in the visible light from a firefly do individually have enough energy to dissociate AgBr. The huge number of 100 MHz TV radiation photons can't compensate for the fact that individually they have too little energy.

40-41

The stopping potential V_0 is given by Eq. (40-4): $eV_0 = hf - \phi = \frac{hc}{\lambda} - \phi$.

Call the two wavelengths λ_1 and λ_2, and the corresponding stopping potentials V_{01} and V_{02}. Thus $eV_{01} = \frac{hc}{\lambda_1} - \phi$ and $eV_{02} = \frac{hc}{\lambda_2} - \phi$. Note that the work function ϕ is a property of the material and is independent of the wavelength of the light.

Subtracting one equation from the other gives

$e(V_{02} - V_{01}) = hc\left(\frac{1}{\lambda_2} - \frac{1}{\lambda_1}\right)$.

The change in stopping potential is $\Delta V_0 = V_{02} - V_{01} = \frac{hc}{e}\left(\frac{\lambda_1 - \lambda_2}{\lambda_1\lambda_2}\right)$.

$$\Delta V_0 = \frac{(6.626\times10^{-34}\ \text{J}\cdot\text{s})(2.998\times10^{8}\ \text{m/s})}{1.602\times10^{-19}\ \text{C}}\left[\frac{400\times10^{-9}\ \text{m} - 320\times10^{-9}\ \text{m}}{(400\times10^{-9}\ \text{m})(320\times10^{-9}\ \text{m})}\right] = 0.775\ \text{V}.$$

The stopping potential will increase by 0.775 V.

40-45

$hf = E_f - E_i$, the energy given to the electron in the atom when a photon is absorbed.

The energy of one photon is $\frac{hc}{\lambda} = \frac{(6.626\times10^{-34}\ \text{J}\cdot\text{s})(2.998\times10^{8}\ \text{m/s})}{40.0\times10^{-9}\ \text{m}}$

$= 4.966\times10^{-18}\ \text{J}\left(\frac{1\ \text{eV}}{1.602\times10^{-19}\ \text{J}}\right) = 31.00\ \text{eV}$.

The final energy of the electron is $E_f = E_i + hf$. In the ground state of the hydrogen atom the energy of the electron is $E_i = -13.60$ eV.

$E_f = -13.60\ \text{eV} + 31.00\ \text{eV} = \underline{17.4\ \text{eV}}$

40-47

a) Eq. (40-19): $m_r = \frac{m_1 m_2}{m_1 + m_2} = \frac{207 m_e m_p}{207 m_e + m_p} = \frac{207(9.109\times10^{-31}\ \text{kg})(1.673\times10^{-27}\ \text{kg})}{207(9.109\times10^{-31}\ \text{kg}) + 1.673\times10^{-27}\ \text{kg}}$

$m_r = 1.695\times10^{-28}\ \text{kg}$

(We have used m_e to clearly identify when we mean the electron mass.)

b) In Eq. (40-17) replace $m = m_e$ by m_r: $E_n = -\frac{1}{\epsilon_0^2}\frac{m_r e^4}{8n^2h^2}$.

Write as $E_n = \left(\frac{m_r}{m_e}\right)\left(-\frac{1}{\epsilon_0^2}\frac{m_e e^4}{8n^2h^2}\right)$, since we know that $\frac{1}{\epsilon_0^2}\frac{m_e e^4}{8h^2} = 13.60$ eV

40-47 (cont)

$$E_n = \frac{m_r}{m_e}\left(-\frac{13.60\text{ eV}}{n^2}\right)$$

$$E_1 = \frac{1.695\times10^{-28}\text{ kg}}{9.109\times10^{-31}\text{ kg}}(-13.60\text{ eV}) = 186.1(-13.60\text{ eV}) = \underline{-2.53\text{ keV}}$$

c) In Eq. (40-13) replace $m=m_e$ by m_r:

$$r_n = \epsilon_0\frac{n^2h^2}{\pi m_r e^2} = \frac{m_e}{m_r}\left(\epsilon_0\frac{n^2h^2}{\pi m_e e^2}\right) = \frac{m_e}{m_r}(n^2)(0.5292\times10^{-10}\text{ m})$$

For $n=1$, $r_1 = \frac{1}{186.1}(0.5292\times10^{-10}\text{ m}) = 2.84\times10^{-13}\text{ m} = \underline{0.284\text{ pm}}$

Note that Eq.(40-13) is for the separation between the two particles. For the hydrogen atom where the center of mass very nearly coincides with the proton this is also equal to the orbit radius. For the atom consisting of a muon and a proton both particles orbit the center of mass. (See Fig. 40-19.)

40-51

From Chapter 16 (Eq. 16-17) the average kinetic energy of a gas atom or molecule is $\frac{3}{2}kT$.

$$\frac{3}{2}kT = 1.0\text{ eV} = 1.602\times10^{-19}\text{ J} \Rightarrow T = \frac{2(1.602\times10^{-19}\text{ J})}{3k} = \frac{2(1.602\times10^{-19}\text{ J})}{3(1.381\times10^{-23}\text{ J/K})} = 7.73\times10^{3}\text{ K}$$

If the temperature of the metal is raised to a value of this order of magnitude, the random kinetic energy of the electrons is large enough for them to escape from the metal (thermionic emission).

40-53

Let E_{tr} be the transition energy, E_{ph} be the energy of the photon with wavelength λ', and E_r be the kinetic energy of the recoiling atom.

$$E_{ph} + E_r = E_{tr}$$

$$E_{ph} = \frac{hc}{\lambda'} \Rightarrow \frac{hc}{\lambda'} = E_{tr} - E_r \Rightarrow \lambda' = \frac{hc}{E_{tr}-E_r}$$

If the recoil energy is neglected then the photon wavelength is $\lambda = \frac{hc}{E_{tr}}$.

$$\Delta\lambda = \lambda' - \lambda = hc\left(\frac{1}{E_{tr}-E_r} - \frac{1}{E_{tr}}\right) = \frac{hc}{E_{tr}}\left(\frac{1}{1-E_r/E_{tr}} - 1\right)$$

$$\frac{1}{1-E_r/E_{tr}} = \left(1-\frac{E_r}{E_{tr}}\right)^{-1} \simeq 1 + \frac{E_r}{E_{th}} \text{ since } \frac{E_r}{E_{th}} \ll 1$$ (We have used the binomial theorem, Appendix B.)

$$\Rightarrow \Delta\lambda = \frac{hc}{E_{tr}}\left(\frac{E_r}{E_{tr}}\right), \text{ or since } E_{tr} = \frac{hc}{\lambda},\ \Delta\lambda = \left(\frac{E_r}{hc}\right)\lambda^2$$

Use conservation of linear momentum to find E_r:
Assuming that the atom is initially at rest, the momentum p_r of the recoiling atom must equal in magnitude and be opposite in direction to the momentum $p_{ph} = \frac{h}{\lambda}$ of the emitted photon: $\frac{h}{\lambda} = p_r$

$E_r = \frac{p_r^2}{2M}$, where M is the mass of the atom $\Rightarrow E_r = \frac{h^2}{2M\lambda^2}$.

40-53 (cont)

Use this result in the above equation:

$$\Delta\lambda = \left(\frac{E_r}{hc}\right)\lambda^2 = \left(\frac{h^2}{2M\lambda^2}\right)\left(\frac{\lambda^2}{hc}\right) = \frac{h}{2Mc}$$; note that this result for $\Delta\lambda$ is independent of the atomic transition energy.

For a hydrogen atom $M = m_p$ and

$$\Delta\lambda = \frac{h}{2m_pc} = \frac{6.626\times10^{-34}\ \text{J.s}}{2(1.673\times10^{-27}\ \text{kg})(2.998\times10^{8}\ \text{m/s})} = \underline{6.60\times10^{-16}\ \text{m}}$$

Note: For the n=5 to n=1 transition the wavelength of the emitted photon is

$$\frac{1}{\lambda} = R\left(\frac{1}{1^2} - \frac{1}{5^2}\right) = \frac{24R}{25} \Rightarrow \lambda = \frac{25}{24R} = 9.50\times10^{-8}\ \text{m, so } \frac{\Delta\lambda}{\lambda} = \frac{6.60\times10^{-16}\ \text{m}}{9.50\times10^{-8}\ \text{m}} = 6.9\times10^{-9};$$

the correction is extremely small.

40-55

a) $\Delta\lambda = \frac{h}{mc}(1-\cos\phi) = \lambda_c(1-\cos\phi)$

Largest $\Delta\lambda$ is for $\phi = 180°$: $\Delta\lambda = 2\lambda_c = 2(2.426\ \text{pm}) = \underline{4.85\ \text{pm}}$

b) $\lambda' - \lambda = \lambda_c(1-\cos\phi)$

Wavelength doubles $\Rightarrow \lambda' = 2\lambda$, so $\lambda' - \lambda = \lambda$. Thus $\lambda = \lambda_c(1-\cos\phi)$

$E = \frac{hc}{\lambda}$, so smallest-energy photon means largest wavelength photon $\Rightarrow \phi = 180°$ and $\lambda = 2\lambda_c = 4.85\ \text{pm}$.

Then $$E = \frac{hc}{\lambda} = \frac{(6.626\times10^{-34}\ \text{J.s})(2.998\times10^{8}\ \text{m/s})}{4.85\times10^{-12}\ \text{m}} = 4.096\times10^{-14}\ \text{J}\left(\frac{1\ \text{eV}}{1.602\times10^{-19}\ \text{J}}\right) = \underline{0.256\ \text{MeV}}$$

40-59

Find the wavelengths of the incident and scattered photons:

$$\lambda = \frac{hc}{E} = \frac{(6.626\times10^{-34}\ \text{J.s})(2.998\times10^{8}\ \text{m/s})}{(2.00\times10^{6}\ \text{eV})\left(\frac{1.602\times10^{-19}\ \text{J}}{1\ \text{eV}}\right)} = 6.200\times10^{-13}\ \text{m}$$

$$\lambda' = \frac{hc}{E'} = \frac{(6.626\times10^{-34}\ \text{J.s})(2.998\times10^{8}\ \text{m/s})}{(0.500\times10^{6}\ \text{eV})\left(\frac{1.602\times10^{-19}\ \text{J}}{1\ \text{eV}}\right)} = 2.48\times10^{-12}\ \text{m}$$

$$\lambda' - \lambda = \lambda_c(1-\cos\phi)$$

$$1-\cos\phi = \frac{\lambda'-\lambda}{\lambda_c} \Rightarrow \cos\phi = 1 - \frac{\lambda'-\lambda}{\lambda_c} = 1 - \frac{2.480\times10^{-12}\ \text{m} - 6.200\times10^{-13}\ \text{m}}{2.426\times10^{-12}\ \text{m}} = 0.2331$$

$$\Rightarrow \phi = \underline{76.5°}$$

CHAPTER 41

Exercises 3, 5, 7, 11, 13, 15

Problems 21, 23, 25, 29, 31, 33

Exercises

41-3

a) The visible spectrum is approximately from 400 nm to 700 nm.

$$\lambda = 700\,\text{nm} \Rightarrow E = \frac{hc}{\lambda} = \frac{(6.626\times10^{-34}\,\text{J}\cdot\text{s})(2.998\times10^{8}\,\text{m/s})}{400\times10^{-9}\,\text{m}} = 4.966\times10^{-19}\,\text{J}\left(\frac{1\,\text{eV}}{1.602\times10^{-19}\,\text{J}}\right) = 3.10\,\text{eV}$$

$$\lambda = 400\,\text{nm} \Rightarrow E = \frac{hc}{\lambda} = \frac{(6.626\times10^{-34}\,\text{J}\cdot\text{s})(2.998\times10^{8}\,\text{m/s})}{700\times10^{-9}\,\text{m}} = 2.838\times10^{-19}\,\text{J}\left(\frac{1\,\text{eV}}{1.602\times10^{-19}\,\text{J}}\right) = 1.77\,\text{eV}$$

The visible spectrum corresponds to photon energies between 1.77 eV and 3.10 eV.

b) $\lambda = \frac{h}{p}$

$E = \frac{1}{2}mv^2 = \frac{p^2}{2m} \Rightarrow p = \sqrt{2mE}$

Thus $\lambda = \frac{h}{\sqrt{2mE}}$.

$$E = 1.77\,\text{eV} = 2.838\times10^{-19}\,\text{J} \Rightarrow \lambda = \frac{6.626\times10^{-34}\,\text{J}\cdot\text{s}}{\sqrt{2(9.109\times10^{-31}\,\text{kg})(2.838\times10^{-19}\,\text{J})}} = 9.21\times10^{-10}\,\text{m} = 0.921\,\text{nm}$$

$$E = 3.10\,\text{eV} = 4.966\times10^{-19}\,\text{J} \Rightarrow \lambda = \frac{6.626\times10^{-34}\,\text{J}\cdot\text{s}}{\sqrt{2(9.109\times10^{-31}\,\text{kg})(4.966\times10^{-19}\,\text{J})}} = 6.97\times10^{-10}\,\text{m} = 0.697\,\text{nm}$$

The range of electron wavelengths is 0.697 nm to 0.921 nm.

41-5

$$\lambda = \frac{h}{p} = \frac{h}{mv} = \frac{6.626\times10^{-34}\,\text{J}\cdot\text{s}}{(2000\,\text{kg})(30.0\,\text{m/s})} = \underline{1.10\times10^{-38}\,\text{m}}$$

This wavelength is extremely short; the car will not exhibit wavelike properties.

41-7

Since the alpha particles are scattered from the surface plane of the crystal Eq. (41-4) applies: $d\sin\theta = m\lambda$.

$$m = 1 \Rightarrow \sin\theta = \frac{\lambda}{d}$$

Use the deBroglie relation to calculate the wavelength of the particles: $\lambda = \frac{h}{p}$.

$$E = \frac{1}{2}mv^2 = \frac{p^2}{2m} \Rightarrow p = \sqrt{2mE} \text{ so } \lambda = \frac{h}{\sqrt{2mE}} = \frac{6.626\times10^{-34}\,\text{J}\cdot\text{s}}{\sqrt{2(6.64\times10^{-27}\,\text{kg})(1.2\times10^{3}\,\text{eV})\left(\frac{1.602\times10^{-19}\,\text{J}}{1\,\text{eV}}\right)}}$$

$\lambda = 4.147\times10^{-13}\,\text{m}$

$$\sin\theta = \frac{\lambda}{d} = \frac{4.147\times10^{-13}\,\text{m}}{72.0\times10^{-12}\,\text{m}} = 5.760\times10^{-3} \Rightarrow \theta = \underline{0.330^\circ}$$

41-11

a) Use the Heisenberg uncertainty principle to compute the uncertainty in the x-component of the electron's velocity.

$$\Delta x \, \Delta p_x \geq \frac{h}{2\pi}, \text{ so } \Delta p_x \geq \frac{h}{2\pi \Delta x}$$

$$p_x = m v_x \Rightarrow \Delta v_x \geq \frac{h}{2\pi m \Delta x}$$

The minimum Δv_x is $\Delta v_x = \frac{h}{2\pi m \Delta x} = \frac{6.626\times10^{-34}\text{ J}\cdot\text{s}}{2\pi(9.109\times10^{-31}\text{ kg})(0.20\times10^{-3}\text{ m})} = 0.5789\text{ m/s}$

$\frac{\Delta v_x}{v_x} = 1.0\% \Rightarrow v_x = \frac{\Delta v_x}{0.0100} = \frac{0.5789\text{ m/s}}{0.0100} = \underline{57.9\text{ m/s}}$

b) The minimum $\Delta v_x = \frac{h}{2\pi m \Delta x} = \frac{6.626\times10^{-34}\text{ J}\cdot\text{s}}{2\pi(1.673\times10^{-27}\text{ kg})(0.20\times10^{-3}\text{ m})} = 3.152\times10^{-4}\text{ m/s}$

$\frac{\Delta v_x}{v_x} = 1.0\% \Rightarrow v_x = \frac{\Delta v_x}{0.0100} = \frac{3.152\times10^{-4}\text{ m/s}}{0.0100} = \underline{0.0315\text{ m/s}}$

Note that the more massive proton must be traveling slower to have the same magnitude of uncertainty principle effects.

41-13

Find the uncertainty ΔE in the particle's energy:

$E = mc^2 \Rightarrow \Delta E = (\Delta m)c^2$

$m = 5m_p$; $\Delta m = (2.0\%)m = 0.020(5m_p) = (0.10)m_p = (0.10)(1.673\times10^{-27}\text{ kg}) = 1.673\times10^{-28}\text{ kg}$

Then $\Delta E = (\Delta m)c^2 = (1.673\times10^{-28}\text{ kg})(2.998\times10^{8}\text{ m/s})^2 = 1.504\times10^{-11}\text{ J}$.

Then use the energy uncertainty principle $\Delta E \, \Delta t \geq \frac{h}{2\pi}$ to estimate the lifetime: $\Delta t \simeq \frac{h}{2\pi \Delta E} = \frac{6.626\times10^{-34}\text{ J}\cdot\text{s}}{2\pi(1.504\times10^{-11}\text{ J})} = \underline{7.0\times10^{-24}\text{ s}}$

41-15

a) The relation between the kinetic energy K of the particle of charge q and the potential difference ΔV through which it has been accelerated is $K = |q|\Delta V$.

And $K = \frac{p^2}{2m} \Rightarrow p = \sqrt{2mK} = \sqrt{2m|q|\Delta V}$.

The deBroglie wavelength is $\lambda = \frac{h}{p} = \frac{h}{\sqrt{2m|q|\Delta V}} = \frac{6.626\times10^{-34}\text{ J}\cdot\text{s}}{\sqrt{2(9.109\times10^{-31}\text{ kg})(1.602\times10^{-19}\text{ C})(400\text{ V})}}$

$\lambda = 6.13\times10^{-11}\text{ m} = \underline{61.3\text{ pm}}$

b) The only change is the mass m of the particle:

$\lambda = \frac{h}{\sqrt{2m|q|\Delta V}} = \frac{6.626\times10^{-34}\text{ J}\cdot\text{s}}{\sqrt{2(1.673\times10^{-27}\text{ kg})(1.602\times10^{-19}\text{ C})(400\text{ V})}} = 1.43\times10^{-12}\text{ m} = \underline{1.43\text{ pm}}$

(The proton wavelength is smaller than the electron wavelength by a factor of $\sqrt{\frac{m_e}{m_p}} = \frac{1}{\sqrt{1836}}$.)

Problems

41-21

$$\Delta x\, \Delta p_x \geq \frac{h}{2\pi} \Rightarrow \text{minimum } \Delta p_x = \frac{h}{2\pi \Delta x}$$

$$\Delta x \simeq \lambda = \frac{h}{p} \Rightarrow \Delta p_x \simeq \frac{h}{2\pi}\frac{p}{h} = \frac{p}{2\pi} \simeq p$$

(Δp_x is the same order of magnitude as p; in making uncertainty principle <u>estimates</u> factors like 2π can be neglected within the accuracy to which the estimate is being done.)

41-23

a) $\Delta x\, \Delta p_x \geq \frac{h}{2\pi}$

Estimate Δx as $\Delta x \simeq 1.0 \times 10^{-15}\text{ m}$. Then the minimum allowed Δp_x is

$$\Delta p_x \simeq \frac{h}{2\pi \Delta x} = \frac{6.626\times10^{-34}\text{ J}\cdot\text{s}}{2\pi\,(1.0\times10^{-15}\text{ m})} = \underline{1.0\times10^{-19}\text{ kg}\cdot\text{m/s}}$$

b) Assume $p \simeq 1.0\times10^{-19}\text{ kg}\cdot\text{m/s}$

Eq. (39-41): $E = \sqrt{(mc^2)^2 + (pc)^2}$

$mc^2 = (9.109\times10^{-31}\text{ kg})(2.998\times10^8\text{ m/s})^2 = 8.187\times10^{-14}\text{ J}$

$pc = (1.0\times10^{-19}\text{ kg}\cdot\text{m/s})(2.998\times10^8\text{ m/s}) = 2.998\times10^{-11}\text{ J}$

$E = \sqrt{(8.187\times10^{-14}\text{ J})^2 + (2.998\times10^{-11}\text{ J})^2} = 2.998\times10^{-11}\text{ J}$

$$K = E - mc^2 = 2.998\times10^{-11}\text{ J} - 8.187\times10^{-14}\text{ J} = 2.99\times10^{-11}\text{ J}\left(\frac{1\text{ eV}}{1.602\times10^{-19}\text{ J}}\right) = \underline{190\text{ MeV}}$$

c) $$|U| = \frac{1}{4\pi\epsilon_0}\frac{e^2}{r} = 8.987\times10^9\text{ N}\cdot\text{m}^2/\text{C}^2\,\frac{(1.602\times10^{-19}\text{ C})^2}{1.0\times10^{-15}\text{ m}} = 2.31\times10^{-13}\text{ J}\left(\frac{1\text{ eV}}{1.602\times10^{-19}\text{ J}}\right) = 1.44\text{ MeV}$$

The estimate from the uncertainty principle of the kinetic energy of an electron confined within the nucleus is much larger than the binding energy from the attractive Coulomb interaction with a proton, so there cannot be electrons within the nucleus.

41-25

Y

screen

aperture

X

0.300 m

Take the origin of coordinates at the aperture, with the x-axis in the direction toward the screen.

a) The electrons are accelerated to a kinetic energy of $K = e\Delta V = (1.602\times10^{-19}\text{ C})(1.20\times10^3\text{ V}) = 1.922\times10^{-16}\text{ J}$

$K = \frac{1}{2}mv^2$ so the speed of the electrons is

$$v = \sqrt{\frac{2K}{m}} = \sqrt{\frac{2(1.922\times10^{-16}\text{ J})}{9.109\times10^{-31}\text{ kg}}} = 2.05\times10^7\text{ m/s}$$

Calculate the time t it takes the electrons to travel 0.300 m from the aperture to the screen:

41-25 (cont)

$$x - x_0 = v_{0x} t + \tfrac{1}{2} a_x t^2 \rightarrow 0$$

$$t = \frac{x - x_0}{v_{0x}} = \frac{0.300\ m}{2.05 \times 10^{7}\ m/s} = 1.46 \times 10^{-8}\ s$$

Use the uncertainty principle to estimate the uncertainty Δv_y in the y-component of the velocity of the electrons introduced when they pass through the aperture:

$$\Delta y \Delta p_y \geq \frac{h}{2\pi}$$

$$\Delta p_y = m \Delta v_y \Rightarrow \Delta v_y \simeq \frac{h}{2\pi m \Delta y}$$

Estimate Δy as $\Delta y \simeq 0.50 \times 10^{-3}\ m \Rightarrow \Delta v_y \simeq \frac{6.626 \times 10^{-34}\ J \cdot s}{2\pi (9.109 \times 10^{-31}\ kg)(0.50 \times 10^{-3}\ m)} = 0.23\ m/s$

The uncertainty in the y-coordinate of the electrons when they reach the screen introduced by Δv_y is $\Delta y \simeq t \Delta v_y \simeq (1.46 \times 10^{-8} s)(0.23\ m/s) = 3.4 \times 10^{-9}\ m = \underline{3.4\ nm}$

b) Δy is only about a factor of 10 times the size of an individual atom, so uncertainty principle effects have absolutely no effect on picture clarity.

41-29

a) $\lambda = \frac{h}{mv} \Rightarrow v = \frac{h}{m \lambda} = \frac{6.626 \times 10^{-34}\ J \cdot s}{(60.0\ kg)(1.0\ m)} = \underline{1.1 \times 10^{-35}\ m/s}$

b) $t = \frac{\text{distance}}{\text{velocity}} = \frac{0.80\ m}{1.1 \times 10^{-35}\ m/s} = 7.3 \times 10^{34}\ s \left(\frac{1\ y}{3.156 \times 10^{7}\ s}\right) = \underline{2.3 \times 10^{27}\ y}$

41-31

$$h = 6.63 \times 10^{-22}\ J \cdot s$$

$$\Delta E \Delta t \geq \frac{h}{2\pi} \Rightarrow \Delta E \simeq \frac{h}{2\pi \Delta t} = \frac{6.63 \times 10^{-22}\ J \cdot s}{2\pi (1.50 \times 10^{-3}\ s)} = 7.03 \times 10^{-20}\ J \left(\frac{1\ eV}{1.602 \times 10^{-19}\ J}\right) = \underline{0.44\ eV}$$

41-33

a) Let the y-direction be from the thrower to the catcher, and let the x-direction be horizontal and perpendicular to the y-direction.

A cube with volume $V = 1000\ cm^3 = 1.00 \times 10^{-3}\ m^3$ has side length $l = V^{1/3} = (1.00 \times 10^{-3}\ m^3)^{1/3} = 0.100\ m$. Thus estimate Δx as $\Delta x \simeq 0.10\ m$.

$\Delta x \Delta p_x \geq \frac{h}{2\pi}$ and $\Delta p_x = m \Delta v_x \Rightarrow \Delta v_x \simeq \frac{h}{2\pi m \Delta x} = \frac{0.0663\ J \cdot s}{2\pi (0.10\ kg)(0.10\ m)} = \underline{1.1\ m/s}$

(The value of h in this other universe has been used.)

b) The time it takes the ball to travel to the second student is

$$t = \frac{20\ m}{5.0\ m/s} = 4.0\ s$$

The uncertainty in the x-coordinate of the ball when it reaches the second student introduced by Δv_x is $\Delta x = (\Delta v_x) t = (1.1\ m/s)(4.0\ s) = 4.4\ m$.

The ball could miss the second student by about 4 m.

CHAPTER 42

Exercises 3, 5, 9, 13, 15, 17, 19

Problems 21, 23, 27, 31, 33

Exercises

42-3

Eq. (42-5): $E_n = \frac{n^2h^2}{8mL^2}$

Ground state energy $E_1 = \frac{h^2}{8mL^2}$; first excited state energy $E_2 = \frac{4h^2}{8mL^2}$.

$\Delta E = E_2 - E_1 = \frac{3h^2}{8mL^2}$

$$L = h\sqrt{\frac{3}{8m\,\Delta E}} = 6.626\times10^{-34}\,\text{J}\cdot\text{s}\sqrt{\frac{3}{8(9.109\times10^{-31}\,\text{kg})(4.0\,\text{eV})\left(\frac{1.602\times10^{-19}\,\text{J}}{1\,\text{eV}}\right)}} = 5.3\times10^{-10}\,\text{m} = \underline{0.53\,\text{nm}}$$

42-5

For the $n=2$ first excited state the normalized wave function is given by Eq. (42-9): $\psi_2(x) = \sqrt{\frac{2}{L}}\sin\left(\frac{2\pi x}{L}\right)$

$$|\psi_2(x)|^2dx = \frac{2}{L}\sin^2\left(\frac{2\pi x}{L}\right)dx$$

a) $|\psi_2|^2dx = 0 \Rightarrow \sin\left(\frac{2\pi x}{L}\right) = 0 \Rightarrow \frac{2\pi x}{L} = m\pi$, $m = 0, 1, 2, \ldots$; $x = m\left(\frac{L}{2}\right)$

$m = 0 \Rightarrow x = 0$

$m = 1 \Rightarrow x = \frac{L}{2}$

$m = 2 \Rightarrow x = L$

The probability of finding the particle is zero at $x = 0$, $\frac{L}{2}$, and L.

b) $|\psi_2|^2dx$ is maximum when $\sin\left(\frac{2\pi x}{L}\right) = \pm1 \Rightarrow \frac{2\pi x}{L} = m\left(\frac{\pi}{2}\right)$, $m = 1, 3, 5, \ldots$; $x = m\left(\frac{L}{4}\right)$

$m = 1 \Rightarrow x = \frac{L}{4}$

$m = 3 \Rightarrow x = \frac{3L}{4}$

The probability of finding the particle is largest at $x = \frac{L}{4}$ and $\frac{3L}{4}$.

42-9

$\psi(x)$ is a solution of Eq. (42-16): $-\frac{\hbar^2}{2m}\frac{d^2\psi(x)}{dx^2} + U(x)\psi(x) = E\psi(x)$

$\psi'(x) = A\psi(x)$

$\frac{d^2\psi'}{dx^2} = A\frac{d^2\psi}{dx^2}$

Thus $-\frac{\hbar^2}{2m}\frac{d^2\psi'}{dx^2} + U(x)\psi' = A\left(-\frac{\hbar^2}{2m}\frac{d^2\psi}{dx^2} + U(x)\psi\right) = A(E\psi) = E(A\psi) = E\psi'$

Therefore ψ' is a solution to Eq. (42-16), with the same E.

42-13

$U_0 = 6E_\infty$, as in Fig. 42-8, so $E_1 = 0.625\,E_\infty$ and $E_3 = 5.09\,E_\infty$ with $E_\infty = \frac{\pi^2\hbar^2}{2mL^2}$.
In this problem the particle bound in the well is a proton, so $m = 1.673\times10^{-27}$ kg.
Then $E_\infty = \frac{\pi^2\hbar^2}{2mL^2} = \frac{\pi^2(1.054\times10^{-34}\,\text{J}\cdot\text{s})^2}{2(1.673\times10^{-27}\,\text{kg})(5.0\times10^{-15}\,\text{m})^2} = 1.311\times10^{-12}$ J.

The transition energy is $\Delta E = E_3 - E_1 = (5.09 - 0.625)E_\infty = 4.465\,E_\infty$

$\Delta E = 4.465(1.311\times10^{-12}\,\text{J}) = 5.854\times10^{-12}$ J

The wavelength of the photon that is absorbed is related to the transition energy by

$\Delta E = \frac{hc}{\lambda} \Rightarrow \lambda = \frac{hc}{\Delta E} = \frac{(6.626\times10^{-34}\,\text{J}\cdot\text{s})(2.998\times10^8\,\text{m/s})}{5.854\times10^{-12}\,\text{J}} = 3.4\times10^{-14}\,\text{m} = \underline{34\,\text{fm}}$

42-15

The probability is $T = Ae^{-2KL}$, with $A = 16\frac{E}{U_0}\left(1 - \frac{E}{U_0}\right)$ and $K = \frac{\sqrt{2m(U_0 - E)}}{\hbar}$.
$E = 50$ eV; $U_0 = 55$ eV; $L = 0.10\times10^{-9}$ m

a) $A = 16\frac{E}{U_0}\left(1 - \frac{E}{U_0}\right) = 16\,\frac{50}{55}\left(1 - \frac{50}{55}\right) = 1.3223$

$K = \frac{\sqrt{2m(U_0 - E)}}{\hbar} = \frac{\sqrt{2(9.109\times10^{-31}\,\text{kg})(55\,\text{eV} - 50\,\text{eV})(1.602\times10^{-19}\,\text{J/eV})}}{1.054\times10^{-34}\,\text{J}\cdot\text{s}} = 1.146\times10^{10}\,\text{m}^{-1}$

$T = Ae^{-2KL} = (1.3223)\,e^{-(2)(1.146\times10^{10}\,\text{m}^{-1})(0.10\times10^{-9}\,\text{m})} = 1.3223\,e^{-2.292} = \underline{0.13}$

b) The only change is the mass m, which appears in K.

$K = \frac{\sqrt{2m(U_0 - E)}}{\hbar} = \frac{\sqrt{2(1.673\times10^{-27}\,\text{kg})(55\,\text{eV} - 50\,\text{eV})(1.602\times10^{-19}\,\text{J/eV})}}{1.054\times10^{-34}\,\text{J}\cdot\text{s}} = 4.912\times10^{11}\,\text{m}^{-1}$

Then $T = Ae^{-2KL} = (1.3223)\,e^{-2(4.912\times10^{11})(0.10\times10^{-9}\,\text{m})} = (1.3223)e^{-98.2} = \underline{3\times10^{-43}}$

42-17

$\omega = \sqrt{\frac{k}{m}} = \sqrt{\frac{80.0\,\text{N/m}}{0.200\,\text{kg}}} = 20.0$ rad/s

The ground state energy is given by Eq. (42-27):
$E = \frac{1}{2}\hbar\omega = \frac{1}{2}(1.054\times10^{-34}\,\text{J}\cdot\text{s})(20.0\,\text{rad/s}) = 1.05\times10^{-33}\,\text{J}\left(\frac{1\,\text{eV}}{1.602\times10^{-19}\,\text{J}}\right) = \underline{6.55\times10^{-15}\,\text{eV}}$

$E_n = (n + \frac{1}{2})\hbar\omega$
$E_{n+1} = (n + 1 + \frac{1}{2})\hbar\omega$
The energy separation between these adjacent levels is
$\Delta E = E_{n+1} - E_n = \hbar\omega = 2E_0 = 2(6.55\times10^{-15}\,\text{eV}) = \underline{1.31\times10^{-14}\,\text{eV}}$

These energies are extremely small; quantum effects are not important for this oscillator.

42-19

The photon wavelength λ is related to the transition energy by $\Delta E = \frac{hc}{\lambda}$. Ground state $E_0 = \frac{1}{2}\hbar\omega$; first excited state $E_1 = \frac{3}{2}\hbar\omega$, so the transition energy is $\Delta E = \left(\frac{3}{2} - \frac{1}{2}\right)\hbar\omega = \hbar\omega = \frac{h}{2\pi}\sqrt{\frac{k}{m}}$.

Equate these two expressions for $\Delta E \Rightarrow \frac{hc}{\lambda} = \frac{h}{2\pi}\sqrt{\frac{k}{m}}$

$\Rightarrow k = m\left(\frac{2\pi c}{\lambda}\right)^2 = (4.7\times10^{-26}\,\text{kg})\left(\frac{2\pi(2.998\times10^8\,\text{m/s})}{220\times10^{-6}\,\text{m}}\right)^2 = \underline{3.4\ \text{N/m}}$

Problems

42-21

The energy levels are given by Eq. (42-5): $E_n = \frac{n^2h^2}{8mL^2}$

Ground state, $n=1 \Rightarrow E_1 = \frac{h^2}{8mL^2}$

First excited state, $n=2 \Rightarrow E_2 = \frac{4h^2}{8mL^2}$

The transition energy is $\Delta E = E_2 - E_1 = \frac{3h^2}{8mL^2}$.

The transition energy is related to the wavelength of the emitted photon by $\Delta E = \frac{hc}{\lambda}$. Combine these two equations $\Rightarrow \frac{hc}{\lambda} = \frac{3h^2}{8mL^2}$

$\lambda = \frac{8mcL^2}{3h} = \frac{8(9.109\times10^{-31}\,\text{kg})(2.998\times10^8\,\text{m/s})(3.78\times10^{-9}\,\text{m})^2}{3(6.626\times10^{-34}\,\text{J}\cdot\text{s})} = 1.57\times10^{-5}\,\text{m} = \underline{15.7\,\mu\text{m}}$

42-23

a) The normalized wave function for the ground state is $\psi_1 = \sqrt{\frac{2}{L}}\sin\left(\frac{\pi x}{L}\right)$.

The probability P of the particle being between $x = \frac{L}{4}$ and $x = \frac{3L}{4}$ is

$P = \int_{L/4}^{3L/4} |\psi_1|^2\,dx = \frac{2}{L}\int_{L/4}^{3L/4} \sin^2\left(\frac{\pi x}{L}\right)dx$

Let $y = \frac{\pi x}{L}$; $dx = \frac{L}{\pi}dy$ and the integration limits become $\frac{\pi}{4}$ and $\frac{3\pi}{4}$

$P = \frac{2}{L}\left(\frac{L}{\pi}\right)\int_{\pi/4}^{3\pi/4}\sin^2 y\,dy = \frac{2}{\pi}\left[\frac{1}{2}y - \frac{1}{4}\sin 2y\right]\Big|_{\pi/4}^{3\pi/4} = \frac{2}{\pi}\left[\frac{3\pi}{8} - \frac{\pi}{8} - \frac{1}{4}\sin\frac{3\pi}{2} + \frac{1}{4}\sin\frac{\pi}{2}\right]$

$P = \frac{2}{\pi}\left[\frac{\pi}{4} - \frac{1}{4}(-1) + \frac{1}{4}(1)\right] = \frac{1}{2} + \frac{1}{\pi} = \underline{0.818}$

(Note: The integral formula $\int \sin^2 y\,dy = \frac{1}{2}y - \frac{1}{4}\sin 2y$ was used.)

b) The normalized wave function for the first excited state is $\psi_2 = \sqrt{\frac{2}{L}}\sin\left(\frac{2\pi x}{L}\right)$.

$P = \int_{L/4}^{3L/4}|\psi_2|^2\,dx = \frac{2}{L}\int_{L/4}^{3L/4}\sin^2\left(\frac{2\pi x}{L}\right)dx$

Let $y = \frac{2\pi x}{L}$; $dx = \frac{L}{2\pi}dy$ and the integration limits become $\frac{\pi}{2}$ and $\frac{3\pi}{2}$.

$P = \frac{2}{L}\left(\frac{L}{2\pi}\right)\int_{\pi/2}^{3\pi/2}\sin^2 y\,dy = \frac{1}{\pi}\left[\frac{1}{2}y - \frac{1}{4}\sin 2y\right]\Big|_{\pi/2}^{3\pi/2} = \frac{1}{\pi}\left[\frac{3\pi}{4} - \frac{\pi}{4} - \frac{1}{4}\sin 3\pi + \frac{1}{4}\sin\pi\right] = \frac{1}{2}$

$P = \underline{0.500}$

42-27

The tunneling probability is given by $T = Ae^{-2KL}$, $A = 16\frac{E}{U_0}\left(1-\frac{E}{U_0}\right)$, $K = \frac{\sqrt{2m(U_0-E)}}{\hbar}$

$\frac{T}{A} = e^{-2KL}$

Take the natural log of both sides of the equation

$\Rightarrow -2KL = \ln\left(\frac{T}{A}\right) = -\ln\left(\frac{A}{T}\right)$

$L = \frac{\ln(A/T)}{2K}$

$A = 16\frac{E}{U_0}\left(1-\frac{E}{U_0}\right) = 16\frac{8}{10}\left(1-\frac{8}{10}\right) = 2.56$

$T = 0.10\% = 1.0\times10^{-3}$

$K = \frac{\sqrt{2m(U_0-E)}}{\hbar} = \frac{\sqrt{2(9.109\times10^{-31}\text{ kg})(10.0\text{ eV}-8.0\text{ eV})(1.602\times10^{-19}\text{ J/eV})}}{1.054\times10^{-34}\text{ J}\cdot\text{s}} = 7.249\times10^{9}\text{ m}^{-1}$

$L = \frac{\ln\left(\frac{2.56}{1\times10^{-3}}\right)}{2(7.249\times10^{9}\text{ m}^{-1})} = 5.4\times10^{-10}\text{ m} = \underline{0.54\text{ nm}}$

42-31

Calculate the angular frequency ω of the pendulum:

$\omega = \frac{2\pi}{T} = \frac{2\pi}{1.00\text{ s}} = 2\pi\text{ s}^{-1}$

The ground-state energy is $E_0 = \frac{1}{2}\hbar\omega = \frac{1}{2}(1.054\times10^{-34}\text{ J}\cdot\text{s})(2\pi\text{ s}^{-1}) = \underline{3.31\times10^{-34}\text{ J}}$

$E_0 = 3.31\times10^{-34}\text{ J}\left(\frac{1\text{ eV}}{1.602\times10^{-19}\text{ J}}\right) = \underline{2.07\times10^{-15}\text{ eV}}$

$E_n = (n+\frac{1}{2})\hbar\omega$

$E_{n+1} = (n+1+\frac{1}{2})\hbar\omega$

The energy difference between these adjacent energy levels is

$\Delta E = E_{n+1} - E_n = \hbar\omega = 2E_0 = \underline{6.62\times10^{-34}\text{ J}} = 4.\underline{14\times10^{-15}\text{ eV}}$

These energies are much too small to detect.

42-33

a) Eq. (42-31): $-\frac{\hbar^2}{2m}\left(\frac{\partial^2\psi}{\partial x^2}+\frac{\partial^2\psi}{\partial y^2}+\frac{\partial^2\psi}{\partial z^2}\right) + U\psi = E\psi$

$\psi_{n_x}, \psi_{n_y}, \psi_{n_z}$ are each solutions of Eq. (42-24)

$\Rightarrow -\frac{\hbar^2}{2m}\frac{d^2\psi_{n_x}}{dx^2} + \frac{1}{2}kx^2\psi_{n_x} = E_{n_x}\psi_{n_x}$

$-\frac{\hbar^2}{2m}\frac{d^2\psi_{n_y}}{dy^2} + \frac{1}{2}ky^2\psi_{n_y} = E_{n_y}\psi_{n_y}$

$-\frac{\hbar^2}{2m}\frac{d^2\psi_{n_z}}{dz^2} + \frac{1}{2}kz^2\psi_{n_z} = E_{n_z}\psi_{n_z}$

$\psi = \psi_{n_x}(x)\,\psi_{n_y}(y)\,\psi_{n_z}(z)$, $U = \frac{1}{2}kx^2 + \frac{1}{2}ky^2 + \frac{1}{2}kz^2$

42-33 (cont)

$$\frac{\partial^2\psi}{\partial x^2} = \left(\frac{\partial^2\psi_{n_x}}{\partial x^2}\right)\psi_{n_y}\psi_{n_z}$$

$$\frac{\partial^2\psi}{\partial y^2} = \psi_{n_x}\left(\frac{\partial^2\psi_{n_y}}{\partial y^2}\right)\psi_{n_z}$$

$$\frac{\partial^2\psi}{\partial z^2} = \psi_{n_x}\psi_{n_y}\left(\frac{\partial^2\psi_{n_z}}{\partial z^2}\right)$$

So $$-\frac{\hbar^2}{2m}\left(\frac{\partial^2\psi}{\partial x^2}+\frac{\partial^2\psi}{\partial y^2}+\frac{\partial^2\psi}{\partial z^2}\right)+U\psi = \left(-\frac{\hbar^2}{2m}\frac{\partial^2\psi_{n_x}}{\partial x^2}+\frac{1}{2}kx^2\psi_{n_x}\right)\psi_{n_y}\psi_{n_z}$$
$$\left(-\frac{\hbar^2}{2m}\frac{\partial^2\psi_{n_y}}{\partial y^2}+\frac{1}{2}ky^2\psi_{n_y}\right)\psi_{n_x}\psi_{n_z}+\left(-\frac{\hbar^2}{2m}\frac{\partial^2\psi_{n_z}}{\partial z^2}+\frac{1}{2}kz^2\psi_{n_z}\right)\psi_{n_x}\psi_{n_y}$$

$$-\frac{\hbar^2}{2m}\left(\frac{\partial^2\psi}{\partial x^2}+\frac{\partial^2\psi}{\partial y^2}+\frac{\partial^2\psi}{\partial z^2}\right)+U\psi = E_{n_x}\psi_{n_x}\psi_{n_y}\psi_{n_z}+E_{n_y}\psi_{n_x}\psi_{n_y}\psi_{n_z}+E_{n_z}\psi_{n_x}\psi_{n_y}\psi_{n_z}$$
$$= (E_{n_x}+E_{n_y}+E_{n_z})\,\psi$$

Therefore, we have shown that this ψ is a solution to Eq. (42-31), with energy $E_{n_x n_y n_z} = E_{n_x}+E_{n_y}+E_{n_z} = (n_x+n_y+n_z+\frac{3}{2})\hbar\omega$

b) Ground state: $n_x = n_y = n_z = 0$; $E_{000} = \frac{3}{2}\hbar\omega$

First-excited state: $n_x = 1,\ n_y = n_z = 0$ or $n_y = 1,\ n_x = n_z = 0$ or $n_z = 1,\ n_x = n_y = 0$

$E_{100} = E_{010} = E_{001} = \frac{5}{2}\hbar\omega$

(There are three different quantum states that have this same energy.)

CHAPTER 43

Exercises 3, 5, 11, 13, 15, 19, 21, 25

Problems 27, 29, 31, 35, 39

Exercises

43-3

Eq. (43-4): $L = \sqrt{l(l+1)}\,\hbar$, $l = 0, 1, 2, \ldots$

$l(l+1) = \left(\frac{L}{\hbar}\right)^2 = \left(\frac{2.583\times10^{-34}\ \text{kg}\cdot\text{m/s}}{1.054\times10^{-34}\ \text{J}\cdot\text{s}}\right)^2 = 6.00$

$l(l+1) = 6 \Rightarrow \underline{l = 2}$

43-5

$L = \sqrt{l(l+1)}\,\hbar$

The maximum l, l_{max}, for a given n is $l_{max} = n-1$.

$n = 1 \Rightarrow l_{max} = 0 \Rightarrow L = 0$

$n = 10 \Rightarrow l_{max} = 9 \Rightarrow L = \sqrt{9(10)}\,\hbar = 9.49\hbar$

$n = 100 \Rightarrow l_{max} = 99 \Rightarrow L = \sqrt{99(100)}\,\hbar = 99.5\hbar$

As n increases the maximum L gets closer to the value $n\hbar$ postulated in the Bohr model.

43-11

a) 4f level $\Rightarrow l = 3 \Rightarrow 2l+1 = 7$ different m_l states.
The 4f level is split into <u>7 levels</u> by the magnetic field.

b) Each m_l level is shifted in energy an amount given by Eq. (43-18): $U = m_l \mu_B B$.
Adjacent levels differ in m_l by one, so $\Delta U = \mu_B B$.

$\mu_B = \frac{e\hbar}{2m} = \frac{(1.602\times10^{-19}\ \text{C})(1.054\times10^{-34}\ \text{J}\cdot\text{s})}{2(9.109\times10^{-31}\ \text{kg})} = 9.274\times10^{-24}\ \text{A}\cdot\text{m}^2$

$\Delta U = \mu_B B = (9.274\times10^{-24}\ \text{A}\cdot\text{m}^2)(2.50\ \text{T}) = 2.318\times10^{-23}\ \text{J}\left(\frac{1\ \text{eV}}{1.602\times10^{-19}\ \text{J}}\right) = \underline{1.45\times10^{-4}\ \text{eV}}$

c) The level of highest energy is for the largest m_l, which is $m_l = l = 3$;
$U_3 = 3\mu_B B$.
The level of lowest energy is for the smallest m_l, which is $m_l = -l = -3$;
$U_{-3} = -3\mu_B B$.

The energy separation between these two levels is $U_3 - U_{-3} = 6\mu_B B$
$= 6(1.45\times10^{-4}\ \text{eV}) = \underline{8.70\times10^{-4}\ \text{eV}}$.

43-13

For a classical particle $L = I\omega$.
For a uniform sphere with mass m and radius R, $I = \frac{2}{5}mR^2$, so $L = \left(\frac{2}{5}mR^2\right)\omega$.
$L = \hbar \Rightarrow \frac{2}{5}mR^2\omega = \hbar$

$$\omega = \frac{5\hbar}{2mR^2} = \frac{5(1.054\times10^{-34}\,J\cdot s)}{2(9.109\times10^{-31}\,kg)(1.0\times10^{-17}\,m)^2} = \underline{2.89\times10^{30}\,rad/s}$$

43-15

j can have the values $l+\frac{1}{2}$ and $l-\frac{1}{2}$. Thus if j takes the values $\frac{5}{2}$ and $\frac{7}{2}$ it must be that $l-\frac{1}{2} = \frac{5}{2} \Rightarrow l = \frac{6}{2} = 3$.

43-19

Eq. (43-27): $E_n = -\frac{Z_{eff}^2}{n^2}(13.6\,eV)$

$n=5$ and $Z_{eff} = 2.771 \Rightarrow E_5 = -\frac{(2.771)^2}{5^2}(13.6\,eV) = -4.18\,eV$
The ionization energy is $\underline{4.18\,eV}$.

43-21

$$E_n = -\frac{Z_{eff}^2}{n^2}(13.6\,eV)$$

a) The element Be has nuclear charge $Z=4$. The ion Be^+ then has 3 electrons. The outermost electron sees the nuclear charge screened by the other two electrons so $Z_{eff} = 4-2 = 2$.

$$E_2 = -\frac{2^2}{2^2}(13.6\,eV) = \underline{-13.6\,eV}$$

b) The outermost electron in Ca^+ sees a $Z_{eff} = 2$.

$$E_4 = -\frac{2^2}{4^2}(13.6\,eV) = \underline{-3.40\,eV}$$

43-25

$f = \frac{E}{h}$ and $f = (2.48\times10^{15}\,Hz)(Z-1)^2$ (Moseley's law, Eq. (43-29))

$\Rightarrow \frac{E}{h} = (2.48\times10^{15}\,Hz)(Z-1)^2$

$$Z-1 = \sqrt{\frac{E}{h(2.48\times10^{15}\,Hz)}} = \sqrt{\frac{9.89\times10^3\,eV(1.602\times10^{-19}\,J/eV)}{(6.626\times10^{-34}\,J.s)(2.48\times10^{15}\,Hz)}} = 31.0$$

Thus $Z = 32$ (germanium).

Problems

43-27 $\psi_{2s} = \frac{1}{\sqrt{32\pi a_0^3}}(2 - r/a_0)\,e^{-r/2a_0}$

43-27 (cont)

a) Let $I = \int_0^\infty |\psi_{2s}|^2 dV = 4\pi \int_0^\infty |\psi_{2s}|^2 r^2 dr$. If ψ_{2s} is normalized then we will find that $I = 1$.

$$I = 4\pi\left(\frac{1}{32\pi a_0^3}\right)\int_0^\infty \left(2-\frac{r}{a_0}\right)^2 e^{-r/a_0} r^2 dr = \frac{1}{8a_0^3}\int_0^\infty \left(4r^2 - 4\frac{r^3}{a_0} + \frac{r^4}{a_0^2}\right)e^{-r/a_0} dr$$

Use the integral formula $\int_0^\infty x^n e^{-\alpha x} dx = \frac{n!}{\alpha^{n+1}}$, with $\alpha = \frac{1}{a_0}$.

$$I = \frac{1}{8a_0^3}\left(4(2!)(a_0)^3 - \frac{4}{a_0}(3!)(a_0)^4 + \frac{1}{a_0^2}(4!)(a_0^5)\right) = \frac{1}{8}(8-24+24) = 1;$$

ψ_{2s} is normalized.

b) $P(r<4a_0) = \int_0^{4a_0} |\psi_{2s}|^2 dV = 4\pi \int_0^{4a_0} |\psi_{2s}|^2 r^2 dr = \frac{1}{8a_0^3}\int_0^{4a_0}\left(4r^2 - 4\frac{r^3}{a_0} + \frac{r^4}{a_0^2}\right)e^{-r/a_0} dr$

Let $P(r<4a_0) = \frac{1}{8a_0^3}(I_1 + I_2 + I_3)$.

$$I_1 = 4\int_0^{4a_0} r^2 e^{-r/a_0} dr$$

Use the integral formula $\int r^2 e^{-\alpha r} dr = -e^{-\alpha r}\left(\frac{r^2}{\alpha} + \frac{2r}{\alpha^2} + \frac{2}{\alpha^3}\right)$ with $\alpha = \frac{1}{a_0}$.

$$\Rightarrow I_1 = -4\left[e^{-r/a_0}(r^2 a_0 + 2ra_0^2 + 2a_0^3)\right]\Big|_0^{4a_0} = (-104e^{-4} + 8)a_0^3$$

$$I_2 = -\frac{4}{a_0}\int_0^{4a_0} r^3 e^{-r/a_0} dr$$

Use the integral formula $\int r^3 e^{-\alpha r} dr = -e^{-\alpha r}\left(\frac{r^3}{\alpha} + 3\frac{r^2}{\alpha^2} + 6\frac{r}{\alpha^3} + 6\frac{1}{\alpha^4}\right)$ with $\alpha = \frac{1}{a_0}$.

$$\Rightarrow I_2 = \frac{4}{a_0}\left[e^{-r/a_0}(r^3 a_0 + 3r^2 a_0^2 + 6ra_0^3 + 6a_0^4)\right]\Big|_0^{4a_0} = (568e^{-4} - 24)a_0^3$$

$$I_3 = \frac{1}{a_0^2}\int_0^{4a_0} r^4 e^{-r/a_0} dr$$

Use the integral formula $\int r^4 e^{-\alpha r} dr = -e^{-\alpha r}\left(\frac{r^4}{\alpha} + 4\frac{r^3}{\alpha^2} + 12\frac{r^2}{\alpha^3} + 24\frac{r}{\alpha^4} + 24\frac{1}{\alpha^5}\right)$

$$\Rightarrow I_3 = -\frac{1}{a_0^2}\left[e^{-r/a_0}(r^4 a_0 + 4r^3 a_0^2 + 12r^2 a_0^3 + 24ra_0^4 + 24a_0^5)\right]\Big|_0^{4a_0} = a_0^3(-824e^{-4} + 24)$$

Then $P(r<4a_0) = \frac{1}{8a_0^3}(I_1 + I_2 + I_3) = \frac{1}{8a_0^3}a_0^3\left([8-24+24] + e^{-4}[-104+568-824]\right)$

$= \frac{1}{8}(8 - 360e^{-4}) = 1 - 45e^{-4} = \underline{0.176}$

43-29

a) $E_{1s} = -\frac{1}{(4\pi\epsilon_0)^2}\frac{me^4}{2\hbar^2}$; $U(r) = -\frac{1}{4\pi\epsilon_0}\frac{e^2}{r}$

$E_{1s} = U(r) \Rightarrow -\frac{1}{(4\pi\epsilon_0)^2}\frac{me^4}{2\hbar^2} = -\frac{1}{4\pi\epsilon_0}\frac{e^2}{r}$

$r = \frac{(4\pi\epsilon_0)2\hbar^2}{me^2} = 2a_0$.

b) For the 1s state the probability that the electron is in the classically forbidden region is $P(r>2a_0) = \int_{2a_0}^\infty |\psi_{1s}|^2 dV = 4\pi\int_{2a_0}^\infty |\psi_{1s}|^2 r^2 dr$

The normalized wave function of the 1s state of hydrogen is given in Example 43-3:

$\psi_{1s}(r) = \frac{1}{\sqrt{\pi a_0^3}} e^{-r/a_0}$

$P(r>2a_0) = 4\pi\left(\frac{1}{\pi a_0^3}\right)\int_{2a_0}^\infty r^2 e^{-2r/a_0} dr$

Use the integral formula $\int r^2 e^{-\alpha r} dr = -e^{-\alpha r}\left(\frac{r^2}{\alpha} + \frac{2r}{\alpha^2} + \frac{2}{\alpha^3}\right)$, with $\alpha = \frac{2}{a_0}$.

43-29 (cont)

$$\Rightarrow P(r>2a_0) = -\frac{4}{a_0^3}\left[e^{-2r/a_0}\left(\frac{a_0 r^2}{2} + \frac{a_0^2 r}{2} + \frac{a_0^3}{4}\right)\right]\Big|_{2a_0}^{\infty} = +\frac{4}{a_0^3}e^{-4}\left(2a_0^3 + a_0^3 + \frac{a_0^3}{4}\right)$$

$$P(r>2a_0) = 4e^{-4}\left(\frac{13}{4}\right) = 13e^{-4} = \underline{0.238.}$$

43-31

a) The energy of the photon equals the transition energy of the atom:

$$\Delta E = \frac{hc}{\lambda}.$$

$$E_n = -\frac{13.60\text{ eV}}{n^2} \Rightarrow E_2 = -\frac{13.60\text{ eV}}{4}, \; E_1 = -\frac{13.60\text{ eV}}{1}$$

$$\Delta E = E_2 - E_1 = 13.60\text{ eV}\left(-\frac{1}{4}+1\right) = \frac{3}{4}(13.60\text{ eV}) = 10.20\text{ eV}\,(1.602\times10^{-19}\text{ J/eV}) = 1.634\times10^{-18}\text{ J}$$

$$\lambda = \frac{hc}{\Delta E} = \frac{(6.626\times10^{-34}\text{ J}\cdot\text{s})(2.998\times10^{8}\text{ m/s})}{1.634\times10^{-18}\text{ J}} = 1.216\times10^{-7}\text{ m} = \underline{121.6\text{ nm}}$$

b) The shift of a level due to the energy of interaction with the magnetic field in the z-direction is $U = m_l \mu_B B$.
The ground state has $m_l = 0$ so is unaffected by the magnetic field.
The $n=2$ initial state has $m_l = -1$ so its energy is shifted downward an amount

$$U = m_l \mu_B B = (-1)(9.274\times10^{-24}\text{ A}\cdot\text{m}^2)(2.50\text{ T}) = -2.318\times10^{-23}\text{ J}\,(1\text{ eV}/1.602\times10^{-19}\text{ J}) = 1.447\times10^{-4}\text{ eV}$$

Note that the shift in energy due to the magnetic field is a very small fraction of the 10.2 eV transition energy. Problem 41-28 (c) shows that in this situation $\left|\frac{\Delta\lambda}{\lambda}\right| = \left|\frac{\Delta E}{E}\right|$.
This gives $|\Delta\lambda| = \lambda\left|\frac{\Delta E}{E}\right| = 121.6\text{ nm}\left(\frac{1.447\times10^{-4}\text{ eV}}{10.2\text{ eV}}\right) = 1.73\times10^{-3}\text{ nm} = \underline{1.73\text{ pm}}.$

The upper level in the transition is lowered in energy so the transition energy is decreased. A smaller ΔE means a larger λ; the magnetic field increases the wavelength.

43-35

Eq. (40-20): $\frac{n_1}{n_0} = e^{-(E_1-E_0)/kT}$

The interaction energy with the magnetic field is
$U = \frac{e}{m}S_z B = \frac{e\hbar}{m}m_s B$ (Example 43-6).
The energy of the $m_s = +\frac{1}{2}$ level is increased and the energy of the $m_s = -\frac{1}{2}$ level is decreased.

$$\frac{n_{1/2}}{n_{-1/2}} = e^{-(U_{1/2}-U_{-1/2})/kT}$$

$$U_{\frac{1}{2}} - U_{-\frac{1}{2}} = \frac{e\hbar}{m}B\left(\frac{1}{2} - \left(-\frac{1}{2}\right)\right) = \frac{e\hbar}{m}B = 2\left(\frac{e\hbar}{2m}\right)B = 2\mu_B B$$

$$\frac{n_{1/2}}{n_{-\frac{1}{2}}} = e^{-2\mu_B B/kT}$$

a) $B = 5.00\times10^{-5}\text{ T} \Rightarrow \frac{n_{\frac{1}{2}}}{n_{-\frac{1}{2}}} = e^{-2(9.274\times10^{-24}\text{ A}\cdot\text{m}^2)(5.00\times10^{-5}\text{ T})/[(1.381\times10^{-23}\text{ J/K})(300\text{ K})]}$

$$\frac{n_{1/2}}{n_{-\frac{1}{2}}} = e^{-2.238\times10^{-7}} = \underline{0.99999978}$$

43-35 (cont)

b) $B = 0.100\,T \Rightarrow \frac{n_{1/2}}{n_{-1/2}} = e^{-4.476\times10^{-4}} = \underline{0.999553}$

c) $B = 10.0\,T \Rightarrow \frac{n_{1/2}}{n_{-1/2}} = e^{-0.04476} = \underline{0.956}$

<u>43-39</u>

a) titanium, $Z = 32$

<u>minimum wavelength</u> $\Rightarrow$ largest transition energy

The highest occupied shell is the N shell ($n=4$). The highest energy transition is $N \to K$ with transition energy $\Delta E = E_N - E_K$. As in Example 43-11 neglect E_N relative to E_K, so $\Delta E = E_K = (Z-1)^2(13.6\text{ eV}) = (22-1)^2(13.6\text{ eV}) = 5.998\times10^3\text{ eV}$ $= 9.608\times10^{-16}\text{ J}$.

The energy of the emitted photon equals this transition energy, so the photon's wavelength is given by $\Delta E = \frac{hc}{\lambda} \Rightarrow \lambda = \frac{hc}{\Delta E}$.

$$\lambda = \frac{(6.626\times10^{-34}\text{ J}\cdot\text{s})(2.998\times10^8\text{ m/s})}{9.608\times10^{-16}\text{ J}} = 2.07\times10^{-10}\text{ m} = \underline{0.207\text{ nm}}$$

<u>maximum wavelength</u> $\Rightarrow$ smallest transition energy $\Rightarrow K_\alpha$ transition

The frequency of the photon emitted in this transition is given by Moseley's law (Eq. 43-29):

$$f = (2.48\times10^{15}\text{ Hz})(Z-1)^2 = (2.48\times10^{15}\text{ Hz})(22-1)^2 = 1.094\times10^{18}\text{ Hz}$$

$$\lambda = \frac{c}{f} = \frac{2.998\times10^8\text{ m/s}}{1.094\times10^{18}\text{ Hz}} = 2.74\times10^{-10}\text{ m} = \underline{0.274\text{ nm}}$$

b) molybdenum, $Z = 42$

Apply the analysis of part (a), just with this different value of Z.

<u>minimum wavelength</u>

$$\Delta E = E_K = (Z-1)^2(13.6\text{ eV}) = (42-1)^2(13.6\text{ eV}) = 2.286\times10^4\text{ eV} = 3.662\times10^{-15}\text{ J}$$

$$\lambda = \frac{hc}{\Delta E} = \frac{(6.626\times10^{-34}\text{ J}\cdot\text{s})(2.998\times10^8\text{ m/s})}{3.662\times10^{-15}\text{ J}} = 5.42\times10^{-11}\text{ m} = \underline{0.0542\text{ nm}}$$

<u>maximum wavelength</u>

$$f = (2.48\times10^{15}\text{ Hz})(Z-1)^2 = (2.48\times10^{15}\text{ Hz})(42-1)^2 = 4.169\times10^{18}\text{ Hz}$$

$$\lambda = \frac{c}{f} = \frac{2.998\times10^8\text{ m/s}}{4.169\times10^{18}\text{ Hz}} = 7.19\times10^{-11}\text{ m} = \underline{0.0719\text{ nm}}$$

CHAPTER 44

Exercises 7, 9, 11, 13, 15, 17, 19, 23, 25, 27

Problems 29, 33, 35, 37, 41

Exercises

44-7

The energy of a rotational level with quantum number l is
$E_l = l(l+1)\frac{\hbar^2}{2I}$ (Eq. 44-3)

$I = m_r r^2$

$m_r = \frac{m_1 m_2}{m_1+m_2} = \frac{m_{Li} m_H}{m_{Li}+m_H} = \frac{(1.17\times10^{-26}\text{ kg})(1.67\times10^{-27}\text{ kg})}{1.17\times10^{-26}\text{ kg}+1.67\times10^{-27}\text{ kg}} = 1.461\times10^{-27}\text{ kg}$

$I = m_r r^2 = (1.461\times10^{-27}\text{ kg})(0.159\times10^{-9}\text{ m})^2 = 3.694\times10^{-47}\text{ kg}\cdot\text{m}^2$

$l=2:\ E_2 = 2(3)\frac{\hbar^2}{2I} = 3\left(\frac{\hbar^2}{I}\right)$

$l=1:\ E_1 = 1(2)\frac{\hbar^2}{2I} = \frac{\hbar^2}{I}$

$\Delta E = E_2 - E_1 = 2\frac{\hbar^2}{I} = 2\frac{(1.054\times10^{-34}\text{ J}\cdot\text{s})^2}{3.694\times10^{-47}\text{ kg}\cdot\text{m}^2} = 6.015\times10^{-22}\text{ J} = \underline{3.75\times10^{-3}\text{ eV}}$

44-9

a) $f = \frac{\omega}{2\pi} = \frac{1}{2\pi}\sqrt{\frac{k}{m_r}} \Rightarrow k = m_r(2\pi f)^2$

$m_r = \frac{m_1 m_2}{m_1+m_2} = \frac{m_H m_{Cl}}{m_H+m_{Cl}} = \frac{(1.67\times10^{-27}\text{ kg})(5.81\times10^{-26}\text{ kg})}{1.67\times10^{-27}\text{ kg}+5.81\times10^{-26}\text{ kg}} = 1.623\times10^{-27}\text{ kg}$

$k = m_r(2\pi f)^2 = 1.623\times10^{-27}\text{ kg}\left(2\pi[8.97\times10^{13}\text{ Hz}]\right)^2 = \underline{516\text{ N/m}}$

b) $E_n = (n+\frac{1}{2})\hbar\omega = (n+\frac{1}{2})hf$, since $\hbar\omega = \frac{h}{2\pi}\omega$ and $\frac{\omega}{2\pi} = f$.
The energy spacing between adjacent levels is $\Delta E = E_{n+1} - E_n = (n+1+\frac{1}{2}-n-\frac{1}{2})hf = hf$, independent of n.

$\Delta E = hf = (6.626\times10^{-34}\text{ J}\cdot\text{s})(8.97\times10^{13}\text{ Hz}) = 5.94\times10^{-20}\text{ J} = \underline{0.371\text{ eV}}$

c) The photon energy equals the transition energy $\Rightarrow \Delta E = \frac{hc}{\lambda}$

$hf = \frac{hc}{\lambda} \Rightarrow \lambda = \frac{c}{f} = \frac{2.998\times10^8\text{ m/s}}{8.97\times10^{13}\text{ Hz}} = 3.34\times10^{-6}\text{ m} = \underline{3.34\ \mu\text{m}}$

This photon is in the <u>infrared</u>.

44-11

Eq. 44-10: $N(n) = NAe^{-E_n/kT} \Rightarrow \frac{N(m_s=+\frac{1}{2})}{N(m_s=-\frac{1}{2})} = e^{-(E_{1/2}-E_{-1/2})/kT}$

0.10% in the $m_s = +\frac{1}{2}$ state $\Rightarrow N(m_s=+\frac{1}{2}) = 1.0\times10^{-3}N = 1.0\times10^{-3}\left(N(m_s=+\frac{1}{2}) + N(m_s=-\frac{1}{2})\right)$

$0.999\,N(m_s=+\frac{1}{2}) = 1.0\times10^{-3}\,N(m_s=-\frac{1}{2})$

$\Rightarrow \frac{N(m_s=+\frac{1}{2})}{N(m_s=-\frac{1}{2})} = 1.0\times10^{-3}$

44-11 (cont)
(The number $N(m_s=-\frac{1}{2})$ in the $m_s=-\frac{1}{2}$ state is very nearly equal to the total number of atoms, N.)

From Example 43-6 the interaction energy between the spin and the magnetic field shifts an m_s level by an energy $U=\frac{e}{m}S_zB=\frac{e\hbar}{m}m_sB=2\mu_B m_s B$.

$E_{1/2}=\mu_B B,\ E_{-1/2}=-\mu_B B \Rightarrow E_{1/2}-E_{-1/2}=2\mu_B B$

Thus $1.0\times10^{-3}=e^{-2\mu_B B/kT}$

Take natural logs of both sides of the equations

$\Rightarrow\ 6.908=2\mu_B B/kT$

$$B=\frac{(6.908)kT}{2\mu_B}=\frac{(6.908)(1.381\times10^{-23}\text{ J/K})(0.300\text{ K})}{2(9.274\times10^{-24}\text{ J/T})}=\underline{1.54\text{ T}}$$

44-13

Eq. (44-19): $C=R\left(\frac{\hbar\omega}{kT}\right)^2\frac{e^{\hbar\omega/kT}}{(e^{\hbar\omega/kT}-1)^2}$

Energy separation between adjacent vibrational levels is $\Delta E=\hbar\omega$

$\Rightarrow \hbar\omega=0.54\text{ eV}=8.65\times10^{-20}\text{ J}.$

a) $T=50\text{ K}$

$$\frac{\hbar\omega}{kT}=\frac{8.65\times10^{-20}\text{ J}}{(1.381\times10^{-23}\text{ J/K})(50\text{ K})}=125.4$$

$$C=R(125.4)^2\frac{e^{125.4}}{(e^{125.4}-1)^2}=(5.4\times10^{-51})R$$

(The vibrational contribution is extremely small, since $kT \ll \hbar\omega$.)

b) $T=300\text{ K}$

$$\frac{\hbar\omega}{kT}=\frac{8.65\times10^{-20}\text{ J}}{(1.381\times10^{-23}\text{ J/K})(300\text{ K})}=20.88$$

$$C=R(20.88)^2\frac{e^{20.88}}{(e^{20.88}-1)^2}=(3.7\times10^{-7})R$$

(The vibrational contribution is still quite small.)

c) $T=2000\text{ K}$

$$\frac{\hbar\omega}{kT}=\frac{8.65\times10^{-20}\text{ J}}{(1.381\times10^{-23}\text{ J/K})(2000\text{ K})}=3.132$$

$$C=R(3.132)^2\frac{e^{3.132}}{(e^{3.132}-1)^2}=\underline{0.468R}$$

(At this high temperature the vibrational levels make an important contribution to C.)

44-15

Each atom occupies a cube with side length 0.282 nm. Therefore, the volume occupied by each atom is $V=(0.282\times10^{-9}\text{ m})^3=2.24\times10^{-29}\text{ m}^3$.

44-15 (cont)

In NaCl there are equal numbers of Na and Cl atoms, so the average mass of the atoms in the crystal is

$m = \frac{1}{2}(m_{Na} + m_{Cl}) = \frac{1}{2}(3.82\times10^{-26}\text{ kg} + 5.89\times10^{-26}\text{ kg}) = 4.855\times10^{-26}\text{ kg}$

The density then is $\rho = \frac{m}{V} = \frac{4.855\times10^{-26}\text{ kg}}{2.24\times10^{-29}\text{ m}^3} = \underline{2.16\times10^3\text{ kg/m}^3}$.

44-17

The photon energy must be at least $E = 1.14\text{ eV} = 1.827\times10^{-19}\text{ J}$.

The wavelength is given by $E = \frac{hc}{\lambda} \Rightarrow \lambda = \frac{hc}{E} = \frac{(6.626\times10^{-34}\text{ J}\cdot\text{s})(2.998\times10^{8}\text{ m/s})}{1.827\times10^{-19}\text{ J}} = 1.09\times10^{-6}\text{ m}$

$\lambda = \underline{1.09\,\mu\text{m}}$

44-19

a) The three-dimensional Schrodinger equation is $-\frac{\hbar^2}{2m}\left(\frac{\partial^2\psi}{\partial x^2} + \frac{\partial^2\psi}{\partial y^2} + \frac{\partial^2\psi}{\partial z^2}\right) + U\psi = E\psi$ (Eq. 44-22): $\psi = A\sin\frac{n_1\pi x}{L}\sin\frac{n_2\pi y}{L}\sin\frac{n_3\pi z}{L}$

Free electrons $\Rightarrow U = 0$

$\frac{\partial\psi}{\partial x} = \frac{n_1\pi}{L}A\cos\left(\frac{n_1\pi x}{L}\right)\sin\left(\frac{n_2\pi y}{L}\right)\sin\left(\frac{n_3\pi z}{L}\right)$

$\frac{\partial^2\psi}{\partial x^2} = -\left(\frac{n_1\pi}{L}\right)^2 A\sin\left(\frac{n_1\pi x}{L}\right)\sin\left(\frac{n_2\pi y}{L}\right)\sin\left(\frac{n_3\pi z}{L}\right) = -\left(\frac{n_1\pi}{L}\right)^2\psi$

Similarly $\frac{\partial^2\psi}{\partial y^2} = -\left(\frac{n_2\pi}{L}\right)^2\psi$ and $\frac{\partial^2\psi}{\partial z^2} = -\left(\frac{n_3\pi}{L}\right)^2\psi$

Therefore, $-\frac{\hbar^2}{2m}\left(\frac{\partial^2\psi}{\partial x^2} + \frac{\partial^2\psi}{\partial y^2} + \frac{\partial^2\psi}{\partial z^2}\right) + \cancel{U\psi}^{\,0} = -\frac{\hbar^2}{2m}\left(-\frac{\pi^2}{L^2}\right)(n_1^2+n_2^2+n_3^2)\psi$

$= \frac{(n_1^2+n_2^2+n_3^2)\pi^2\hbar^2}{2mL^2}\psi$

This equals $E\psi$, with $E = \frac{(n_1^2+n_2^2+n_3^2)\pi^2\hbar^2}{2mL^2}$, which is Eq. (44-33).

b) <u>Ground level</u> $\Rightarrow$ lowest $E \Rightarrow n_1 = n_2 = n_3 = 1$ and $E = \frac{3\pi^2\hbar^2}{2mL^2}$.

No other combination of n_1, n_2, and n_3 gives this same E so the only degeneracy is the degeneracy of two due to spin.

<u>First excited level</u> $\Rightarrow$ next lowest E $\Rightarrow$ one n equal 2 and the others equal 1.

$E = (2^2+1^2+1^2)\frac{\pi^2\hbar^2}{2mL^2} = \frac{6\pi^2\hbar^2}{2mL^2}$

There are three different sets of n_1, n_2, n_3 values that give this E:

$n_1 = 2, n_2 = 1, n_3 = 1$; $n_2 = 2, n_1 = 1, n_3 = 1$; and $n_3 = 2, n_1 = 1, n_2 = 1$.

This gives a degeneracy of 3 so the total degeneracy, with the factor of 2 from spin, is 6.

<u>Second excited level</u> $\Rightarrow$ next lowest E $\Rightarrow$ two n equal 2 and the others equal 1.

44-19 (cont)

$$E = (2^2+2^2+1^2)\frac{\pi^2\hbar^2}{2mL^2} = \frac{9\pi^2\hbar^2}{2mL^2}$$

There are three different sets of n_1, n_2, n_3 values that give this E:

$n_1=2, n_2=2, n_3=1$; $n_1=2, n_3=2, n_2=1$; $n_2=2, n_3=2, n_1=1$.

Thus as in part (b) the total degeneracy, including spin, is 6.

44-23

The electron contribution to the heat capacity of a metal is $C=\left(\frac{\pi^2 kT}{2E_F}\right)R$.

$$C = \frac{\pi^2(1.381\times10^{-23}\,J/K)(300\,K)}{2(7.06\,eV)(1.602\times10^{-19}\,J/eV)}R = \underline{0.0181\,R}.$$

44-25

Eq.(44-29): $f(E) = \frac{1}{e^{(E-E_F)/kT}+1}$, where $f(E)$ is the probability that a state with energy E is occupied. The donor levels are 0.01 eV below the bottom of the conduction band and E_F is midway between the donor levels and the bottom of the conduction band

$\Rightarrow E-E_F = 0.005\,eV$ and $\frac{E-E_F}{k} = \frac{(0.005\,eV)(1.602\times10^{-19}\,J/eV)}{1.381\times10^{-23}\,J/K} = 58.00\,K$

$T = 250\,K$

$$\frac{E-E_F}{kT} = \frac{58.00\,K}{250\,K} = 0.2320$$

$$f = \frac{1}{e^{0.2320}+1} = \underline{0.44}$$

$T = 300\,K$

$$\frac{E-E_F}{kT} = \frac{58.00\,K}{300\,K} = 0.1933$$

$$f = \frac{1}{e^{0.1933}+1} = \underline{0.45}$$

$T = 350\,K$

$$\frac{E-E_F}{kT} = \frac{58.00\,K}{350\,K} = 0.1657$$

$$f = \frac{1}{e^{0.1657}+1} = \underline{0.46}$$

f is large (near 50%) and not very sensitive to temperature, for temperatures near room temperature.

44-27

The voltage-current relation is given by Eq.(44-35): $I = I_0\left(e^{eV/kT}-1\right)$

a) Find I_0: $V = +15.0\times10^{-3}\,V$ gives $I = 12.0\times10^{-3}\,A$

$$\frac{eV}{kT} = \frac{(1.602\times10^{-19}\,C)(15.0\times10^{-3}\,V)}{(1.381\times10^{-23}\,J/K)(300\,K)} = 0.5800$$

$$I_0 = \frac{I}{e^{eV/kT}-1} = \frac{12.0\times10^{-3}\,A}{e^{0.5800}-1} = 1.527\times10^{-2}\,A = 15.27\,mA$$

44-27 (cont)

Then can calculate I for V = 10.0 mV:

$$\frac{eV}{kT} = \frac{(1.602\times10^{-19}\,C)(10.0\times10^{-3}\,V)}{(1.381\times10^{-23}\,J/K)(300\,K)} = 0.3867$$

$$I = I_0\left(e^{eV/kT}-1\right) = (15.27\,mA)\left(e^{0.3867}-1\right) = \underline{7.21\,mA}$$

b) $\frac{eV}{kT}$ has the same magnitudes as in part (a) but now V is negative so $\frac{eV}{kT}$ is negative.

$\underline{V = -15.0\,mV} \Rightarrow \frac{eV}{kT} = -0.5800$

$$I = I_0\left(e^{eV/kT}-1\right) = (15.27\,mA)\left(e^{-0.5800}-1\right) = \underline{-6.72\,mA}$$

$\underline{V = -10.0\,mV} \Rightarrow \frac{eV}{kT} = -0.3867$

$$I = I_0\left(e^{eV/kT}-1\right) = (15.27\,mA)\left(e^{-0.3867}-1\right) = \underline{-4.90\,mA}$$

Problems

44-29

a) U must make up for the difference between the ionization energy of Na (5.1 eV) and the electron affinity of Cl (3.6 eV).

$$\Rightarrow U = -(5.1\,eV - 3.6\,eV) = -1.5\,eV = -2.40\times10^{-19}\,J$$

$$-\frac{1}{4\pi\epsilon_0}\frac{e^2}{r} = -2.40\times10^{-19}\,J$$

$$r = \frac{1}{4\pi\epsilon_0}\frac{e^2}{2.40\times10^{-19}\,J} = 8.987\times10^9\,N\cdot m^2/C^2\,\frac{(1.602\times10^{-19}\,C)^2}{2.40\times10^{-19}\,J} = 9.60\times10^{-10}\,m = \underline{0.96\,nm}$$

b) Ionization energy of K = 4.3 eV

Electron affinity of Br = 3.5 eV

$$\Rightarrow U = -(4.3\,eV - 3.5\,eV) = -0.8\,eV = -1.28\times10^{-19}\,J$$

$$-\frac{1}{4\pi\epsilon_0}\frac{e^2}{r} = -1.28\times10^{-19}\,J$$

$$r = \frac{1}{4\pi\epsilon_0}\frac{e^2}{1.28\times10^{-19}\,J} = 8.987\times10^9\,N\cdot m^2/C^2\,\frac{(1.602\times10^{-19}\,C)^2}{1.28\times10^{-19}\,J} = 1.8\times10^{-9}\,m = \underline{1.8\,nm}$$

44-33

a) $E_\ell = \ell(\ell+1)\frac{\hbar^2}{2I}$

Calculate I for $Na^{35}Cl$:

$$m_r = \frac{m_1 m_2}{m_1+m_2} = \frac{m_{Na}\,m_{Cl}}{m_{Na}+m_{Cl}} = \frac{(3.8176\times10^{-26}\,kg)(5.8068\times10^{-26}\,kg)}{3.8176\times10^{-26}\,kg + 5.8068\times10^{-26}\,kg} = 2.303\times10^{-26}\,kg$$

$$I = m_r r^2 = 2.303\times10^{-26}\,kg\,(0.2361\times10^{-9}\,m)^2 = 1.284\times10^{-45}\,kg\cdot m^2$$

$\underline{\ell = 2 \rightarrow \ell = 1 \text{ transition}}$

$$\Delta E = E_2 - E_1 = [6-2]\frac{\hbar^2}{2I} = \frac{2\hbar^2}{I} = \frac{2(1.054\times10^{-34}\,J\cdot s)^2}{1.284\times10^{-45}\,kg\cdot m^2} = 1.730\times10^{-23}\,J$$

$$\Delta E = \frac{hc}{\lambda} \Rightarrow \lambda = \frac{hc}{\Delta E} = \frac{(6.626\times10^{-34}\,J\cdot s)(2.998\times10^8\,m/s)}{1.730\times10^{-23}\,J} = 1.148\times10^{-2}\,m = \underline{1.148\,cm}$$

44-33 (cont)

$l=1 \rightarrow l=0$ transition

$$\Delta E = E_1 - E_0 = [2-0]\frac{\hbar^2}{2I} = \frac{\hbar^2}{I} = \frac{1}{2}(1.730\times10^{-23}\,\text{J}) = 8.650\times10^{-24}\,\text{J}$$

$$\lambda = \frac{hc}{\Delta E} = \frac{(6.626\times10^{-34}\,\text{J}\cdot\text{s})(2.998\times10^{8}\,\text{m/s})}{8.650\times10^{-24}\,\text{J}} = 2.297\times10^{-2}\,\text{m} = \underline{2.297\,\text{cm}}$$

b) Calculate I for Na ^{37}Cl:

$$m_r = \frac{m_{Na}\,m_{Cl}}{m_{Na}+m_{Cl}} = \frac{(3.8176\times10^{-26}\,\text{kg})(6.1384\times10^{-26}\,\text{kg})}{3.8176\times10^{-26}\,\text{kg}+6.1384\times10^{-26}\,\text{kg}} = 2.354\times10^{-26}\,\text{kg}$$

$$I = m_r r^2 = 2.354\times10^{-26}\,\text{kg}\,(0.2361\times10^{-9}\,\text{m})^2 = 1.312\times10^{-45}\,\text{kg}\cdot\text{m}^2$$

$l=2 \rightarrow l=1$ transition

$$\Delta E = \frac{2\hbar^2}{I} = \frac{2(1.054\times10^{-34}\,\text{J}\cdot\text{s})^2}{1.312\times10^{-45}\,\text{kg}\cdot\text{m}^2} = 1.693\times10^{-23}\,\text{J}$$

$$\lambda = \frac{hc}{\Delta E} = \frac{(6.626\times10^{-34}\,\text{J}\cdot\text{s})(2.998\times10^{8}\,\text{m/s})}{1.693\times10^{-23}\,\text{J}} = \underline{1.173\,\text{cm}}$$

$l=1 \rightarrow l=0$ transition

$$\Delta E = \frac{\hbar^2}{I} = \frac{1}{2}(1.693\times10^{-23}\,\text{J}) = 8.465\times10^{-24}\,\text{J}$$

$$\lambda = \frac{hc}{\Delta E} = \frac{(6.626\times10^{-34}\,\text{J}\cdot\text{s})(2.998\times10^{8}\,\text{m/s})}{8.465\times10^{-24}\,\text{J}} = \underline{2.347\,\text{cm}}$$

The differences in the wavelengths for the two isotopes are:

$l=2 \rightarrow l=1$ transition: $1.173\,\text{cm} - 1.148\,\text{cm} = \underline{0.025\,\text{cm}}$

$l=1 \rightarrow l=0$ transition: $2.347\,\text{cm} - 2.297\,\text{cm} = \underline{0.050\,\text{cm}}$

44-35

a) Problem 44-34 gives $I = 2.71\times10^{-47}\,\text{kg}\cdot\text{m}^2$.

$$I = m_r r^2$$

$$m_r = \frac{m_H\,m_{Cl}}{m_H+m_{Cl}} = \frac{(1.67\times10^{-27}\,\text{kg})(5.81\times10^{-26}\,\text{kg})}{1.67\times10^{-27}\,\text{kg}+5.81\times10^{-26}\,\text{kg}} = 1.623\times10^{-27}\,\text{kg}$$

$$r = \sqrt{\frac{I}{m_r}} = \sqrt{\frac{2.71\times10^{-47}\,\text{kg}\cdot\text{m}^2}{1.623\times10^{-27}\,\text{kg}}} = 1.29\times10^{-10}\,\text{m} = \underline{0.129\,\text{nm}}$$

b) Each transition is between the level l to the level $l-1$.

$E_l = l(l+1)\frac{\hbar^2}{2I}$, so $\Delta E = E_l - E_{l-1} = [l(l+1) - l(l-1)]\frac{\hbar^2}{2I} = l\frac{\hbar^2}{I}$.

The transition energy is related to the photon wavelength by $\Delta E = \frac{hc}{\lambda}$.

Combine these two equations for $\Delta E \Rightarrow l\frac{\hbar^2}{I} = \frac{hc}{\lambda}$

$$l = \frac{2\pi c I}{\hbar\lambda} = \frac{2\pi(2.998\times10^{8}\,\text{m/s})(2.71\times10^{-47}\,\text{kg}\cdot\text{m}^2)}{(1.054\times10^{-34}\,\text{J}\cdot\text{s})\,\lambda} = \frac{4.843\times10^{-4}\,\text{m}}{\lambda}$$

$$\lambda = 60.4\,\mu\text{m} \Rightarrow l = \frac{4.843\times10^{-4}\,\text{m}}{60.4\times10^{-6}\,\text{m}} = 8$$

$$\lambda = 69.0\,\mu\text{m} \Rightarrow l = \frac{4.843\times10^{-4}\,\text{m}}{69.0\times10^{-6}\,\text{m}} = 7$$

$$\lambda = 80.4\,\mu\text{m} \Rightarrow l = \frac{4.843\times10^{-4}\,\text{m}}{80.4\times10^{-6}\,\text{m}} = 6$$

44-35 (cont)

$\lambda = 96.4\,\mu m \Rightarrow l = \dfrac{4.843\times10^{-4}\,m}{96.4\times10^{-6}\,m} = 5$

$\lambda = 120.4\,\mu m \Rightarrow l = \dfrac{4.843\times10^{-4}\,m}{120.4\times10^{-6}\,m} = 4$

c) Longest $\lambda \Rightarrow$ smallest $\Delta E \Rightarrow$ transition from $l=1$ to $l=0$.

$\Delta E = l\dfrac{\hbar^2}{I} = (1)\dfrac{(1.054\times10^{-34}\,J\cdot s)^2}{2.71\times10^{-47}\,kg\cdot m^2} = 4.099\times10^{-22}\,J$

$\lambda = \dfrac{hc}{\Delta E} = \dfrac{(6.626\times10^{-34}\,J\cdot s)(2.998\times10^{8}\,m/s)}{4.099\times10^{-22}\,J} = 4.85\times10^{-4}\,m = \underline{485\,\mu m}$

d) What changes is m_r, the reduced mass of the molecule. The transition energy is $\Delta E = l\dfrac{\hbar^2}{I}$ and $\Delta E = \dfrac{hc}{\lambda} \Rightarrow \lambda = \dfrac{2\pi c I}{l\hbar}$ (part (b)).

$I = m_r r^2$, so λ is directly proportional to m_r.

$\dfrac{\lambda(HCl)}{m_r(HCl)} = \dfrac{\lambda(DCl)}{m_r(DCl)} \Rightarrow \lambda(DCl) = \lambda(HCl)\dfrac{m_r(DCl)}{m_r(HCl)}$

The mass of a deuterium atom is approximately twice the mass of a hydrogen atom $\Rightarrow m_D = 3.34\times10^{-27}\,kg$.

$m_r(DCl) = \dfrac{m_D m_{Cl}}{m_D + m_{Cl}} = \dfrac{(3.34\times10^{-27}\,kg)(5.81\times10^{-26}\,kg)}{3.34\times10^{-27}\,kg + 5.81\times10^{-26}\,kg} = 3.158\times10^{-27}\,kg$

$\lambda(DCl) = \lambda(HCl)\dfrac{3.158\times10^{-27}\,kg}{1.623\times10^{-27}\,kg} = (1.946)\,\lambda(HCl)$

$l=8 \rightarrow l=7$: $\lambda = (60.4\,\mu m)(1.946) = 118\,\mu m$

$l=7 \rightarrow l=6$: $\lambda = (69.0\,\mu m)(1.946) = 134\,\mu m$

$l=6 \rightarrow l=5$: $\lambda = (80.4\,\mu m)(1.946) = 156\,\mu m$

$l=5 \rightarrow l=4$: $\lambda = (96.4\,\mu m)(1.946) = 188\,\mu m$

$l=4 \rightarrow l=3$: $\lambda = (120.4\,\mu m)(1.946) = 234\,\mu m$

44-37

a) $I = m_r r^2$

$m_r = \dfrac{m_C m_O}{m_C + m_O} = \dfrac{(1.99\times10^{-26}\,kg)(2.66\times10^{-26}\,kg)}{1.99\times10^{-26}\,kg + 2.66\times10^{-26}\,kg} = 1.138\times10^{-26}\,kg$

$I = m_r r^2 = (1.138\times10^{-26}\,kg)(0.113\times10^{-9}\,m)^2 = \underline{1.45\times10^{-46}\,kg\cdot m^2}$

b) The energy levels are $E_{nl} = l(l+1)\dfrac{\hbar^2}{2I} + (n+\tfrac{1}{2})\hbar\sqrt{\dfrac{k}{m}}$ (Eq. 44-9)

$\sqrt{\dfrac{k}{m}} = \omega = 2\pi f \Rightarrow E_{nl} = l(l+1)\dfrac{\hbar^2}{2I} + (n+\tfrac{1}{2})hf$

(i) transition, $n=1 \rightarrow n=0$, $l=1 \rightarrow l=0$

$\Delta E = (2-0)\dfrac{\hbar^2}{2I} + (1+\tfrac{1}{2}-\tfrac{1}{2})hf = \dfrac{\hbar^2}{I} + hf$

$\Delta E = \dfrac{hc}{\lambda} \Rightarrow \lambda = \dfrac{hc}{\Delta E} = \dfrac{hc}{\frac{\hbar^2}{I} + hf} = \dfrac{c}{\frac{\hbar}{2\pi I} + f}$

$\dfrac{\hbar}{2\pi I} = \dfrac{1.054\times10^{-34}\,J\cdot s}{2\pi(1.45\times10^{-46}\,kg\cdot m^2)} = 1.157\times10^{11}\,Hz$

$\lambda = \dfrac{c}{\frac{\hbar}{2\pi I} + f} = \dfrac{2.998\times10^{8}\,m/s}{1.157\times10^{11}\,Hz + 6.498\times10^{13}\,Hz} = \underline{4.606\,\mu m}$

44-37 (cont)

(ii) transition $n=1 \rightarrow n=0$, $l=2 \rightarrow l=1$

$$\Delta E = (6-2)\frac{\hbar^2}{2I} + hf = 2\frac{\hbar^2}{I} + hf$$

$$\lambda = \frac{c}{2\left(\frac{\hbar}{2\pi I}\right) + f} = \frac{2.998\times10^8\ \text{m/s}}{2(1.157\times10^{11}\ \text{Hz}) + 6.498\times10^{13}\ \text{Hz}} = \underline{4.597\ \mu\text{m}}$$

(iii) transition $n=2 \rightarrow n=0$, $l=1 \rightarrow l=0$

$$\Delta E = (2-0)\frac{\hbar^2}{2I} + 2hf = \frac{\hbar^2}{I} + 2hf$$

$$\lambda = \frac{c}{\frac{\hbar}{2\pi I} + 2f} = \frac{2.998\times10^8\ \text{m/s}}{1.157\times10^{11}\ \text{Hz} + 2(6.498\times10^{13}\ \text{Hz})} = \underline{2.305\ \mu\text{m}}$$

44-41

a) $$U_{tot} = -\frac{\alpha e^2}{4\pi\epsilon_0 r} + \frac{A}{r^8}$$

$$\frac{dU_{tot}}{dr} = \frac{\alpha e^2}{4\pi\epsilon_0 r^2} - \frac{8A}{r^9}$$

$$\left.\frac{dU_{tot}}{dr}\right|_{r=r_0} = 0 \Rightarrow \frac{\alpha e^2}{4\pi\epsilon_0 r_0^2} = \frac{8A}{r_0^9}$$

$$A = \frac{r_0^7 \alpha e^2}{32\pi\epsilon_0}$$

Then $U_{tot} = -\frac{\alpha e^2}{4\pi\epsilon_0 r} + \frac{r_0^7 \alpha e^2}{32\pi\epsilon_0 r^8}$.

$$r = r_0 \Rightarrow U_{tot} = -\frac{\alpha e^2}{4\pi\epsilon_0 r_0} + \frac{\alpha e^2}{32\pi\epsilon_0 r_0} = -\frac{7\alpha e^2}{32\pi\epsilon_0 r_0} = -\left(\frac{1}{4\pi\epsilon_0}\right)\left(\frac{7\alpha e^2}{8 r_0}\right)$$

For NaCl,

$$U_{tot} = -(8.987\times10^9\ \text{N}\cdot\text{m}^2/\text{C}^2)\left(\frac{7(1.75)(1.602\times10^{-19}\ \text{C})^2}{8(0.281\times10^{-9}\ \text{m})}\right) = -1.257\times10^{-18}\ \text{J} = -7.85\ \text{eV}$$

b) To remove a Na^+Cl^- ion pair from the crystal requires 7.85 eV. When neutral Na and Cl atoms are formed from the Na^+ and Cl^- atoms there is a net release of energy $-5.14\ \text{eV} + 3.61\ \text{eV} = -1.53\ \text{eV}$, so the net energy required to remove a neutral Na, Cl pair from the crystal is $7.85\ \text{eV} - 1.53\ \text{eV} = \underline{6.32\ \text{eV}}$.

CHAPTER 45

Exercises 3, 5, 9, 11, 13, 15, 19, 21, 25, 27

Problems 31, 33, 35, 37, 41, 43

Exercises

45-3

When the spin component is parallel to the field the interaction energy is $U = -\mu_z B$. When the spin component is antiparallel to the field the interaction energy is $U = +\mu_z B$. The transition energy for a transition between these two states is $\Delta E = 2\mu_z B$, where $\mu_z = 2.7928\,\mu_n$.

The transition energy is related to the photon frequency by $\Delta E = hf$ $\Rightarrow 2\mu_z B = hf$.

$$B = \frac{hf}{2\mu_z} = \frac{(6.626\times10^{-34}\,\text{J}\cdot\text{s})(90.8\times10^{6}\,\text{Hz})}{2(2.7928)(5.051\times10^{-27}\,\text{J/T})} = \underline{2.13\,\text{T}}$$

45-5

The text calculates that the binding energy of the deuteron is 2.23 MeV. A photon that breaks the deuteron up into a proton and a neutron must have at least this much energy.

$$E = \frac{hc}{\lambda} \Rightarrow \lambda = \frac{hc}{E} = \frac{(6.626\times10^{-34}\,\text{J}\cdot\text{s})(2.998\times10^{8}\,\text{m/s})}{(2.23\times10^{6}\,\text{eV})(1.602\times10^{-19}\,\text{J/eV})} = 5.56\times10^{-13}\,\text{m} = \underline{0.556\,\text{pm}}$$

45-9

a) A ${}^{9}_{4}\text{Be}$ atom has 4 protons, 9−4 = 5 neutrons, and 4 electrons. The mass defect therefore is $\Delta M = 4m_p + 5m_n + 4m_e - M({}^{9}_{4}\text{Be})$

$\Delta M = 4(1.0072765\,\text{u}) + 5(1.0086649\,\text{u}) + 4(0.0005485799\,\text{u}) - 9.012182\,\text{u} = 0.062443\,\text{u}$

The energy equivalent is $E_B = (0.062443\,\text{u})(931.5\,\text{MeV/u}) = \underline{58.2\,\text{MeV}}$.

b) Eq. (45-9): $E_B = C_1 A - C_2 A^{2/3} - C_3 \frac{Z(Z-1)}{A^{1/3}} - C_4 \frac{(A-2Z)^2}{A}$

$$E_B = (15.7\,\text{MeV})(9) - (17.8\,\text{MeV})(9)^{2/3} - (0.71\,\text{MeV})\frac{4(3)}{9^{1/3}} - (23.7\,\text{MeV})\frac{(9-8)^2}{9}$$

$$E_B = +141.3\,\text{MeV} - 77.0\,\text{MeV} - 4.1\,\text{MeV} - 2.6\,\text{MeV} = 57.6\,\text{MeV}$$

The percentage difference between the calculated and measured E_B is

$$\frac{57.6\,\text{MeV} - 58.2\,\text{MeV}}{58.2\,\text{MeV}} = \underline{-1.0\%}.$$

Eq. (45-9) has a greater percentage accuracy for ${}^{56}\text{Fe}$.

45-11

a) α decay : Z decreases by 2, $A = N + Z$ decreases by 4 (α particle is a ${}^4_2\text{He}$ nucleus)

$${}^{238}_{94}\text{Pu} \rightarrow {}^4_2\text{He} + {}^{234}_{92}\text{U}$$

b) β^- decay : Z increases by 1, $A = N + Z$ remains the same (β^- particle is an electron ${}^{0}_{-1}e$)

$${}^{19}_{8}\text{O} \rightarrow {}^{0}_{-1}e + {}^{19}_{9}\text{F}$$

c) β^+ decay: Z decreases by 1, $A = N + Z$ remains the same (β^+ particle is a positron ${}^{0}_{+1}e$)

$${}^{25}_{13}\text{Al} \rightarrow {}^{0}_{+1}e + {}^{25}_{12}\text{Mg}$$

Note: In each case the total charge and total number of nucleons for the decay products equals the charge and number of nucleons for the parent nucleus; these two quantities are conserved in the decay.

45-13

a) The β^- decay reaction is ${}^3_1\text{H} \rightarrow {}^{0}_{-1}e + {}^3_2\text{He}$. ${}^3_1\text{H}$ is unstable with respect to β^- decay since the mass of the ${}^3_2\text{He}$ nucleus plus the emitted electron is less than the mass of the ${}^3_1\text{H}$ nucleus:

The mass of the ${}^3_1\text{H}$ nucleus is $3.016049\ u - 0.0005486\ u = 3.015500\ u$.

The masses in Table 45-2 are for the neutral atoms, so the mass of the ${}^3_2\text{He}$ nucleus is $3.01603\ u - 2(0.0005486\ u) = 3.01493\ u$.

The mass decrease in the decay is $3.015500\ u - 3.01493\ u - 0.0005486\ u = 0.00002\ u$.

b) Find the energy equivalent of the mass decrease found in part (a):

$0.00002\ u\ (931.5\ \text{MeV}/u) = 0.02\ \text{MeV} = \underline{20\ \text{keV}}$

45-15

a) Eq. (45-17): $\lambda = \dfrac{0.693}{T_{1/2}} = \dfrac{0.693}{(4.47\times10^9\ y)\left(\dfrac{3.156\times10^7\ s}{1\ y}\right)} = \underline{4.91\times10^{-18}\ s^{-1}}$

b) Activity of 1.00 Ci $\Rightarrow \left|\dfrac{dN}{dt}\right| = 3.70\times10^{10}$ decays/s

Eq. (45-14): $\left|\dfrac{dN}{dt}\right| = \lambda N \Rightarrow N = \dfrac{|dN/dt|}{\lambda} = \dfrac{3.70\times10^{10}\ \text{decays/s}}{4.91\times10^{-18}\ s^{-1}} = 7.54\times10^{27}$ nuclei

The sample must contain 7.54×10^{27} uranium nuclei.

The mass of one uranium ${}^{238}\text{U}$ atom is approximately 238 u, so the mass of the sample is $m = N(238\ u) = (7.54\times10^{27})(238)(1.66054\times10^{-27}\ kg) = \underline{2.98\times10^3\ kg}$

c) One α particle is emitted in each decay, so the problem is asking for the activity, in decays/s, of this sample.

Compute N, the number of nuclei in the sample. The mass of one ${}^{238}\text{U}$ atom is approximately 238 u, so $N = \dfrac{20.0\times10^{-3}\ kg}{238\ u} = \dfrac{20.0\times10^{-3}\ kg}{238(1.66054\times10^{-27}\ kg)} = 5.061\times10^{22}$ nuclei.

45-15 (cont) $\left|\frac{dN}{dt}\right| = \lambda N = (4.91\times10^{-18}\,s^{-1})(5.061\times10^{22}) = \underline{2.48\times10^{5}\text{ decays/s}}$

45-19

a) $\left|\frac{dN}{dt}\right| = \lambda N$

$$\lambda = \frac{0.693}{T_{1/2}} = \frac{0.693}{(1.28\times10^{9}\,y)\left(\frac{3.156\times10^{7}\,s}{1\,y}\right)} = 1.715\times10^{-17}\,s^{-1}$$

The mass of one ^{40}K atom is approximately 40 u, so the number of ^{40}K nuclei in the sample is $N = \frac{5.00\times10^{-9}\,kg}{40\,u} = \frac{5.00\times10^{-9}\,kg}{40(1.66054\times10^{-27}\,kg)} = 7.528\times10^{16}$.

$\left|\frac{dN}{dt}\right| = \lambda N = (1.715\times10^{-17}\,s^{-1})(7.528\times10^{16}) = \underline{1.29\text{ decays/s}}$

b) $\left|\frac{dN}{dt}\right| = (1.29\text{ decays/s})\left(\frac{1\,Ci}{3.70\times10^{10}\text{ decay/s}}\right) = \underline{3.49\times10^{-11}\,Ci}$

45-21

a) The energy E of each photon is $E = \frac{hc}{\lambda} = \frac{(6.626\times10^{-34}\,J\cdot s)(2.998\times10^{8}\,m/s)}{0.0200\times10^{-9}\,m} = 9.932\times10^{-15}\,J$

The total energy absorbed is the number of photons absorbed times the energy of each photon: $(8.00\times10^{10}\text{ photons})(9.932\times10^{-15}\text{ J/photon}) = \underline{7.95\times10^{-4}\,J}$

b) The absorbed dose is the energy absorbed divided by the mass of the absorbing tissue:

$$\text{absorbed dose} = \frac{7.95\times10^{-4}\,J}{0.600\,kg} = 1.32\times10^{-3}\,J/kg\left(\frac{1\,rad}{0.01\,J/kg}\right) = 0.132\,rad$$

The equivalent dose in rem equals the RBE times the absorbed dose in rad. For x rays RBE = 1, so the equivalent dose is $\underline{0.132\text{ rem}}$.

45-25

$${}^{2}_{1}H + {}^{9}_{4}Be \rightarrow {}^{7}_{3}Li + {}^{4}_{2}He$$

a) If we use the neutral atom masses then there are the same number of electrons (five) in the reactants as in the products. Their masses cancel, so we get the same mass defect whether we use nuclear masses or neutral atom masses.

The neutral atom masses are given in Table 45-2.
${}^{2}_{1}H + {}^{9}_{4}Be$ has mass $2.01410\,u + 9.01218\,u = 11.02628\,u$.
${}^{7}_{3}Li + {}^{4}_{2}He$ has mass $7.01600\,u + 4.00260\,u = 11.01860\,u$.
The mass decrease is $11.02628\,u - 11.018600\,u = 0.00768\,u$.
this corresponds to an energy release of $0.00768\,u\left(\frac{931.5\,MeV}{1\,u}\right) = \underline{7.15\,MeV}$.

b) Estimate the threshold energy by calculating the Coulomb potential energy when the ${}^{2}_{1}H$ is at the radius of the ${}^{9}_{4}Be$ nucleus.

45-25 (cont)

$r = r_0 A^{1/3}$ (Eq. 45-1) $\Rightarrow r = 1.2\times10^{-15}\,\text{m}\,(9)^{1/3} = 2.5\times10^{-15}\,\text{m}$ is the ${}^{9}_{4}\text{Be}$ nuclear radius.

$U = \frac{1}{4\pi\epsilon_0}\frac{q_1 q_2}{r} \Rightarrow U = \frac{1}{4\pi\epsilon_0}\frac{(1)(4)e^2}{r} = (9.0\times10^{9}\,\text{N}\cdot\text{m}^2/\text{C}^2)\frac{4(1.60\times10^{-19}\,\text{C})^2}{2.5\times10^{-15}\,\text{m}} = 3.7\times10^{-13}\,\text{J}$

$U = (3.7\times10^{-13}\,\text{J})\left(\frac{1\,\text{eV}}{1.602\times10^{-19}\,\text{J}}\right) = 2.3\times10^{6}\,\text{eV} = \underline{2.3\,\text{MeV}}$

This is a rough estimate of the threshold energy for the reaction.

45-27

a) ${}^{2}_{1}\text{H} + {}^{14}_{7}\text{N} \rightarrow {}^{6}_{3}\text{Li} + {}^{10}_{5}\text{B}$

The neutral atoms on each side of the reaction equation have a total of 8 electrons, so the electron masses cancel when neutral atom masses are used. The neutral atom masses are found in Table 45-2.

mass of ${}^{2}_{1}\text{H} + {}^{14}_{7}\text{N}$ is $2.01410\,\text{u} + 14.00307\,\text{u} = 16.01717\,\text{u}$

mass of ${}^{6}_{3}\text{Li} + {}^{10}_{5}\text{B}$ is $6.01512\,\text{u} + 10.01294\,\text{u} = 16.02806\,\text{u}$

The mass increases, so energy is absorbed by the reaction.

The Q value is $(16.01717\,\text{u} - 16.02806\,\text{u})(931.5\,\text{MeV/u}) = \underline{-10.1\,\text{MeV}}$

b) The kinetic energy that must be available to cause the reaction is 10.1 MeV. Thus in Eq. (45-24) $K_{cm} = 10.1$ MeV. The mass M of the stationary target (${}^{14}_{7}\text{N}$) is $M = 14\,\text{u}$. The mass m of the colliding particle (${}^{2}_{1}\text{H}$) is 2u. Then by Eq. (45-24) the minimum kinetic energy K that the ${}^{2}_{1}\text{H}$ must have is

$K = \frac{M+m}{M}K_{cm} = \left(\frac{14\,\text{u}+2\,\text{u}}{14\,\text{u}}\right)(10.1\,\text{MeV}) = \underline{11.5\,\text{MeV}}$

Problems

45-31

${}^{198}_{79}\text{Au} \rightarrow {}^{198}_{80}\text{Hg} + {}^{0}_{-1}e^{-}$

The mass change is $197.968217\,\text{u} - 197.966743\,\text{u} = 1.474\times10^{-3}\,\text{u}$.
(The neutral atom masses include 79 electrons before the decay and 80 electrons after the decay. This one additional electron in the products accounts correctly for the electron emitted by the nucleus.)

The total energy released in the decay is $(1.474\times10^{-3}\,\text{u})(931.5\,\text{MeV/u}) = 1.373\,\text{MeV}$. This energy is divided between the energy of the emitted photon and the kinetic energy of the β^- particle. Thus the β^- has kinetic energy $1.373\,\text{MeV} - 0.412\,\text{MeV} = \underline{0.961\,\text{MeV}}$.

45-33

We have to be careful; after ^{87}Rb has undergone radioactive decay it is no longer a rubidium atom.
Let N_{85} be the number of ^{85}Rb atoms; this number doesn't change.
Let N_0 be the number of ^{87}Rb atoms on earth when the solar system was formed.
Let N be the present number of ^{87}Rb atoms.

The present measurements say that

$$0.2783 = \frac{N}{N+N_{85}} \Rightarrow (N+N_{85})(0.2783) = N \Rightarrow \boxed{N = 0.3856\, N_{85}}$$

The percentage we are asked to calculate is $\frac{N_0}{N_0+N_{85}}$.
N and N_0 are related by $N = N_0 e^{-\lambda t} \Rightarrow N_0 = e^{+\lambda t} N$

$$\text{So } \frac{N_0}{N_0+N_{85}} = \frac{N e^{\lambda t}}{N e^{\lambda t} + N_{85}} = \frac{(0.3856\, e^{\lambda t}) \cancel{N_{85}}}{0.3856 e^{\lambda t} \cancel{N_{85}} + \cancel{N_{85}}} = \frac{0.3856\, e^{\lambda t}}{0.3856 e^{\lambda t} + 1}$$

$$t = 4.6\times10^{9}\,y \;;\; \lambda = \frac{0.693}{T_{1/2}} = \frac{0.693}{4.89\times10^{10}\,y} = 1.417\times10^{-11}\,y^{-1}$$

$$e^{\lambda t} = e^{(1.417\times10^{-11}y^{-1})(4.6\times10^{9}y)} = e^{0.06518} = 1.0674$$

$$\Rightarrow \frac{N_0}{N_0+N_{85}} = \frac{(0.3856)(1.0674)}{(0.3856)(1.0674)+1} = 0.2916 = \underline{29.16\%}$$

45-35

a) First find the number of decays each second:

$$35.0\times10^{-6}\,Ci\left(\frac{3.70\times10^{10}\,\text{decays/s}}{1\,Ci}\right) = 1.295\times10^{6}\,\text{decays/s}$$

The average energy per decay is 1.25 MeV, and one-half of this energy is deposited in the tumor. The energy delivered to the tumor per second then is

$$\tfrac{1}{2}(1.295\times10^{6}\,\text{decays/s})(1.25\times10^{6}\,\text{eV/decay})(1.602\times10^{-19}\,\text{J/eV}) = \underline{1.30\times10^{-7}\,\text{J/s}}$$

b) The absorbed dose is the energy absorbed divided by the mass of the tissue:

$$\frac{1.30\times10^{-7}\,\text{J/s}}{0.500\,\text{kg}} = 2.60\times10^{-7}\,\text{J/kg}\cdot\text{s}\left(\frac{1\,\text{rad}}{0.01\,\text{J/kg}}\right) = \underline{2.60\times10^{-5}\,\text{rad/s}}$$

c) equivalent dose (rem) = RBE × absorbed dose (rad)
In one second the equivalent dose is $(0.70)(2.60\times10^{-5}\,\text{rad}) = \underline{1.82\times10^{-5}\,\text{rem}}$

d) $$\frac{200\,\text{rem}}{1.82\times10^{-5}\,\text{rem/s}} = 1.10\times10^{7}\,s\left(\frac{1\,h}{3600\,s}\right) = 3050\,h = \underline{127\,d.}$$

45-37

$N = N_0 e^{-\lambda t}$

The problem says $\frac{N}{N_0} = \frac{1}{8}$; solve for t.

$\frac{1}{8} = e^{-\lambda t} \Rightarrow \ln 8 = \lambda t \Rightarrow t = \frac{\ln 8}{\lambda}$

Example 45-8 gives $\lambda = 1.21\times10^{-4}\,y^{-1}$ for ^{14}C.

45-37 (cont)

Thus $t = \frac{\ln 8}{1.21\times10^{-4}\,y} = \underline{1.7\times10^{4}\,y.}$

45-41

a) The radius of ${}^{2}_{1}H$ is $R = (1.2\times10^{-15}\,m)(2)^{1/3} = 1.51\times10^{-15}\,m$.
The barrier energy is the Coulomb potential energy of two ${}^{2}_{1}H$ nuclei separated by this distance:
$U = \frac{1}{4\pi\epsilon_0}\frac{e^2}{r} = (8.987\times10^{9}\,N\cdot m^2/C^2)\frac{(1.602\times10^{-19}C)^2}{1.51\times10^{-15}\,m} = 1.527\times10^{-13}\,J = \underline{0.95\,MeV}$

b) ${}^{2}_{1}H + {}^{2}_{1}H \rightarrow {}^{3}_{2}He + {}^{1}_{0}n$
If we use neutral atom masses there are two electrons on each side of the reaction equation, so their masses cancel. The neutral atom masses are given in Table 45-2.

${}^{2}_{1}H + {}^{2}_{1}H$ has mass $2(2.01410\,u) = 4.02820\,u$
${}^{3}_{2}He + {}^{1}_{0}n$ has mass $3.01603\,u + 1.008665\,u = 4.024695\,u$
The mass decrease is $4.02820\,u - 4.024695\,u = 3.505\times10^{-3}\,u$. This corresponds to a liberated energy of $(3.505\times10^{-3}u)(931.5\,MeV/u) = \underline{3.26\,MeV}$, or $(3.26\times10^{6}eV)(1.602\times10^{-19}J/eV) = \underline{5.22\times10^{-13}\,J.}$

c) Each reaction takes two ${}^{2}_{1}H$ nuclei. Each mole of D_2 has 6.022×10^{23} molecules so 6.022×10^{23} pairs of atoms. The energy liberated when one mole of deuterium undergoes fusion is $(6.022\times10^{23})(5.22\times10^{-13}J) = \underline{3.14\times10^{11}\,J/mol}$

45-43

Consider 1.00 kg of body tissue. The mass of ${}^{40}K$ in 1.00 kg of tissue is $(0.21\times10^{-2})(0.012\times10^{-2})(1.00\,kg) = 2.52\times10^{-7}kg$.

The mass of a ${}^{40}K$ atom is approximately 40 u, so the number of ${}^{40}K$ nuclei in 1.00 kg of tissue is $\frac{2.52\times10^{-7}kg}{40u} = \frac{2.52\times10^{-7}\,kg}{40u\,(1.66054\times10^{-27}kg/u)} = 3.79\times10^{18}$

The activity is $\left|\frac{dN}{dt}\right| = \lambda N$.
$\lambda = \frac{0.693}{T_{1/2}} = \frac{0.693}{1.25\times10^{9}\,y} = 5.544\times10^{-10}\,y^{-1}$

$\left|\frac{dN}{dt}\right| = (5.544\times10^{-10}y^{-1})(3.794\times10^{18}) = 2.103\times10^{9}$ decays/y
In 70 y there are $(70y)(2.103\times10^{9}$ decays/y$) = 1.47\times10^{11}$ decays.

For each decay an average of 0.50 MeV of energy is absorbed.
The energy absorbed by 1.00 kg of tissue is $(1.47\times10^{11}$ decays$)(0.50$ MeV/decay$)$
$= 7.35\times10^{10}MeV = 7.35\times10^{16}eV\,(1.602\times10^{-19}\,J/eV) = 0.012\,J.$

The absorbed dose is $0.012\,J/kg\left(\frac{1\,rad}{0.01\,J/kg}\right) = \underline{1.2\,rad}$
RBE = 1.0, so the equivalent dose is $\underline{1.2\,rem.}$

CHAPTER 46

Exercises 3, 5, 7, 9, 13, 15, 19, 23, 25

Problems 27, 29, 35, 37

Exercises

46-3

By momentum conservation the two photons must have equal and opposite momenta. Then $E = pc$ says the photons must have equal energies.

The rest energy of the pion is $(0.145\text{ u})(931.5\text{ MeV/u}) = 135.1\text{ MeV}$. Each photon has half this energy, or 67.5 MeV.

$$E = hf \Rightarrow f = \frac{E}{h} = \frac{(67.5\times10^{6}\text{ eV})(1.602\times10^{-19}\text{ J/eV})}{6.626\times10^{-34}\text{ J}\cdot\text{s}} = \underline{1.63\times10^{22}\text{ Hz}}$$

$$\lambda = \frac{c}{f} = \frac{2.998\times10^{8}\text{ m/s}}{1.63\times10^{22}\text{ Hz}} = 1.84\times10^{-14}\text{ m} = \underline{18.4\text{ fm}}$$

46-5

a) The nuclear charge of Na is 11 and of Ne is 10. The charge of the β^+ (positron) is $+e$, so the charge is $+11e$ on both sides of the reaction equation.

b) To be sure to treat the electron masses correctly work with nuclear masses. Get these from the atomic masses by subtracting the masses of the electrons.

Nuclear mass of ${}^{22}_{11}\text{Na}$ is $21.994435\text{ u} - 11(0.0005485799\text{ u}) = 21.988401\text{ u}$.
Nuclear mass of ${}^{22}_{10}\text{Ne}$ is $21.991384\text{ u} - 10(0.0005485799\text{ u}) = 21.985898\text{ u}$.

The mass decrease in the decay is $m({}^{22}_{11}\text{Na}) - m({}^{22}_{10}\text{Ne}) - m(\beta^+)$
$= 21.988401\text{ u} - 21.985898\text{ u} - 0.0005486\text{ u} = 0.001954\text{ u}$.
The energy released is the energy equivalent of the mass decrease
$(0.001954\text{ u})(931.5\text{ MeV/u}) = \underline{1.82\text{ MeV}}$.

46-7

a) Eq. (46-7): $\omega = \frac{|q|B}{m} \Rightarrow B = \frac{m\omega}{|q|}$

$\omega = 2\pi f \Rightarrow B = \frac{2\pi m f}{|q|}$

A deuteron is a deuterium nucleus $({}^{2}_{1}\text{H})$. Its charge is $q = +e$. Its mass is the mass of the neutral ${}^{2}_{1}\text{H}$ atom (Table 45-2) minus the mass of the one atomic electron:

$$m = 2.01410\text{ u} - 0.0005486\text{ u} = 2.013551\text{ u}\left(\frac{1.66054\times10^{-27}\text{ kg}}{1\text{ u}}\right) = 3.344\times10^{-27}\text{ kg}$$

$$B = \frac{2\pi m f}{|q|} = \frac{2\pi(3.344\times10^{-27}\text{ kg})(6.00\times10^{6}\text{ Hz})}{1.602\times10^{-19}\text{ C}} = \underline{0.787\text{ T}}$$

46-7 (cont)

b) Eq. (46-8): $K = \frac{q^2B^2R^2}{2m} = \frac{[(1.602\times10^{-19}\,C)(0.787\,T)(0.320\,m)]^2}{2(3.344\times10^{-27}\,kg)} = 2.434\times10^{-13}\,J$

$K = 2.434\times10^{-13}\,J\left(\frac{1\,eV}{1.602\times10^{-19}\,J}\right) = \underline{1.52\times10^{6}\,MeV}$

$K=\frac{1}{2}mv^2 \Rightarrow v=\sqrt{\frac{2K}{m}} = \sqrt{\frac{2(2.434\times10^{-13}\,J)}{3.344\times10^{-27}\,kg}} = 1.21\times10^{7}\,m/s$

Note: $\frac{v}{c} = 0.04$, so it is ok to use the nonrelativistic expression for kinetic energy.

<u>46-9</u>

a) The masses of the target and projectile particles are equal, so Eq. (46-10) can be used: $E_a^2 = 2mc^2(E_m + mc^2)$

$E_m = \frac{E_a^2}{2mc^2} - mc^2$

The mass of the alpha particle can be calculated by subtracting two electron masses from the ^4_2He atomic mass:

$m = m_\alpha = (4.002603\,u - 2(0.0005486\,u)) = 4.001506\,u$

Then $mc^2 = (4.001506\,u)(931.5\,MeV/u) = 3.727\,GeV$

$E_m = \frac{E_a^2}{2mc^2} - mc^2 = \frac{(12.0\,GeV)^2}{2(3.727\,GeV)} - 3.727\,GeV = \underline{15.6\,GeV}$

b) Each beam must have $\frac{1}{2}E_a = \underline{6.0\,GeV}$

<u>46-13</u>

The mass decrease is $m(\Sigma^+) - m(n) - m(\pi^+)$ and the energy released is $mc^2(\Sigma^+) - mc^2(n) - mc^2(\pi^+) = 1189\,MeV - 939.6\,MeV - 139.6\,MeV = \underline{110\,MeV}$. (The mc^2 values for each particle were taken from Table 46-3.)

<u>46-15</u>

a) $n \rightarrow p + e^- + \bar{\nu}_e$; $B_n = 1$, $B_p = 1$, $B_{e^-} = 0$, $B_{\bar{\nu}_e} = 0$; $B = 1$ on both sides of the reaction equation $\Rightarrow$ B is conserved.

b) $p + n \rightarrow p + \pi^0$; $B_p = 1$, $B_n = 1$, $B_{\pi^0} = 0$; $B = 2$ initially, $B = 1$ for the products $\Rightarrow$ B is not conserved.

c) $p \rightarrow \pi^+ + \pi^0$; $B_p = 1$, $B_{\pi^+} = 0$, $B_{\pi^0} = 0$; $B = 1$ initially, $B = 0$ for the products $\Rightarrow$ B is not conserved.

d) $p + p \rightarrow p + p + \pi^0$; $B_p = 1$, $B_{\pi^0} = 0$; $B = 2$ on both sides of the reaction equation $\Rightarrow$ B is conserved.

46-19

Each value for the combination is the sum of the values for each quark.

a) uus

$Q = \frac{2}{3}e + \frac{2}{3}e - \frac{1}{3}e = +e$

$B = \frac{1}{3} + \frac{1}{3} + \frac{1}{3} = 1$

$S = 0 + 0 - 1 = -1$

$C = 0 + 0 + 0 = 0$

b) $c\bar{s}$

The values for $\bar{s}$ are the negative of those for s.

$Q = \frac{2}{3}e + \frac{1}{3}e = +e$

$B = \frac{1}{3} - \frac{1}{3} = 0$

$S = 0 + 1 = +1$

$C = +1 + 0 = +1$

c) $\overline{ddu}$

$Q = +\frac{1}{3}e + \frac{1}{3}e - \frac{2}{3}e = 0$

$B = -\frac{1}{3} - \frac{1}{3} - \frac{1}{3} = -1$

$S = 0 + 0 + 0 = 0$

$C = 0 + 0 + 0 = 0$

d) $c\bar{b}$

$Q = \frac{2}{3}e + \frac{1}{3}e = +e$

$B = \frac{1}{3} - \frac{1}{3} = 0$

$S = 0 + 0 = 0$

$C = +1 + 0 = +1$

46-23

a) Eq. (46-14): $v = \left[\frac{(\lambda/\lambda_0)^2 - 1}{(\lambda/\lambda_0)^2 + 1}\right]c = \left[\frac{\left(\frac{1030\text{nm}}{590\text{nm}}\right)^2 - 1}{\left(\frac{1030\text{nm}}{590\text{nm}}\right)^2 + 1}\right]c$

$v = 0.5059c = (0.5059)(2.998 \times 10^8 \text{ m/s}) = \underline{1.52 \times 10^8 \text{ m/s}}$

b) Eq. (46-15): $v = H_0 r$ (Hubble's law)

$r = \frac{v}{H_0} = \frac{1.52 \times 10^8 \text{ m/s}}{17 \times 10^{-3} (\text{m/s})/\text{light-yr}} = \underline{8.9 \times 10^9 \text{ light-yr}}$

46-25

a) $p + {}^2_1H \rightarrow {}^3_2He$, or ${}^1_1H + {}^2_1H \rightarrow {}^3_2He$

If neutral atom masses are used then the masses of the two atomic electrons on each side of the reaction equation will cancel.

46-25 (cont)

Taking the atomic masses from Table 45-2 the mass decrease is $m(^1_1H) + m(^2_1H) - m(^3_2He) = 1.00783\,u + 2.01410\,u - 3.01603\,u = 0.00590\,u$.

The energy released is the energy equivalent of this mass decrease:

$(0.00590\,u)(931.5\ \text{MeV/u}) = \underline{5.50\ \text{MeV}}$

b) $n + ^3_2He \rightarrow ^4_2He$

If neutral helium masses are used then the masses of the two atomic electrons on each side of the reaction equation will cancel.

The mass decrease is $m(n) + m(^3_2He) - m(^4_2He) = 1.0086649\,u + 3.01603\,u - 4.00260\,u = 0.02209\,u$.

The energy released is the energy equivalent of this mass decrease:

$(0.02209\,u)(931.5\ \text{MeV/u}) = \underline{20.6\ \text{MeV}}$

Problems

46-27

a) $E_a = 2(20\ \text{TeV}) = 40\ \text{TeV}$

b) The beam energy of 20 TeV is much larger than the rest mass energy of a proton (938 MeV) so the beam is highly relativistic. Use the relativistic version of Eq. (46-7):

$$\frac{v}{r} = \frac{|q|B}{m}\sqrt{1 - v^2/c^2} \Rightarrow B = \frac{mv}{|q|r}\frac{1}{\sqrt{1 - v^2/c^2}} = \frac{mv\gamma}{|q|r}$$

Can determine γ from the energy:

$$E = mc^2\gamma \Rightarrow \gamma = \frac{E}{mc^2} = \frac{20 \times 10^6\ \text{MeV}}{938\ \text{MeV}} = 2.13 \times 10^4$$

(This large γ means that v is very close to c.)

$$2\pi r = 85\ \text{km} \Rightarrow r = \frac{85 \times 10^3\ \text{m}}{2\pi} = 1.35 \times 10^4\ \text{m}$$

$$\text{Then } B = \frac{mv\gamma}{|q|r} = \frac{(1.673 \times 10^{-27}\ \text{kg})(2.998 \times 10^8\ \text{m/s})(2.13 \times 10^4)}{(1.602 \times 10^{-19}\ \text{C})(1.35 \times 10^4\ \text{m})} = \underline{4.94\ \text{T}}$$

c) Need $E_a = 40\ \text{TeV} = 40 \times 10^6\ \text{MeV}$

Since the target and projectile particles are both protons, Eq. (46-10) can be used:

$$E_a^2 = 2mc^2(E_m + mc^2)$$

$$E_m = \frac{E_a^2}{2mc^2} - mc^2 = \frac{(40 \times 10^6\ \text{MeV})^2}{2(938.3\ \text{MeV})} - 938.3\ \text{MeV} = 8.53 \times 10^{11}\ \text{MeV} = \underline{8.53 \times 10^5\ \text{TeV}}$$

(This shows the great advantage of colliding beams at relativistic energies.)

46-29

The total available energy must be at least the total rest energy of the product particles: $E_a = mc^2(\Lambda) + mc^2(\eta^0) = 1116\ \text{MeV} + 548.8\ \text{MeV} = 1664.8\ \text{MeV}$.

46-29 (cont)

Since the target and projectile particles are different we must use Eq.(46-9):

$$E_a^2 = 2Mc^2E_m + (Mc^2)^2 + (mc^2)^2$$

$M = m_p$, $m = m_{K^-}$

$$E_a^2 = 2m_pc^2E_{K^-} + (m_pc^2)^2 + (m_{K^-}c^2)^2$$

$$E_{K^-} = \frac{E_a^2 - (m_pc^2)^2 - (m_{K^-}c^2)^2}{2m_pc^2} = \frac{(1664.8\text{ MeV})^2 - (938.3\text{ MeV})^2 - (493.7\text{ MeV})^2}{2(938.3\text{ MeV})} = 877.9\text{ MeV}$$

This is the total energy of the K^- particle. Its kinetic energy is

$K = E - m_{K^-}c^2 = 877.9\text{ MeV} - 493.7\text{ MeV} = \underline{384\text{ MeV}}$

46-35

$$\phi \rightarrow K^+ + K^-$$

a) The mass decrease is $m(\phi) - m(K^+) - m(K^-)$. The energy equivalent of the mass decrease is $mc^2(\phi) - mc^2(K^+) - mc^2(K^-)$. The rest mass energy mc^2 for the ϕ meson is given in Problem 46-34, and the values for K^+ and K^- are given in Table 46-3. The energy released then is 1020 MeV − 493.7 MeV − 493.7 MeV = 32.6 MeV. The K^+ gets half this, 16 MeV.

b) $\phi \xrightarrow{?} K^+ + K^- + \pi^0$

The energy equivalent of the $K^+ + K^- + \pi^0$ masses is 493.7 MeV + 493.7 MeV + 135.0 MeV = 1121 MeV. This is greater than the energy equivalent of the ϕ mass. The mass of the decay products would be greater than the mass of the parent particle; the decay is energetically forbidden.

c) $\phi \xrightarrow{?} K^+ + \pi^-$

The reaction $\phi \rightarrow K^+ + K^-$ is observed. K^+ has strangeness +1 and K^- has strangeness −1, so the total strangeness of the decay products is zero. But strangeness must be conserved, so we deduce that the ϕ particle has strangeness zero.

π^- has strangeness 0, so $K^+ + \pi^-$ has strangeness −1. The decay $\phi \rightarrow K^+ + \pi^-$ then would violate conservation of strangeness.

$\phi \xrightarrow{?} K^+ + \mu^-$

μ^- has strangeness 0, so this decay would also violate conservation of strangeness.

<u>46-37</u>

a) The energy equivalent of the mass decrease is
$mc^2(\Omega^-) - mc^2(\Lambda^0) - mc^2(K^-) = 1672\text{ MeV} - 1116\text{ MeV} - 493.7\text{ MeV} = \underline{62\text{ MeV}}$

b) The Ω^- is at rest means the initial linear momentum is zero. Conservation of linear momentum then says that the Λ^0 and K^- must have equal and opposite momenta $\Rightarrow m_{\Lambda^0} v_{\Lambda^0} = m_{K^-} v_{K^-}$

$$\boxed{v_{K^-} = \left(\frac{m_{\Lambda^0}}{m_{K^-}}\right) v_{\Lambda^0}}$$

Also, the sum of the kinetic energies of the Λ^0 and K^- must equal the total kinetic energy $K_{tot} = 62$ MeV calculated in part (a):

$$K_{tot} = K_{\Lambda^0} + K_{K^-}$$

$$K_{\Lambda^0} + \tfrac{1}{2} m_{K^-} v_{K^-}^2 = K_{tot}$$

Use the momentum conservation result:

$$K_{\Lambda} + \tfrac{1}{2} m_{K^-} \left(\frac{m_{\Lambda^0}}{m_{K^-}}\right)^2 v_{\Lambda^0}^2 = K_{tot}$$

$$K_{\Lambda} + \left(\frac{m_{\Lambda^0}}{m_{K^-}}\right)\left(\tfrac{1}{2} m_{\Lambda^0} v_{\Lambda^0}^2\right) = K_{tot}$$

$$K_{\Lambda^0}\left(1 + \frac{m_{\Lambda^0}}{m_{K^-}}\right) = K_{tot}$$

$$K_{\Lambda^0} = \frac{K_{tot}}{1 + \frac{m_{\Lambda^0}}{m_{K^-}}} = \frac{62\text{ MeV}}{1 + \frac{1116\text{ MeV}}{493.7\text{ MeV}}} = 19\text{ MeV}$$

$K_{\Lambda^0} + K_{K^-} = K_{tot} \Rightarrow K_{K^-} = K_{tot} - K_{\Lambda^0} = 62\text{ MeV} - 19\text{ MeV} = 43\text{ MeV}$

The fraction for the Λ^0 is $\frac{19\text{ MeV}}{62\text{ MeV}} = 31\%$.
The fraction for the K^- is $\frac{43\text{ MeV}}{62\text{ MeV}} = 69\%$.

The lighter particle carries off more of the kinetic energy that is released in the decay.